Bacterial Conjugation

Bacterial Conjugation

Edited by

Don B. Clewell
The University of Michigan
Ann Arbor, Michigan

Plenum Press • New York and London

Library of Congress Cataloging-in-Publication Data

Bacterial conjugation / edited by Don B. Clewell.
    p.   cm.
  Includes bibliographical references and index.
  ISBN 0-306-44376-7
  1. Conjugation (Biology) 2. Plasmids.   I. Clewell, Don B.
QR86.5.B33 1993
589.9'01662--dc20
                                            93-7458
                                             CIP

ISBN 0-306-44376-7

© 1993 Plenum Press, New York
A Division of Plenum Publishing Corporation
233 Spring Street, New York, N.Y. 10013

Printed in the United States of America

Contributors

Gordon L. Archer • Departments of Microbiology and Immunology and Medicine, Medical College of Virginia, Virginia Commonwealth University, Richmond, Virginia 23298

Don B. Clewell • Department of Biologic and Materials Sciences, School of Dentistry and Department of Microbiology and Immunology, School of Medicine, The University of Michigan, Ann Arbor, Michigan 48109-0402

Walter B. Dempsey • University of Texas Health Science Center and Department of Veterans Affairs, Veterans Affairs Medical Center, Dallas, Texas 75216

Stephen K. Farrand • Department of Plant Pathology and Microbiology, University of Illinois at Urbana-Champaign, Urbana, Illinois 61801

Susan E. Flannagan • Biologic and Materials Sciences, School of Dentistry, The University of Michigan, Ann Arbor, Michigan 48109-0402

Laura S. Frost • Department of Microbiology, M330 Biological Sciences Building, University of Alberta, Edmonton, Alberta T6G 2E9, Canada

Donald G. Guiney • Department of Medicine and Center for Molecular Genetics, University of California, San Diego, San Diego, California 92103

Dieter Haas • Mikrobiologisches Institut, Eidgenössische Technische Hochschule, CH-8092 Zürich, Switzerland

Donald R. Helinski • Department of Biology and Center for Molecular Genetics, University of California, San Diego, La Jolla, California 92093-0634

David A. Hopwood • John Innes Institute, John Innes Centre, Norwich NR4 7UH, England

Karin Ippen-Ihler • Medical Microbiology and Immunology, Health Science Center, Texas A & M University, College Station, Texas 77843-1114

Clarence I. Kado • Davis Crown Gall Group, Department of Plant Pathology, University of California, Davis, Davis, California 95616

Tobias Kieser • John Innes Institute, John Innes Centre, Norwich NR4 7UH, England

Barbara Lewis Kittell • Department of Biology and Center for Molecular Genetics, University of California, San Diego, La Jolla, California 92093-0634

Erich Lanka • Max-Planck-Institut für Molekulare Genetik, Abteilung Schuster, D-1000 Berlin 33, Federal Republic of Germany

Francis L. Macrina • Department of Microbiology and Immunology, Medical College of Virginia, Virginia Commonwealth University, Richmond, Virginia 23298-0678

Cornelia Reimmann • Mikrobiologisches Institut, Eidgenössische Technische Hochschule, CH-8092 Zürich, Switzerland

Ronald A. Skurray • School of Biological Sciences, University of Sydney, Sydney, New South Wales 2006, Australia

Brian Wilkins • Department of Genetics, University of Leicester, Leicester LE1 4RH, United Kingdom

Neil Willetts • Biotech Australia Proprietary, Limited, Roseville, New South Wales 2069, Australia

Preface

Bacterial plasmids originating in a wide range of genera are being studied from a variety of perspectives in hundreds of laboratories around the globe. These elements are well known for carrying "special" genes that confer important survival properties, frequently necessary under atypical conditions. Classic examples of plasmid-borne genes are those providing bacterial resistance to toxic substances such as antibiotics, metal ions, and bacteriophage. Often included are those determining bacteriocins, which may give the bacterium an advantage in a highly competitive environment. Genes offering metabolic alternatives to the cell under nutritionally stressed conditions are also commonly found on plasmids, as are determinants important to colonization and pathogenesis.

It is likely that in many, if not most, cases plasmids and their passenger determinants represent DNA acquired recently by their bacterial hosts, and it is the characteristic mobility of these elements that enables their efficient establishment in new bacterial cells by the process known as conjugation. Whereas many plasmids are fully capable of promoting their own conjugal transfer, others move only with help from coresident elements. The ability of a plasmid to establish itself in a variety of different species is common, and recent studies have shown that transfer can in some cases occur from bacterial cells to eukaryotes such as yeast. Interkingdom transfer in nature is indeed evident from the findings that certain diseases in plants are directly related to the introduction of bacterial DNA by a conjugationlike process. It is noteworthy that the genes commonly found on plasmids often exhibit their own form of mobility by being located on transposons able to move from one DNA molecule to another intracellularly. Some transposons even have self-conjugative potential and can transfer directly to another bacterial cell without having to hitchhike on a plasmid; these are the so-called conjugative transposons.

Considering the general significance of conjugation, from a basic as well as clinical, environmental, and evolutionary perspective, a volume that brings together a comprehensive collection of reviews focusing primarily on this subject would seem timely. It is hoped that this book will serve as a useful resource for individuals interested in the mysteries of DNA transfer in the bacterial world.

Don B. Clewell

Ann Arbor, Michigan

Contents

3. Key Regulatory Aspects of Transfer of F-Related Plasmids

Walter B. Dempsey

13. **Conjugal Transfer in Anaerobic Bacteria**

Francis L. Macrina

14. **Sex Pheromones and the Plasmid-Encoded Mating Response in**
Enterococcus faecalis

Don B. Clewell

15. The Conjugative Transposons of Gram-Positive Bacteria

Don B. Clewell and Susan E. Flannagan

Chapter 1

Bacterial Conjugation
A Historical Perspective

NEIL WILLETTS

1. Introduction

The phenomenon of conjugation in bacteria provides one of the cornerstones of bacterial genetics. Use of the technology led to our initial understanding of the circular bacterial chromosome carrying *Escherichia coli*'s array of genes and of the recombination process whereby these genes could be reassorted.

The discovery of bacterial conjugation by Lederberg and Tatum (82) ranks with the other key discoveries of the 1940s and early 1950s, including the Luria and Delbrück (86) fluctuation test for the origin of mutations, the demonstration that DNA is the "transforming principle" (10), and the description of both lysogeny (87) and transduction (151).

The discovery illustrates, perhaps better than any other, the principle of scientific serendipity, since we now realize that the odds against chromosome transfer being observed in what was essentially a random strain of *E. coli* were enormous. Lederberg's good fortune depended on an almost miraculous combination of circumstances:

- The strain chosen, *E. coli* K12, carried a conjugative plasmid, F. This is relatively rare, particularly in strains of the preantibiotic predrug resistance factor era.
- Expression of the conjugation system of this plasmid was derepressed. F is an exception, along with only one or two others, among F-like plasmids in this regard. Without this the frequency of recombinants would have been too low to observe.

This chapter is dedicated to Professor William Hayes for his recognition of the first conjugative plasmid, F, and in gratitude for his support and friendship over the past 30 years.

NEIL WILLETTS • Biotech Australia Proprietary Limited, Roseville, New South Wales 2069, Australia. *Bacterial Conjugation*, edited by Don B. Clewell. Plenum Press, New York, 1993.

- F is unusual among plasmids in carrying a collection of insertion sequences, allowing reciprocal recombination with a variety of similar sequences on the bacterial chromosome. This is required for chromosome transfer and for Hfr formation.
- The *thr leu* recipient strain had spontaneously lost its F plasmid, thus avoiding the 300-fold decrease that would otherwise have occurred because of surface exclusion.
- The two markers used to reduce the frequency of prototrophic revertants in the recipient population, *thr* and *leu*, are closely linked, hence coselection of $thr^+ leu^+$ recombinants did not reduce the observed recombination frequency as would have happened if the markers had been on the opposite sides of the chromosome.

During the 1940s and 1950s, the major focus was on the application of conjugation to elucidation of the fundamental genetic characteristics of the *E. coli* genome, and this largely distracted attention from the curious and interesting properties of F itself. Although the agent responsible for conjugation was recognized as an "infectious vector," dubbed the F factor (26, 55), F was essentially considered as the "point" from which transfer of the bacterial chromosome was initiated. This story has been told in detail by Hayes (56) and recently by Brock (22).

I therefore plan to pick up the story of plasmid-determined conjugation in the early 1960s, when consideration of the genetic properties of F began and when it was discovered that F itself is only one representative among a broad spectrum of conjugative plasmids in a wide range of bacterial genera. I need make no apologies for my bias toward the conjugation process determined by F itself, which historically has served as the paradigm, as well as the focus of conjugation studies in my own laboratory. I will conclude my history in 1984, the year in which my own research on conjugation terminated because of my definitive emigration to Australia. As I think this chapter will demonstrate, discoveries even in this relatively recent period can easily and truly be accounted as historic, in the rapidly advancing field of molecular biology.

2. Recognition of F as a Physical and Genetic Entity

Proof that F was in fact a DNA molecule came at the beginning of this period from experiments in which the sex factor was transferred from *E. coli* (50%GC) to *Serratia marcescens* (58%GC). After density gradient centrifugation of F$^+$ *S. marcescens* DNA extracts, a small new peak corresponding to 50%GC density, and therefore to F, was visible (91).

Genetic (70) and later physical (23) evidence suggested that the bacterial chromosome is circular. F itself was initially thought to have a linear structure (70), but Campbell's (24) elegant model for integration of λ prophage (and by analogy, F) into the chromosome proposed that F was also a circular DNA molecule. Earlier genetic evidence for circularity of the Hfr chromosome (124) was consistent with this model. Circularity was also a major feature of the replicon model of Jacob et al (71). The concept of circularity was soon supported by physical data for several plasmids, including F (50). The ethidium bromide-cesium chloride density gradient technique for purification of covalently closed circular DNA molecules had been introduced in 1967 (114) and rapidly became the method of choice for separating plasmid DNA from linear chromosomal fragments.

Experiments during the 1960s gave estimates for the size of the F factor ranging from 68 to 250 kb (reviewed in 56, 57). It was therefore realized early on that F is about 1 to 2% of the size of the bacterial chromosome and so must contain about 100 genes (56). By that time, many genes had been identified in the similarly sized λ genome, and as Hayes (56) then commented, "It is surprising that more sex factor functions have not yet become apparent."

3. Mutational Studies of Conjugation

An Hfr mutant retaining F incompatibility but unable to transfer its chromosome was described by Lederberg and Lederberg (81). However, the first clear transfer-deficient mutant of F (in fact isolated in F*lac* carrying the convenient *lac* "handle") was $F_{D5}lac$, identified after ultraviolet (UV) mutation followed by replica plating of single donor colonies onto a lawn of an F^-lac^- recipient strain (32, 70). This mutant had a transfer level reduced 10^4 to 10^5 times in comparison with F*lac*, and it was shown that the mutation was indeed carried by the F*lac* element itself, not by the bacterial chromosome. $F_{D5}lac$-carrying strains were resistant to the newly discovered male-specific bacteriophages (83, 84), providing a method whereby further transfer-deficient F*lac* mutants could be selected, as male-specific phage-resistant strains (33, 34). A series of mutants resistant to one or another of the phages was made in Ray Valentine's laboratory (120, 121). Most of these mutants were resistant to all such phages and had lost the pilus, but a few were sensitive to some but not to others; all had also lost transfer ability.

More systematic studies of the genetic basis of conjugation were initiated in the late 1960s by isolation of the large collections of transfer-deficient mutants necessary to carry out a complementation analysis. This took place both in Hirota and Nishimura's laboratory in Osaka, Japan, and in John Clark's laboratory in Berkeley, California. The Japanese research (60) was subsequently continued and expanded by Eichii Ohtsubo, while the Berkeley mutants (5) were isolated and analyzed by Mark Achtman and myself, in Berkeley and in Edinburgh. Many of these mutants were isolated not by the simple method of selecting F-specific phage-resistant strains but rather by screening mutagenized populations of F prime strains (or strains carrying the related F-like transfer-derepressed R factor, R100-1) for their donor abilities by replica-plating techniques. This was important, since F encodes other functions required for transfer in addition to the F pilus, and mutants in such genes still retain the pilus and hence their sensitivity to F-specific bacteriophages.

4. Complementation Analysis and Mapping Studies

Complementation analysis of F mutants deficient in transfer was complicated at that time by two important factors. First, because the transfer function was defective, it could no longer be used to transfer the F factor to another cell to give the necessary heterozygote; second, because of incompatibility resulting from the control of replication, two autonomous F factors will not coexist stably in the same cell. Early attempts were made to study complementation between transfer-deficient Hfr and F*lac* mutants (33) but were confused by the belief at that time that transfer and replication were related (see below).

In the Japanese studies, incompatibility was avoided by carrying out the complementation analysis using stable "heterozygotes" containing both a transfer-deficient F*gal* mutant and a transfer-deficient mutant of the related F-like plasmid R100-1 (109). These strains can be constructed by P1 transduction of the entire R100-1 *tra* mutant into the F*gal tra* strain, and the two plasmids are compatible. However, interpretation of the data depends on the transfer gene products being interchangeable between F and R100-1. Although earlier studies in that laboratory had shown this is often the case (60), it later proved not to be so for all *tra* genes, particularly those involved in conjugative DNA metabolism and in regulation (137).

The Berkeley/Edinburgh studies overcame the complications in totally different ways. They relied on the construction of *transient* heterozygotes containing two F*lac tra* mutants, whose transfer ability was immediately assayed, prior to segregation of the F*lac tra* elements away from each other during subsequent cell growth. The transient heterozygotes were produced either by initial transfer of a suppressible F*lac tra* mutant from a *sup* host to an "F⁻ phenocopy" culture of a T6^R strain carrying the second mutant (6), followed by lysis of the T6^S donor strain with excess T6 phage, or by P1 transduction of one F*lac tra* mutant into a strain carrying the second (139, 141, 142).

Taken together, these initial studies identified 12 cistrons (*traA, B, C, D, E, F, G, H, I, J, K, L*) required for conjugation. Mutants in most of these genes were defective in F-pilus formation, but those in some, notably *traI*, were not. Some mutants in *traG* still produced the F pilus whereas others did not, and *traD* mutants made a pilus and were sensitive to some F pilus-specific phages but resistant to others. It was therefore deemed likely that the *traD, G*, and *I* gene products were required for the conjugational DNA metabolism that was known to be associated with transfer.

Between 1973 and 1984, seven more genes were added to this list: *traM* (7); *traN, U, V, W* (96); *traY* (89); and *traQ* (101).

Mapping of these *tra* genes was carried out by the classical technique of complementation analysis between representative mutants in each gene and a series of *tra* deletion mutants. At that time, such deletion mutants could be obtained only by in vivo techniques. Ohtsubo (108) used P1 transduction of F*gal*, which was itself too long to be accommodated in the phage head. Ippen-Ihler et al (67) selected deletions as temperature-resistant derivatives of a λcI857 lysogen (117) in a strain in which F$_{ts114}$*lac* had been integrated within the nearby *gal* genes. Taken together, these analyses showed the order of the genes then identified to be *traJALEKBCFHGDI*.

Ohtsubo's F*gal* deletion mutants were used in electron microscope heteroduplex studies by Sharp et al (118). This landmark paper not only gave approximate locations for some of the *tra* genes but also placed the transfer region on the first physical map of F and identified "hot spots" (later, insertion sequence or "IS" elements) responsible for Hfr and F prime formation.

The first indications of the operon structure of the *tra* genes came from observations of the polarity of some of the suppressible (chain-terminating) mutations in complementation tests with other point mutants. In particular, *traK4* was polar on *traBCFHG*, indicating that all six genes were in a single operon (6, 141).

In the 1970s, the mutator phage Mu, with its ability to insert randomly into *E. coli* genes, became popular as a means of carrying out in vivo genetic manipulations. In particularly, such insertions were shown to produce strongly polar mutations. Helmuth and Achtman (58) took advantage of this to derive a series of polar Mu insertions into *tra* genes, and complementation tests showed that almost all of these genes constituted a single large polycistronic operon (*traALEKBCFHGSDI*), with only *traJ* being transcribed separately. Later studies suggested that, in fact, *traI* had its own promoter as well (7, 147).

Another in vivo genetic tool popular in the 1970s was the insertion of λ*cI*857 prophage into abnormal chromosomal locations after deletion of the *att*λ site (119). McIntire and Willetts (88, 89, 146) adapted this technique to isolate F*lac* plasmids into which the prophage had been inserted, either within or close to the *tra* genes. These derivatives could be used in a variety of ways: λ insertions into the *tra* genes were polar, giving information about operon structures; they could be used to derive overlapping deletion mutants for mapping studies; and they provided a source of λ*tra* transducing phages (in vivo cloning, preceding the ready availability of today's in vitro techniques). These λ*tra* transducing phages were themselves quite useful. First, they could be used for complementation analysis and for genetic mapping; they were used, for example, to map *traV, W, U*, and *N* (96). Second, they were used to provide physical mapping data, obtained at that time by a combination of the now old-fashioned electron microscopic measurements of λ/λ*tra* phage heteroduplexes and the newly invented restriction fragment agarose gel length measurements (72, 144, 145, 147). Third, they were used to identify *tra* proteins, and fourth, they were used as a source of *tra* DNA for hybridization studies (see below).

Cloning of *tra* genes using today's standard in vitro restriction enzyme-based techniques also began in the 1970s with the studies of Skurray et al (122) of *Eco*RI fragments. Reports of cloning and subcloning and of restriction enzyme/agarose gel fragment size analysis since that time are too numerous to mention, but as for other complex biological systems, the *tra* gene clones were exploited, along with the allied DNA sequencing technology first described in 1977, to give a detailed genetic and physical analysis of the *tra* region (see Chapter 2).

5. *tra* Gene Proteins

In the late 1970s, *tra* genes cloned by in vivo λ*tra* methods, or by in vitro techniques, were used to visualize and to size the *tra* gene products. Key studies of that era were by Kennedy et al (74), and Thompson and Achtman (125, 126), using in vitro clones expressed in minicells or in vitro protein synthesizing systems, and by Ippen-Ihler (66), Willetts and McIntire (147), and Willetts and Maule (144, 145), using λ*tra* phages expressed in heavily UV-irradiated cells. Ultimately, these studies were supplemented and supplanted by DNA sequencing information giving precise predicted protein sizes (see Chapter 2).

Cell fractionation studies combined with these *tra* protein labeling techniques also gave the first indications of the cellular locations of the proteins: this was expected to be instructive, since the F pilus is a surface structure, and during conjugation, DNA efficiently crosses two cell surfaces. Consistent with this, many of the proteins were found in the inner and outer membranes (8, 99, 100). However, early assignations using minicells or over-producing systems have sometimes proved misleading.

6. The Pilus

Davis's (39) classic experiment, separating donor and recipient cells with a fritted glass filter, showed that cell-to-cell contact was essential for transfer of the bacterial chromosome. This led to the idea that conjugation required some sort of "mating bridge" between the cells. Attempts were made to visualize this in the electron microscope, but whether the "conjugation tubes" seen by Anderson et al (9) were indeed such, or drying artifacts (19), remains unproven.

In 1960, altered properties of the surface of F$^+$ donor cells were demonstrated: the surface carried a new antigen (110) and allowed infection by a new class of RNA-containing bacteriophages (83). At that time, chromosomally determined pili or fimbriae on the surface of enterobacteria had been well studied, and the tubular structure of type I pili had been elucidated (reviewed by Brinton [19]). However, the presence of such pili did not parallel donor ability (18). The idea that a pilus serves as the organ of DNA transport was revived in 1964 when the electron microscopic studies of Crawford and Gesteland (30) and of Brinton et al (21) showed that the male-specific RNA phages adsorbed to the sides of a new type of pilus present on F-carrying cells. In contrast to the numerous type I pili (to which the phages did not adsorb), there was, on average, only one F pilus per cell (19). Two years later, Caro and Schnös (25) identified a new class of male-specific phages carrying single-stranded DNA, that attached to the tip of the pilus.

Physiological studies showed that stationary phase cultures of F$^+$ or Hfr cells, known to be poor donors, also had very few pili, and that blending donor cultures both stripped off pili and severely reduced donor ability (19, 106). This pointed to a requirement for the pilus in the mating process. Competition between F pilus-specific DNA bacteriophage infection and conjugation supported this view (65, 105).

The question that exercised the minds of conjugation researchers during the 1960s and 1970s was, does the pilus provide the "conjugation tube" through which DNA passes from donor to recipient cell, or does it simply provide the means of bringing the cells into close contact? The first hypothesis was attractive, since the F pilus is 8.5 to 9.0 nm across, is 1 to 2 μm long, and appears to have an axial hole 2.0 to 2.5 nm across (19, 78, 129); this hole is just large enough to accommodate single-stranded DNA. However, it could not be proved by simple electron microscope observation of *apparently* mating cells *apparently* joined by a pilus.

Initial light microscopic observations suggested that the cells of many mating pairs often did not come into close contact and led Brinton (19) to propose the "epivirus" model whereby DNA is conducted from donor to recipient via the F pilus. Analogies were drawn with the tail structures of the T-even phages and with the role of the pilus in conducting the RNA and DNA of F pilus-specific phages into the cell. In agreement with this hypothesis, micromanipulation experiments by Ou and Anderson (111) suggested that DNA could be transferred between cells that had never been in contact. However, disbelievers thought that perhaps the cells had sneaked close together while the observer's attention was temporarily distracted. Indeed, Ou and Anderson found that F pili often did bring the mating cells into intimate contact and that DNA transfer between such pairs was apparently more efficient. Pilus retraction leading to wall-to-wall contact of mating cells had been proposed to be a step in mating pair formation (31) as well as in pilus-specific phage infection (92). Consistent with this, sodium dodecyl sulfate at concentrations that dissociate the pilus did not inhibit DNA transfer once mating pairs had formed (4).

Achtman (2) enlarged the concept of mating pairs to that of mating aggregates, of up to 20 cells in close contact. He and his coworkers made extensive studies of these aggregates and of the nature of the interaction between donor and recipient cells (reviewed in 3). In particular, he distinguished aggregates stable to shear forces or sodium dodecyl sulfate (SDS) treatment from unstable aggregates and showed that formation of stable aggregates required the F *traG* (C-terminal segment) and *traN* products that are not required for pilus formation per se. Achtman and Skurray (3) provided a detailed model for the cell-to-cell interactions occurring during conjugation, which they denoted the "mating cycle."

Although Brinton thought in 1965 that the F pilus was tubular, he later proposed instead that it had a double-fiber structure, making a role as the pathway for DNA transfer less appealing although not impossible (20). Finally, however, Folkhard et al (47) used X-ray diffraction and electron microscopy to show that the structure was indeed tubular as first thought, reviving the potential for its role as a conjugation tube. Convincing evidence that it can or does normally serve in this role, however, remained elusive even in 1984.

Genetic analysis in the 1970s showed that at least 12 genes were required for pilus formation. This was in notable contrast to the biochemical simplicity of the F pilus itself, which, as far as could be determined, was composed of a single protein subunit, F pilin (20). Willetts (137) carried out complementation tests showing that the gene responsible for the reduced efficiency of plating of F pilus-specific RNA phages on cells carrying the F-like plasmid R100-1, as opposed to F itself, was *traA*. He therefore proposed that *traA* was the pilin gene. This was confirmed by Minkley et al (97) by examining the number of labeled tyrosine peptides obtained from F pili synthesized by serine- or tyrosine-suppressed *traA* amber mutants.

In 1984, Frost et al showed that *traA* encoded pilin in a pre-, perhaps pre-pro, form (51). Their DNA sequencing and biochemical studies gave the full amino acid sequence, identified the point at which cleavage occurs to give mature pilin, and showed that the N-terminal amino acid was acetylated. Studies of the biochemical pathway whereby the initial *traA* product is processed into the completed pilus, and the involvement of the other *tra* gene products in this or in any membrane-located "basal structure," began in the early 1980s in Karin Ippen-Ihler's laboratory (99, 100, 101), but much remained to be learned in 1984.

7. Conjugational DNA Metabolism

Implicit in the "epidemic spread" of an F factor through an F population are the concepts that conjugation involves F replication and that this can occur independent of chromosome replication (26, 55).

In 1963, Jacob and Brenner (68) proposed the replicon model to explain regulation of the initiation of replication of any "unit of replication": this model was formally similar to the operon model for gene regulation but involved positive control via an initiator molecule rather than repression. Jacob et al (71) then proposed a mechanism for DNA transfer during conjugation based on the replicon model. On donor-recipient cell contact, F replication in the donor cell was initiated, and the geometry of the contact region resulted in one of the double-stranded replicas being driven into the recipient cell. This neatly accounted for the polarity of chromosome transfer by Hfr strains, for its initiation at the site of F integration, and for "epidemic spread."

Despite its intellectual appeal, several aspects of this model for conjugation proved to be incorrect. In particular, studies in the late 1960s and early 1970s showed that the origin of conjugative transfer was distinct from the origin of vegetative replication and that the mechanism of conjugative DNA replication was quite different from that proposed. Jacob et al (71) were misled by two of the pieces of evidence they adduced. First, some of their temperature-sensitive F*lac* replication mutants were simultaneously conjugation deficient. They themselves found that these F_{ts}*lac* mutants were temperature sensitive for replication but not for transfer (but they ignored this), and Willetts and Achtman (141) later found that one such mutant F_{ts62}*lac* contained a *traG* mutation; double mutations are therefore the likely explanation. Second, they found that acridine orange almost completely inhibited conjugation, and drew a parallel with its inhibition of F replication, which leads to "curing" (59). However, four other laboratories found acridine orange to have a relatively minor effect on conjugation (quoted in 142).

DNA labeling experiments carried out by Jacob et al (71) confirmed that DNA replication does occur during conjugation, and Gross and Caro (54) showed that only one preexisting DNA strand is transferred. The mechanism involved was further elucidated in the landmark papers of Ohki and Tomizawa (107) and of Rupp and Ihler (115), both based on zygotic induction of phage λ followed by separation and identification of the transferred λ DNA strands by a procedure published the previous year by Hradecna and Szybalski (63). These experiments showed that a unique preexisting strand of F is transferred starting from the 5' end and that during transfer, the complementary strand remaining in the donor cell is replicated. However, contrary to the predictions of Jacob et al (71), the complementary strand to the transferred strand was synthesized in the *recipient* cell (107, 130). The similarity of this replication mechanism to the rolling circle mechanism that had just been proposed for φX174 replication (53) was clearly apparent.

Although it was known that DNA replication in the donor cell normally accompanies conjugation, the question remained as to whether it is *essential* to drive DNA transfer as proposed in the model by Jacob et al. Early experiments on this point using *thy* mutants (14, 113) nalidixic acid (11, 61), and *dnaB*$_{ts}$ mutants (13, 17) to inhibit DNA replication were difficult to interpret unequivocally. Identification of single-stranded transferred DNA in minicell recipients implied that replication was not essential and provided further evidence for transfer of single-stranded DNA (29). Curtiss (31) and Brinton (20) discussed all of these experiments at length. The matter was not finally resolved until Sarathy and Siddiqi (116) used a *thy dnaB*$_{ts}$ double mutant to block donor DNA synthesis totally and demonstrated that this did *not* prevent transfer.

Experiments using *dnaE*$_{ts}$ mutants later showed that the complement to the transferred strand is synthesized by DNA polymerase III in both the recipient (136) and donor (75) cells. Kingsman and Willetts (75) also confirmed that donor cell conjugative DNA synthesis was not required for conjugative DNA transfer.

Kingsman and Willetts (75) proposed a detailed model for processing of DNA during conjugation, based on a modification of the rolling circle model as applied to replication of the single-stranded DNA bacteriophage φX174. This modification was necessary to explain their observation that the predicted 3'OH terminus at the nick was insufficient for donor conjugal DNA synthesis, since de novo synthesis of RNA primers was also required. A similar observation had been made the year before for φX174 RF DNA replication (90). Furthermore, Kingsman and Willetts (75) suggested that a unit length of plasmid DNA might be transferred and recircularized by a mechanism analogous to that described for φX174 cistron A endonuclease-ligase, which binds to the 5' terminus of the initial nick

(42). The later observation that a unit length of F *oriT*-containing plasmid DNA was transferred and recircularized by a *recA*- and *tra*-independent event was consistent with this (44). Initially, a rolling circle mechanism without a defined terminus had been assumed, based on Fulton's evidence (52) that early and late chromosomal markers in Hfr transfer show some linkage. However, this could have been due to multiple mating events (3) or to occasional recircularization reactions between the early and late DNA prior to recombination (75).

Transfer genes not required for pilus formation had been presumed to be needed for DNA processing during conjugation, and in line with this hypothesis, Kingsman and Willetts (75) found that the *traI* and *traM* products were required for both donor conjugative DNA synthesis and physical transfer of plasmid DNA. In contrast, piliated *traG* and *traN* mutants allowed donor conjugative DNA synthesis but without physical DNA transfer. This was surprising, since Achtman and Skurray (3) had found that these genes are required for stabilization of mating pairs, and it is still not clear what happens to the replicated DNA.

An early observation was that the sex factor F (or a part of it) is always transferred last by the Hfr chromosome (150); the origin of transfer was therefore presumed to be located at an F-chromosome junction or within F. Following the original suggestion of Rupp and Ihler (115), Willetts (138) proposed that the origin of transfer (*oriT*) was the site at which a specific endonuclease nicked the DNA strand to be transferred and began replication of the complementary strand remaining in the donor cell. Using a series of RecA⁻ Hfr deletions extending into the integrated F factor, he located the position of a site required in *cis* for transfer (that is, *oriT*) at one extreme of the transfer gene region and determined its orientation to be such that the transfer genes are transferred last. Later in vitro cloning studies located *oriT* more precisely within a 600-bp fragment close to *traM* (125).

An ingenious assay for endonucleolytic nicking at *oriT* was invented by Everett and Willetts (43), further exploiting λ transducing phages carrying part of the transfer region, in this case *oriT*. The DNA of such λ*oriT* transducing phages was nicked in the predicted strand when grown in F*tra*⁺, but not F⁻, cells. In fact, the transfer genes required for the nicking were *traY* and *traZ* (more recently found to be synonymous with a segment of *traI*; see Chapter 2), as well as *traJ*, probably acting indirectly in its regulatory role. The *traM* gene was not required for nicking, even though it is not required for pilus synthesis either; it was therefore hypothesized that its product may serve to transmit the "mating signal" generated by mating pair formation and required to initiate DNA transfer (43, 148).

Nicked λ*oriT* genomes were used in sequencing studies to determine a more precise location for *oriT*, outside the transfer gene region and close to *traM* (128). A serendipitous finding was that pBR-based plasmids carrying *oriT* in a particular orientation were unstable in the presence of F*lac*: stability therefore provided a selection for mutants in the *oriT* sequence (44). These mutations were located by sequencing to within 15 to 20 base pairs of the location of the *oriT* nick site (128).

Everett and Willetts (43) also observed that while the 3′ terminus at the nick was a hydroxyl group, the 5′ terminus was apparently blocked. This was reminiscent of earlier findings for the small nonconjugative but mobilizable plasmid ColE1. Clewell and Helinski (27, 28) had observed that a superhelical plasmid DNA-protein complex isolated by gentle cell lysis was converted to the open circular form by treatment with detergents or proteases. They therefore named this a "relaxation complex." Furthermore, on relaxation, a large protein was covalently linked to the 5' terminus at the strand-specific nick (85). Initially, it was thought that this phenomenon was related to replication of ColE1, but it was later implicated in conjugation (40, 64). The relaxation nick site is therefore the *oriT* site of ColE1.

Similar relaxation complexes were reported for other plasmids, including F (76). However in a later study, Johnson et al (73) could not repeat this observation for plasmids containing a cloned F *oriT* region in Tra+ cells. The concept that the nicked F DNA strand is transferred by a "pilot protein" by analogy with DNA phages (77) could not therefore be confirmed. Of the two transfer genes required for nicking *oriT*, the product of *traI* (of which "*traZ*" is now known to be part) was shown to be the DNA-unwinding enzyme helicase I (1); such an enzyme is clearly necessary for conjugation and might even provide the driving force for DNA transfer (148). Whether this large protein (180 kD) has an additional role and how and if it interacts with the *traY* product remained the subject of speculation in 1984, fueled only by indirect evidence based on the plasmid specificities among related F-like plasmids of *oriT* and of the *traM, Y, I* (or *Z*) products (137, reviewed in 148).

The processing of plasmid DNA during conjugation was reviewed in detail by Willetts and Wilkins (148), at the end of the period under discussion. This review contains a model for conjugative DNA transfer, which on comparison with that of Jacob et al (71) illustrates progress over the 20-year period (Figure 1.1).

8. Regulation of Conjugation

As mentioned in the introduction in this chapter, F expresses its conjugation system constitutively and transfers very efficiently. However, it became apparent in the 1960s that this was not the norm and that many naturally occurring plasmids, identified by virtue of the colicin or antibiotic resistance determinants that they carried, transferred at a level about 1000-fold less than that of F.

In the early 1960s, Japanese researchers began to study the interaction between the newly discovered drug resistance transfer factors and the F factor. They noted that some R factors that themselves transferred at only a low level inhibited F transfer and pilus formation (41, 103, 133, 134). This inhibitory ability was named i^+ by Hirota and fi^+ by Watanabe, and Egawa and Hirota (41) drew the analogy with Jacob and Monod's (69) recently proposed operon model for genetic regulation.

Shortly afterward, the phenomenon of high-frequency transfer (HFT) by cultures of cells that had recently received a colicinogenic or resistance transfer factor, and that then transferred the plasmid at high frequency for a limited period, was reported (123, 132). This type of "zygotic induction" of transfer ability was analogous to that previously seen for expression of the *lac* and λ prophage genes after bacterial chromosome transfer. This led Meynell and Datta (93) to suggest that the conjugation systems of the F and R factors were closely related and that an R factor-determined repressor molecule inhibited expression of both. Coordinate derepression of conjugation, F-like pilus formation, and RNA phage sensitivity in HFT cultures of R factor-carrying strains were indeed observed (37). The isolation of transfer-deregulated mutants from the fi^+ R factors R100 and R1 that transferred at high frequency, that made a pilus related to that of F in pilus-specific phage sensitivity, and that no longer inhibited F transfer or pilus formation confirmed this idea (41, 94, 104).

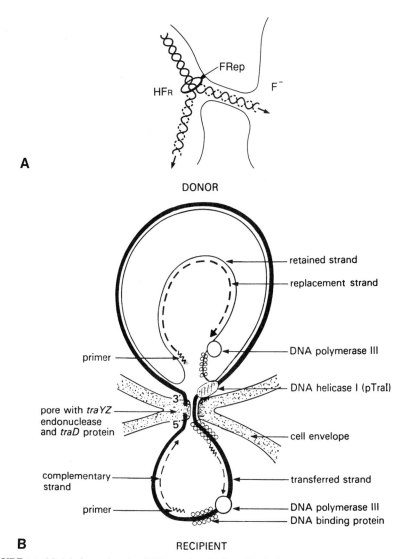

FIGURE 1.1. Models for conjugative DNA transfer. (A) From Jacob, Brenner, and Cuzin, 1963 (72). (B) From Willetts and Wilkins, 1984 (148), with permission.

The system for regulation of conjugation by F or F-like plasmids turned out to be more complex than originally thought, however. When my graduate student David Finnegan isolated a series of supposedly "operator-constitutive" mutants of F, not subject to transfer repression by the *fin* product of the F-like plasmid R100, he was surprised to find that some of these were *recessive* in transient heterozygote tests. We therefore hypothesized that these were mutant in a second gene required for regulation, *traP* (later *finP*), carried by F itself, whose product could not be replaced by the equivalent but plasmid-specific product of R100 (45). The *finP* gene was mapped close to *traJ* (149). Initially, it was thought that the *finP* gene product might be a small protein (73, 127), but later sequencing studies indicated that it was in fact an antisense RNA molecule, acting in conjunction with the R100 *finO* product to regulate *traJ* expression (102). F was right up to date in participating in this RNA-based regulatory mechanism discovered in the early 1980s!

Genetic studies of the dominant operator-constitutive mutations showed that the immediate target of the *finO* and *finP* gene products was *traJ* (46). Consistent with this, one such mutation mapped very close to *traJ*. It had previously been observed that both transfer inhibition and *traJ* mutations prevented not only conjugation per se but also the allied properties of pilus formation and surface exclusion: This led to the concept that the *traJ* product was a positive regulator of other transfer genes (45, 141, 143). The genetic data of Finnegan and Willetts (46) showed that this was indeed the case, and the *traALEKBCFHGSDI* operon structure (58) provided the basis for a simple model of the mechanism of action of *traJ*. Ultimately, hybridization experiments using the tedious filter hybridization technology that predated blotting techniques, with in vivo λ*tra*-cloned DNA, allowed formal proof of the assumption that control of *traJ* by the *finO* and *finP* products and of the major transfer operon by the *traJ* product was indeed at the level of transcription (140). The sequence locations of the *traJ* and transfer operon promoters were later described, and in vitro *galK* fusions were used for detailed study of the control regions (48, 102, 127).

This regulatory system is of surprising complexity compared, say, to that of the *lac* gene. This may be an indication of its importance in preserving two otherwise incompatible properties: a high level of transfer to allow "epidemic spread" through a new cell population and (because pilus formation is reduced along with transfer itself) resistance to the pilus-specific bacteriophages common in the environment. It is curious that although the unrelated transfer systems of some other plasmids are also regulated (for example, those of IncI plasmids) (94, 123), those of others (for example, those of IncP and IncN plasmids) are not.

9. Conjugation: A Common Phenomenon

Even though studies of conjugation through the 1940s and 1950s concentrated almost solely on F, realization that the conjugative F plasmid was not unique came early on. In studies subsequent to those with *E. coli* K12, Lederberg (80) found that 50 out of 2000 naturally occurring *E. coli* strains were able to give rise to prototrophic streptomycin-resistant recombinants after their culture with an auxotrophic streptomycin-resistant *E. coli* K12 F⁻ recipient strain. A number of these appeared to carry sex factors distinct from F. Hayes (56) reported that analysis of these sex factors remained rudimentary, and as far as I am aware, this is still the case. In a more recent study, Datta (36) found that 59 out of 375 enterobacterial strains collected during the 1917 to 1954 period carried distinct conjugative plasmids.

The identification of a sex factor in *E. coli* led to similar studies in other bacterial genera, resulting in the discovery of the FP factor in *Pseudomonas* (62) and of a transmissible colicinogenic factor in *Vibrio* (12). In fact, colicinogenic factors transmissible among the enterobacteria formed the first major group of conjugative plasmids to be described (49, 112). This group also included the first representative to be described of the interesting class of nonconjugative but mobilizable plasmids, ColE1.

A second even larger and more important group of transmissible plasmids was discovered in the late 1950s and early 1960s in Japan: the resistance transfer factors. These were, and still are, responsible for the spread of antibiotic resistance genes among the enterobacteria, particularly in hospital situations. Transmission by conjugation was demonstrated using the Davis U-tube by Mitsuhashi et al (98). Initial publications were in Japanese, and the interest of English-speaking scientists was first aroused by publications in English language journals by Watanabe and his colleagues in the early 1960s, culminating in his 1963 review (131).

A large number of R factors (as they came to be called) and plasmids carrying other determinants such as K antigen genes were identified and characterized during the 1960s and 1970s in many countries. As a consequence, efforts to group and to classify them began. Initially, two groups were suggested: fi^+ plasmids making F-like pili and fi^- plasmids with I-like pili, the latter being related to the prototype plasmid ColIb (79, 95, 135). However, it was soon realized that the fi^- group was heterogeneous and that both groups included plasmids with different replication and therefore incompatibility properties.

Classification via incompatibility grouping was therefore introduced, in particular through the pioneering efforts of Naomi Datta (reviewed in 35). Classification relied on the interrelationship between the regulatory systems controlling plasmid replication (perhaps a more "fundamental" property than conjugation), and its detailed description is beyond the scope of this chapter. However, superimposed on this classification was a consideration of the plasmid's conjugation system. Hence plasmids in different incompatibility groups but with conjugation systems related to that of F, particularly in the type of pilus formed and therefore in susceptibility to the same pilus-specific bacteriophages, were designated IncFI, IncFII, and so on. Other incompatibility groups were identified by different letters. Plasmids in the same incompatibility group often proved to have related conjugation systems, whereas plasmids in different groups usually had different conjugation systems, determining different pili adsorbing different bacteriophages.

In all cases, conjugation by these plasmids proved to be related in a mechanistic sense to that of F, with a pilus being required for mating pair formation. However, the pili differed, not only in adsorbing different types of the pilus-specific bacteriophages but in their morphological and serological properties. Correspondingly, the various transfer regions were unrelated both genetically and in their DNA homologies. Studies of these new types of pili were undertaken by David Bradley during the late 1970s (see his 1980 review [15]). Three morphological types were identified: thin flexible, thick flexible, and rigid. A significant related discovery was that rigid pili allow conjugation to proceed at up to four orders of magnitude higher frequency on solid, as compared to liquid, medium (16).

In general, these plasmids could transfer among the enterobacterial species but not to other gram-negative organisms. However, R factors were also found in due course in clinical *Pseudomonas* infections (38). The IncP1 group was of particular note, as plasmids belonging to it were able both to transfer to and to replicate in essentially all gram-negative bacteria. The phenomenon of conjugation is also found (though with quite different, nonpilus mechanisms) in gram-positive bacteria and *Streptomyces*. (See Chapters 11 and 15 for more detailed descriptions of these.)

In summary, by 1984 it was realized that conjugative plasmids are a common phenomenon in bacteria, that they determine an unexpected diversity of conjugation systems, and that they play an important role not only with regard to the practical issue of antibiotic resistance transfer but probably also in the process of bacterial evolution. Bill Hayes's view expressed in a 1964 lecture that "F in bacteria is an evolutionary backwater" has been comprehensively disproved.

10. New Techniques for Old

It is instructive to compare the techniques used for the gradual understanding of the F conjugation system beginning in the 1960s, with those that can be applied today. This provides a further illustration of the principle that advances in science depend on advances in technology.

Mutants are still required for study of the genetics of a multigene system but can now be obtained by methods that avoid the tediousness of screening 100,000 colonies individually. At the Berkeley laboratory, a person was employed for the sole purpose of carrying out the patching and replica plating required to establish our collection of *tra* mutants. Instead, transposon insertion or in vitro techniques offer less tedious alternatives, and the advantage of physically locating the gene into which the insertion has taken place at the same time.

Now that plasmid conjugation and replication have been clearly separated, both genetically and physically, the choice for complementation analysis might be to clone the entire transfer region under analysis, perhaps in segments, into a small nontransmissible, compatible plasmid. Two sets of transfer-deficient mutants could then be established, allowing tests to be carried out in *stable* heterozygotes. Plasmid transformation techniques in *E. coli* provide a much easier route to heterozygotes, avoiding the tedious isolation of relatively rare suppressible *tra* mutations and inefficient P1 transduction.

Deletion mutants for mapping studies of the transfer region were obtained in the early 1970s by in vivo techniques, either the use of P1 transduction to select F*gal* deletion mutants that could now be accommodated in the phage head or insertion of λcI857 prophage near F on an Hfr chromosome, followed by selection of temperature-resistant mutants, some of which carried deletions extending into the F genome. Physical mapping studies of these deletions were carried out by electron microscopic length measurements, since agarose gel analysis of the sizes of DNA fragments was still to come.

In contrast, in vitro cloning techniques and BAL-31 digestion now allow production of deletions with a much greater degree of facility and control. Similarly, in vivo cloning techniques requiring abnormal insertion of λ into or close to the conjugation region, followed by identification of rare λ*tra* transducing phages formed by aberrant excision, were useful at the time but have also now been replaced by in vitro cloning.

The initial studies of transcriptional regulation were unable to take advantage of Northern or dot blotting techniques, as yet uninvented. Instead, a ponderous filter hybridization technique was used. The source of *tra* DNA was a large scale λ*tra* phage preparation rather than a restriction fragment. If one wanted to measure the size of the transcript, sucrose gradient fractionation followed by interminable filter hybridizations was the only technique available. In Edinburgh, Ron Skurray tried to carry this out in 1977 for the expected large F transcript and discovered only the attendant frustrations (the experiment never was successful).

The outcome of all of the technological advances over the 20 years or so, from the mid-1960s to the mid-1980s, is that these days elucidation of the genetics, including physical studies, of a plasmid conjugation system might make up one or two respectable Ph.D. theses rather than keep several research groups around the world occupied for 10 to 20 years. Nowadays, one might adopt entirely the reverse approach to that used for F, by cloning and sequencing the conjugation region first, identifying the genes from the predicted open reading frames, and producing site-specific mutations in them to examine the functions of their products.

11. A Personal Odyssey

The preceding sections describe in rather impersonal scientific terms the gaining of an understanding of conjugation and of the overall phenomenon of conjugative plasmids between the early 1960s and the mid-1980s. However, the other side of this story concerns the people involved, from many countries worldwide, and their interactions over this 25-year period. As always, science provided the opportunity and the catalyst for many collaborations and close personal relationships.

My own interest in F dates back to lectures received from Sydney Brenner in 1961 in Cambridge but more especially to participation in 1964 in the course in microbial genetics, given by Bill Hayes and his colleagues at the MRC Genetics Unit at Hammersmith Hospital in London. There I was initiated into the mysteries of Hfr and F$^+$ strains by Bill and Julian Gross, and I learned of "HFT" colicinogen transfer from Roy Clowes. This experience resulted in my determination to change from biochemistry to the exciting new field of molecular genetics.

Then followed an interlude in Mexico, during which I and my Mexican colleagues Luis Cañedo and Jaime Martuscelli isolated our first set of transfer-deficient F*lac* mutants and performed trials of the P1 transductional complementation test. In 1967 I moved to California for a postdoctorate on recombination genetics (to John Clark's chagrin—he had wanted me to study its biochemistry), completing my transmutation to a molecular geneticist. At this time, Mark Achtman was a postgraduate student in his laboratory, vainly trying to construct diploid *E. coli* cells by transferring the entire Hfr chromosome into one of John Clark's newly discovered RecA$^-$ hosts. For reasons that are now clear, this project failed, and Mark therefore took up the F conjugation project that I had begun in Mexico, but he proceeded in a much larger and more ambitious way.

In 1966, Bill Hayes's MRC Unit moved to Edinburgh University, where he and Martin Pollock became joint professors in the new Department of Molecular Biology, Britain's first. I returned to join his unit and the university in 1968 and took up my interest in conjugation systems full time. After completion of his thesis, Mark Achtman came to spend 2 years (1969 to 1971) in my laboratory as a postdoctoral fellow, to continue our collaboration.

During the 1970s a succession of scientists from around the world with an interest in conjugation came to spend a period in Edinburgh. Visitors on sabbatical leave included John Clark (1969) and Walt Dempsey (1973 to 1974) from the United States. Bill Paranchych

(1972 to 1973) from Canada had a long-term interest in pilus biochemistry and came to learn about F molecular genetics; this ultimately catalyzed two further visits by Laura Frost (1981, 1983) from his laboratory, to sequence the *traA* pilus gene. Among the postdoctoral fellows, Karen Ippen (1969 to 1970) from the United States was involved (along with our congenial Russian visitor, Yura Fomitchev) in making and analyzing *tra* deletion mutants, and Sarah McIntire (1973 to 1976), also from the United States, played a large part in the λ*tra* studies. Alan Kingsman (1975 to 1977) and Roger Everett (1977 to 1980), both from the United Kingdom, made major contributions to studies of conjugal DNA metabolism and *oriT*, and Doug Johnson (1976 to 1978) from Canada was involved in our *tra* cloning studies.

There was also a strong Australian contingent. One of my early Ph.D. students was David Finnegan (1969 to 1972), who researched the mechanism of regulation of conjugation, discovering *finP*, and he was followed by Peter Reeves (1972 to 1973), who came for a sabbatical year and worked on *oriT*. Last, but not least, Ron Skurray (1976 to 1978) came to collaborate on studies of the regulation of conjugation; this led in due course to the two reviews (1980, 1987) that we coauthored.

The network of collaborations and friendships was maintained during this period, with regular discussions at conferences and during laboratory visits. During the 1970s there were about a dozen international conferences dedicated to plasmids, reflecting the upsurge of interest in these entities and the growing recognition of their widespread occurrence and importance. Between 1980 and 1984, three Gordon Conferences on Extrachromosomal Elements were held, which were particularly influential. Everyone who attended those meetings will remember, as well as the formal presentations, the discussions at the lake or around the pool and the mandatory performance of the Conjugation Song. This song was instigated in 1978 at a plasmid conference in Berlin by Stuart Levy. Every year, the conjugational "hard-core" stayed up until 4 or 5 A.M. to write new verses for this, with well-lubricated imaginations. I will cite just the refrain:

> Conjugation, penetration, consummation
> Oh that's great
> Derepress your inhibition
> Find a mate and aggregate
> Now penetrate the membrane
> Right on to the chromosome
> If you miss you're just a plasmid
> Try to be an episome

In summary, I wish also to record the memorable and enjoyable interactions and friendships between this group of enthusiasts. This is the way science *really* works.

12. Conclusion

What will the next historical perspective of conjugation look like in 25 years? Judging by the events during the 25 years between 1960 and 1984, this is almost entirely unpredictable. However, there is every reason to hope that its major topic will be the elucidation of the molecular biochemistry of the DNA transfer process, including the precise architecture of the conjugation proteins in and on the cell surface, and of their interactions with the transferred DNA. Dissection of such complex structural systems still eluded us in 1984. Along with this dissection of the specifics should be a greater understanding of the generalities, in terms of the evolutionary relationships between different conjugation systems, and an appreciation of how they have contributed to the evolution of microorganisms.

This prediction, however, may be optimistic. The heyday of the genetics of bacteria and their plasmids is already past, and even during this period of enthusiasm, conjugation was regarded as the esoteric retreat of relatively few aficionados. The mechanism of conjugation is discussed only briefly, if at all, in genetics texts written during the past 10 to 20 years, usually with the superficial presentation of an outdated model. Perhaps, then, conjugation will instead join other historic areas in the evolution of biological science, to be relegated to obscurity in the excitement of pursuing the molecular biology of higher eukaryotes.

In either case, this book has a place—as a basis for the deepening and broadening of our understanding of conjugation and its mechanisms or as a requiem for the efforts of the past 50 years.

References

1. Abdel-Monem, M., Taucher-Scholz, G., and Klinkert, M.Q., 1983, Identification of *Escherichia coli* DNA helicase I as the *traI* gene product of the F sex factor, *Proc. Natl. Acad. Sci. USA* 80:4659–4663.
2. Achtman M., 1975, Mating aggregates in *Escherichia coli* conjugation, *J. Bacteriol.* 123:505–515.
3. Achtman, M., and Skurray, R., 1977, A redefinition of the mating phenomenon in bacteria, in: *Microbial Interactions, Receptors and Recognition, Series B*, Vol. 3 (J.L. Reissig, ed.), Chapman & Hall, London, pp. 233–279.
4. Achtman, M., Morelli, G., and Schwuchow, S., 1978, Cell-cell interactions in conjugating *E. coli*: role of F pili and fate of mating aggregates, *J. Bacteriol.* 135:1053–1062.
5. Achtman, M., Willetts, N., and Clark, A. J., 1971, Beginning a genetic analysis of conjugational transfer determined by the F factor in *Escherichia coli* by isolation and characterisation of transfer-deficient mutants, *J. Bacteriol.* 106:529–538.
6. Achtman, M., Willetts, N., and Clark, A.J., 1972, Conjugational complementation analysis of transfer-deficient mutants of F*lac* in *Escherichia coli*, *J. Bacteriol.* 110:831–842.
7. Achtman, M., Skurray, R.A., Thompson, R., Helmuth, R., Hall, S., Beutin, L., and Clark, A.J., 1978, Assignment of *tra* cistrons to *Eco*RI fragments of F sex factor DNA, *J. Bacteriol.* 133:1383–1392.
8. Achtman, M., Manning, P., Edelbluth, C., and Herrlich, P., 1979, Export without proteolytic processing of inner and outer membrane proteins encoded by F sex factor *tra* cistrons in *E. coli* minicells, *Proc. Natl. Acad. Sci. USA* 76:4837–4841.
9. Anderson, T.F., Wollman, E.L., and Jacob, F., 1957, Sur les procéssus de conjugaison et de recombinaison chez *E. coli*. III. Aspects morphologiques en microscopie électronique, *Ann. Inst. Pasteur* 93:450.
10. Avery, O.T., Macleod, C.M., and McCarthy, M., 1944, Studies on the chemical nature of the substance inducing transformation of pneumonococcal types. I. Induction of transformation by a desoxyribonucleic acid fraction isolated from pneumonococcus type III, *J. Exptl. Med.* 79:137.
11. Barbour, S.D., 1967, Effect of nalidixic acid on conjugational transfer and expression of episomal Lac genes in *Escherichia coli* K-12, *J. Mol. Biol.* 28:373–376.
12. Bhaskaran, K., 1958, Genetic recombination in *Vibrio cholerae*, *J. Gen. Microbiol.* 19:71.
13. Bonhoeffer, F., 1966, DNA transfer and DNA synthesis during bacterial conjugation, *Z. Vererbungslehre* 98:141.
14. Bouck, N., and Adelberg, E.A., 1963, The relationship between DNA synthesis and conjugation in *Escherichia coli*, *Biochem. Biophys. Res. Comm.* 11:24.
15. Bradley, D.E., 1980, Morphological and serological relationships of conjugative pili, *Plasmid* 4:155–169.

16. Bradley, D.E., Taylor, D.E., and Cohen, D.R., 1980, Specification of surface mating systems among conjugative drug resistance plasmids in *Escherichia coli* K-12, *J. Bacteriol.* 143:1466–1470.

17. Bresler, S.E., Lanzov, V.A., and Lukjaniec-Blinkova, A.A., 1968, On the mechanism of conjugation in *Escherichia coli* K12, *Mol. Gen. Genet.* 102:269–284.

18. Brinton, C.C. Jr., 1959, Non-flagellar appendages of bacteria, *Nature* 183:782–786.

19. Brinton, C.C. Jr., 1965, The structure, function, synthesis and genetic control of bacterial pili and a molecular model for DNA and RNA transport in gram negative bacteria, *Trans. NY Acad. Sci.* 27:1003–1054.

20. Brinton, C.C. Jr., 1971, The properties of sex pili, the viral nature of "conjugal" genetic transfer systems, and some possible approaches to the control of bacterial drug resistance, *Crit. Rev. Microbiol.* 1:105–160.

21. Brinton, C.C., Gemski, P. Jr., and Carnahan, J., 1964, A new type of bacterial pilus genetically controlled by the fertility factor of *Escherichia coli* K12 and its role in chromosome transfer, *Proc. Natl. Acad. Sci. USA* 52:776–783.

22. Brock, T.D., 1990, Mating, in: *The Emergence of Bacterial Genetics*, Cold Spring Harbor Laboratory Press, pp. 75–112.

23. Cairns, J., 1963, The bacterial chromosome and its manner of replication as seen by autoradiography, *J. Mol. Biol.* 6:208–213.

24. Campbell, A.M., 1962, Episomes, *Advances Genet.* 11:101–145.

25. Caro, L.G., and Schnös, M., 1966, The attachment of male-specific bacteriophage fl to sensitive strains of *E. coli*, *Proc. Natl. Acad. Sci. USA* 56:126–132.

26. Cavalli, L.L., Lederberg, J., and Lederberg, E.M., 1953, An infective factor controlling sex compatibility in *Bacterium coli*, *J. Gen. Microbiol.* 8:89–103.

27. Clewell, D.B., and Helinski, D.R., 1969, Supercoiled circular DNA-protein complex in *E. coli*: purification and induced conversion to an open circular DNA form, *Proc. Natl. Acad. Sci. USA* 62:1159–1166.

28. Clewell, D., and Helinski, D., 1970. Properties of a supercoiled deoxyribonucleic acid-protein relaxation complex and strand specificity of the relaxation event, *Biochemistry* 9:4428–4440.

29. Cohen, A., Fisher, W.D., Curtiss, R. III, and Adler, H.I., 1968, DNA isolated from *Escherichia coli* minicells mated with F⁺ cells, *Proc. Natl. Acad. Sci. USA* 61:68.

30. Crawford, E.M., and Gesteland, R.G., 1964, The adsorption of bacteriophage R17, *Virology* 22:165–167.

31. Curtiss, R., 1969, Bacterial conjugation, *Ann. Rev. Microbiol.* 23:69–136.

32. Cuzin, F., 1962, Mutants défectifs de l'épisome sexuel chez *Escherichia coli* K-12, *C.R. Acad. Sci. Paris* 225:1149.

33. Cuzin, F., and Jacob, F., 1965, Génétique Microbienne. Analyse génétique fonctionnelle de l'épisome sexuel d'*Escherichia coli* K12, *C.R. Acad. Sci. Paris* 260:2087–2090.

34. Cuzin, F., and Jacob, F., 1967, Mutations de l'épisome F d'*Escherichia coli* K12, *Ann. Inst. Pasteur* 112:1–9.

35. Datta, N., 1975, Epidemiology and classification of plasmids, in: *Microbiology—1974* (D. Schlessinger, ed.), American Society for Microbiology, Washington DC, pp. 9–15.

36. Datta, N., 1985, Plasmids as organisms, in: *Plasmids in Bacteria* (D.R. Helinski, S.N. Cohen, D.B. Clewell, D.A. Jackson, and A. Hollaender, eds.), Plenum Press, New York, pp. 3–16.

37. Datta, N., Lawn, A.M., and Meynell, E., 1966, The relationship of F type piliation and F phage sensitivity to drug resistance transfer in R+F⁻ *Escherichia coli* K12, *J. Gen. Microbiol.* 45:365–376.

38. Datta, N., Hedges, R.W., Shaw, E.J., Sykes, R.B., and Richmond, M.H., 1971, Properties of an R factor from *Pseudomonas aeruginosa*, *J. Bacteriol.* 108:1244–1249.

39. Davis, B.D., 1950, Non-filterability of the agents of genetic recombination in *E. coli*, *J. Bacteriol.* 60:507.

40. Dougan, G., and Sherratt, D., 1977, The transposon Tn1 as a probe for studying ColE1 structure and function, *Mol. Gen. Genet.* 151:151–160.

41. Egawa, R., and Hirota, Y., 1962, Inhibition of fertility by multiple drug-resistance factors in *Escherichia coli* K12, *Japan J. Genet.* 37:66–69.

42. Eisenberg, S., Scott, J.F., and Kornberg, A., 1976, Enzymatic replication of viral and complementary strands of duplex DNA of phage φX174 proceeds by separate mechanisms, *Proc. Natl. Acad. Sci. USA* 73:3151–3155.

43. Everett, R., and Willetts, N., 1980, Characterisation of an *in vivo* system for nicking at the origin of conjugal DNA transfer of the sex factor F, *J. Mol. Biol.* 136:129–150.

44. Everett, R., and Willetts, N., 1982, Cloning, mutation and location of the origin of conjugal transfer, *EMBO J.* 1:747–753.

45. Finnegan, D., and Willetts, N., 1971, Two classes of F*lac* mutants insensitive to transfer inhibition by an F-like R factor, *Mol. Gen. Genet.* 111:256–264.
46. Finnegan, D., and Willetts, N., 1973, The site of action of the F transfer inhibitor, *Mol. Gen. Genet.* 127:307–316.
47. Folkhard, W., Leonard, K. R., Malsey, S., Marvin, D.A., Dubochet, J., Engel, A., Achtman, M., and Helmuth R., 1979, X-ray diffraction and electron microscope studies on the structure of bacterial F pili, *J. Mol. Biol.* 130:145–160.
48. Fowler, T., Taylor, T., and Thompson, R., 1983, The control region of the F plasmid transfer operon: DNA sequence of the *traJ* and *traY* genes and characterization of the *traYZ* promoter, *Gene* 26:79–89.
49. Frédéricq, P., 1954, Transduction génétique des propriétés colicinogénes chez *E. coli* et *S. sonnei, Compt. Rend. Séanc. Soc. Biol.* 148:399–402.
50. Freifelder, D., 1968, Studies with *E. coli* sex factors, *Cold Spring Harbor Symp. Quant. Biol.* 33:425–434.
51. Frost, L.S., Paranchych, W., and Willetts, N.S., 1984, DNA sequence of the F *traALE* region that includes the gene for F pilin, *J. Bacteriol.* 160:395–401.
52. Fulton, C., 1965, Continuous chromosome transfer in *Escherichia coli, Genetics* 116:885–892.
53. Gilbert, W., and Dressler, D., 1968, DNA replication: the rolling circle model, *Cold Spring Harbor Symp. Quant. Biol.* 33:437–484.
54. Gross, J.D., and Caro, L.G., 1966, DNA transfer in bacterial conjugation, *J. Mol. Biol.* 16:269–284.
55. Hayes, W., 1953, Observations on a transmissible agent determining sexual differentiation in *Bacterium coli, J. Gen. Microbiol.* 8:72–88.
56. Hayes, W., 1964, *The Genetics of Bacteria and Their Viruses*, John Wiley & Sons, New York, pp. 740.
57. Helinski, D.R., and Clewell, D.B., 1971, Circular DNA, *Ann. Rev. Biochem.* 25:899–942.
58. Helmuth, R., and Achtman, M., 1975, Operon structure of DNA transfer cistrons on the F sex factor, *Nature* 257:652–656.
59. Hirota, Y., 1960, The effect of acridine dyes on mating type factors in *Escherichia coli, Proc. Natl. Acad. Sci. USA* 46:57.
60. Hirota, Y., Fujii, T., and Nishimura, Y., 1966, Loss and repair of conjugal fertility and infectivity of the resistance factor and sex factor in *Escherichia coli, J. Bacteriol.* 91:1298–1304.
61. Hollom, S., and Pritchard, R.H., 1965, Effect of inhibition of DNA synthesis on mating in *Escherichia coli* K12, *Genet. Res. Cambridge* 6:479–483.
62. Holloway, B.W., 1955, Genetic recombination in *Pseudomonas aeruginosa, J. Gen. Microbiol.* 13:572.
63. Hradecna, Z., and Szybalski, W., 1967, Fractionation of the complementary strands of coliphage λ DNA based on the asymmetric distribution of the poly I.G. binding sites, *Virology* 32:633.
64. Inselburg, J., 1977, Studies of colicin E1 plasmid functions by analysis of deletions and TnA insertions of the plasmid, *J. Bacteriol.* 132:332–340.
65. Ippen, K.A., and Valentine, R.C., 1967, The sex hair of *E. coli* as a sensory fiber, conjugation tube, or mating arm? *Biochem. Biophys Res. Comm.* 27:674–680.
66. Ippen-Ihler, K., 1978, Isolation of a lambda transducing bacteriophage carrying F*traJ* activity, in: *Microbiology—1978* (D. Schlessinger, ed.), American Society for Microbiology, Washington, DC, pp. 146–149.
67. Ippen-Ihler, K., Achtman, M., and Willetts, N., 1972, Deletion map of the *Escherichia coli* K-12 sex factor F: the order of eleven transfer cistrons, *J. Bacteriol.* 110:857–863.
68. Jacob, F., and Brenner, S., 1963, Sur la regulation de la synthése du DNA chez les bactéries: l'hypothése du réplicon, *C.R. Acad. Sci. Paris* 256:298.
69. Jacob, F., and Monod, J., 1961, Genetic regulatory mechanisms in the synthesis of proteins, *J. Mol. Biol.* 3:318–356.
70. Jacob, F., and Wollman, E.L., 1961, *Sexuality and the Genetics of Bacteria*, Academic Press, New York, p. 374.
71. Jacob, F., Brenner, S., and Cuzin, F., 1963, On the regulation of DNA replication in bacteria, *Cold Spring Harbor Symp. Quant. Biol.* 28:329–348.
72. Johnson, D., and Willetts, N., 1983, Lambda-transducing phages carrying transfer genes isolated from an abnormal prophage insertion into the *traY* gene of F, *Plasmid* 9:71–85.
73. Johnson, D., Everett, R., and Willetts, N., 1981, Cloning of F DNA fragments carrying the origin of transfer *oriT* and the fertility inhibition gene *finP, J. Mol. Biol.* 153:187–202.
74. Kennedy, N., Beutin, L., Achtman, M., Skurray, R., Rahmsdorf, U., and Herrlich, P., 1977, Conjugation proteins encoded by the F sex factor, *Nature* 270:580–585.

75. Kingsman, A., and Willetts, N., 1978, The requirements for conjugal DNA synthesis in the donor strain during the F*lac* transfer, *J. Mol. Biol.* 122:287–300.
76. Kline, B.C., and Helinski, D.R., 1971, F1 sex factor of *Escherichia coli*. Size and purification in the form of a strand-specific relaxation complex of supercoiled deoxyribonucleic acid and protein, *Biochemistry* 10:4975–4980.
77. Kornberg, A., 1974, *DNA Synthesis*, W.H. Freeman & Co., San Francisco.
78. Lawn, A.M., 1966, Morphological features of the pili associated with *E. coli* K12 carrying R factors or the F factor, *J. Gen. Microbiol.* 45:377–383.
79. Lawn, A.M., Meynell, M., Meynell, G.G., and Datta, N., 1967, Sex pili and the classification of sex factors in the *Enterobacteriaceae*, *Nature* 216:343–346.
80. Lederberg, J., 1951, Prevalence of *E. coli* strains exhibiting genetic recombination, *Science* 114:68.
81. Lederberg, J., and Lederberg, E.M., 1956, Infection and heredity, in: *Cellular Mechanisms in Differentiation and Growth* (D. Rudrick, ed.), Princeton University Press, Princeton, New Jersey, pp. 110–124.
82. Lederberg, J., and Tatum, E.L., 1946, Gene recombination in *Escherichia coli*, *Nature* 158:558.
83. Loeb, T., 1960, Isolation of a bacteriophage specific for the F+ and Hfr mating types of *Escherichia coli* K12, *Science* 131:932–933.
84. Loeb, T., and Zinder, N.D., 1961, A bacteriophage containing RNA, *Proc. Natl. Acad. Sci. USA* 47:282.
85. Lovett, M.A., and Helinski, D.R., 1975, Relaxation complexes of plasmid DNA and protein. II. Characterisation of the proteins associated with the unrelaxed and relaxed complexes of plasmid ColE1, *J. Biol. Chem.* 250:8790–8797.
86. Luria, S.E., and Delbrück, M., 1943, Mutations of bacteria from virus sensitivity to virus resistance, *Genetics* 28:491–511.
87. Lwoff, A., Gutmann, A., 1950, Recherches sur un *Bacillus megathérium* lysogéne, *Ann. Inst. Pasteur* 78:711.
88. McIntire, S., and Willetts, N., 1978, Plasmid cointegrates of F*lac* and lambda prophage, *J. Bacteriol.* 134:184–192.
89. McIntire, S., and Willetts, N., 1980, Transfer-deficient cointegrates of F*lac* and lambda prophage, *Mol. Gen. Genet.* 178:165–172.
90. Machida, Y., Okazaki, T., and Okazaki, R., 1977, Discontinuous replication of replicative form DNA from bacteriophage ΦX174, *Proc. Natl. Acad. Sci. USA* 74:2776–2779.
91. Marmur, J., Rownd, R., Falkow, S., Baron, L.S., Schildkraut, C., and Doty, P., 1961, The nature of intergeneric episomal infection, *Proc. Natl. Acad. Sci. USA* 47:972–979.
92. Marvin, D.A., and Hohn, B., 1969, Filamentous bacterial viruses, *Bacteriol Rev.* 33:172–209.
93. Meynell, E., and Datta, N., 1965, Functional homology of the sex-factor and resistance transfer factors, *Nature* 207:884–885.
94. Meynell, E., and Datta, N., 1967, Mutant drug resistance factors of high transmissibility, *Nature* 214:885–887.
95. Meynell, E., Meynell, G.G., and Datta, N., 1968, Phylogenetic relationships of drug-resistance factors and other transmissible bacterial plasmids, *Bacteriol Rev.* 32:55–83.
96. Miki, T., Horiuchi, T., and Willetts, N., 1978, Identification and characterization of four new *tra* cistrons on the *E. coli* K-12 sex factor F, *Plasmid* 1:316–323.
97. Minkley, E.G., Polen, S., Brinton, C.C. Jr., and Ippen-Ihler, K., 1976, Identification of the structural gene for F-pilin, *J. Mol. Biol.* 108:111–121.
98. Mitsuhashi, S., Harada, K., and Hashimito, H., 1960, Multiple resistance of enteric bacteria and transmission of drug resistance to other strains by mixed cultivation, *Japan J. Exp. Med.* 30:179–184.
99. Moore, D. Sowa, B.A., and Ippen-Ihler, K., 1981, The effect of *tra* mutations on the synthesis of the F pilin membrane polypeptide, *Mol. Gen. Genet.* 184:260–264.
100. Moore, D. Sowa, B.A., and Ippen-Ihler, K., 1981, Location of an F-pilin pool in the inner membrane, *J. Bacteriol.* 146:251–252.
101. Moore, D., Sowa, B.A., and Ippen-Ihler, K., 1982, A new activity in the F*tra* operon which is required for F-pilin synthesis, *Mol. Gen. Genet.* 188:459–464.
102. Mullineaux, P., and Willetts, N., 1985, Promoters in the transfer region of plasmid F, in: *Plasmids in Bacteria* (D.R. Helinski, S.N. Cohen, D.B. Clewell, D.A. Johnson, and A. Hollaender, eds.), Plenum, New York, pp. 605–614.

103. Nakaya, R., Nakamura, A., and Murata, Y., 1960, Resistance transfer agents in *Shigella*, *Biochim. Biophys. Res. Comm.* 3:654–659.

104. Nishimura, Y., Ishibashi, M., Meynell, E., and Hirota, Y., 1967, Specific piliation directed by a fertility factor and a resistance factor of *Escherichia coli*, *J. Gen. Microbiol.* 49:89–98.

105. Novotny, C., Knight, W.S., and Brinton, C.C., 1968, Inhibition of bacterial conjugation by ribonucleic acid and deoxyribonucleic acid male-specific bacteriophages, *J. Bacteriol.* 95:314–326.

106. Novotny, C., Raizen, E., Knight, W., and Brinton, C., 1969, Functions of F pili in mating-pair formation and male bacteriophage infection studied by blending spectra and reappearance kinetics, *J. Bacteriol.* 98:1307.

107. Ohki, M., and Tomizawa, J., 1968, Asymmetric transfer of DNA strands in bacterial conjugation, *Cold Spring Harbor Symp. Quant. Biol.* 3:651–657.

108. Ohtsubo, E., 1970, Transfer-defective mutants of sex factors in *Escherichia coli*. II. Deletion mutants of an F-prime and deletion mapping of cistrons involved in genetic transfer, *Genetics* 64:189–197.

109. Ohtsubo, E., Nishimura, Y., and Hirota, Y., 1970, Transfer defective mutants of sex factors in *Escherichia coli*. I. Defective mutants and complementation analysis, *Genetics* 64:173–188.

110. Ørskov, I., and Ørskov, F., 1960, An antigen termed f^+ occurring in $f^+E.$ *coli* strains, *Acta. Pathol. Microbiol. Scand.* 48:37.

111. Ou, J.T., and Anderson, T.F., 1970, Role of pili in bacterial conjugation, *J. Bacteriol.* 102:648–654.

112. Ozeki, H., Stocker, B.A.D., and Smith, S.M., 1962, Transmission of colicinogeny between strains of *Salmonella typhimurium* grown together, *J. Gen. Microbiol.* 28:671.

113. Pritchard, R.H., 1963, The relationship between conjugation, recombination and DNA synthesis in *E. coli*, in: *Genetics Today: Proceedings of the XI International Congress of Genetics*, Vol. 2, The Hague, Pergamon Press, London, pp. 55–78.

114. Radloff, R., Bauer, W., and Vinograd, J., 1967, A dye-buoyant-density method for the detection and isolation of closed circular duplex DNA: the closed circular DNA in HeLa cells, *Proc. Natl. Acad. Sci. USA* 57:1514–1522.

115. Rupp, W.D., and Ihler, G., 1968, Strand selection during bacterial mating, *Cold Spring Harbor Symp. Quant. Biol.* 33:647–650.

116. Sarathy, P.V., and Siddiqi, O., 1973, DNA synthesis during bacterial conjugation. II. Is DNA replication in the Hfr obligatory for chromosome transfer? *J. Mol. Biol.* 78:443–451.

117. Shapiro, J.A., and Adhya, S.L., 1969, The galactose operon of *E. coli* K12. II. A deletion analysis of operon structure and polarity, *Genetics* 62:249–264.

118. Sharp, P.A., Hsu, M.T., Ohtsubo, E., and Davidson, N., 1972, Electron microscope heteroduplex studies of sequence relations among plasmids of *Escherichia coli*. I. Structure of F-prime factors, *J. Mol. Biol.* 71:471–497.

119. Shimada, K., Weisberg, R.A., Gottesman, M.E., 1972, Prophage lambda at unusual chromosomal locations. I. Location of the secondary attachment sites and the properties of the lysogens, *J. Mol. Biol.* 63:483–503.

120. Silverman, P., Mobach, W.W., and Valentine, R.C., 1967, Sex hair (F-pili) mutants of *E. coli*, *Biochem. Biophys. Res. Comm.* 27:412–416.

121. Silverman, P., Rosenthal, S., and Valentine, R.C., 1968, Two new classes of F pili mutants of *E. coli* resistant to infection by the male specific bacteriophage f2, *Virology* 36:142–146.

122. Skurray, R.A., Nagaishi, H., and Clark, A.J., 1976, Molecular cloning of DNA from F sex factor of *Escherichia coli* K-12, *Proc. Natl. Acad. Sci. USA* 73:64–68.

123. Stocker, B.A.D., Smith, S.M., and Ozeki, H., 1963, High infectivity of *Salmonella typhimurium* newly infected by the ColI factor, *J. Gen. Microbiol.* 30:201–221.

124. Taylor, A.L., and Adelberg, E.A., 1961, Evidence for a closed linkage group in Hfr males of *Escherichia coli* K-12, *Biochem. Biophys. Res. Comm.* 5:400.

125. Thompson, R., and Achtman, M., 1978, The control region of the F sex factor DNA transfer cistrons: restriction mapping and DNA cloning, *Mol. Gen. Genet.* 165:295–304.

126. Thompson, R., and Achtman, M., 1979, The control region of the F sex factor DNA transfer cistrons: physical mapping by deletion analysis, *Mol. Gen. Genet.* 169:49–57.

127. Thompson, R., and Taylor, L., 1982, Promoter mapping and DNA sequencing of the F plasmid transfer genes *traM* and *traJ*, *Mol. Gen. Genet.* 188:513–518.

128. Thompson, R., Taylor, L., Kelly, K., Everett, R., and Willetts, N., 1984, The F plasmid origin of transfer: DNA sequence of wild-type and mutant origins and location of origin-specific nicks, *EMBO J.* 3:1175–1180.

129. Valentine, R.C., and Strand, M., 1965, Complexes of F-pili and RNA bacteriophage, *Science* 148:511–513.

130. Vapnek, D., and Rupp, W.D., 1970, Asymmetric segregation of the complementary sex-factor DNA strands during conjugation in *Escherichia coli, J. Mol. Biol.* 53:287–303.

131. Watanabe, T., 1963a, Infective heredity of multiple drug resistance in bacteria, *Bacteriol. Rev.* 27:87–115.

132. Watanabe, T., 1963b, Episome-mediated transfer of drug resistance in *Enterobacteriaceae*. VI. High-frequency resistance transfer system in *Escherichia coli, J. Bacteriol.* 85:788–794.

133. Watanabe, T., and Fukasawa, T., 1962, Episome-mediated transfer of drug resistance in *Enterobacteriaceae*. IV. Interactions between resistance transfer factor and F-factor in *Escherichia coli* K12, *J. Bacteriol.* 83: 727–735.

134. Watanabe, T., Fukasawa, T., and Takano, T., 1962, Conversion of male bacteria of *Escherichia coli* K12 to resistance of f phages by infection with the episome "resistance transfer factor," *Virology* 17:218–219.

135. Watanabe, T., Nishida, H., Ogata, C., Arai, T., and Sato, S., 1964, Episome mediated transfer of drug resistance in *Enterobacteriaceae*. VII. Two types of naturally occurring R-factors, *J. Bacteriol.* 88:716–726.

136. Wilkins, B.M., and Hollom, S.E., 1974, Conjugational synthesis of F*lac*⁺ and ColI DNA in the presence of rifampicin and in *Escherichia coli* K-12 mutants defective in DNA synthesis, *Mol. Gen. Genet.* 134:143–156.

137. Willetts, N.S., 1971, Plasmid specificity of two proteins required for conjugation in *E. coli* K-12, *Nature, London New Biol.* 230:183–185.

138. Willetts, N.S., 1972, Location of the origin of transfer of the sex factor F, *J. Bacteriol.* 112:773–778.

139. Willetts, N.S., 1973, Characterization of the F transfer cistron, *traL, Genet. Res.* 21:205–213.

140. Willetts, N.S., 1977, The transcriptional control of fertility in F-like plasmids, *J. Mol. Biol.* 112:141–148.

141. Willetts, N.S., and Achtman, M., 1972, Genetic analysis of transfer by the *Escherichia coli* sex factor F, using P1 transductional complementation, *J. Bacteriol.* 110:843–851.

142. Willetts, N., and Broda, P., 1969, The *Escherichia coli* sex factor, in: *Ciba Foundation Symposium on Bacterial Episomes and Plasmids* (G.E.W. Wolstenholm and M. O'Connor, eds.), J. and A. Churchill Ltd., London, pp. 32–48.

143. Willetts, N. S., and Finnegan, D.J., 1970, Characteristics of *E. coli* K12 strains carrying both an F prime and an R factor, *Genet. Res.* 16:113–122.

144. Willetts, N.S., and Maule, J., 1979, Investigation of the F conjugation gene *traI:traI* mutants and *traI* transducing phages, *Mol. Gen. Genet.* 169:325–336.

145. Willetts, N.S., and Maule, J., 1980, Characterization of a transducing phage carrying the F conjugation gene *traG, Mol. Gen. Genet.* 178:675–680.

146. Willetts, N., and McIntire, S., 1977, λ*tra* transducing phages, in: *Plasmids: Medical and Theoretical Aspects* (S. Mitsuhashi, L. Rosival, V. Kreméry, eds.), Avicemum—Springer Verlag, Prague and Berlin, pp. 131–140.

147. Willetts, N.S., and McIntire, S., 1978, Isolation and characterization of λ*tra* transducing phages from EDFL223 (F*lac traB*::EDλ4), *J. Mol. Biol.* 126:525–549.

148. Willetts, N.S., and Wilkins, B., 1984, Processing of plasmid DNA during bacterial conjugation, *Microbiol. Rev.* 48:24–41.

149. Willetts, N., Maule, J., and McIntire, S., 1976, The genetic locations of *traO, finP*, and *tra-4* on the *E. coli* K-12 sex factor F, *Genet. Res.* 26:255–263.

150. Wollman, E.L., and Jacob, F., 1955, Sur le méchanisme du transfert de matériel génétique au cours de la recombinaison chez *E. coli* K-12, *C.R. Acad. Sci. Paris* 240:2449.

151. Zinder, N.D., and Lederberg, J., 1952, Genetic exchange in *Salmonella, J. Bacteriol.* 64:679.

Chapter 2

Genetic Organization of Transfer-Related Determinants on the Sex Factor F and Related Plasmids

KARIN IPPEN-IHLER and RONALD A. SKURRAY

1. Introduction

The bacterial conjugation system encoded by the F sex factor was the first to be described and subjected to detailed genetic and molecular analysis (for recent reviews see 88 and 167, and for earlier reviews and articles see Chapter 1 of this volume). Because of this, F-mediated conjugation has dominated much of our thinking about conjugal DNA transfer mechanisms and has certainly taken "pride of place" in textbooks as *the* model to present to undergraduates in microbial genetics. The F-conjugation system gained even further significance when it was recognized that F was only one of a large group of plasmids that encode a common transfer mechanism. The F-like plasmids, many of which carry determinants of importance in human and veterinary medicine, for example, for antibiotic resistance and for toxin and hemolysin production (Table 2.1), were originally grouped together on the basis of the sex pili that hosts carrying them elaborated. These pili were morphologically and serologically similar to those encoded by F and rendered the host susceptible to F- or "male-specific" phages. The plasmids were later classified on the basis of incompatibility (Inc) (Table 2.1); two plasmids in the same Inc group could not both stably coexist in the same host cell (33).

Where examined, molecular and genetic analyses have confirmed the relatedness of the transfer (*tra*) systems encoded by the F-like group of plasmids. For example, genetic

KARIN IPPEN-IHLER • Medical Microbiology and Immunology, Health Science Center, Texas A & M University, College Station, Texas 77843-1114.   RONALD A. SKURRAY • School of Biological Sciences, University of Sydney, Sydney, New South Wales 2006, Australia.
Bacterial Conjugation, edited by Don B. Clewell. Plenum Press, New York, 1993.

TABLE 2.1 F-like Plasmids

Incompatibility Group and Plasmid	Associated Phenotype[a]	References
IncFI		
F		16, 20
F'*lac*	Lac	20, 28
ColV	Cva	20, 28
ColV2-K94 (= ColV-K94 = ColV2)	Cva	20, 28
ColV3-K30	Cva	16, 20, 28
P307 (= EntP307)	Ent	16, 28
pA'::Tn5		16
pHH507	Tc	28
pIP162		20, 28
pIP162-1 (= RIP162-1)	Ap, Cm, Sm, Su	20, 28
pIP162-2 (= RIP162-2)	Ap, Tc	20, 28
pIP174	Ap, Cm, Sm, Su, Tc	28
pIP180	Ap, Cm, Km, Sm, Su, Tc	28
pIP234		28
R129	Ap, Cm, Sm, Su	20
R129-1	Cm, Sm, Su	20
R386	Tc	16, 20, 28
R453	Ap, Cm, Hg, Sm, Sp, Su, Tc	20, 28
R455	Ap, Cm, Hg, Sm, Sp, Su, Tc	20, 28
R455-2	Ap, Cm, Sm, Su	20
R456	Ap, Cm, Sm, Sp, Su, Tc	20, 28
R714a	Ap, Cm, Sm, Sp, Su, Tc	20
R773	Asa, Asi, Sm, Tc	20, 28
R978b	Ap, Cm, Sm, Su, Tc, Lac	20
RGN238	Ap, Cm, Sm, Su, Tc	20, 28
IncFII		
334	Ap, Cm, Sm, Su	20
ColB-CA18	Cba	20
ColB2	Cba	20
ColB2-K77	Cba	20
ColVB-K260	Cba, Cva	20
ENT$_I$	Ent	20
JR66 (= JR66b)	Cm, Gm, Km, Sm, Su, Tc	16, 20
JR70	Cm, Fa, Km, Sm, Su, Tc	20
JR72	Cm, Fa, Km, Nm, Sm, Su, Tc	20
JR73	Cm, Fa, Km, Sm, Su, Tc	20
pIP24	Tc	28
pIP100	Cm, Km, Su	28
pIP187 (= RIP187)	Sm, Tc, Tp	20, 28
pIP214		28
pSC50	Ap, Cm, Sm, Su	20
pSC206	Ap, Cm, Km, Sm, Su	20
R1	Ap, Cm, Km, Sm, Sp, Su	16, 20, 28
R1*drd*-19 (= R1-19)	Ap, Cm, Km, Sm, Sp, Su	20, 28
R1-1	Ap, Cm, Fa, Km, Sm, Sp, Su	20
R1-16	Km	28
R1-K^s	Ap, Cm, Sm, Su	20
R6	Cm, Fa, Hg, Km, Nm, Pm, Sm, Sp, Su, Tc	20, 157

TABLE 2.1 (Continued.)

Incompatibility Group and Plasmid	Associated Phenotype[a]	References
R6-5	Cm, Fa, Hg, Km, Nm, Pm, Sm, Sp, Su	16, 20, 157
R28	Fa, Hg, Sm, Sp, Su, Tc	20
R52	Hg, Sm, Sp, Su, Tc	20
R82	Sm, Sp, Su, Tc	20
R100 (= NR1 = R222Jap = 222)	Cm, Fa, Hg, Sm, Sp, Su, Tc	20, 28
R100-1*drd* (= R100-1)	Cm, Fa, Hg, Sm, Sp, Su, Tc	16, 20
R136 (= 240)	Tc	20, 28
R136*i*−1 (= pED241)	Tc	27
R192	Cm, Hg, Sm, Sp, Su, Tc	20
R192-1	Tc	20
R192*drd7*	Cm, Sm, Sp, Su, Tc	20
R196-1	Cm, Sm, Su, Tc	20
R348F	Cm, Hg, Km, Sm, Su, Tc	20
R429	Ap, Cm, Km, Tc	20
R444	Ap, Cm, Sm, Sp, Su, Tc	20
R445	Ap, Cm, Sm, Sp, Su, Tc	20
R452	Ap, Cm, Sm, Sp, Su, Tc	20
R457	Ap, Cm, Sm, Sp, Su, Tc	20
R458	Ap, Cm, Km, Sm, Sp, Su, Tc	20
R459	Ap, Cm, Km, Sm, Sp, Su, Tc	20
R459-4	Ap, Cm, Hg, Sm, Su, Tc	20
R459-5	Tc	20
R459-7	Ap, Cm, Hg, Su, Tc	20
R459-13	Cm, Su, Tc	20
R494	Ap, Hg, Km, Tc	20, 28
R496	Ap	20
R538-1	Cm, Hg, Sm, Sp, Su	16, 20, 28
R704a	Ap, Cm, Km, Sm, Su	20
R813	Ap, Cm, Sm, Su, Tc	20
R814	Ap, Cm, Sm, Su, Tc	20
R815	Ap, Cm, Sm, Su, Tc	20
R830	Cm, Sm, Su, Tc	20
RTF1		20
IncFIII		
ColB-K98 (= ColB4-K98)	Cba	16, 20
ColB-K166	Cba	16, 20
MIP240 (= pIP240)	Hly	20, 33a
pSU306	Ap	16
pSU316	Hly	110
IncFIV		
ColVB*trp*	Cba, Cva, Trp	165
R124	Tc	16, 20
R124*drd*-2	Tc	28
IncFV[b]		
F$_O$*lac*[c]	Lac	20, 28
28pED208[c] (= EDP208)	Lac	50
pGL109		16

(continued)

TABLE 2.1 (*Continued.*)

Incompatibility Group and Plasmid	Associated Phenotype[a]	References
IncFVI		
pSU1	Hly	33a
pSU104		28
pSU105	Hly	33a
pSU212 (= Hly-212)	Hly	33a
IncFVII		
pSU233	Hly	110a
Unclassified F-like		
ColV3	Cva	20
R36	Tc	20
R51	Tc	20
R269F	Ap, Km, Sm	20
R312	Hg	20
pSU233	Hly	110a

[a]Phenotypic symbols are Ap, ampicillin resistance; Asa, arsenate resistance; Asi, arsenite resistance; Cm, chloramphenicol resistance; Cba, colicin B production; Cva, colicin V production; Ent, enterotoxin production; Fa, fusidic acid resistance; Hg, mercury resistance; Hly, hemolysin production; Km, kanamycin resistance; Lac, lactose utilization; Nm, neomycin resistance; Pm, paromomycin resistance; Sm, streptomycin resistance; Sp, spectinomycin resistance; Su, sulfonamide resistance; Tc, tetracycline resistance; Tp, trimethoprim resistance.
[b]IncFV has previously been named IncF$_o$lac and IncFO.
[c]F$_o$lac and pED208 have also been classified as members of the IncS complex (27). Since the other plasmids of the IncS complex (R71, TP224, and pPLS) encode pili that resemble F$_o$lac pili in morphology, serology, and phage adsorption properties, these plasmids might also be considered to define new IncF complex incompatibility groups. Plasmids of the IncD group (R771b, R778b, and R687) similarly encode pili with an F-like morphology that confers sensitivity to F-specific filamentous DNA phages and could represent an additional IncF group.

complementation studies between *tra* genes from F and F-like plasmids did demonstrate that a number of them, particularly those associated with pilus production, were inter-changeable (see 88, 166 for reviews). In other cases, variation between products prevented complementation, leading to the identification of plasmid-specific alleles of some *tra* genes implicated in conjugal DNA metabolism, surface exclusion, and regulation of *tra* gene expression (165, 166).

The genetic homology implied by the complementation studies was supported by DNA:DNA heteroduplex analyses of F and selected F-like plasmids (147), which demonstrated that extensive physical homology also exists across the *tra* regions of these plasmids and that this homology extended through what is now termed the leading region of transfer, that is, the segment 5' to *oriT* (see section 10). In some instances, plasmid-specific alleles were associated with particular regions of microheterogeneity seen in heteroduplex analyses (see 88, 165, 166, 167 for reviews).

This chapter concentrates on the genetic determinants and products of the transfer and leading regions of the F group of plasmids, with particular attention to the information derived from the recently assembled nucleotide sequence of the entire F plasmid *tra* region (L.S. Frost, K. Ippen-Ihler, and R. Skurray, in preparation).

2. Genetic Characterization of the Transfer Region

Considerable work has been focused on the identification and characterization of *tra* region genes and gene products. On the F plasmid, all of the genes required for F-directed conjugation are located within the 34-kb F *tra* region shown in Figure 2.1. The majority of the F protein products stemming from the region have been identified and, in many cases, their subcellular location has been determined experimentally. Selected segments of the *tra* regions carried by F-like plasmids have also been sequenced, allowing comparison of a number of different alleles (30, 40, 44, 46, 48, 49, 50, 51, 58, 68, 73, 84, 99, 100, 110, 119, 120, 133, 174, 175).

The functions of 19 F *tra* products (from genes *traA, B, C, D, E, F, G, H, I, J, K, L, M, N, S, T, U, V, W*) were initially deduced from the characteristics of F*lac* transfer-deficient or surface-exclusion-deficient derivatives carrying amber, frameshift, or missense mutations that affect individual *tra* cistrons (166). Mutations affecting the capacity of F genes to be expressed in the presence of repressed F-like plasmids, such as R100, similarly led to identification of *finP* and *finO* loci (52). In addition, protein product and DNA sequence analyses revealed the existence of 16 transfer region loci that were not represented in earlier mutant collections. Those transcribed in the *tra* gene direction (Figure 2.1) were named *traP, Q, R, Y, X* and (since the alphabet was exhausted) *trbA, B, C, D, E, F, G, H, I, J* (Frost et al, in preparation); one open reading frame on the opposite strand is known as *artA* (171). Through assays of transfer-related functions and the construction of F derivatives carrying kanamycin (*kan*) resistance insertions, the participation of many of these loci in conjugative transfer has also been investigated. Nevertheless, the role of a number of *tra* region gene products remains to be discovered.

3. Transfer Gene Functions

As indicated in Figure 2.1, most F transfer region genes can be grouped according to the conjugation stage or transfer-related property that requires their expression (for reviews of the stages, structures, and mechanistic processes of conjugation see 7, 87, 88, 138, 148, 166, 167, 168). These groups include the following:

1. *Regulation:* Expression of transfer functions is dependent on the positive regulatory product, TraJ. TraJ expression is repressed by the *finP* product when a *finO* product is available.
2. *Synthesis of pili:* Donors carrying F-like plasmids elaborate pili that extend from the donor surface and initiate the mating contact with recipient cells. Donors that are unable to express pili do not form mating aggregates and are transfer deficient.
3. *Aggregate stabilization:* Some transfer-deficient donors express pili and initiate mating contacts but do not accumulate in mating aggregates measurable in a Coulter counter. Such mutants are thought to be deficient in steps required to stabilize the mating contact.
4. *Conjugative DNA metabolism:* Other transfer-defective mutants express pili and form stable mating aggregates but are, nevertheless, unable to transfer DNA to recipient cells. This group includes mutants defective in genes required for nicking

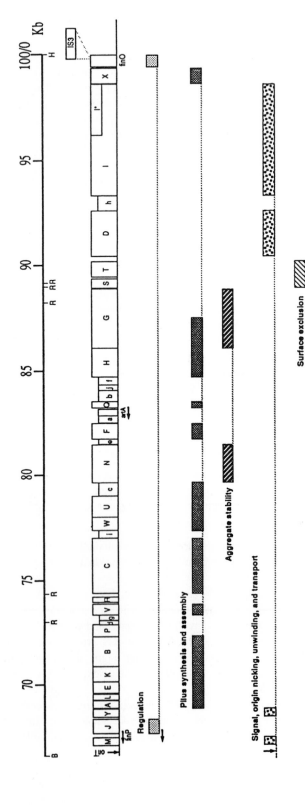

FIGURE 2.1. Map of the F plasmid transfer region. Boxes indicate the position and size of F transfer region genes, transcribed from left to right. Capital letters indicate *tra* genes; small letters indicate *trb* genes. The *finP* RNA product and *artA* product are encoded by the anti-*tra* strand. Where known, functional groups are indicated by the character of the shaded areas below the map. A line above the map gives F plasmid coordinates from the 100-kb map (167); H, R indicate *Hind*III and *Eco*RI sites, respectively, and B indicates the *Bgl*II site at 66.6F, for correspondence with Figure 2.2. Drawn according to compiled DNA sequence data for *traM-traJ-traY* (55, 84, 153), *traA-traL-traE* (57), *traK-traB-traP-trbD* (L.S. Frost, personal communication), *trbG-traV-traR* (T. Doran et al, unpublished data), *traC* (145), *rbI-traW* (114a), *traU* (128), *trbC-traN-trbE* (113, 114), *traF-trbA-artA-traQ-trbB-trbJ* (171, 173), *trbF-traH-traG* (53, 78), *traS-traT-traD* (91, 92), *trbH-traI* (18), *traX-finO* (30), *traD → finO* (174, 175).

the DNA at the origin of transfer or for displacement or transport of single-stranded DNA.

5. *Surface exclusion:* The *traS* and *traT* products are not required for transfer; these proteins interfere with the entry of F-like plasmid DNA into a cell that already contains the same F-like plasmid.

4. Transfer Loci Involved in Gene Regulation and Expression

Transfer region regulatory products are listed in Table 2.2. Important promoters located near the beginning of the *tra* region include P_M and P_J, which initiate individual transcripts for *traM* and *traJ*, respectively (35, 37, 63, 129, 153), and the major *tra* operon promoter P_Y, located just upstream from the *traY* gene. The majority of *tra* region loci are expressed from P_Y. Activation of this promoter requires the plasmid product, TraJ, and the host protein, SfrA (63, 129, 149, 150); the precise mechanism by which TraJ acts remains to be established. The chromosomally encoded integration host factor (IHF) has also been implicated in the efficiency of expression from P_Y, P_M, and P_J (34, 38, 64, 149, 150). Because the TraJ product is the positive regulator for expression of the majority of downstream *tra* genes (*traY–traX*) (Figure 2.1), plasmid-specific regulation of transfer functions can be effected through control of TraJ synthesis. This becomes repressed if two "F inhibitory" or "fertility inhibition" loci, *finO* and *finP*, are active.

4.1. FinOP Regulation of *traJ* Expression

The transcript initiated from the *traJ* promoter includes an untranslated leader sequence that can interact with FinP. The FinP repressor is a small RNA transcript, complementary to the *traJ* mRNA leader sequence and the translational start site for *traJ* (35, 44, 51, 73, 110, 129). FinP is expressed from a promoter located near the beginning of the *traJ* open reading frame but on the antisense strand (Figure 2.1). As *finP* RNA inhibits *traJ* expression by annealing to its complementary sequence on the *traJ* mRNA, *finP* genes are plasmid specific. The RNA in the *finP* region can form two stem-loop secondary structures, one of which could interact with the *traJ* mRNA ribosome biding site; *finP*-specific differences occur within these loops, and mutations affecting *finP* activity have been traced to alterations that change the stability of a stem or inactivate the *finP* promoter

TABLE 2.2   F Transfer Region Regulatory Products

Product	Size[a] (kDa)	Comment, References
FinP	–	An RNA molecule complementary to the *traJ* mRNA leader sequence (51, 60)
FinO	21.2	Predicted cytoplasmic protein. F plasmid FinO protein is not made because of IS3 insertion in *finO*. Predicted size is taken from sequence analysis of *finO* from R100 and R6-5 (23, 30, 174, 175)
TraJ	27.1	Cytoplasmic, positive control protein (31, 32, 153)

[a]Size predicted from DNA sequence.

(51, 60). Mutation of the FinP loop sequence has also been shown to affect transfer repression of a wild-type plasmid *in trans* (98). Excess expression of the R100 *traJ* leader sequence leads to derepression of R100 transfer, and it has been suggested that transcripts initiated in the R100 *traM* gene may play a similar role (36, 37).

Although *finP* RNA synthesis is constitutive, inhibition of *traJ* expression occurs only when a *finO* gene has also been expressed in the host. All F-like plasmids examined carry *finO* sequences, and while the majority are FinO⁺, for example R1, R6-5, and R100, others have a mutation in *finO* and are derepressed for transfer, for example R1*drd*-19, R100-1*drd*, and ColV2-K94 (Table 2.1) (26, 121). Because the *finO* gene of F is inactive through the insertion of IS*3* (23, 174) (Figure 2.1), F *tra* genes are expressed constitutively unless FinO is expressed from a coresident plasmid. Recent studies suggest that the *finO* product may be required to stabilize *finP* RNA (105a). Polypeptide analyses and predictions from the nucleotide sequence indicate that FinO is a 21.2-kDa protein (24, 30, 39, 156, 161, 174, 175). The degree of F transfer inhibition that results appears to correlate with the amount of the FinO protein available and, interestingly, cotranscription with an additional locus, ORF286 (or ORFC), seems to affect *finO* expression (161). Plasmids such as R100 and R6-5, which carry an ORF286 locus encoding a 32-kDa protein and mapping just upstream from *finO*, repress F transfer 100- to 1000-fold; ColB2 and other FinO⁺ plasmids that lack an ORF286 sequence repress F transfer only 20- to 50-fold (24, 30, 161, 165). The F plasmid is deleted for ORF286 sequences with the possible exception of the most distal few nucleotides (30), suggesting that in the FinOP⁺ ancestral form of the F plasmid, *finO* was also subject to the *cis* effect of the adjacent ORF286 locus.

4.2. Other Fin Systems

Five other F transfer inhibition systems (FinC, Q, U, V, W) have been described (63, 166), and, for all, the exact site of action and mechanism of inhibition remain to be identified. To date, only the sequences for genes involved in FinC (130) and FinQ (76) have been reported. FinC is specified by copy number mutants of plasmid CloDF13; high cellular levels of the products of two genes, *mobA* and *rpi*, are able to act independently to reduce both conjugal DNA transfer and the efficiency of plating of f2 RNA male-specific phage (but not pilus formation) by a F*lac* donor (130). FinC has been proposed to act to interfere with the function of *traD* (163). FinQ is specified by IncI₁ plasmids, such as R62 and R820a, and affects conjugal DNA transfer, pilus formation, and surface exclusion (166). The *finQ* gene product, a 40-kDa protein, is proposed to be a rho-independent transcriptional terminator acting predominantly at a site, *fisQ*, located between *traC* and *traN* (63, 76, L. Ham and R. Skurray, unpublished results).

4.3. Expression of *tra* Operon Genes

Because full expression of even such distal transfer genes as *traD* and *traI* is dependent on expression from P_Y, it has been suggested that transcription from P_Y continues through these loci (63, 81, 111b). If so, transcription presumably continues through to *traX*, and P_Y

controls a *traY-X* operon. However, the extent of the transcript initiated at P_Y remains to be determined, and other models are possible (77). The P_Y transcript, like other long transcripts, is expected to be processed into mRNA species that vary in stability. The transcript from which pilin subunits are synthesized has been shown, in fact, to be a stable core RNA that results from such processing. The 5′ ends of R1 plasmid *traA* transcripts vary, but the 3′ end is constant, apparently protected from degradation by stem and loop structures formed within the *traL* sequence (97).

Low-level, P_Y-independent expression of distal *tra* loci has also been detected. For example, the product of the surface exclusion gene, *traT*, can be detected even in *traJ* mutants or under conditions of transfer repression (25, 125, 140, 141). Evidence that constitutive promoter sequences precede the F *traS* and *traT* genes and the R100 *traT* locus has been described (77, 92, 116, 133). Transcripts initiated upstream to and encoding the F surface exclusion gene, *traT*, end at a strong, transcript-terminating, stem-loop structure in the *traT-traD* interval (77). Additional evidence suggests that the distal genes, *traD* and *traI*, are constitutively expressed (63, 77, 92, 129). Characterization of a promoter sequence that precedes the F *trbF* gene indicates it has TraJ-dependent activity (78). The activity of a promoter in the anti-*tra* orientation, for *artA*, has also been reported (171). The extent to which such promoters influence transfer properties is presently unknown.

In hosts expressing T7 RNA polymerase, F derivatives carrying a bacteriophage T7 late gene promoter sequence instead of the P_Y promoter sequence are able to express F pili and transfer to recipients (111b). In this case, F transfer operon expression is TraJ- and SfrA-independent, and transfer operon products can be specifically labeled in the presence of rifampicin.

5. Genes Involved in the Expression of Pilus Filament

Expression of the transfer region of an F-like plasmid leads to the elaboration of long, flexible, plasmid-specific filaments. Usually only one or two F pili are visible per donor cell; these hollow, cylindrical filaments, 8 nm in diameter with an axial hole of 2 nm, commonly extend 1 to 2 μm from the cell surface (117, 138, 166). F pili act as receptors for "male-specific" bacteriophages: RNA phages such as f2, MS2, R17, and Qβ adsorb to the circumference of pili, while filamentous DNA phages such as f1, fd, and M13 attach to the tip (see Chapter 7 in this volume). The pili of other F-like plasmids are essentially similar but exhibit variations in phage adsorption, phage sensitivity, and serological properties (104, 137, 138). It is thought that mating is initiated by an interaction between the pilus tip and the recipient surface. F hosts that are unable to express F pili, or from which pili have been removed, do not form mating aggregates or initiate conjugal DNA metabolism.

F pili are assembled from an inner membrane pool of F-pilin subunits (124). Synthesis of these subunits involves expression of three *tra* gene products, and at least 11 additional products are required for subunit assembly (see below). Although the genetic loci affecting pilus number and outgrowth/retraction kinetics remain uncharacterized, the assembly process does appear to be reversible (54, 132, 151). Thus, the F pilus acts as an external receptor; once phages and recipient cells attach to the pilus, they can be brought to the cell surface when the pilin subunits disassemble and the pilus retracts.

5.1. The F-Pilin Subunit

Synthesis of a mature F-pilin subunit requires the products (Table 2.3) of three genes, *traA*, *traQ*, and *traX*, located at the beginning, middle, and far end of the transfer operon. Although the mRNA species from which the three genes are expressed have not been characterized, the disparate location of these genes suggests that synthesis of mature pilin subunits requires most of the *tra* region to be expressed. Whereas F pilin is a 70 amino acid polypeptide with an acetylated amino terminus, the F-pilin structural gene, *traA*, encodes a 121 amino acid pilin precursor protein from which a 51 amino acid signal sequence must be removed (57). The *traQ* product is required for efficient utilization and processing of this precursor (89, 103, 111a, 126, 172), while *traX* is required for acetylation of the pilin polypeptide amino terminus (128a).

The pilin proteins and pilin structural (*traA*) genes of F and 11 other IncFI and IncFII plasmids, including R100-1, R1, and ColB2, have been characterized (57, 58). Both the signal sequences and processed products of these *traA* genes are very similar. As in the F *traA* product signal sequence, hydrophobic residues at the amino terminus precede a 19 amino acid segment that is punctuated with positively charged residues. This charged region is followed by a more classic signal peptidase I signal recognition sequence; a hydrophobic core precedes an Ala-Met-Ala | Ala processing site. However, while the pilins of F-like plasmids also have a blocked (Ac-Ala) amino terminus, variation in the amino acid sequence immediately following the amino terminus confers group-specific serological differences among these pili. The acetylated amino-terminal amino acid sequence has been shown to be the major antigen for F and F-like pilins (47, 59, 170). Thus, polyclonal antisera raised against purified wild-type F pili reacts well with mature F-pilin subunits but poorly with the unprocessed or unacetylated precursor products from *traA* (103). Analysis of the IncFV plasmid pED208 (Table 2.1), a transfer-derepressed derivative of F_o *lac* formed through the insertion of IS2 near the *tra* operon promoter, has revealed that, while the sequences of its *traA* gene and pilin product show very little correspondence to F-like sequences, pED208 pilin also has an aceytlated N-terminus and is predicted to have a similar structure (50, 169, 170). Interestingly, elaboration of pED208 pili does confer sensitivity to filamentous DNA phages, such as M13, but not to the RNA phages, f2, R17, MS2, or Qβ. A different group of RNA phages infect IncFV plasmid hosts (27).

The *traQ* gene was located through the discovery that membranes from Hfr deletion mutants lacking the *tra* region distal to *traF* do not contain the 7.2-kDa pilin polypeptide (126). Analysis of products from *traA* clones further showed that removal of the *traA*

TABLE 2.3   F Transfer Region Products Involved in Pilin Synthesis

Product	Size[a] (kDa)	Comment, References
TraA	12.8	Processed to the 7.2-kDa N-acetylated F-pilin subunit; unassembled pilin detected as an inner-membrane protein (57, 103, 111a, 126, 128a)
TraQ	10.9	Inner-membrane protein required for processing TraA (111a, 171, 172)
TraX	27.5	Required for pilin subunit modification (N-acetylation) but not for filament assembly; predicted to be inner-membrane protein (30, 61, 79, 128a)

[a]Size predicted from DNA sequence.

product signal peptide is dependent on TraQ, a 94 amino acid, inner-membrane protein (171, 172). In the absence of *traQ* expression, the 12.8-kDa product of *traA* is unstable, and very little 7.2-kDa polypeptide is detectable; when TraQ is available, expression and processing of the *traA* product occurs efficiently and membrane pilin polypeptides accumulate (111a). A *traQ::kan* insertion mutant that has been constructed was found to be extremely transfer deficient and to lack both F pili and membrane F-pilin subunits. However, it could be complemented to a normal phenotype by a clone that expresses the *traQ* gene (94).

The 7.2-kDa *traA* product synthesized in hosts that express *traA* and *traQ* clones was found to react very poorly with antipilin sera and to migrate slightly faster than mature pilin (103). Identification of the F gene required for the N-terminal acetylation of F pilin took advantage of the availability of two monoclonal antibodies: JEL93, specific for the acetylated F-pilin amino-terminal sequence, and JEL92, specific for an internal epitope (59). Western blots of inner-membrane preparations showed that only *traA* and *traQ* were required to obtain expression of a membrane-pilin polypeptide reaction with JEL92. However, reaction with JEL93 required expression of the *traX* sequence (128a). Interestingly, pilus filaments are elaborated by strains that lack *traX*. Indeed, in standard *Escherichia coli* matings, the phenotype of a *traX::kan* insertion mutant is relatively normal, and plasmid transfer and F-pilus-specific phage infection frequencies approximate that of the wild type (K. Maneewannakul and K. Ippen-Ihler, unpublished data). Nevertheless, pili purified from a *traX::kan* mutant contain pilin subunits with an unblocked Ala-Gly-Ser amino terminus (D. Moore and K. Ippen-Ihler, unpublished data). These *traX*-mutant pilin subunits may be arranged differently in the pilus filament. JEL93 and JEL92 do not usually bind to pilus filaments (59), but Grossman et al (75) found JEL92 bound along the length of the numerous pili expressed by pTG801, an F *tra* region clone that lacks sequences distal to *traG* (74).

After assembly, pili contains F-pilin subunits related by a fivefold rotation axis around the helix axis (117). Because the monoclonal antibodies specific for epitopes at and near the pilin amino terminus typically bind to disorganized subunits present in the vesiclelike basal knobs seen at the ends of purified free pili, but not to the sides or tips of normal F pili (59), it appears that the amino-terminal end of the F-pilin polypeptide is usually not exposed on the filament surface. It cannot be ruled out, however, that under some circumstances amino-terminal sequences are exposed at the tip of pili (59, 170). Analysis of F *traA* mutations that alter RNA phage sensitivity, as well as the altered *traA* sequences and phage sensitivities associated with different F-like plasmids, has suggested that F pilin domains at residues 14 to 17 and at the carboxy terminus are exposed along the length of the filament (56, 58).

5.2. Pilus Assembly Proteins

F derivatives carrying mutations in genes *traL*, *traE*, *traK*, *traB*, *traV*, *traC*, *traW*, *trbC*, *traF*, *traH*, or in the N-terminal region of *traG* do not elaborate F pili, although their membranes contain F-pilin subunits (113, 125). Mutations in *traU* also cause a reduction in pilus number and affect RNA phage susceptibility (128). Thus, all 12 products of these genes (Table 2.4) are thought to be involved in pilus assembly. A defect in any of these products causes a profound reduction in transfer frequency.

TABLE 2.4   F Transfer Region Products Involved in Pilus Assembly

Product	Size[a] (kDa)	Comment, References
TraL	10.4	Inner-membrane association predicted (57, 61)
TraE	21.2	Inner-membrane protein (11, 57, 89, 103)
TraK	25.6	Predicted to be processed to 23.3-kDa periplasmic protein (61, L.S. Frost, personal communication)
TraB	50.5	Inner-membrane association predicted (61, L.S. Frost, personal communication)
TraV	18.6	Predicted to be processed as a lipoprotein of 16.6 kDa (mature, unmodified) attached to outer membrane; observed size 21 kDa (61, 127, T. Doran et al, unpublished data)
TraC	99.2	Cytoplasmic protein; possibly inner-membrane associated in presence of other *tra* membrane proteins (145, 145a)
TraW	23.6	Processed to a 21.7-kDa periplasmic protein (112, 114a)
TraU	36.8	Processed to a 34.3-kDa periplasmic protein; *traU* mutants make pili, but in reduced numbers (128)
TrbC	23.4	Processed to a 21.2-kDa periplasmic protein (113)
TraF	28.0	Processed to a 25.9-kDa periplasmic protein (173)
TraH	50.2	Predicted to be processed to a 47.8-kDa periplasmic protein (53, 78, 115)
TraG	102.4	Inner-membrane protein; deletion of C-terminus does not affect piliation (53, 115)

[a]Size predicted from DNA sequence.

5.2.1. Inner-Membrane Proteins

Although TraC can be purified as a soluble protein, recent studies with anti-TraC serum suggest that it is normally associated with the membrane through interactions with other *tra* proteins (145, 145a). This association may involve the *traL*, *traE*, *traB*, and *traG* products, pilus assembly proteins that all contain membrane-spanning regions consistent with insertion in the cytoplasmic membrane (53, 57, 61, L.S. Frost, personal communication). Preliminary results show that *trbI* and *traP* mutants make pili but exhibit increased male-specific phage resistance (114a, P. Kathir, D. Moore, and K. Ippen-Ihler, unpublished data), so the *trbI* and *traP* products, also inner-membrane proteins (114a, L.S. Frost, personal communication), could also be involved in pilus assembly. The actual conformation of the *traL, E, B, P*, and *G* products and their individual interactions with F pilin and the *tra* proteins on the cytoplasmic and periplasmic side of the membrane remain to be investigated.

5.2.2. Periplasmic Proteins

The products of genes *traW*, *traU*, *trbC*, and *traF* have been shown to undergo signal-sequence processing and to be enriched in periplasmic protein fractions (114a, 113, 128, 173). The sequence of *traK* and the sequence and polypeptide analysis of *traH* suggest that these gene products are also processed to periplasmic proteins (53, 61, 78, 115, L.S. Frost, personal communication). It seems likely, then, that these six products interact with each other to form a periplasmic complex that, in association with the *tra* membrane proteins, achieves the assembly of pilus filaments. The *trbB* product could be a seventh member of the complex, since this is also a periplasmic protein (171). However, *trbB* is apparently dispensable for F-specific phage sensitivity and transfer under the standard conditions that have been tested (94).

5.2.3. Outer-Membrane Proteins

Interaction with the *traV* product, also required for elaboration of the pilus filament, might possibly provide passage of the assembled filament through the outer membrane. The DNA sequence of this protein indicates TraV may undergo lipoprotein processing, in which case it would be likely to be attached to the outer membrane by lipid modification (T. Doran, S. Loh, N. Firth, and R. Skurray, unpublished data). The two other *tra* region products (TraN and TraT) that localize in the outer membrane have no known involvement with pilus expression.

6. Products Required for Aggregate Stabilization

Mutations in *traN* or in the C-terminal segment of *traG* result in F-piliated, transfer-deficient strains. Studies of mating mixtures containing *traN* and piliated *traG* mutant donors have led to the suggestion that these two gene products (Table 2.5) may be required to advance or stabilize an interaction between the surfaces of conjugating donor and recipient cells (115). Observations with *E. coli* have indicated that mixtures of piliated F donors and recipient (F$^-$) cells typically aggregate into clusters of cells in surface contact (6). Treatment with 0.01% sodium dodecyl sulphate (SDS) to disrupt F pili can prevent the aggregates from forming but does not block DNA transfer once they have been established (9, 136). Thus, although there is some evidence that DNA transfer can occur through extended F-pilus filaments (79, 135), transfer typically occurs between cells held close together. Indeed, a recent report has described the formation of wall-to-wall contacts between donor and recipient cells, as observed by video microscopy, and has correlated these contacts with conjugative junctions between the outer membranes of the cells using thin-section electron microscopy (41). Interestingly, no fusion or channel between the inner membranes of conjugating cells could be detected in these studies. For discussions of the recipient cell surface and its role in conjugation, see reviews by Achtman and Skurray (7) and Willetts and Skurray (166).

TABLE 2.5 F Transfer Region Products Required for Mating Aggregate Stabilization

Product	Size[a] (kDa)	Comment, References
TraN	65.7	Processed to 63.8-kDa outer-membrane protein (114)
TraG	102.4	C-terminal portion of protein essential for stable aggregates (53, 115)
TraG*	50.0[b]	Periplasmic-located product corresponding to C-terminus of TraG; probably results from internal cleavage. Function unknown but contains sequences shown to be required for stabilization (53)

[a]Size predicted from DNA sequence.
[b]Size observed; exact N-terminus to be determined.

Manning et al (115) tested a number of piliated transfer-deficient F*lac* mutants for aggregation defects. Their Coulter counter analyses indicated that, whereas *traM*, *traD*, and *traI* mutations had no effect on aggregation properties, *traN* and *traG* mutations blocked aggregate accumulation. However, the work of Kingsman and Willetts (95) had shown that these same *traN* and *traG* mutations did not block triggering of donor conjugal DNA synthesis, a pilus-dependent event requiring contact with recipient cells. Thus, Manning et al (115) suggested that *traN* and *traG* products might be required to stabilize the initial cell contacts made after F pili retract. Defects in these two proteins would cause cells to quickly fall apart again.

Product analysis indicates that, together, the *traN* and *traG* proteins span the donor envelope. TraN is in a position to interact directly with the recipient envelope. A part of this 66-kDa outer-membrane protein has been shown to be exposed on the cell surface (114). Other domains of TraN might interact with a periplasmic *traG*-product component. The 102-kDa bifunctional *traG* product has two large hydrophilic domains that are punctuated by membrane-spanning segments responsible for the association of the protein with the inner membrane. Both large hydrophilic portions of TraG are believed to reside on the periplasmic side of the inner membrane (53). Only the N-terminal domain of TraG is required for pilus assembly. The C-terminal segment of TraG, which includes one of the large hydrophilic domains and apparently extends into the periplasm, may be released by processing. Using anti-TraG serum, Firth and Skurray (53) detected not only the 102-kDa TraG protein but also, in a periplasmic fraction, a 50-kDa C-terminal polypeptide, TraG*. These studies raise the question of whether aggregate stabilization depends on the entire *traG* product or only on the activity of TraG*.

7. Surface Exclusion Genes

On F-like plasmids a pair of surface exclusion genes, *traS* and *traT*, is located just distal to the large group of *tra* region genes required for formation and stabilization of mating aggregates. Although expression of these genes increases when the major *tra* promoter is active, promoters preceding *traS* and *traT* loci may allow the surface exclusion (Sfx) proteins to be synthesized even under conditions of transfer repression (25, 77, 92, 116, 125, 133, 140, 141, 152). The two surface exclusion products (Table 2.6) erect a barrier, but not a total block, to entry of a like plasmid into the cell. For example, F transfer frequencies are reduced several hundredfold when F Sfx[+] recipients are used. The barrier is specific and may permit normal entry of a different F-like plasmid. Willetts and Maule (164) identified four surface exclusion specificity groups represented by (a) F; (b) ColV2, R538-1;

TABLE 2.6 F Transfer Region Products Required for Surface Exclusion

Product	Size[a] (kDa)	Comment, References
TraS	16.9	Inner-membrane protein (92)
TraT	26.0	Processed to 25.9-kDa (observed) outer-membrane lipoprotein; predicted, mature, unmodified size 23.8 kDa (92, 123, 139)

[a]Size predicted from DNA sequence.

(c) ColVB*trp*, R1-19; and (d) R100-1, pED241 (= R136*i*⁻1). A fifth group is represented by the F₀*lac* derivative, pED208 (46).

7.1. *traT*

The *traT* product is a lipoprotein, detectable as a major outer-membrane component of an F plasmid host (92, 122, 123, 139). TraT proteins are also known to enhance the survival of pathogenic bacteria in serum and may be separately encoded by virulence plasmids. The properties of this family of lipoproteins were reviewed recently by Sukupolvi and O'Connor (152), who have also compared known *traT* product sequences. Interestingly, TraT group specificity is derived from a single amino acid differences in a five-amino acid region (residues 116–120) of the polypeptide chain (80).

When F donors are mixed with recipient cells that express F *traT*, mating aggregates fail to accumulate; however, transfer is depressed only about 10- to 20-fold (8). TraT effects are stoichiometric, and TraT specificities seem to correlate with the group specificities found at the amino terminus of pilins (58, 164, 165). Minkley and Willetts (123) found that purified F TraT protein could also inhibit mating, reducing F transfer more severely than transfer of R100-1. They suggested that TraT interacts with pilus tips, competitively inhibiting an interaction with their conjugal receptor and reducing the rate of functional pairing. Harrison et al (80) have similarly shown that purified R6-5 TraT protein has a group-specific transfer inhibitory effect. However, Riede and Eschbach (144) have suggested that TraT masks a region of the outer-membrane protein, OmpA, that would normally be exposed on recipient cell surfaces. OmpA is implicated in donor-recipient surface interactions (for reviews see 7, 166). F donor aggregates formed with OmpA-deficient recipient cells are unstable, although transfer occurs normally if these cells are mated on an agar surface (10).

7.2. *traS*

Characterization of *traS* function may eventually provide an important key to distinguishing preliminary-pairing from functional-pairing events. Although the presence of this small inner-membrane protein does not significantly impede mating aggregate formation, F DNA entry into TraS⁺ recipients is reduced 100- to 200-fold (8). Kingsman and Willetts (95) found that recipients expressing F surface exclusion proteins failed to trigger conjugative DNA events in F donors. Thus, Manning et al (115) suggested that TraS may interfere with transmission of the mating signal. The two sequences of *traS* products that are available (F and pED208) vary considerably (46, 92). TraS might therefore interact with plasmid-specific components, such as those thought to be associated in the *oriT* complex.

8. Genes Required for DNA Nicking, Displacement, and Transport

DNA-related events in F-like plasmid donors include the introduction of a single-stranded nick at the origin of transfer, *oriT*; 5′ → 3′ displacement of the nicked strand; and transport of the displaced DNA strand into the recipient cell. Complementary strand synthesis, replacing the transmitted strand, usually also occurs in mating donors. An in vivo prerequisite to conjugal DNA metabolism is that a functional, intercellular conjugal contact must first be made. Kingsman and Willetts (95) found that donor replacement-strand synthesis was undetectable in mixtures containing either mutant donors that lacked F pili or recipient cells that expressed F surface exclusion proteins. They suggested that cell contact generates a "mating signal" that precipitates DNA strand displacement and replacement synthesis. Details of *oriT*-related events are discussed more fully in Chapter 5. The plasmid-specific *tra* proteins TraM, TraY, and TraI (Helicase I) have been postulated to perform *oriT*-related functions (Table 2.7). TraI exhibits the enzymatic activities required for both nicking and strand displacement, while TraM and TraY appear to participate in the nicking complex in vivo, perhaps by contributing to complex formation or by mediating the mating signal that triggers events at *oriT*. Although the structures of TraM and TraY indicate that, like TraI, they are cytosolic proteins, they could normally interact with inner-membrane *tra* products. The inner-membrane protein, TraD, is thought to be required for DNA transport to recipients (Table 2.7).

8.1. *oriT*

DNA sequencing studies have defined the *oriT* regions of F (154), R1 (134), R100 (119), ColB4-K98 (49), P307 (68), and pED208 (40). The sites of the TraI-dependent F *oriT* nick (118, 143, 155) and R100 *oriT* nick (86) have also been characterized. Deletion analysis has identified sequences on both sides of the nick site that contribute to nicking and transfer from *oriT* (62). F *oriT* region features include two intrinsic bends and binding sites for integration host factor (IHF), TraY, and TraM (40a, 102, 160). Similar features have been identified in the R100 *oriT* region (85, 86, 119). Plasmic specificities reflect differences in the *tra* protein sequences and their corresponding binding sites.

8.2. *traI*

Although located 26.5 kb from *oriT*, the *traI* gene plays a central role in *oriT*-related events. F *traI* encodes both a full-length, 192-kDa product (TraI or Helicase I) and a smaller,

TABLE 2.7 F Transfer Region Products Involved in Conjugal DNA Metabolism

Product	Size[a] (kDa)	Comment, References
TraM	14.5	F-specific; multiple binding sites in *oriT-traM* interval. Soluble protein, possibly associates with membrane. Required for mating signal? (40, 40a, 146, 153)
TraY	15.2	F-specific; soluble protein required to nick *oriT* in vivo. Binds near *oriT*. F *traY* results from duplication; *traY* genes of F-like plasmids are smaller (43, 55, 57, 83, 84, 102)
TraD	81.7	Inner-membrane protein required for DNA transport (91, 136a, 175)
TraI	192.0	Helicase I; introduces nick at *oriT* and unwinds $5' \rightarrow 3'$ (4, 18, 118, 143)
TraI*	87.8	Cytoplasmic protein arising by translational restart in *traI*; function unknown (18, 158)

[a]Size predicted from DNA sequence.

antigenically related, 87.8-kDa polypeptide (TraI*) stemming from a second translational start site within *traI* (18, 158, 159). Helicase I is a well-characterized ATP-dependent helicase (1–4, 15, 101, 162). Its capacity to processively unwind DNA in the $5' \rightarrow 3'$ direction is consistent with the direction in which the conjugally transferred DNA strand is displaced. Recent work has shown that TraI/Helicase I also has an additional function. A distal *tra* activity (originally named TraZ) had been known to be required for nicking λ-F *oriT* transducing phages in vivo (43). Traxler and Minkley (159) showed that this activity stemmed from *traI*. In vitro analyses have further demonstrated that,in the presence of Mg^{+2}, Helicase I forms a complex with superhelical F *oriT*-plasmid DNA, and, when heated with SDS or (ethylenedinitrilo) tetraacetic acid (EDTA) or treated with protease K, introduces a site- and strand-specific nick at *oriT* and remains bound to the 5' nick terminus (118, 143). Interestingly, the amino-terminal region of TraI/Helicase I is not required for DNA unwinding but is required for *oriT* nicking (15, 143, 159).

8.3. *traY*

While the in vitro TraI-dependent, F *oriT*-nicking reactions described can proceed in the absence of F TraY protein, Inamoto et al (86) have recently described an in vitro R100 *oriT*-nicking reaction dependent on both the R100 *traY* and *traI* products. This result is consistent with the report of Everett and Willetts (43) that in vivo nicking of λ F *oriT* phages depended on expression of F *traY* as well as a distal (F *traI*) locus. Both *traY* and *traI* are also required for *oriT*-enhanced recombination (21). Thus, although F TraI, alone, can nick *oriT* in vitro, in vivo conditions may require TraY.

The *traY* sequences of F (55), R100 (84), RI and R1-19 (48, 99), ColB4-K98 (49), P307 (73), and pED208 (50) have been characterized, and the TraY proteins of F and R100 have been purified (83, 102). Interestingly, *traY* sequences begin with unusual amino-terminal start codons, UUG (84, 102) or GUG (73, 99). The *traY* gene of F appears to result from tandem duplication, since both the amino-terminal region and the carboxy-terminal region of its 131 amino acid product exhibit homology to the smaller TraY products of other plasmids, such as the 75 amino acid R100 TraY product (84). Two TraY protein binding sites have been identified: *sbyA* is between the two intrinsic bends and IHF binding sites near *oriT*, while *sbyB* is near the P_Y promoter (83, 102). Thus, association of TraY polypeptides bound to these two sites might effect formation of a loop of the intervening DNA. Studies with *oriT* and P_Y plasmids that include both *sby* sites will be of interest in exploring the contribution of TraY to origin nicking, as well as the possibility that TraY plays a regulatory role in *tra* transcription. With respect to the latter, TraY has been identified as a member of the Mnt and Arc family of repressor proteins (17).

8.4. *traM*

The 14.5-kDa F *traM* product is encoded by *oriT* distal sequences preceding *traJ* and the major *tra* region promoter (153). The *traM* sequences of R100 (44), R1 and R1-19 (48, 100), ColB4-K98 (49), P37 (73), pSU316 (110), and pED208 (40) are available for comparison. Experiments with purified TraM proteins have shown that TraM binds to a series of *oriT* region sites located between *sbyA* and the *traM* open reading frame (5, 40a, 146). The TraM binding region extends to and overlaps the *traM* promoter indicating that *traM* transcription can be autoregulated.

Kingsman and Willetts (95) found that donor DNA replacement synthesis required *traM* as well as *traI*; a similar requirement for *traY* is predicted from the *traY* dependence of in vivo nicking assays. Because no requirement for TraM was apparent in λ-*oriT* nicking assays, everett and Willetts (43) suggested that TraM might convey the mating signal to the F plasmid, precipitating one of the events that leads to strand displacement. Further studies are, however, needed to determine exactly how TraM acts. As for TraY, possibilities include an influence on the formation or stability of TraI-*oriT* binding, or activation of TraI nicking or unwinding activities.

8.5. *traD*

The F and R100 *traD* products are predicated to be highly congruent except that R100 *traD* appears to contain an insert including an unusual sequence of repeated Gln-Gln-Pro codons in the C-terminal region of the protein (18, 91, 175). A consensus ATP binding sequence is in the amino-terminal region. The 81.7-kDa F TraD protein has been purified and shown to be an inner-membrane protein that exhibits in vitro DNA binding activity (136a). Strains carrying F *traD* mutants are very transfer deficient but elaborate an increased number of F pili and are sensitive to the filamentous DNA phages and RNA phage Qβ. However, phages of the f2, R17, MS2 family adsorb to, but do not infect, these cells. F mobilization of the ColE1 plasmid requires F *traD*, although CloDf13 mobilization is *traD* independent (168). Several additional findings suggest that TraD plays a role in nucleic acid transport. DNA replacement synthesis occurred but was reduced when mating mixtures contained F *traD* mutant donors (95). Panicker and Minkley (136) used a temperature-sensitive *traD* mutant to show that *traD* is required for transfer only after SDS-resistant mating pairs have formed. Unfortunately, the nature of the connection through which DNA transport occurs during conjugation and the identity of other components that might participate in the process are unknown. A recent study suggesting that TraD might accompany transported DNA into the recipient cell warrants further investigation (142).

9. Genes of Unknown Function in the Transfer Region

Table 2.8 lists F genes encoded within the transfer region that have been identified through either sequence analysis or sequence and product analysis. The phenotypic changes accompanying mutation of the *traP*, *traR*, *trbD*, *trbF*, *trbG*, and *trbI* loci are still under investigation, and the functional category to which these genes might belong is still unclear, although the long pili expressed by some *trbI* mutants suggest this gene product may have a role in regulating pilus assembly and/or retraction (114a). Recent studies have shown, however, that *kan* insertions that inactivate any one of the loci *traR*, *trbA*, *trbB*, *trbE*, *trbH*, *trbJ*, or *artA* do not cause any marked alteration in the male-specific phage sensitivity or plasmid transfer frequency of cultures tested under standard conditions (94, 111, 114). Such genes, like *traX* and the leading region loci, may express transfer-related functions that are

TABLE 2.8 F Transfer Region Products of Unknown Function

Product	Size[a] (kDa)	Comment, References
TraP	22.0	Inner-membrane location predicted (61, 127, L.S. Frost, personal communication)
TrbD	7.1	Product unidentified; predicted to be cytoplasmic (61, L.S. Frost, personal communication)
TrbG	9.1	Product unidentified; predicted to be cytoplasmic (61, T. Doran et al, unpublished data)
TraR	8.3	Predicted to be cytoplasmic (111, 127, T. Doran et al, unpublished data)
TrbI	14.1	Intrinsic inner-membrane protein (114a)
TrbE	9.9	Intrinsic inner-membrane protein (114)
TrbA	12.9	Inner-membrane protein (171, 172)
TrbB	19.5	Processed to a 17.4-kDa periplasmic protein (171)
TrbJ	10.2	Initiated at GUG codon; predicted to be inner-membrane protein (111, 171)
TrbF	14.5	Predicted to be intrinsic inner-membrane protein (78, Y.N. Lin, J. Tennent, and R. Skurray, unpublished data)
TrbH	26.2	Product unidentified; predicted to be inner-membrane protein (18, 111)
ArtA	12.1	Product unidentified; predicted to be an inner-membrane protein; encoded by the anti-*tra* DNA strand (171)

[a]Size predicted from DNA sequence.

dispensable in *E. coli* broth grown cells. Studies of alternate hosts and nonstandard physiological conditions may eventually reveal that these proteins also contribute to transfer. Alternatively, such proteins may express activities unrelated to conjugation. It is noteworthy that R100 does not encode an ORF corresponding to the F *trbH* sequence, and the function of an additional ORF (ORFE) located between the R100 *traT* and *traD* genes is also unknown (18, 175).

10. Genes in the Leading Region of Transfer

The sequence of the leading region, 5′ to *oriT*, is highly conserved among F-like plasmids (108). Although this 13-kb region of DNA, extending from *oriT*, at 66.7 F to the primary replication-incompatibility region at 53.8 F (Figure 2.2), is not essential to F transfer, it is considered to be the first to enter a recipient cell during conjugation.

At least two of the leading region genes encode products (Table 2.9) that would be expected to be advantageous after conjugal entry of single-stranded DNA into recipient cells. One, the F *ssb* gene, encodes a single-stranded DNA binding protein capable of complementing defects in *E. coli ssb* (22, 96). Interestingly, the IncI₁ conjugative plasmid, CoIIb-P9, also carries an *ssb* locus (82), and homology to F *ssb* has been detected on transmissible plasmids in diverse Inc groups (69). An additional locus, *psiB*, originally characterized on R6-5, also seems to be a conserved sequence among both IncF and IncI₁ plasmids (71, 82). The *psiB* product interferes with induction of the cellular SOS response by inhibiting the RecA protein coprotease activity; *psiB* also inhibits homologous recombination but not in a recipient cell following conjugation (14, 42). Notably, both *ssb* and *psiB* on CoIIb-P9 and *psiB* on F have been shown to be zygotically induced in recipient cells following conjugation, and it has been suggested that these gene products are important in

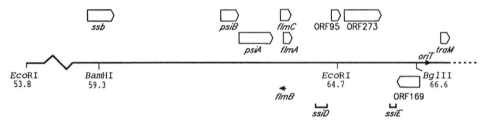

FIGURE 2.2. Map of the F plasmid leading region. The thick linear line represents the physical map of the leading region with relevant restriction sites and F coordinates shown; the arrowhead at *oriT* indicates the position of the origin of transfer and the direction of DNA transfer. The arrowed boxes represent coding regions and directions of transcription of the genes, as determined by DNA sequence analysis. The thin arrowed line represents the extent and direction of transcription of *flmB* antisense RNA and the bracketed lines indicate *ssi* sequences. Genes and the products they encode are described in the text and in Table 2.9. Three other leading region products have been identified by polypeptide analysis (29). The locations of protein 3a and 3d have been mapped to the region between *Eco*RI (53.8F) and *ssb*, and the location of protein 3e has been mapped between *ssb* and *psiB*; these regions are yet to be sequenced.

the establishment of a newly transferred plasmid (13, 93). A gene, *psiA*, located immediately downstream of *psiB* (Figure 2.2) appears not to be involved in the Psi function, although its 24-kDa protein product, PsiA, does seem to be coexpressed with PsiB (12, 14, 109).

A third locus, *flm*, on the F leading region encodes a function, Flm (Table 2.9), capable of extending the maintenance of unstable plasmids (107). This locus was previously referred

TABLE 2.9   F Leading Region Products

Product	Size[a] (kDa)	Comment, References
Ssb	30.0	Single-stranded DNA binding protein (22, 96). Zygotically induced in recipient cells following ColIb plasmid transfer and may be involved in establishment of plasmid in the recipient cell (93)
PsiB	15.7	Predicted cytoplasmic protein. Inhibits generation of signal to induce SOS genes (14, 42). Zygotically induced in recipient cells following F or ColIb plasmid transfer and may be involved in establishment of plasmid in the recipient cell (13, 93)
PsiA	28.7	Predicted cytoplasmic protein; function unknown (12, 14, 109)
FlmA	6.1	Predicted to be inner-membrane associated protein; lethal product that acts post-segregationally to kill cells that have lost F plasmid (70, 107); homologous to Hok product from R1 (66)
FlmC	7.9	Product unidentified; predicted cytoplasmic protein; FlmC and FlmA are encoded by overlapping genes; *flmA* expression may be dependent on translation of *flmC* (107, S. Loh and R. Skurray, unpublished data); homologous to Mok from R1 (67)
FlmB	–	100-nt regulatory antisense RNA that binds to *flmA* mRNA to inhibit its translation (107); homologous to Sok from R1 (66)
ORF95	11.0	Product unidentified; predicted cytoplasmic protein; function unknown (108)
ORF273	31.6	Predicted cytoplasmic protein; function unknown; equivalent to previously described protein 6d (29, 108)
ORF169	19.3	Product unidentified; predicted inner-membrane protein; function unknown (108)

[a]Size predicted from DNA sequence.

to as *parL* (106) and has also been designated *stm* (70). Like many other leading region sequences, the *flm* locus is highly conserved among F-like plasmids (S. Loh and R. Skurray, unpublished data), with the most extensive studies reported for the *parB* or *hok/sok* locus of R1 (65, 66, 67). In this maintenance system, the FlmA protein is almost identical to the 52 amino acid Hok polypeptide, a lethal membrane-associated protein that is expressed only in plasmid-free segregants, thereby ensuring the predominance of plasmid-carrying cells in a population (65, 107). Both *flmA* and *hok* sequences are overlapped by a 70 amino acid ORF designated FlmC (107, S. Loh and R. Skurray, unpublished data) or Mok (67), respectively. It is likely that *flmC* and *flmA* (*mok* and *hok*) are translationally coupled (107). The third gene of the maintenance systems of F and R1, designated *flmB* and *sok*, respectively, encodes a 100-nt-long antisense RNA that posttranscriptionally regulates expression of *flmC* (*mok*) RNA, which in turn regulates *flmA* (*hok*) translation (67, 107). The expression of the FlmA protein has been shown to rely on differential rates of decay of the stable *flmC/flmA* message compared to the unstable *flmB* RNA (107, S. Loh and R. Skurray, unpublished data). Daughter cells that fail to obtain a copy of the plasmid at cell division but do inherit *flmB*:*flmC/flmA* RNA duplexes would be killed by the subsequent expression of the lethal FlmA product after decay of the less stable *flmB* antisense RNA. The role of a plasmid maintenance function, such as Flm, following conjugal transfer could be in the killing of plasmid-free segregants that may arise through cell division after reception of the conjugative plasmid but prior to its replication. Therefore, Flm would be involved in ensuring the establishment of the plasmid within a population of cells rather than just the recipient cell per se.

The leading region also encodes a number of ORFs (Figure 2.2, Table 2.9) and products for which no function has been ascribed (29, 108). Two of these, ORF169 and ORF273, appear highly conserved among F-like and some non-F-like conjugative plasmids (108), and sequences for them from R1 are now available (72). ORF169, also designated gene *X*, has been sequenced from R100 (45). Potential IHF recognition sequences and primosome assembly sites have also been detected in the leading region sequence by computer search (108). Three of the recognized primosome assembly sites are located within two single-strand initiation sequences, *ssiD* and *ssiE*, recently identified in the leading region (131) (Figure 2.2). Primers initiated at these sequences could be involved in synthesis, in the recipient, of the complement to the transferred DNA strand.

11. Conclusions

Descriptions of the history of bacterial conjugation (see 19 and Chapter 1 in this volume) attest to the impact that the discovery and use of the F-plasmid conjugation system has had on modern molecular genetics, a theme also pursued by Francois Jacob in his account of early days in the field (90). The F system provided the first evidence that bacteria could engage in gene transfer involving communication between sexually differentiated cells, a mechanism now recognized to be widely distributed in gram-negative and gram-positive bacteria and to occur between bacteria and organisms from other kingdoms. Indeed, what is known for F has often provided a basis for analysis of these systems. Further, the F sex factor was the first plasmid to be identified, thereby providing bacterial genetics with the subdiscipline of plasmid biology.

Ever since the report by Lederberg and Tatum (105) of chromosomal gene transfer in *E. coli* K12, studies have been directed toward understanding the precise mechanistic details of the F conjugal process. The combined information from these numerous studies has allowed models for the stages in conjugation to be devised, and, with the availability of the entire *tra* region nucleotide sequence, the pace of analysis of the various *tra* gene products and their roles in these stages has quickened, particularly for those products associated with conjugal DNA metabolism. Yet, despite the sequence information, so many aspects of F conjugation remain a virtual mystery. For example, we know the cellular locations (real or predicted) of some 15 proteins involved in F-pilus biogenesis, but the exact processes whereby this structure is assembled and transported through the cell envelope to decorate the surface of the donor cell remain to be established. Equally obscure are the processes involved in the transport of DNA between donor and recipient cells and the nature of the barrier to this transport exerted by surface exclusion. The availability of antisera specific for an increasing array of *tra* proteins may provide the key toward understanding these interactions.

It is also encouraging to note that attention is again returning to fundamental questions regarding donor and recipient cell surfaces and surface interactions. The recognition of the importance of the recipient cell is also undergoing a resurgence through the finding that certain leading region genes appear to be expressed postconjugally in the recipient. The conservation of leading region sequences among F-like and non-F-like conjugative plasmids supports the contention that its continued analysis will reveal further conjugation-associated functions.

The sequence data have so far given only minimal insights into the evolution and origins of the *tra* region. One reason for this is the very limited homology that F and F-like *tra* gene products have with other sequences in the databases. Such limited homology hints at the early divergence of the F transfer genes from other bacterial and/or phage sequences. The F plasmid *tra* region is therefore likely to specify an "ancient" mechanism of gene transfer. Even after some 46 years of study, this mechanism still awaits detailed resolution.

ACKNOWLEDGMENTS. We wish to thank Sumit Maneewannakul and Sue Loh for help in preparing the figures, Neville Firth for assembling Table 2.1, both Nadim Majdalani and Neville Firth for assistance with electronic communications, and all four for many valuable discussions, suggestions, and critical reading of the manuscript. The help of Deanna Moore, Kesmanee Maneewannakul and Jane Lantz is also much appreciated. The work from our laboratories was supported by grants from the U.S. Public Health Service, the National Institutes of Health (K.I.-I.; AI14426), and the Australian Research Council (R.A.S.).

References

1. Abdel-Monem, M., and Hoffmann-Berling, H., 1976, Enzymatic unwinding of DNA. I. Purification and characterization of a DNA-dependent ATPase from *Escherichia coli*, *Eur. J. Biochem.* 65:431–440.
2. Abdel-Monem, M., Durwald, H., and Hoffmann-Berling, H., 1976, Enzymatic unwinding of DNA. II. Chain separation by an ATP-dependent DNA unwinding enzyme, *Eur. J. Biochem.* 65:441–449.
3. Abdel-Monem, M., Lauppe, H.F., Kartenbeck, J., Durwald, H., and Hoffmann-Berling, H., 1977,

Enzymatic unwinding of DNA. III. Mode of action of *Escherichia coli* DNA unwinding enzyme, *J. Mol. Biol.* 110:667–685.

4. Abdel-Monem, M., Taucher-Scholz, G., and Klinkert, M.Q., 1983, Identification of *Escherichia coli* DNA helicase I as the *traI* gene product of the F sex factor, *Proc. Natl. Acad. Sci. USA* 80:4659–4663.

5. Abo, T., Inamoto, S., and Ohtsubo, E., 1991, Specific DNA binding of the TraM protein to the *oriT* region of plasmid R100, *J. Bacteriol.* 173:6347–6354.

6. Achtman, M., 1975, Mating aggregates in *Escherichia coli* conjugation, *J. Bacteriol.* 123:505–515.

7. Achtman, M., and Skurray, R., 1977, A redefinition of the mating phenomenon in bacteria, in: *Microbial Interactions, Receptors and Recognition* Series B, Vol. 3 (J.L. Reissig, ed.), Chapman & Hall, London, pp. 233–279.

8. Achtman, M., Kennedy, N., and Skurray, R., 1977, Cell-cell interactions in conjugating *Escherichia coli*: role of *traT* protein in surface exclusion, *Proc. Natl. Acad. Sci. USA* 74:5104–5108.

9. Achtman, M., Morelli, G., and Schwuchow, S., 1978, Cell-cell interactions in conjugating *Escherichia coli*: role of F pili and fate of mating aggregates, *J. Bacteriol.* 135:1053–1061.

10. Achtman, M., Schwuchow, S., Helmuth, R., Morelli, G., and Manning, P.A., 1978, Cell-cell interactions in conjugating *Escherichia coli*: Con⁻ mutants and stabilization of mating aggregates, *Mol. Gen. Genet.* 164:171–183.

11. Achtman, M., Manning, P.A., Edelbluth, C., and Herrlich, P., 1979, Export without proteolytic processing of inner and outer membrane proteins encoded by F sex factor *tra* cistrons in *Escherichia coli* minicells, *Proc. Natl. Acad. Sci. USA* 76:4837–4841.

12. Bagdasarian, M., Bailone, A., Bagdasarian, M.M., Manning, P.A., Lurz, R., Timmis, K.N., and Devoret, R., 1986, An inhibitor of SOS induction, specified by a plasmid locus in *Escherichia coli*, *Proc. Natl. Acad. Sci. USA* 83:5723–5726.

13. Bagdasarian, M., Bailone, A., Angulo, J., Scholz, P., Bagdasarian, M., and Devoret, R., 1992, PsiB, an anti-SOS protein, is transiently expressed by the F sex factor during its transmission to an *Escherichia coli* K-12 recipient, *Mol. Microbiol.* 6:885–894.

14. Bailone, A., Bäckman, A., Sommer, S., Célérier, J., Bagdasarian, M.M., Bagdasarian, M., and Devoret, R., 1988, PsiB polypeptide prevents activation of RecA protein in *Escherichia coli*, *Mol. Gen. Genet.* 214: 389–395.

15. Benz, I., and Müller, H., 1990, *Escherichia coli* DNA helicase I. Characterization of the protein and of its DNA-binding properties, *Eur. J. Biochem.* 189:267–276.

16. Bergquist, P.L., 1987, Incompatibility, in: *Plasmids: A Practical Approach* (K.G. Hardy, ed.), IRL Press, Oxford, pp. 37–78.

17. Bowie, J.U., and Sauer, R.T., 1990, TraY proteins of F and related episomes are members of the Arc and Mnt repressor family, *J. Mol. Biol.* 211:5–6.

18. Bradshaw, H.D., Jr., Traxler, B.A., Minkley, E.G., Jr., Nester, E.W., and Gordon, M.P., 1990, Nucleotide sequence of the *traI* (Helicase I) gene from the sex factor F, *J. Bacteriol.* 172:4127–4131.

19. Brock, T.D., 1990, *The Emergence of Bacterial Genetics*, Cold Spring Harbor Laboratory Press, Cold Spring Harbor, New York.

20. Bukhari, A. I., Shapiro, J.A., and Adhya, S.L., 1977, *DNA Insertion Elements, Plasmids and Episomes*, Cold Spring Harbor Laboratory, Cold Spring Harbor, New York.

21. Carter, J.R., and Porter, R.D., 1991, *traY* and *traI* are required for *oriT*-dependent enhanced recombination between *lac*-containing plasmids and λ*plac5*, *J. Bacteriol.* 173:1027–1034.

22. Chase, J.W., Merrill, B.M., and Williams, K.R., 1983, F sex factor encodes a single-stranded DNA binding protein (SSB) with extensive sequence homology to *Escherichia coli* SSB, *Proc. Natl. Acad. Sci. USA* 80:5480–5484.

23. Cheah, K.-C., and Skurray, R., 1986, The F plasmid carries an IS3 insertion within *finO*, *J. Gen. Microbiol.* 132:3269–3275.

24. Cheah, K.-C., Ray, A., and Skurray, R., 1984, Cloning and molecular analysis of the *finO* region from the antibiotic-resistance plasmid R6-5, *Plasmid* 12:222–226.

25. Cheah, K-C., Ray, A., and Skurray, R., 1986, Expression of F plasmid traT: independence of *traY*→*Z* promoter and *traJ* control, *Plasmid* 16:101–107.

26. Cheah, K-C., Hirst, R., and Skurray, R., 1987, *finO* sequences on conjugally repressed and derepessed F-like plasmids, *Plasmid* 17:233–239.

27. Coetzee, J.N., Bradley, D.E., Hedges, R.W., Hughes, V.M., McConnell, M.M., Du Toit, L., and

Tweehuysen, M., 1986, Bacteriophages $F_olac\ h$, SR, SF: phages which adsorb to pili encoded by plasmids of the S-complex, *J. Gen. Microbiol.* 132:2907–2917.

28. Couturier, M., Bex, F., Bergquist, P.L., and Maas, W.K., 1988, Identification and classification of bacterial plasmids, *Microbiol. Rev.*, 52:375–395.

29. Cram, D., Ray, A., O'Gorman, L., and Skurray, R., 1984, Transcriptional analysis of the leading region in F plasmid DNA transfer, *Plasmid* 11:221–233.

30. Cram, D.S., Loh, S.M., Cheah, K-C., and Skurray, R.A., 1991, Sequence and conservation of genes at the distal end of the transfer region on plasmids F and R6-5, *Gene* 104:85–90.

31. Cuozzo, M., and Silverman, P.M., 1986, Characterization of the F plasmid TraJ protein synthesized in F' and Hfr strains of *Escherichia coli* K12, *J. Biol. Chem.* 261:5175–5179.

32. Cuozzo, M., Silverman, P.M., and Minkley, E.G., Jr., 1984, Overproduction in *Escherichia coli* K12 and purification of the TraJ protein encoded by the conjugative plasmid F, *J. Biol. Chem.* 259:6659–6666.

33. Datta, N., 1975, Epidemiology and classification of plasmids, in: *Microbiology–1974* (D. Schlessinger, ed.), American Society for Microbiology, Washington, DC, pp. 9–15.

33a. de la Cruz, F., Zabala, J.C., and Ortíz, J.M., 1979, Incompatibility among α-hemolytic plasmids studied after inactivation of the α-hemolysin gene by transposition of Tn802, *Plasmid* 2:507–519.

34. Dempsey, W.B., 1987, Integration host factor and conjugative transfer of the antibiotic resistance plasmid R100, *J. Bacteriol.* 169:4391–4392.

35. Dempsey, W.B., 1987, Transcript analysis of the plasmid R100 *traJ* and *finP* genes, *Mol. Gen. Genet.* 209:533–544.

36. Dempsey, W.B., 1989, Derepression of conjugal transfer of the antibiotic resistance plasmid R100 by antisense RNA, *J. Bacteriol.* 171:2886–2888.

37. Dempsey, W.B., 1989, Sense and antisense transcripts of *traM*, a conjugal transfer gene of the antibiotic resistance plasmid R100, *Mol. Microbiol.* 3:561–570.

38. Dempsey, W.B., and Fee, B.E., 1990, Integration host factor affects expression of two genes at the conjugal transfer origin of plasmid R100, *Mol. Microbiol.* 4:1019–1028.

39. Dempsey, W., and McIntire, S., 1983, The *finO* gene of antibiotic resistance plasmid R100, *Mol. Gen. Genet.* 190:444–451.

40. Di Laurenzio, L., Frost, L.S., Finlay, B.B., and Paranchych, W., 1991, Characterization of the *oriT* region of the IncFV plasmid pED208, *Mol. Microbiol.* 5:1779–1790.

40a. Di Laurenzio, L., Frost, L.S., and Paranchych, W., 1992, The TraM protein of the conjugative plasmid F binds to the origin of transfer of the F and ColE1 plasmids. *Mol. Microbiol.* 6:2952–2959.

41. Dürrenberger, M.B., Villiger, W., and Bächi, T., 1991, Conjugational junctions: morphology of specific contacts in conjugating *Escherichia coli* bacteria, *J. Struct. Biol.* 107:146–156.

42. Dutreix, M., Bäckman, A., Célérier, J., Bagdasarian, M.M., Sommer, S., Bailone, A., Devoret, R., and Bagdasarian, M., 1988, Identification of *psiB* genes of plasmids F and R6-5. Molecular basis for *psiB* enhanced expression in plasmid R6-5, *Nucl. Acids Res.* 16:10669–10679.

43. Everett, R., and Willetts, N., 1980, Characterization of an in vivo system for nicking at the origin of conjugal DNA transfer of the sex factor F, *J. Mol. Biol.* 136:129–150.

44. Fee, B.E., and Dempsey, W.B., 1986, Cloning, mapping, and sequencing of plasmid R100 *traM* and *finP* genes, *J. Bacteriol.* 167:336–345.

45. Fee, B.E., and Dempsey, W.B., 1988, Nucleotide sequence of gene X of antibiotic resistance plasmid R100, *Nucl. Acids Res.* 16:4726.

46. Finlay, B.B., and Paranchych, W., 1986, Nucleotide sequence of the surface exclusion genes *traS* and *traT* from the $IncF_olac$ plasmid pED208, *J. Bacteriol.* 166:713–721.

47. Finlay, B.B., Frost, L.S., Paranchych, W., Parker, J.M., and Hodges, R.S., 1985, Major antigenic determinants of F and ColB2 pili, *J. Bacteriol.* 163:331–335.

48. Finlay, B.B., Frost, L.L., and Paranchych, W., 1986, Nucleotide sequences of the R1-19 plasmid transfer genes *traM*, *finP*, *traJ*, and *traY* and the *traYZ* promoter, *J. Bacteriol.* 166:368–374.

49. Finlay, B.B., Frost, L.S., and Paranchych, W., 1986, Origin of transfer of IncF plasmids and nucleotide sequences of the type II *oriT*, *traM*, and *traY* alleles from ColB4-K98 and the type IV *traY* allele from R100-1, *J. Bacteriol.* 168:132–139.

50. Finlay, B.B., Frost, L.S., and Paranchych, W., 1986, Nucleotide sequence of the *traYALE* region from IncFV plasmid pED208, *J. Bacteriol.* 168:990–998.

51. Finlay, B.B., Frost, L.S., Paranchych, W., and Willetts, N.S., 1986, Nucleotide sequences of five IncF plasmid *finP* alleles, *J. Bacteriol.* 167:754–757.

52. Finnegan, D.J., and Willetts, N.S., 1971, Two classes of F*lac* mutants insensitive to transfer inhibition by an F-like R factor, *Mol. Gen. Genet.* 111:256–264.

53. Firth, N., and Skurray, R., 1992, Characterization of the F plasmid bifunctional conjugation gene, *traG*, *Mol. Gen. Genet.* 232:145–153.

54. Fives-Taylor, P., 1978, F pili, in: *Pili* (D.E. Bradley, E. Raizen, P. Fives-Taylor, and J. Ou, eds.), International Conferences on Pili, Washington DC, pp. 145–159.

55. Fowler, T., Taylor, L, and Thompson, R., 1983, The control region of the F plasmid transfer operon: DNA sequence of the *traJ* and *traY* genes and characterization of the *traY-Z* promoter, *Gene* 26:79–89.

56. Frost, L.S., and Paranchych, W., 1988, DNA sequence analysis of point mutations in *traA*, the F pilin gene, reveal two domains involved in F-specific bacteriophage attachment, *Mol. Gen. Genet.* 213:134–139.

57. Frost, L.S., Paranchych, W., and Willetts, N.S., 1984, DNA sequence of the F *traALE* region that includes the gene for F pilin, *J. Bacteriol.* 160:395–401.

58. Frost, L.S., Finlay, B.B., Opgenorth, A., Paranchych, W., and Lee, J.S., 1985, Characterization and sequence analysis of pilin from F-like plasmids, *J. Bacteriol.* 164:1238–1247.

59. Frost, L.S., Lee, J.S., Scraba, D.G., and Paranchych, W., 1986, Two monoclonal antibodies specific for different epitopes within the amino-terminal region of F pilin, *J. Bacteriol.* 168:192–198.

60. Frost, L., Lee, S., Yanchar, N., and Paranchych, W., 1989, *finP* and *fisO* mutations in FinP anti-sense RNA suggest a model for FinOP action in the repression of bacterial conjugation by the F*lac* plasmid JCFLO, *Mol. Gen. Genet.* 218:152–160.

61. Frost, L., Usher, K., and Paranchych, W., 1991, Computer analysis of the F transfer region, *Plasmid* 25:226.

62. Fu, Y.-H.F., Tsai, M.-M., Luo, Y., and Deonier, R.C., 1991, Deletion analysis of the F plasmid *oriT* locus, *J. Bacteriol.* 173:1012–1020.

63. Gaffney, D., Skurray, R., and Willetts, N., 1983, Regulation of the F conjugation genes studied by hybridization and *tra-lacZ* fusion, *J. Mol. Biol.* 168:103–122.

64. Gamas, P., Caro, L., Galas, D., and Chandler, M., 1987, Expression of F transfer functions depends on the *Escherichia coli* integration host factor, *Mol. Gen. Genet.* 207:302–305.

65. Gerdes, K., Bech, F.W., Jørgensen, S.T., Løbner-Olesen, A. Rasmussen, P.B., Atlung, T., Boe, L., Karlstrom, O., Molin, S., and von Meyenburg, K., 1986, Mechanism of postsegregational killing by the *hok* gene product of the *parB* system of plasmid R1 and its homology with the *relF* gene product of the *E. coli relB* operon, *EMBO J.* 5:2023–2029.

66. Gerdes, K., Helin, K., Christensen, O.W., and Løbner-Olesen, A., 1988, Translational control and differential RNA decay are key elements regulating postsegregational expression of the killer protein encoded by the *parB* locus of plasmid R1, *J. Mol. Biol.* 203:119–129.

67. Gerdes, K., Thisted, T., and Martinussen, J., 1990, Mechanism of postsegregational killing by the *hok/sok* system of plasmid R1: *sok* antisense RNA regulates formation of a *hok* mRNA species correlated with killing of plasmid-free cells, *Mol. Microbiol.* 4:1807–1818.

68. Göldner, A., Graus, H., and Högenauer, G., 1987, The origin of transfer of P307, *Plasmid* 18:76–83.

69. Golub, E.I., and Low, K.B., 1986, Unrelated conjugative plasmids have sequences which are homologous to the leading region of the F factor, *J. Bacteriol.* 166:670–672.

70. Golub, E.I., and Panzer, H.A., 1988, The F factor of *Escherichia coli* carries a locus of stable plasmid inheritance *stm*, similar to the *parB* locus of plasmid R1, *Mol. Gen. Genet.* 214:353–357.

71. Golub, E., Bailone, A., and Devoret, R., 1988, A gene encoding an SOS inhibitor is present in different conjugative plasmids, *J. Bacteriol.* 170:4392–4394.

72. Graus, H., Hödl, A., Wallner, P., and Högenauer, G., 1990, The sequence of the leading region of the resistance plasmid R1, *Nucl. Acids Res.* 18:1046.

73. Graus-Göldner, A., Graus, H., Schlacher, T., and Högenauer, G., 1990, The sequences of genes bordering *oriT* in the enterotoxin plasmid P307: comparison with the sequences of plasmids F and R1, *Plasmid* 24:119–131.

74. Grossman, T.H., and Silverman, P.M., 1989, Structure and function of conjugative pili: inducible synthesis of functional F-pili by *Escherichia coli* K-12 containing a *lac-tra* operon fusion, *J. Bacteriol.* 171:650–656.

75. Grossman, T.H., Frost, L.S., Silverman, P.M., 1990, Structure and function of conjugative pili: monoclonal antibodies as probes for structural variants of F pili, *J. Bacteriol.* 172:1174–1179.

76. Ham, L.M., and Skurray, R., 1989, Molecular analysis and nucleotide sequence of *finQ*, a transcriptional inhibitor of the F plasmid transfer genes, *Mol. Gen. Genet.* 216:99–105.

77. Ham, L.M., Cram, D., and Skurray, R., 1989, Transcriptional analysis of the F plasmid surface exclusion region: mapping of *traS*, *traT*, and *traD* transcripts, *Plasmid* 21:1–8.

78. Ham, L.M., Firth, N., and Skurray, R., 1989, Nucleotide sequence of the F plasmid transfer gene, *traH*: identification of a new gene and a promoter within the transfer operon, *Gene* 75:157–165.

79. Harrington, L.C., and Rogerson, A.C., 1990, The F pilus of *Escherichia coli* appears to support stable DNA transfer in the absence of wall-to-wall contact between cells, *J. Bacteriol.* 172:7263–7264.

80. Harrison, J.L., Taylor, I.M., Platt, K., and O'Connor, C.D., 1992, Surface exclusion specificity of the TraT lipoprotein is determined by single alterations in a five-amino-acid region of the protein, *Mol. Microbiol.* 6:2825–2832.

81. Helmuth, R., and Achtman, M., 1975, Operon structure of DNA transfer cistrons on the F sex factor, *Nature* 257:652–656.

82. Howland, C.J., Rees, C.E.D., Barth, P.T., and Wilkins, B.M., 1989, The *ssb* gene of plasmid CollB-P9, *J. Bacteriol.* 171:2466–2473.

83. Inamoto, S., and Ohtsubo, E., 1990, Specific binding of the TraY protein to *oriT* and the promoter region for the *traY* gene of plasmid R100, *J. Biol. Chem.* 265:6461–6466.

84. Inamoto, S., Yoshioka, Y., and Ohtsubo, E., 1988, Identification and characterization of the products from the *traJ* and *traY* genes of plasmid R100, *J. Bacteriol.* 170:2749–2757.

85. Inamoto, S., Abo, T., and Ohtsubo, E., 1990, Binding sites of Integration Host Factor in *oriT* of plasmid R100, *J. Gen. Appl. Microbiol.* 36:287–293.

86. Inamoto, S., Yoshioka, Y., and Ohtsubo, E., 1991, Site- and strand-specific nicking in vitro at *oriT* by the TraY-TraI endonuclease of plasmid R-100, *J. Biol. Chem.* 266:10086–10092.

87. Ippen-Ihler, K., and Maneewannakul, S., 1991, Conjugation among enteric bacteria: mating systems dependent on expression of pili, in: *Microbiol Cell-Cell Interactions* (M. Dworkin, ed.), American Society for Microbiology, Washington, DC, pp. 35–69.

88. Ippen-Ihler, K.A., and Minkley, E.G., Jr., 1986, The conjugation system of F, the fertility factor of *Escherichia coli*, *Ann. Rev. Genet.* 20:593–624.

89. Ippen-Ihler, K., Moore, D., Laine, S., Johnson, D.A., and Willetts, N.S., 1984, Synthesis of F-pilin polypeptide in the absence of F *traJ* product, *Plasmid* 11:116–129.

90. Jacob, F., 1988, *The Statue Within*, Basic Books, New York.

91. Jalajakumari, M.B., and Manning, P.A., 1989, Nucleotide sequence of the *traD* region in the *Escherichia coli* F sex factor, *Gene* 81:195–202.

92. Jalajakumari, M.B., Guidolin, A., Buhk, H.J., Manning, P.A., Ham, L.M., Hodgson, A.L.M., Cheah, K.-C., and Skurray, R.A., 1987, Surface exclusion genes *traS* and *traT* of the F sex factor of *Escherichia coli* K-12: determination of the nucleotide sequence and promoter and terminator activities, *J. Mol. Biol.* 198:1–11.

93. Jones, A.L., Barth, P.T., and Wilkins, B.M., 1992, Zygotic induction of plasmid *ssb* and *psiB* genes following conjugative transfer of IncI1 plasmid CollB-P9, *Mol. Microbiol.* 6:605–614.

94. Kathir, P., and Ippen-Ihler, K., 1991, Construction and characterization of derivatives carrying insertion mutations in F plasmid transfer region genes, *trbA*, *artA*, *traQ*, and *trbB*, *Plasmid* 26:40–54.

95. Kingsman, A., and Willetts, N., 1978, The requirements for conjugal DNA synthesis in the donor strain during F*lac* transfer, *J. Mol. Biol.* 122:287–300.

96. Kolodkin, A.L., Capage, M.A., Golub, E.I., and Low, K.B., 1983, F sex factor of *Escherichia coli* K-12 codes for a single-stranded DNA binding protein, *Proc. Natl. Acad. Sci. USA* 80:4422–4426.

97. Koraimann, G., and Högenauer, G., 1989, A stable core region of the *tra* operon mRNA of plasmid R1-19, *Nucl. Acids Res.* 17:1283–1298.

98. Koraimann, G., Koraimann, C., Koronakis, V., Schlager, S., and Högenauer, G., 1991, Repression and derepression of conjugation of plasmid R1 by wild-type and mutated *finP* antisense RNA, *Mol. Microbiol.* 5:77–87.

99. Koronakis, V., and Högenauer, G., 1986, The sequences of the *traJ* gene and the 5′ end of the *traY* gene of the resistance plasmid R1, *Mol. Gen. Genet.* 203:137–142.

100. Koranakis, V.E., Bauer, E., and Högenauer, G., 1985, The *traM* gene of resistance plasmid R1: comparison with the corresponding sequence of the *Escherichia coli* F factor, *Gene* 36:79–86.

101. Lahue, E.E., and Matson, S.W., 1988, *Escherichia coli* DNA helicase I catalyzes a unidirectional and highly processive unwinding reaction, *J. Biol. Chem.* 263:3208–3215.

102. Lahue, E.E., and Matson, S.W., 1990, Purified *Escherichia coli* F-factor TraY protein binds *oriT*, *J. Bacteriol.* 172:1385–1391.

103. Laine, S., Moore, D., Kathir, P., and Ippen-Ihler, K., 1985, Genes and gene products involved in the synthesis of F-pili, in: *Plasmids in Bacteria* (D.R. Helinski, S.N. Cohen, D.B. Clewell, D.A. Jackson, and A. Hollaender, eds.), Plenum Press, New York, pp. 535–553.

104. Lawn, A.M., and Meynell, E., 1970, Serotypes of sex pili, *J. Hygiene* 68:683–694.

105. Lederberg, J., and Tatum, E.L., 1946, Gene recombination in *Escherichia coli*, *Nature* 158:558.

105a. Lee, S.H., Frost, L.S., and Paranchych, W. 1992, FinOP repression of the F plasmid involves extension of the half-life of FinP antisense RNA by FinO, *Mol. Gen. Genet.* 235:131–139.

106. Loh, S.M., Ray, A., Cram, D.S., O'Gorman, L.E., and Skurray, R.A., 1986, Location of a second partitioning region (ParL) on the F plasmid, *FEMS Microbiol. Lett.* 37:179–182.

107. Loh, S.M., Cram, D.S., and Skurray, R.A., 1988, Nucleotide sequence and transcriptional analysis of a third function (Flm) involved in F-plasmid maintenance, *Gene* 66:259–268.

108. Loh, S., Cram, D., and Skurray, R., 1989, Nucleotide sequence of the leading region adjacent to the origin of transfer on plasmid F and its conservation among conjugative plasmids, *Mol. Gen. Genet.* 219:177–186.

109. Loh, S., Skurray, R., Célérier, J., Bagdasarian, M., Bailone, A., and Devoret, R., 1990, Nucleotide sequence of the *psiA* (plasmid SOS inhibition) gene located on the leading region of plasmids F and R6-5, *Nucl. Acids Res.* 18:4597.

110. López, J., Salazar, L., Andrés, I., Ortíz, J.M., and Rodriguez, J.C., 1991, Nucleotide sequence of the *oriT-traM-finP* region of the haemolytic plasmid pSU316: comparison to F, *Nucl. Acids Res.* 19:3451.

110a. López, J., Rodriguez, J.C., Andrés, I., and Ortíz, J.M., 1989, Characterization of the RepFVII replicon of the haemolytic plasmid pSU233: nucleotide sequence of an IncFVII determinant, *J. Gen. Micro.* 135:1763–1768.

111. Maneewannakul, K., and Ippen-Ihler, K., 1993, Construction and analysis of F plasmid *traR*, *trbJ* and *trbH* mutants, *J. Bacteriol.* 175:1528–1531.

111a. Maneewannakul, K., Maneewannakul, S., and Ippen-Ihler, K., 1993, F pilin synthesis, *J. Bacteriol.* 175:1384–1391.

111b. Maneewannakul, K., Maneewannakul, S., and Ippen-Ihler, K., 1992, Sequence alterations affecting F plasmid transfer gene expression: a conjugation system dependent on transcription by the RNA polymerase of phage T7. *Mol. Microbiol.* 6:2961–2973.

112. Maneewannakul, S., Kathir, P., Moore, D., Le, L.-A., Wu, J.H., and Ippen-Ihler, K., 1987, Location of F plasmid transfer operon genes *traC* and *traW* and identification of the *traW* product, *J. Bacteriol.* 169:5119–5124.

113. Maneewannakul, S., Maneewannakul, K., and Ippen-Ihler, K., 1991, Characterization of *trbC*, a new F plasmid *tra* operon gene that is essential to conjugative transfer, *J. Bacteriol.* 173:3872–3878.

114. Maneewannakul, S., Kathir, P., and Ippen-Ihler, K., 1992, Characterization of the F plasmid mating aggregate gene, *traN*, and of a new F transfer region locus, *trbE*, *J. Mol. Biol.* 225:299–311.

114a. Maneewannakul, S., Maneewannakul, K., and Ippen-Ihler, K., 1992, Characterization, localization, and sequence of F transfer region products: the pilus assembly gene product TraW and a new product, TrbI. *J. Bacteriol.*, 174:5567–5574.

115. Manning, P.A., Morelli, G., and Achtman, M., 1981, *traG* protein of the F sex factor of *Escherichia coli* K-12 and its role in conjugation, *Proc. Natl. Acad. Sci. USA* 78:7487–7491.

116. Manning, P.A., Morelli, G., and Fisseau, C., 1984, RNA-polymerase binding sites within the *tra* region of the F factor of *Escherichia coli* K-12, *Gene* 27:121–123.

117. Marvin, D.A., and Folkhard, W., 1986, Structure of F-pili: reassessment of the symmetry, *J. Mol. Biol.* 191:299–300.

118. Matson, S.W., and Morton, B.S., 1991, *Escherichia coli* DNA helicase I catalyzes a site- and strand-specific nicking reaction at the F plasmid *oriT*, *J. Biol. Chem.* 266:16232–16237.

119. McIntire, S.A., and Dempsey, W.B., 1987, *oriT* sequence of the antibiotic resistance plasmid R100, *J. Bacteriol.* 169:3829–3832.

120. McIntire, S.A., and Dempsey, W.B., 1987, Fertility inhibition gene of plasmid R100, *Nucl. Acids Res.* 15:2029–2042.

121. Meynell, E., Meynell, G.G., and Datta, N., 1968, Phylogenetic relationships of drug resistance factors and other transmissible bacterial plasmids, *Bacteriol. Rev.* 32:55–83.

122. Minkley, E.G., Jr., 1984, Purification and characterization of pro-TraTp, the signal sequence-containing precursor of a secreted protein encoded by the F sex factor, *J. Bacteriol.* 158:464–473.

123. Minkley, E.G., Jr., and Willetts, N.S., 1984, Overproduction, purification and characterization of the F *traT* protein, *Mol. Gen. Genet.* 196:225–235.

124. Moore, D., Sowa, B.A., and Ippen-Ihler, K., 1981, Location of an F-pilin pool in the inner membrane, *J. Bacteriol.* 146:251–259.

125. Moore, D., Sowa, B.A., and Ippen-Ihler, K., 1981, The effect of *tra* mutations on the synthesis of the F-pilin membrane polypeptide, *Mol. Gen. Genet.* 184:260–264.

126. Moore, D., Sowa, B.A., and Ippen-Ihler, K., 1982, A new activity in the F *tra* operon which is required for F-pilin synthesis, *Mol. Gen. Genet.* 188:459–464.

127. Moore, D., Wu, J.H., Kathir, P., Hamilton, C.M., and Ippen-Ihler, K., 1987, Analysis of transfer genes and gene products within the *traB-traC* region of the *Escherichia coli* fertility factor F, *J. Bacteriol.* 169:3994–4002.

128. Moore, D., Maneewannakul, K., Maneewannakul, S., Wu, J.H., Ippen-Ihler, K., and Bradley, D.E., 1990, Characterization of the F-plasmid conjugative transfer gene *traU*, *J. Bacteriol.* 172:4263–4270.

128a. Moore, D., Hamilton, C.H., Maneewannakul, K., Mintz, Y., Frost, L.S., and Ippen-Ihler, K., 1993, The *Escherichia coli* K12 F plasmid gene, *traX* is required for acetylation of F-pilin, *J. Bacteriol.* 175:1375–1383.

129. Mullineaux, P., and Willetts, N., 1985, Promoters in the transfer region of plasmid F, in: *Plasmids in Bacteria* (D.R. Helinski, S.N. Cohn, D.B. Clewell, D.A. Jackson, and A. Hollaender, eds.), Plenum Press, New York, pp. 605–614.

130. Nijkamp, H.J.J., de Lang, R., Stuitje, A.R., van den Elzen, P. J.M., Veltkamp, E., and van Putten, A.J., 1986, The complete nucleotide sequence of the bacteriocinogenic plasmid CloDF13, *Plasmid* 16:135–160.

131. Nomura, N., Masai, H., Inuzuka, M., Miyazaki, C., Ohtsubo, E., Itoh, T., Sasamoto, S., Matsui, M., Ishizaki, R., and Arai, K-I., 1991, Identification of eleven single-strand initiation sequences (*ssi*) for priming of DNA replication in the F, R6K, R100 and ColE2 plasmids, *Gene* 108:15–22.

132. Novotny, C.P., and Fives-Taylor, P., 1974, Retraction of F pili, *J. Bacteriol.* 117:1306–1311.

133. Ogata, R.T., Winters, C., and Levine, R.P., 1982, Nucleotide sequence analysis of the complement resistance gene from plasmid R100, *J. Bacteriol.* 151:819–827.

134. Ostermann, E., Kricek, F., and Högenauer, G., 1984, Cloning the origin of transfer region of the resistance plasmid R1, *EMBO J.* 3:1731–1735.

135. Ou, J.T., and Anderson, T.F., 1970, Role of pili in bacterial conjugation, *J. Bacteriol.* 102:648–654.

136. Panicker, M. M., and Minkley, E.G., Jr., 1985, DNA transfer occurs during a cell surface contact stage of F sex factor-mediated bacterial conjugation, *J. Bacteriol.* 162:584–590.

136a. Panicker, M.M., and Minkley, E.G., Jr., 1992, Purification and properties of the F sex factor TraD protein, an inner membrane conjugal transfer protein, *J. Biol. Chem.* 267:12761–12766.

137. Paranchych, W., 1975, Attachment, ejection and penetration stages of the RNA phage infectious process, in: *RNA Phages* (N. Zinder, ed.), Cold Spring Harbor Laboratory, Cold Spring Harbor, New York, pp. 85–111.

138. Paranchych, W., and Frost, L.S., 1988, The physiology and biochemistry of pili, *Adv. Microbiol. Physiol.* 29:53–114.

139. Perumal, N.B., and Minkley, E.G., Jr., 1984, The product of the F sex factor *traT* surface exclusion gene is a lipoprotein, *J. Biol. Chem.* 259:5357–5360.

140. Rashtchian, A., Crooks, J.H., and Levy, S.B., 1983, *traJ* independence in expression of *traT* on F, *J. Bacteriol.* 154:1009–1012.

141. Ray, A., Cheah, K.-C., Skurray, R., 1986, An F-derived conjugative cosmid: analysis of *tra* polypeptides in cosmid-infected cells, *Plasmid* 16:90–100.

142. Rees, C.E.D., and Wilkins, B.M., 1990, Protein transfer into the recipient cell during bacterial conjugation: Studies with F and RP4, *Mol. Microbiol.* 4:1199–1205.

143. Reygers, U., Wessel, R., Müller, H., and Hoffmann-Berling, H., 1991, Endonuclease activity of *Escherichia coli* DNA helicase I directed against the transfer origin of the F factor, *EMBO J.* 10:2689–2694.

144. Riede, I., and Eschbach, M.-L., 1986, Evidence that TraT interacts with OmpA of *Escherichia coli*, *FEBS Lett.* 205:241–245.

145. Schandel, K.A., Maneewannakul, S., Vonder Haar, R.A., Ippen-Ihler, K., and Webster, R.E., 1990, Nucleotide sequence of the F plasmid gene, *traC*, and identification of its product, *Gene* 96:137–140.

145a. Schandel, K. A., Muller, M., and Webster, R.E., 1992, Localization of TraC: A protein involved in the assembly of the F conjugative pilus, *J. Bacteriol.* 174:3800–3694.

146. Schwab, M., Gruber, H., and Högenauer, G., 1991, The TraM protein of plasmid R1 is a DNA binding protein, *Mol. Microbiol.* 5:439–446.

147. Sharp, P.A., Cohen, S.N., and Davidson, N., 1973, Electron microscope heteroduplex studies of sequence relations among plasmids of *Escherichia coli*. II. Structure of drug resistance (R) factors and F factors, *J. Mol. Biol.* 75:235–255.

148. Silverman, P.M., 1987, The structural basis of prokaryotic DNA transfer, in: *Bacterial Outer Membranes as Model Systems* (M. Inouye, ed.), John Wiley & Sons, New York, pp. 277–310.

149. Silverman, P.M., Wickersham, E., and Harris, R., 1991, Regulation of the F plasmid *traY* promoter in *Escherichia coli* by host and plasmid factors, *J. Mol. Biol.* 218:119–128.

150. Silverman, P.M., Wickersham, E., Rainwater, S., and Harris, R., 1991, Regulation of the F plasmid *traY* promoter in *Escherichia coli* K12 as a function of sequence context, *J. Mol. Biol.* 220:271–279.

151. Sowa, B.A., Moore, D., and Ippen-Ihler, K., 1983, Physiology of F-pilin synthesis and utilization, *J. Bacteriol.* 153:962–968.

152. Sukupolvi, S., and O'Connor, C.D., 1990, TraT lipoprotein, a plasmid-specified mediator of interactions between gram-negative bacteria and their environment, *Microbiol. Rev.* 54:331–341.

153. Thompson, R., and Taylor, L., 1982, Promoter mapping and DNA sequencing of the F plasmid transfer genes, *traM* and *traJ*, *Mol. Gen. Genet.* 188:513–518.

154. Thompson, R., Taylor, L., Kelly, K., Everett, R., and Willetts, N., 1984, The F plasmid origin of transfer: DNA sequence of wild-type and mutant origins and location of origin-specific nicks, *EMBO J.* 3:1175–1180.

155. Thompson, T.L., Centola, M.B., and Deonier, R.C., 1989, Location of the nick at *oriT* of the F plasmid, *J. Mol. Biol.* 207:505–512.

156. Timmis, K.N., Andrés, I., and Achtman, M., 1978, Fertility repression of F-like conjugative plasmids: physical mapping of the R6-5 *finO* and *finP* cistrons and identification of the *finO* protein, *Proc. Natl. Acad. Sci. USA* 75:5836–5840.

157. Timmis, K.N., Cabello, F., and Cohen, S.N., 1978, Cloning and characterization of *Eco*RI and *Hind*III restriction endonuclease-generated fragments of antibiotic resistance plasmids R6-5 and R6, *Mol. Gen. Genet.* 162:121–137.

158. Traxler, B.A., and Minkley, E.G., Jr., 1987, Revised genetic map of the distal end of the F transfer operon: implications for DNA helicase I, nicking at *oriT*, and conjugal DNA transport, *J. Bacteriol.* 169:3251–3259.

159. Traxler, B.A., and Minkley, E.G., Jr., 1988, Evidence that DNA helicase I and *oriT* site-specific nicking are both functions of the F traI protein, *J. Mol. Biol.* 204:205–209.

160. Tsai, M.-M., Fu, Y.-H.F., and Deonier, R.C., 1990, Intrinsic bends and integration host factor binding at F plasmid *oriT*, *J. Bacteriol.* 172:4603–4609.

161. van Biesen, T., and Frost, L.S., 1992, Differential levels of fertility inhibition among F-like plasmids are related to the cellular concentration of *finO* mRNA, *Mol. Microbiol.* 6:771–780.

162. Wessel, R., Müller, H., and Hoffmann-Berling, H., 1990, Electron microscopy of DNA helicase-I complexes in the act of strand separation, *Eur. J. Biochem.* 189:277–285.

163. Willetts, N., 1980, Interactions between the F conjugal transfer system and CloDF13::TnA plasmids, *Mol. Gen. Genet.* 180:213–217.

164. Willetts, N.S., and Maule, J., 1984, Interactions between the surface exclusion systems of some F-like plasmids, *Genet. Res.* 24:81–89.

165. Willetts, N.S., and Maule, J., 1985, Specificities of IncF plasmid conjugation genes, *Genet. Res.* 47:1–11.

166. Willetts, N., and Skurray, R., 1980, The conjugation system of F-like plasmids, *Ann. Rev. Genet.* 14:41–76.

167. Willetts, N., and Skurray, R., 1987, Structure and function of the F factor and mechanism of conjugation, in: *Escherichia coli and Salmonella typhimurium: Cellular and Molecular Biology* (F.C. Neidhart, J.L. Ingraham, K.B. Low, B. Magasanik, M. Schaechter, and H.E. Umbarger, eds.), American Society for Microbiology, Washington DC, pp. 1110–1133.

168. Willetts, N., and Wilkins, B., 1984, Processing of plasmid DNA during bacterial conjugation, *Microbiol. Rev.* 48:24–41.

169. Worobec, E.A., Paranchych, W., Parker, J.M.R., Taneja, A.K., and Hodges, R.S., 1985, Antigen-antibody interaction: the immunodominant region of EDP208 pili, *J. Biol. Chem.* 260:938–943.

170. Worobec, E.A., Frost, L.S., Pieroni, P., Armstrong, G.D., Hodges, R.S., Parker, J.M.R., Finlay, B.B., and Paranchych, W., 1986, Location of the antigenic determinants of conjugative F-like pili, *J. Bacteriol.* 167:660–665.

171. Wu, J.H., and Ippen-Ihler, K., 1989, Nucleotide sequence of *traQ* and adjacent loci in the *Escherichia coli* K-12 F plasmid transfer operon, *J. Bacteriol.* 171:213–221.

172. Wu, J.H., Moore, D., Lee, T., and Ippen-Ihler, K., 1987, Analysis of *Escherichia coli* K12 F factor transfer genes: *traQ*, *trbA* and *trbB*, *Plasmid* 18:54–69.
173. Wu, J.H., Kathir, P., and Ippen-Ihler, K., 1988, The product of the F-plasmid transfer operon gene, *traF*, is a periplasmic protein, *J. Bacteriol.* 170:3633–3639.
174. Yoshioka, Y., Ohtsubo, H., and Ohtsubo, E., 1987, Repressor gene *finO* in plasmids R100 and F: constitutive transfer of plasmid F is caused by insertion of IS*3* into F *finO*, *J. Bacteriol.* 169:619–623.
175. Yoshioka, Y., Fujita, Y., and Ohtsubo, E., 1990, Nucleotide sequence of the promoter-distal region of the *tra* operon of plasmid R100, including *traI* (DNA helicase I) and *traD* genes, *J. Mol. Biol.* 214:39–53.

Chapter 3

Key Regulatory Aspects of Transfer of F-Related Plasmids

WALTER B. DEMPSEY

1. Introduction

Current experiments designed to increase our understanding of the biochemistry of conjugation, as well as earlier experiments that mapped the transfer genes of the F sex factor, owe much of their design to the fact that transfer of the F sex factor is unregulated. Unregulated expression of transfer occurs because one of the critical regulatory genes of the transfer operon, the *finO* gene, was inactivated by the insertion of an IS3 sequence some time before F became a subject of study (6, 60). The existence of the moderately complex regulatory system in F transfer, to which this *finO* gene belongs, was first discovered by analysis of transfer of antibiotic resistance factors related to F (12, 54). In a series of elegant and insightful studies initially in the laboratory of Meynell (36, 37) and then in the laboratories of Anderson (1a, 23, 46) and of Willetts (17–19), the basic control system for both F and related R-factors such as R100 (also known as R222, NR1) was established several years ago. The final model proposed by Willetts (55) has stood against all challenges and is now generally accepted. A biological consequence of the "FinO-P" control system, as the model is now frequently called, is the subject of this chapter.

Regulation of transfer in the "FinO-P" system is primarily transcriptional. In a simplified overview, all but two of the transfer genes are contained in the 30 kb long *tra* operon. The remaining two are next to the 5' end of the *tra* operon. The gene proximal to the *tra* operon, *traJ*, encodes a protein which, if made, activates the adjacent *tra* operon promoter (56). The promoter for *traJ* is constitutively expressed (7). The promotor for the distal gene, *traM*, is regulated. The regulatory genes which give the system its name are

WALTER B. DEMPSEY • University of Texas Health Science Center and Department of Veterans Affairs, Veterans Affairs Medical Center, Dallas, Texas 75216.

Bacterial Conjugation, edited by Don B. Clewell. Plenum Press, New York, 1993.

finO and *finP*. Both of these *fin* genes are absolutely required for regulation. The nature of the *finO* gene product(s) is unknown, but the stable *finP* gene products are antisense RNAs that apparently limit translation of *traJ* transcripts (16).

The relative sizes and positions of *traM* and *traJ* are shown in Figure 3.1. The figure also shows the origin of transfer (*oriT*), the site of the single strand nick (T), gene *X*, and *traY*. *traY* is the first gene of the 30-kb *tra* operon, and transcripts originating at the *traY* promoter can extend 30 kb to the end of the *tra* operon (not shown). The large arrow in Figure 3.1 indicates that gene *X* is the first gene to enter a recipient during transfer. The gene not shown, *finO*, maps just distal to the 3′ end of the *tra* operon.

For the sake of simplicity, this review assumes the ideal case that all transcripts beginning at *traY* are extended to the end of the 30-kb-long *tra* operon (5). For that reason, *traY* in Figure 3.1 is taken to represent both itself and the entire *tra* operon. Precise molecular details of the above model are still the subjects of active investigations by several different groups. This essay will focus primarily on the reactions that initiate conjugal transfer of the fully repressed R100 from *Escherichia coli* K12.

2. Properties of R100

2.1. Early Studies

Early studies with R100 isolated from patients infected with *Shigella flexneri* showed that the rate of transfer of R100 from an established *E. coli* strain was very low when compared to the rate of transfer of F sex factor from the same strain (38). These measurements were generally performed using short mating times before the mixtures were plated onto a selective medium. If the mating times were extended, the kinetics of the mating showed that there was rapid retransfer from each new recipient for several hours. This was called "epidemic spread" or "infectious antibiotic resistance" because it led to conversion of all of the susceptible cells in the mixture to R+ in a very short time (53, 54). This phenomenon was explained satisfactorily as arising from the need for time to establish effective concentrations of the *trans*-acting regulatory elements FinO and FinP in the recipients. The aspect of the phenomenon that has not been explained yet is what causes the low-level transfer by a repressed plasmid in the first place. That is the focus of this chapter.

2.2. Statement of Problem

This problem is a significant one when one considers that the act of conjugation is a major commitment of energy by the donor. It requires synthesis of at least 26 different proteins, including enzymes, structural proteins, and DNA binding proteins. Accordingly, it is likely that conjugation from a fully repressed cell involves the throwing of a molecular switch that lets the entire transfer operon behave for an adequate amount of time as if it were fully derepressed. Only in this way can sufficient amounts of the pilus and its molecular motor be made and put into action together with the 20 or more other proteins required to run the conjugal process. In many respects this aspect of transfer derepression is analogous to induction of a lysogenic phage. The specific question asked here about R-factor transfer

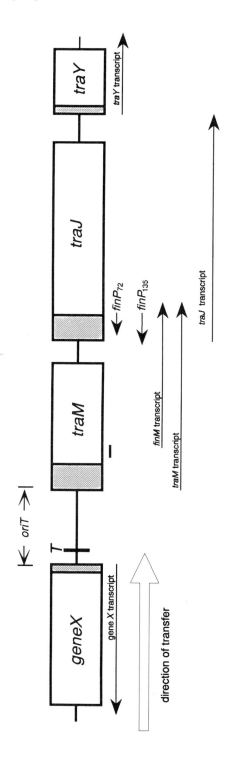

Transcript map of the R100 transfer origin region

FIGURE 3.1. Linear map of the genes near the R100 transfer origin. The map is drawn to scale from the known R100 sequences (13, 14, 25). For each gene the hatched areas represent untranslated mRNA leader regions, and the open boxes represent the ORFs. *T* represents the site of single-stranded nicking on the bottom strand as the map is drawn. Gene X is the gene immediately 3′ to the nick. As such it is the first gene to enter the recipient. This is indicated by the arrow labeled "direction of transfer." The last gene to enter is *traM*. The small horizontal line under *traM* is the site of the promoter called *finM*. Thin lines below *traM* show the origins and best estimate of the 3′ ends of the *traM* and *finM* transcripts. The site of initiation of both the 72-nucleotide-long and the 135-nucleotide-long *finP* transcripts is the same (7).

is similar to asking about the source of the low-level amounts of λ phage seen in uninduced cultures of λ lysogens.

The question for R-factor transfer then is, if there is a switch that occasionally turns transfer on in cultures fully repressed for transfer (R100), what is the nature of that event? One can rule out permanent switches such as up-mutations because the repressed donor phenotype is observed in the new transconjugants after several generations. If mutations are ruled out, some other processes cannot be. From the λ analogy, one could propose that there was a momentary insufficiency of a repressor (FinO?) due to cleavage or a synthetic rate inadequate to saturate a particular cell volume. From what is known about the DNA sequences of the genes involved early in transfer expression, two other possibilities suggest themselves. One possibility might be the rare transcription of a very efficiently translated transcript (TraY) whose transcription is normally blocked. Another possibility might be rare translation of a continuously synthesized transcript (TraJ) whose translation is normally blocked. Regarding both of these suggestions, recent arguments made by de Smit and van Duin (10) suggest that one or two copies of a transcript having a "perfect" Shine-Dalgarno sequence and proper spacing for an initiation codon would be efficiently translated and provide many copies of a protein. If the protein specifically and efficiently bound DNA, it might serve as part of a switching mechanism.

To recapitulate, to initiate transfer in a fully repressed cell, a single, random, spontaneous event occurs that does one of the three things proposed above as initiators of epidemic spread of R100. These events are inactivation of FinO protein, 30S ribosome binding and translation of a single *traJ* transcript, or synthesis and translation of a single *traY* transcript. Once one of these events occurs, ensuing steps in all three cases lead to synthesis of TraY protein. TraY then binds to the *oriT* region to throw the switch. Table 3.1 lists the predictions for each event and how each leads to TraY.

2.3. Proposed Biological Latching Relay

Most of us are familiar with the functioning of mechanical or electronic latching relays. They are at the heart of burglar alarm systems. These are switches that lock "on"

TABLE 3.1 Relationships between Initiating Events and the Genetic Switch

Event	Response	*tra* Operon turned on by	Source of TraY for switch
FinO inactivated	Upstream *traJ* promoters turned on; FinP concentration drops	TraJ synthesis	TraJ activation of *traY* promoter
traJ transcript translated	TraJ synthesized	a) TraJ synthesis	TraJ activation of *traY* promoter
		b) transcription and translation continues through *traY* and *tra* operon	Self
traY transcribed	*traY* translated	a) TraY synthesis	Self
		b) transcription and translation continues through *tra* operon	Self

after a single pulse. A similar switch is proposed here. The heart of the proposal is the following: *traJ* transcripts originating at a tightly regulated promoter(s) *upstream* of the normal, continuously active *traJ* promoter provide additional *traJ* mRNA in the presence of TraY, which by virtue of both its quantity and its different secondary structure serves as a very efficient decoy sense mRNA that captures *finP* RNA. This lowers the concentration of the antisense *finP* RNA to the point that some *TraJ* transcripts get translated. Once the TraJ protein is made, it fully activates the *traY* promoter and transfer begins. I propose that these upstream *traJ* promoters are completely inactive in fully repressed R100. I propose that the activation of these upstream *traJ* promoters is a kind of positive feedback or "switch" that keeps transfer on in R100-containing cells once it has been randomly started. Aspects of this type of regulation are seen during vegetative replication of R100 and R1 (39, 57).

The components of this switch in R100 can be visualized in Figure 3.1. The position of one of the upstream *traJ* promoters (*finM*) is indicated by the small line under the *traM* open reading frame. The other is the *traM* promoter itself. Experiments by the author, presented below, show that transcripts originating at both these promoters cross *traJ*. (To reduce confusion in presenting this model, no distinction is made below regarding which upstream promoter is used.) The first step in activation of the upstream promoters is binding of TraY protein to its site near the *traM* promoter. Experiments are in progress to distinguish between TraY binding alone and a combination of TraY binding and Helicase I (TraI) nicking at *oriT* as sufficient causes for activation of the upstream *traJ* promoters. Activation presumably terminates once the donor strand is fully transferred to the recipient because conjugation of the F-like plasmids involves donation of a unit copy of a preexisting single strand. Donation thus necessarily involves complete unwinding and strand separation. This activity would remove the bound proteins.

The model just proposed, that a random event causes *traY* transcription and that the resulting TraY protein activates an upstream *traJ* promoter(s) by its binding between the nick site and *traM*, is an interpretation of several facts already established, the data presented in Figure 3.4 of this review, and some speculation. In the next section, a brief review of the facts known about the RNA structures, the proteins involved in the model and their binding sites, and *finP* and *finO* is presented.

3. Components of the Biological Relay

3.1. RNA Secondary Structures

The foundation on which this model is built is the set of RNA structures predicted by the latest version of the Zuker program for *traJ* and *finP* transcripts (26, 51, 52, 61). The DNA sequences from the *oriT* region through *traJ* are known for F sex factor, R100, R1, and several other related plasmids. Transcript mapping has identified the 5' ends of the transcripts. When one uses this information to generate the structures of R100 *traJ* transcripts from the normal *traJ* promoter and those of *traJ* from an upstream *traJ* promoter, one finds two critical differences in the region that overlaps the antisense RNA from the *finP* gene. First, the upstream *traJ* RNA has three stem loops that are the complements of the three stem loops in the longer of the two R100 *finP* transcripts, whereas the normal *traJ* RNA matches only two of the *finP* stem loops. The second difference is that the Zuker program predicts that the region of *traJ* mRNA that is the exact complement of the

5' end of *finP* is always a single-stranded loop structure in the upstream *traJ* mRNA, whereas this critical region is frequently double stranded in the *traJ* mRNA from the normal *traJ* promoter. This is discussed further in the *traJ* section below.

Other RNA structures contributing to this model are those related to translational regulation. Inspection of the RNA sequence for translation initiation sites in the region depicted in Figure 3.1 shows that both the *traJ* gene and the *traY* gene have transcripts with strong complementary homology to the Shine-Dalgarno sequence of 3'UCCUCCA5' at the 3' end of the 16S ribosomal RNA (*traJ* = 5'-AGGAGGU-3' *traY* = 5'-AGGAGG-3'). Therefore, both the *traY* transcript and the *traJ* transcript qualify as efficiently translatable transcripts, and either might serve as part of a triggering system. A clear difference between these two transcripts is that the Shine-Dalgarno sequence of *traJ* is largely buried in the stem of a strong stem loop (loop 1 Figure 3.2A), whereas the Shine-Dalgarno sequence of *traY* is single stranded (Figure 3.2E).

3.2. Protein Binding in the *oriT* Region

At least four proteins, TraM (1, 11, 43), TraY (24, 30), Helicase I (TraI) (25b, 34, 41), and IHF (9, 25a, 50), have been shown to bind specifically in the *oriT* region of the F-related plasmids. (I am proposing here that a fifth protein, a FinO protein, also binds here.) Footprint analyses for binding proteins on *oriT* regions are known. Helicase I binds at the nick site T (Figure 3.1) and in purified F sex factor systems has been shown to be sufficient cause for nicking (32, 34, 41). This is not the case for R100 (25b). Figure 3.3 shows footprints. One can see that the F footprints are nearly identical to those of R100 (50). Similar footprints for similar proteins on related plasmids are simply inferred at present.

TraY binds on the *traM* side of *oriT* (Figure 3.3). TraM also binds to four separate sites on the 5' end of *traM*. Some of these TraM sites overlap the *traM* promoter and the *traM* leader (Figure 3.3). IHF also binds in this same region. Studies of the interactions of TraM, TraY, and IHF proteins are incomplete. For example, the known binding sites for IHF in R100 and the predicted TraM protein overlap to a degree, and it is unknown if the proteins are mutually exclusive or synergistic.

Each of the transfer genes involved in the model being presented here is the object of study of one or more labs, and there is not space here to critically review all of the data for each of the genes. Instead, the following is a brief synopsis of each gene.

3.3. *traM*

traM is the gene immediately 5' to *traJ* and immediately 3' to *oriT* (see Figure 3.1). *traM* is expressed continuously in the derepressed (*finO⁻*) R100-1 strain, but it is not expressed in the repressed (*finO⁺*) R100 strain (9). The *traM* transcript encodes three nested ORFs, each of which has an appropriately spaced Shine-Dalgarno sequence (13). The longest is for a protein of 127 residues and a size of 14.5 kDal. The isoelectric point for this TraM protein is 4.78 as calculated by the Chargepro program of PCgene. The other two are for proteins of 10.0 kDal, pI = 4.6, and 8.5 kDal, pI = 4.6. Other programs of PCgene show two potential phosphorylation sites and one glycosylation site in the longest *traM* protein. Proteins identified as TraM protein on polyacrylamide gels show the following sizes: For plasmid R1, 14.4 kD (43) and 8.5 kD (29); for R100, 10.5 kDal (9); and for pED208, 9.6 kD (11). The size differences may be the results of processing of TraM or the use of different start sites.

The *traM* proteins of R100, R1, and pED208 have been purified and shown both by gel-shift assays and by DNase1 footprinting to bind to their respective plasmid DNA in the region between *oriT* and the start of the *traM* ORF (1, 11, 43). The positions of these bindings are shown in Figure 3.3. A similar, but unpublished, finding was made with both F and R100 *traM* sites by R. Thompson in 1983 (personal communication). TraM binding is discussed elsewhere in this volume (see Chapter 5). One should note here that several of the investigators of *traM* have pointed out that some manner of autoregulation of *traM* expression is expected from the position of its binding locus.

traM has a tandem pair of promoters in R100 (8), in R1 (29), and probably in F (47). Figure 3.1 shows that the R100 *traM* transcript begins 100 nucleotides upstream of the *traM* ORF. In contrast to R1, the tandem promoter in R100 is not in this leader but, instead, is contained entirely inside the longest *traM* ORF. To avoid confusion, we named this tandem R100 promoter "*finM*" until its role is known and a more suitable name is suggested. Although the *finM* transcript contains some of the smaller of the ORFs nested inside the *traM* gene, proteins of sizes corresponding to these small ORFs cannot be seen on SDS-polyacrylamide gels made from strains carrying *finM* but lacking the *traM* promoter.

Northern blots of RNA extracted from cells containing R100-1 were compared to blots made with RNA from cells containing the *traM* region and different amounts of accompanying R100 DNA, all in multicopy plasmid vectors (8, 9). When analyzed with a probe that detected both *traM* and *finM* transcripts, these blots showed that extracts from the R100-1 cells contained relatively equal amounts of the *traM* transcript and the *finM* transcript, whereas cells containing the cloned R100 fragments made overwhelming amounts of the *finM* transcript and virtually undetectable amounts of *traM* transcript (8). Although there was a great disproportion between the amounts of *traM* transcript and the amounts of *finM* transcript in these experiments, all clones containing the entire *traM* gene made easily detectable amounts of TraM protein (9). The amount of TraM seemed unrelated to the amount of *traM* transcript. One could interpret the results as supporting either the hypothesis that TraM protein negatively regulated *traM* transcription but not *finM* transcription or the hypothesis that TraM protein somehow was involved in processing the 5′ end of the *traM* transcript to give a *finM*-like transcript.

It is now known that most of the *traM*/*finM* transcripts reach into the *traJ* ORF (see below). Inspection of the DNA sequence between the 3′ end of the *traM* ORF and the 5′ end of the *traJ* ORF shows no consensus transcription terminators of either type (2, 42) before the end of *traJ*. This sequence does show numerous strong inverted repeats, however. Therefore, the discretely sized *traM*/*finM* transcripts that were seen in these experiments are presently as likely to be transcripts processed back to a stable 3′ ends as they are to be transcripts specifically terminated at the observed lengths.

3.4. *traJ*

The *traJ* gene encodes a protein of 223 amino acid residues that has a molecular mass of 26 kDal (25). The protein analysis programs of PCgene identify three potential glycosylation sites in the *traJ* protein, five potential phosphorylation sites, three potential

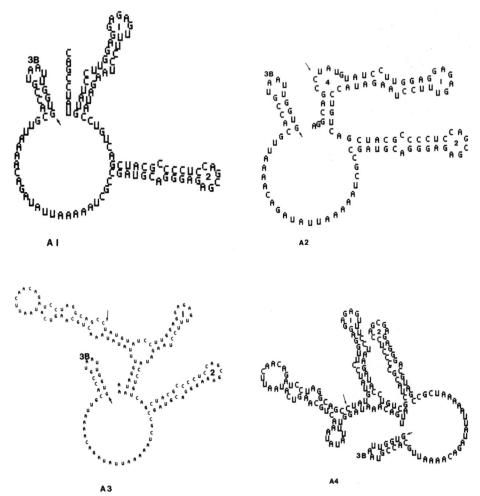

FIGURE 3.2. Structures of RNA molecules predicted by the Zuker PCfold program version 4.0. The folded structures were drawn by the public domain program *Molecule* written by J. Ryan Thompson. Structures A1, A2, A3, and A4 represent *traJ* transcripts initiating at the normal *traJ* promoter. The 5' ends of the transcripts are indicated with a small arrow. The series shows that the sequence 3'CUAU-5', the complement to the 5' end of the *finP* transcript shown first at the far 3' end of A1, becomes part of loop 4 as the transcript lengthens in A2 (long arrow) and then becomes sequestered (long arrow) as transcription continues (A3) and stays sequestered (A4). Structure B represents the *traJ* transcript from an upstream promoter. The first base after the translation stop signal of the *traM* ORF was chosen as the 5' end in order to keep the drawing small. The 5' end is indicated with a small arrow. Note that structure B retains loop 4. Structure C represents the *traJ* transcript made by the FisO mutant of the F sex factor as initiated from the normal *traJ* promoter of F. An arrow points to the single base change in the FisO mutant. A second arrow points to the region complementary to the 5' end of the *finP* transcript of F. Structure D represents the *traJ* transcript of the F sex factor initiated from the normal *traJ* promoter. An arrow points to the base that is changed in the FisO mutant, and a second arrow points to the region complementary to the 5' end of *finP*. Note that the complementary region is sequestered in the normal F *traJ* transcript. Structure E represents the very weak stem loop that might form in the *traY* leader region. Note that the Shine-Dalgarno sequence is fully exposed in the loop of the structure. Structure F represents the short *finP* transcript of R100. An arrow shows the 5' end. The structure shown is 79 bases long. Structure G represents the long *finP* transcript of R100. The 5' end is marked with an arrow. The structure shown is 138 bases long. (Figure continued on next page.)

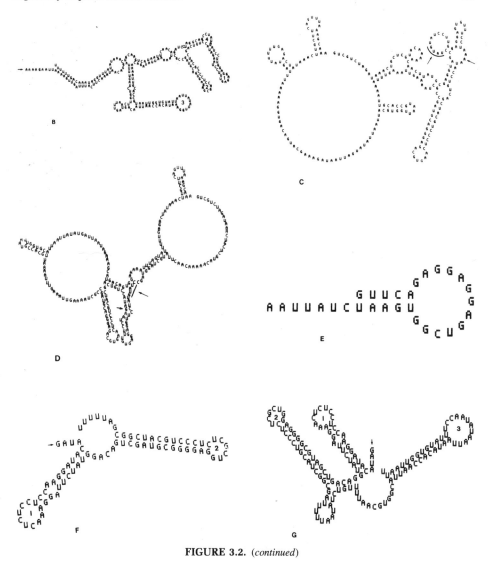

FIGURE 3.2. (*continued*)

membrane-associated helices, and one transmembrane sequence. As noted elsewhere in this chapter, the *traJ* protein appears to function by activating a promoter for the *traY* gene.

The exact site of the normal *traJ* promoter was identified by primer extension experiments (7). The *traJ* transcript originating at the R100 *traJ* promoter begins 103 bases 5′ to the *traJ* ORF. Northern blots probed with *traJ*-specific single-stranded RNA probes showed that this promoter was continuously active in both R100 (repressed) and R100-1 (derepressed (7). The sizes of the *traJ* transcripts detected from R100 by this method varied with different strains. In our common ED2149 strain it was crudely estimated at 235 nucleotides (7). The *traJ* transcripts in R100-1 strains appeared to be full-length. One

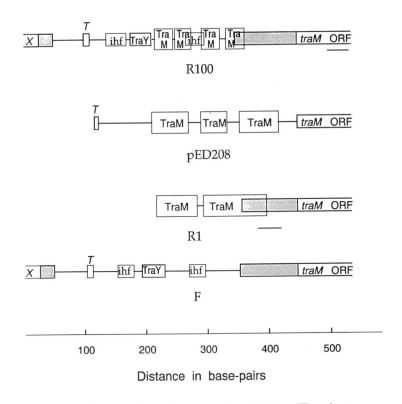

Footprints of proteins that bind to the *oriT* region

of F-related plasmids

FIGURE 3.3. Maps of the regions protected by different proteins in R100 and related plasmids. The linear genetic maps are drawn from the known sequences of the three plasmids illustrated. Alignment was made on the first base of the *traM* ORF. The small rectangles indicate the regions protected in footprinting experiments by the named proteins. The horizontal lines under the *traM* genes represent the sites of the promoters tandem to the *traM* promoters in R100 and R1. *T* represents the site of single-stranded nicking (25a). The complete R1 sequence between gene *X* and *traM* is unavailable, hence unmarked. Sources of the footprinting data for IHF are R100 (9, 23a), F (50); TraY protein: R100 (24), F (S. W. Matson, unpublished); TraM protein: R100 (1a), R1 (43), pED208 (11).

interpretation of this finding is that regulation of *traJ* gene expression from the normal *traJ* promoter must be via translation. The short *traJ* mRNAs seen in R100 are presumed to result from a failure of the RNA to be translated.

The leader region of *traJ* can fold into three separate stem loops. These are shown in Figure 3.2A. The loops are numbered from 3′ to 5′ to allow easy comparison to *finP* transcripts. The strengths of the stem loops as calculated by the PCfold program of Zuker (26, 51, 52, 61) are Structure I (complement of *finP* Structure I) = −5.0 kcal, Structure II (complement of *finP* Structure II) = −15.1 kcal, Structure IIIb (no common *finP* equivalent) = −2.8 kcal.

Experiments described below establish that the *traJ* leader is also transcribed from an upstream promoter in R100-1. The normal *traJ* promoter is in the middle of an extensive region of dyad symmetry. Normal *traJ* transcripts beginning at this promoter will have a different predicted structure than *traJ* transcripts entering this region from the *traM* or *finM* promoters because the latter would contain all of this dyad symmetry and would fold differently from those that originate within it at the *traJ* promoter. (See Figure 3.2A, B.) These longer transcripts would thus have three stem loops that were the exact complements of the longer *finP* molecule. Structures I and II would be the same as those seen in transcripts originating from the *traJ* promoter, but the third structure would be different. The strength of this alternative stem loop, called Structure III (complement of *finP* Structure III), is -17.4 kcal. These are shown in Figure 3.2A and 3.2B.

Another difference between upstream *traJ* transcripts and regular *traJ* transcripts is in the structures predicted for them after transcription enters the *traJ* ORF and the transcripts are not being translated. By following Barry Polisky's personal suggestion to look at potential structures and how they changed with each addition of a single base to the 3' end of a transcript, I found that both types of *traJ* transcripts form a short stem loop (labeled 4 in Figure 3.2) as transcription enters the ORF region. The loop of this structure contains the region that is the complement of the 5' end of the *finP* transcripts. Loop 4 disappears from the regular *traJ* transcript after 51 bases of the ORF have been transcribed. This procession of structures for *traJ* mRNA made from the normal *traJ* promoter is shown in Figure 3.2A1, A2, and A3. Loop 4 is maintained by the transcript from the upstream *traJ* promoter once it forms. These two differences characterize the *traJ* transcript from the upstream promoter as being a real analogue of the kinds of prokaryotic sense-antisense pairs that have been studied by others. They also suggest roles for a FinO protein. For example, the FinO protein could be a repressor of one or both upstream TraJ promoters, or it could be a binding protein that prevented the normal *traJ* transcript from folding into the form without the loop that matches the 5' end of *finP*. In support of this interpretation is the predicted structure of the *traJ* transcript from the FinO-insensitive (FisO) variant of the F sex factor (21). Compare structures C and D in Figure 3.2. This FisO *traJ* RNA structure has a loop just like the RNA from the upstream promoters in R100, and thus it mimics them. One could argue that the normal role of FinO was to form this loop. Because the loop forms independently in this *fisO* mutant, FinO is not required.

The upstream *traJ* transcript is like the sense RNA in the ColE1 sense-antisense system (31, 49) because it has three stem loops that are the exact complements of the three stem loops of the antisense RNA. The finding that it can apparently fold stably into a form that keeps the above complement to the 5' end of *finP* as a single-stranded loop makes it look like it was made to interact rapidly with *finP*. According to the present hypotheses of both Tomizawa (31) and Kleckner (27), maximum *rate* of formation of the double-stranded RNA between a sense-antisense pair occurs when the 5' end of one member of the pair can attack a single-stranded complementary region on the 3' end of the other member rather than a complementary 3' region that is already self-paired.

I assume that this model of rapid sense-antisense interaction that has come out of the independent work of the Kleckner, Polisky (40), and Tomizawa labs represents the most common way that pairs interact. If this assumption is true, then the three-stem-loop structure of the long *finP* transcript and the *traJ* transcript from the upper *traJ* promoter are predicted to be the most frequently successful reactant pair in the *traJ-finP* system because the 3' end of traJ that reacts with the 5' end of *finP* is always single-stranded.

3.5. *traY*

The R100 TraY protein is a small (8543 MW), plasmid-specific protein that has a calculated isoelectric point of 9.6 (25). The TraY protein binds at two sites on R100 (24). One site is centered 61 bases away from the site of the single-stranded nick (discussed above), and the other is at the *traY* promoter. Matson (personal communication) has found analogous binding sites in the sex factor F for TraY. The site of the TraY binding near the nick site, as determined by footprinting, is shown in Figure 3.3. In a purified system using components from the F sex factor, Lahue and Matson have shown that the binding of TraY by itself does not cause nicking (30). It is now known that TraY is not required for nicking to occur in F (34, 41), but it is required for nicking of R100 (25b).

The TraJ protein has only one role in transfer that is currently established with considerable rigor and that is to activate the *traY* promoter (44, 45). The TraY protein binds near the site of nicking that begins transfer. It presently seems possible that escape of either the *traY* or the *traJ* gene from its normal regulation could lead to a few molecules of TraY protein. From what is already published about the *traY* promoter (44, 45), we know that the *traY* promoter of F sex factor can occasionally transcribe *traY* without any TraJ protein being present, so a rare, spontaneous *traY* transcript is a reasonable candidate for the initial trigger. The location of TraY binding sites at the promoters of both *traM* and *traY* in plasmids F and R100 suggests that its role is regulating promoter activity. (We have found an occasional activation of the R100-1 *traY* promoter without TraJ in unpublished observations in the laboratory.)

3.6. *finP*

Probing Northern blots of RNA from R100-containing cells with *finP* specific probes shows two *finP* transcripts (7) (Figure 3.1). Primer extension experiments show that both transcripts initiate at the same site (7). The exact sizes of the two have been determined to be 76 and 135 nucleotides on RNase protection gels with RNA size standards (Dempsey, in preparation). The sizes predicted by the Zuker program are 72 and 135 (Figure 3.2F, G).

An uncomplicated explanation for the existence of two transcripts is that the terminator for the first is a bit leaky. The alternative explanation is that the smaller sized RNA is only the result of processing. Structural analysis with the Zuker RNA-folding program (61), which uses the 1989 data of Turner et al (26, 51, 52), shows that the 72-nucleotide position is at the 3' end of a stem loop that is followed by a string of uridine residues (Fig. 3.2F). In a similar fashion, the 3' end of the 135-nucleotide structure is at the beginning of a run of uridines that are immediately at the 3' end of another stem loop (Figure 3.2G). Because both structures have the general characteristics of RNA ending at rho-independent terminators (2), leakiness of the terminator at 72 bp is presently an adequate explanation of the 135-bp RNA. The area remains under study.

Figure 3.2G shows that the 135-nucleotide-long R100 *finP* can fold into three stem loops. The strengths calculated for the individual loops by the latest PCfold program are −6.1 kcal for Structure I, −22.1 kcal for Structure II, and −12.5 kcal for Structure III.

The sequence of the R100 *finP* gene contains a few small ORFs. There is no evidence that these are translated. A comparative study of the sequences of five different *finP* genes found in the IncF conjugal plasmids led Finlay et al (16) to conclude that it is unlikely to encode a protein. The common elements in all were the secondary structures. Additional conviction about the importance of the structures and not their content comes from comparing *finP* to the number of antisense systems now known.

At present no *finP* mutants have been studied in R100, but a large and perhaps sufficient set of *finP* mutants have been studied in the F sex factor (21) and the plasmid R1 (28). The location of the mutations in these mutants are generally in those places that alter the stability of the stem loops corresponding to Structures I and II (Figure 3.2F, G).

3.7. *finO*

finO is the other control gene that regulates transfer operon expression. It has been mapped just beyond the 3′ end of the transfer operon (48, 59). Its presence is absolutely required for any repressive effect of the antisense gene *finP* to be detected, but its mechanism of action is unknown. The nature of the *trans*-acting product(s) of the *finO* gene has not yet been unequivocally established, even though DNA fragments with full *finO* activity have been cloned from several plasmids and sequenced (3, 4, 35, 60).

Sequence analysis shows that one strand of *finO* contains an ORF for a basic protein consisting of 186 amino acid residues, a predicted size of 21.2 kD, and a predicted isoelectric point of 10.0 to 10.2. The opposite strand of *finO* contains two regions of homology with the *traM-traJ* transcript. One of these is a 12-nucleotide run at the end of the ORF that matches (11/12) the end of the R100 *traM* ORF. The other is a 15-nucleotide run that matches (13/15) the bases in Structure II of *finP*. If either or both of these stretches are normally transcribed, the resultant RNA could alter the folding rates of the *traJ* transcripts (7). Recent Northern blots made in the author's laboratory with RNA from cells containing wild-type R100 or R100-1 and probed with strand-specific *finO* riboprobes show that small but equal amounts of a 540-base-long and a 350-base-long transcript are made from *finO* in the anti-sense direction. Probes specific for the ORF or "sense" strand show that transcripts from this strand are more numerous but are heavily degraded. The maximum size of the predominant species is approximately 500 nucleotides. The intensities of these high molecular weight spots is greatly increased in *finO* mutants, probably because the *finO* gene is being transcribed from *tra* promoters. Northern blots of cells containing cloned *finO* so far shows no trace of transcript from the sense strand.

A protein of 21.5 kD was tentatively identified as the *finO* protein of the related plasmid R6-5 by Timmis et al (48). We showed that R100 *finO* made a 21.2-kD protein when cloned in front of the *tac* promoter in plasmid pKK223-2 (35). We have not yet been able to show that cells actively making this protein are *finO*⁺. What we observe is that these *tac* plasmids, presumably *finO*⁺, are cured at rates too high to allow us to test for *finO* phenotype. Not only is the pKK223-2 cured but the F sex-factor plasmids, introduced to test for *finO*, are also cured at unusually high rates. To date, no one has reported a convincing experiment that shows that the 21.2-kD protein is the active agent in *finO*. These observations notwithstanding, in the model to be presented later, I assume that at least one FinO protein exists and acts as a repressor.

Mutational evidence that suggests that the 558-nucleotide-long ORF of *finO* is an active component of *finO* is found in the sequence of the spontaneously occurring *finO* minus mutation in R100-1. That sequence, by Ohtsubo and co-workers (60), shows the insertion of an extra A into a run of A's in the middle of the ORF. In apparent support of this ORF as *finO*, we showed that an 18-base deletion into the 5′ end of the ORF or a 13-base deletion into the 3′ end of the ORF inactivated the *finO* response (35).

On the other hand, when we made and sequenced internal addition and deletion mutations at the (*Sma*I) *Xma*I site inside the ORF, we found contradictory evidence (35). The interesting mutants were two in which frameshifting insertions or deletions were found. These mutant *finO* fragments retained *finO* activity. Sequence analysis showed that these frameshift mutations would introduce several extra basic amino acids into the "FinO" protein, whereas the R100-1 insertion mutations would not. This could explain the retention of activity. The activity might also be explained if these frameshifts occurred exclusively in intergenic spaces found between genes for the putative small RNAs in the bottom strand that were mentioned previously. The unexpected activity could also be explained by the fact that the mutants were tested only in multicopy plasmids. Rigorous testing of these mutant *finO* fragments in single-copy plasmids has not yet been completed.

A puzzling and still unexplained observation made with *finO* is that the amount of detectable *finP* transcript increases in the presence of *finO* (7). We have not yet been able to show any effect of *finO* on the cloned *finP* promoter. It reamins possible that the *finO* product affects *finP* degradation. In the model presented here, I simply speculate that a FinO protein acts as a repressor of one or both of the upstream *traJ* promoters. By repressing synthesis of "decoy" RNA, which would trap free *finP* RNA, FinO would leave more FinP RNA free to be detected. One expects from the nonplasmid specificity of *finO* that any binding site FinO might bind to would be common among all plasmids sensitive to *finO*.

3.8. IHF and *traI*

Recent reviews of both helicase activity (32, 33) and IHF activity (15, 20, 22, 50, 58) have appeared. The reader is directed there and to Chapter 5 for discussions of how these proteins affect DNA structure.

3.9. Evidence that Upstream Transcripts Cross *traJ*

The transcript of the R100 *traJ* gene includes an untranslated, 103-nucleotide-long leader. The finding that the *traM* and *finM* transcripts generally crossed this *traJ* leader before they terminated was made by RNase protection experiments. Single-stranded RNA probes were constructed that could distinguish between *traJ* transcripts that began at the normal *traJ* promoter and transcripts that began 5′ to it. The structure of one of the relevant probes is shown in the top portion of Figure 3.4. This probe can complement the normal *traJ* leader and an additional 60 bases 5′ to that leader. In an RNase protection experiment, this probe will give a 103-nt-long product from transcripts originating at the *traJ* promoter and a 165-nt-long product from intact transcripts originating at upstream promoters. Thus,

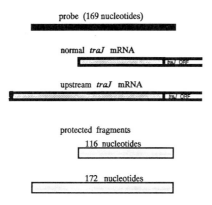

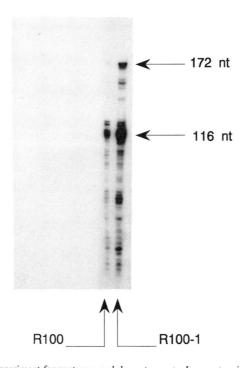

FIGURE 3.4. RNase protection experiment for upstream and downstream *traJ* promoters in R100. The upper part shows the sizes and location of a riboprobe synthesized from a *Sau*3a-*Bam*H1 fragment. The number of nucleotides of homology between the probe and the R100 is shown. Beneath the probe are drawn to scale the relevant parts of the two types of *traJ* mRNA. The fragments of probe protected by the mRNA predicted from the known sequence are shown under "protected fragments." The lower part shows an autoradiograph of the products remaining after hybridization of the RNA to 2× purified probe followed by digestion with RNase A and RNase T1. The mix was run out on a sequencing gel with RNA standard. The first track shows results with RNA extracted from ED2149 containing R100, the second track (no signal) shows the results with RNA extracted from ED2149 without plasmids, and the third track shows results with RNA extracted from ED2149 containing R100-1. The arrows show sizes determined by RNA standards.

if there are two different transcripts that cross the entire *traJ* leader, RNase protection experiments conducted with probe A will show two protected fragments: one of 103 nucleotides represents the transcripts beginning at the normal *traJ* promoter, and one of 165 nucleotides represents transcripts beginning at least 62 bases farther upstream. As noted previously, earlier work had shown that transcription occurs continually from the *traJ* promoters of both the fully repressed R100 and the derepressed R100-1. When RNA was prepared from R100-1 and from R100 and analyzed with probe A in an RNA protection assay, I found, as expected, that the *traJ* transcript originating at the *traJ* promoter was easily detected in extracts of both kinds of cells. This is shown by the band at 103 nucleotides in both tracks of the gel depicted in the lower half of Figure 3.4. Further inspection of the figure shows that the R100-1 track has the additional 165-nt-long band expected if transcripts begin upstream to the end of the probe. This band is missing from the R100 track. The figure shows then that the *traJ* transcript is made continuously from the *traJ* promoter in both repressed and derepressed strains and that an additional transcript from an upstream promoter crosses the entire *traJ* leader region when transfer is derepressed.

4. Recapitulation or Sources of Speculations

Cloned R100 fragments that contain the *traM* gene make TraM protein but only very small amounts of *traM* transcript of a discrete size (8, 9). These same fragments make a large amount of discretely sized transcript from the adjacent *finM* promoter. R100-1 makes roughly equal amounts of both *traM* and *finM* transcripts, while R100, on the other hand, makes neither. Other cloning experiments show that the presence of the *traJ* gene or TraJ protein has no detectable effect on the expression of cloned *traM* and *finM* (9).

One conclusion from these observations is that maximum expression of the *traM* promoter in clones requires an additional component in order to give the amounts of *traM* transcript seen in R100-1. This component could be a protein, such as a processed form of TraM, or a product of another *tra* gene, or it could be a nick in the DNA or a combination of the two. An alternative conclusion is that TraM protein is an autoregulator, and in cells containing multicopy *traM* plasmids, the level of this TraM is much higher than it is in R100-1, and it shuts off the *traM* promoter but not the *finM* promoter.

Because the *finM* promoter is turned off in R100 but is on equally well in both R100-1 and cloned fragments carrying either *finM* alone or in different combinations with *traM*, *traJ*, and *oriT*, it appears that this promoter behaves simply as if it were a negatively regulated promoter. R100 makes something that represses *finM*, and R100-1 does not, namely FinO. Although one possible candidate for a repressor that is present in R100 and absent in R100-1 is a FinO protein, preliminary analyses of RNA from cells containing multicopy *traM* and *finM* and low-copy *finO* clones show no effect of *finO* on *traM* or *finM* transcripts (Dempsey, unpublished observations). It is more likely that FinO inhibits *traM*, and when TraM protein is made it binds to the *traM* promoter and activates the *finM* promoter.

5. Summary

Our current hypothesis is that R100 in its native state is supercoiled, and its *oriT* region is bent by the binding of IHF. The upstream *traJ* promoters are not operating because FinO

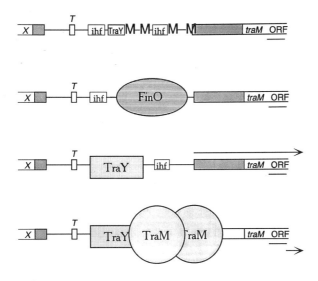

Model for FinO, TraY, TraM regulation of transfer

1. FinO represses *traM*

2. TraY blocks FinO, *traM* is transcribed

3. TraM activates upstream *traJ* promoter

FIGURE 3.5. A speculative model of how the transfer genes of R100 get locked on in response to a transitory signal. The four lines are the linear map of R100 drawn to scale. The hatched areas are the leader regions of gene *X* and *traM*. *T* is the site of single-stranded nicking. The small horizontal line under the ORF is the promoter for the upstream *traJ* transcript that starts at *finM*. The small boxes in the first line show the sites protected in footprint assays for the two proteins IHF and TraY. It is proposed that one role for the FinO protein is as repressor for the upstream *traJ* promoters. This is approximated in line 2, where FinO covers the *traM* promoter. The model speculates that the tandem *finM* promoter is also inactive at this time because of blocking caused by either FinO binding or by the tertiary structure stabilized by the binding. Line 3 shows that the binding of TraY displaces FinO and allows the *traM* promoter to work. The activation of the *traM* promoter and the DNA unwinding associated with it frees the *finM* promoter from its restraints, and it too begins to work. Translation of the *traM* ORF provides TraM protein, which then binds to the *traM* promoter to self-regulate the *traM* transcript and to lock open the *finM* promoter, which then continues to work and provide large amounts of the upstream *traJ* transcript that has the effect of inactivating the antisense *finP* and allowing the *tra* operon to express.

protein is bound as a repressor (Figure 3.5). FinO binding requires the native, supercoiled, three-dimensional structure in order to bind. Loss of FinO protein by either cleavage or dissociation alters the three-dimensional structure, and allows at least one of the upstream *traJ* promoters to work. Binding of a rare TraY protein also alters the three-dimensional structure of the region and results in the displacement of FinO protein, and again an upstream *traJ* promoter is activated. Activation of upstream *traJ* promoters works as a switch because transcripts from these promoters cross into the regulatory region of *traJ*. These transcripts have the three complementary stem loops that allow rapid alignment ("kissing") with *finP* RNA as well as the free 3′ end that allows rapid hybridization with the free 5' end of *finP* RNA with its consequent rapid unwinding of the stems and consequent rewinding to form the double-stranded FinP-TraJ complex. This drops the concentration of free *finP* RNA and allows some *traJ* transcripts to be translated to TraJ protein. Once TraJ is made, its activation of the *traY* promoter ensures adequate *tra* expression for one round of transfer.

Nicking at *oriT* occurs either because of the synthesis of a *traI* protein from the same transcript that made the original TraY protein or from the synthesis of such a protein from a later transcript. Nicking may be necessary for the upstream *traJ* promoter to start, or instead it may be the final lock on the switch that prevents the normally silent promoter from reassuming the off state.

The phenomenon of some rare initiating event triggering the full expression of the *tra* operon of R100 has been the subject of this chapter. Reversal of the process and restoration of the donor strain to the fully repressed state is of equal interest, but space and data relevant to that process limit what can be said at this time. Several factors may prove to be important. Some of these factors are as follows: The strand donated is preexisting and single stranded; the *finO* gene is intact and presumably making its product(s) throughout the larger part of the transfer; once disrupted by helicase, the *finO* gene is the first transfer-related gene to be restored in the donor; completion of transfer includes removal of helicase, ligation of the 3' end of the newly made strand to its 5' end, and supercoiling of the newly completed R100.

The single-strandedness of the donor strand means that the unwinding catalyzed by the helicase will remove most of the binding proteins. A protein like TraY that bound to *tra* because of a random pulse in its concentration will have to compete with proteins like FinO that are still being made by constitutive genes. The location of *finO* seems to give it a special advantage in this regard. Restoration of R100 to its supercoiled form at the end of transfer also restores the putative FinO binding site to its most competitive native form. The ideas presented here are stimulating work in the author's lab, and he is hopeful that they will provide ideas for critical experiments in other labs as well.

ACKNOWLEDGMENTS. The author wishes to thank Don Clewell both for inviting him to write this chapter and for his useful critiques of earlier drafts of it. The author also wishes to thank Richard Deonier for extended discussions about earlier phases of the model and Steve Matson for making available the footprinting data for F TraY before their publication. The excellent technical assistance of both Teresa Thompson and Vera Lelianova is gratefully acknowledged. Research in the author's laboratory is supported by the U.S. Department of Veteran's Affairs.

References

1. Abo, T., Inamoto S., and Ohtsubo, E., 1991, Specific DNA binding of the TraM protein to the *oriT* region of plasmid R100, *J. Bacteriol.* **173:**6347–6354.
1a. Anderson, E. S., and Smith, H. R., 1972, Fertility inhibition in strains of *Salmonella typhimurium*, *Mol. Gen. Genet.* **118:**79–84.
2. d'Aubenton-Carafa, Y., Brody, E., and Thermes, C., 1990, Prediction of rho-independent *Escherichia coli* transcription terminators. A statistical analysis of their RNA stem-loop structures, *J. Mol. Biol.* **216:**835–858.
3. Cheah, K-C., Hirst, R., and Skurray, R., 1987, *finO* sequences on conjugally repressed and derepressed F-like plasmids, *Plasmid* **17:**233–239.

4. Cheah, K-C., Ray, A., and Skurray, R., 1984, Cloning and molecular analysis of the *finO* region from the antibiotic resistance plasmid R6-5, *Plasmid* **12**:222–226.

5. Cheah, K-C., Ray, A., and Skurray, R., 1986, Expression of F plasmid *traT*: independence of *traY-Z* promoter and *traJ* control, *Plasmid* **16**:101–107.

6. Cheah, K-C., and Skurray, R., 1986, The F plasmid carries an IS*3* insertion within *finO*, *J. Gen Microbiol.* **132**:3269–3275.

7. Dempsey, W. B., 1987, Transcript analysis of the plasmid R100 *traJ* and *finP* genes, *Mol. Gen. Genet.* **209**:533–544.

8. Dempsey, W. B., 1989, Sense and antisense transcripts of *traM*, a conjugal transfer gene of the antibiotic resistance plasmid R100, *Mol. Microbiol.* **3**:561–570.

9. Dempsey, W. B., and Fee, B. E., 1990, Integration host factor affects expression of two genes at the conjugal transfer origin of plasmid R100, *Mol. Microbiol.* **4**:1019–1028.

10. de Smit, M. H., and van Duin, J., 1990, Control of prokaryotic translational initiation by mRNA secondary structure, in: *Progress in Nucleic Acid Research and Molecular Biology*, Vol. 38 (J. Davidson), Academic Press, New York, pp. 1–35.

11. Di Laurenzio, L., Frost, L. S., Finlay, B. B., and Paranchych, W., 1991, Characterization of the *oriT* region of the IncFV plasmid pED208, *Mol. Microbiol.* **5**:1779–1790.

11a. Di Laurenzio, L., Frost, L. S., and Paranchych, W., 1992, The TraM protein of the conjugative plasmid F binds to the origin of transfer of the F and ColE1 plasmids, *Mol. Microbiol.* **6**:2951–2959. .

12. Egawa, R., and Hirota, Y., 1962, Inhibition of fertility by multiple drug resistance factor (R) in *Escherichia coli* K-12, *Jap. J Genet.* **37**:66–69.

13. Fee, B. E., and Dempsey, W. B., 1986, Cloning, mapping, and sequencing of plasmid R100 *traM* and *finP* genes. *J. Bacteriol.* **167**:336–345.

14. Fee, B. E., and Dempsey, W. B., 1988, Nucleotide sequence of gene *X* of antibiotic resistance plasmid R100, *Nucl. Acids Res.* **16**:4726.

15. Figueroa, N., Wills, N., and Bossi, L., 1991, Common sequence determinants of the response of a prokaryotic promoter to DNA bending and supercoiling, *EMBO J.* **10**:941–949.

16. Finlay, B. B., Frost, L. S., Paranchych, W., and Willetts, N. S., 1986, Nucleotide sequences of five IncF plasmid *finP* alleles, *J. Bacteriol.* **167**:754–757.

17. Finnegan, D., and Willetts, N., 1972, The nature of the transfer inhibitor of several F-like plasmids, *Mol. Gen. Genet.* **119**:57–66.

18. Finnegan, D., and Willetts, N., 1973, The site of action of the F transfer inhibitor, *Mol. Gen. Genet.* **127**:307–316.

19. Finnegan, D. J., and Willetts, N. S., 1971, Two classes of F*lac* mutants insensitive to transfer inhibition by an F-like R factor, *Mol. Gen. Genet.* **111**:256–264.

20. Friedman, D. I., 1988, Integration host factor: a protein for all reasons, *Cell* **55**:545–554.

21. Frost, L., Lee, S., Yanchar, N., and Paranchych, W., 1989, *finP* and *fisO* mutations in *finP* anti-sense RNA suggest a model for FinOP action in the repression of bacterial conjugation by the F*lac* plasmid JCFLO, *Mol. Gen. Genet.* **218**:152–160.

22. Goodrich, J. A., Schwartz, M. L., and McClure, W. R., 1990, Searching for and predicting the activity of sites for DNA binding proteins: compilation and analysis of the binding sites for *Escherichia coli* integration host factor (IHF), *Nucl. Acids Res.* **18**:4993–5000.

23. Grindley, N. D., Grindley, J. N., Smith, H. R., and Anderson, E. S., 1973, Characterisation of derepressed mutants of an F-like R factor, *Mol. Gen. Genet.* **120**:27–34.

24. Inamoto, S., and Ohtsubo, E., 1990, Specific binding of the TraY protein to *oriT* and the promoter region for the *traY* gene of plasmid R100, *J. Biol. Chem.* **265**:6461–6466.

25. Inamoto, S., Yoshioka, Y., and Ohtsubo, E., 1988, Identification and characterization of the products from the *traJ* and *traY* genes of plasmid R100, *J. Bacteriol.* **170**:2749–2757.

25a. Inamoto, S., Abo, T., and Ohtsubo, E., 1990, Binding sites of integration host factor in *oriT* of plasmid R100, *J. Gen. Appl. Microbiol.* **36**:287–293.

25b. Inamoto, S., Yoshioko, Y., and Ohtsubo, E., 1991, Site- and strand-specific nicking in vitro at *oriT* by the TraY-TraI endonuclease of plasmid R100, *J. Biol. Chem.* **266**:10086–10092.

26. Jaeger, J. A., Turner, D. H., and Zuker, M., 1989, Improved predictions of secondary structures for RNA, *Proc. Natl. Acad. Sci. USA* **86**:7706–7710.

27. Kittle, J. D., Simons, R. W., Lee, J., and Kleckner, N., 1989, Insertion sequence IS*10* anti-sense pairing

initiates by an interaction between the 5' end of the target RNA and a loop in the anti-sense RNA, *J. Mol. Biol.* **210:**561–572.

28. Koraimann, G., Koraimann, C., Koronakis, V., Schlager, S., and Högenauer, G., 1991, Repression and derepression of conjugation of plasmid R1 by wild-type and mutated *finP* antisense RNA, *Mol. Microbiol.* **5:**77–87.

29. Koronakis, V. E., Bauer, E., and Högenauer, G., 1985, The *traM* gene of the resistance plasmid R1: comparison with the corresponding sequence of the *Escherichia coli* F factor, *Gene* **36:**79–86.

30. Lahue, E. E., and Matson, S. W., 1990, Purified *Escherichia coli* F-factor TraY protein binds *oriT*, *J. Bacteriol.* **172:**1385–1391.

31. Masukata, H., and Tomizawa, J., 1990, A mechanism of formation of a persistent hybrid between elongating RNA and template DNA, *Cell* **62:**331–338.

32. Matson, S. W., 1991, DNA helicases of *Escherichia coli*, in: *Progress in Nucleic Acid Research and Molecular Biology*, Vol. 40 (J. Davidson), Academic Press, New York, pp. 289–326.

33. Matson, S. W., and Kaiser-Rogers, K. A., 1990, DNA helicases, *Ann. Rev. Biochem.* **59:**289–329.

34. Matson, S. W., and Morton, B. W., 1991, *Escherichia coli* DNA helicase I catalyzes a site- and strand-specific nicking reaction at the F plasmid *oriT*, *J. Biol. Chem.* **266:**16232–16237.

35. McIntire, S. A., and Dempsey, W. B., 1987, Fertility inhibition gene of plasmid R100, *Nucl. Acids Res.* **15:**2029–2042.

36. Meynell, E., and Cooke, M., 1969, Repressor-minus and operator constitutive de-repressed mutants of F-like R factors: their effect on chromosomal transfer by HfrC, *Genet. Res.* **14:**309–313.

37. Meynell, E., and Datta, N., 1967, Mutant drug resistance factors of high transmissibility, *Nature* **214:** 885–887.

38. Nakaya, R., Nakamura, A., and Murata, Y., 1960, Resistance transfer agents in *Shigella*, *Biochem. Biophys. Res. Commun.* **3:**654–659.

39. Persson, C., Wagner, E. G. H., and Nordström, K., 1990, Control of replication of plasmid R1: formation of an initial transient complex is rate limiting for antisense RNA-target RNA pairing, *EMBO J.* **9:**3777–3785.

40. Polisky, B., Zhang, X. Y., and Fitzwater, T., 1990, Mutations affecting primer RNA interaction with the replication repressor RNA I in plasmid ColE1: potential RNA folding pathway mutants, *EMBO J.* **9:**295–304.

41. Reygers, U., Wessel, R., Müller, H., and Hoffmann-Berling, H., 1991, Endonuclease activity of *Escherichia coli* DNA helicase I directed against the transfer origin of the F factor, *EMBO J.* **10:**2689–2694.

42. Richardson, J. P., 1990, Rho-dependent transcription termination, *Biochim. Biophys. Acta.* **1048:**127–138.

43. Schwab, M., Gruber, H., and Högenauer, G., 1991, The TraM protein of plasmid R1 is a DNA-binding protein, *Mol. Microbiol.* **5:**439–446.

44. Silverman, P. M., Wickersham, E., and Harris, R., 1991, Regulation of the F plasmid *traY* promoter in *Escherichia coli* by host and plasmid factors, *J. Mol. Biol.* **218:**119–128.

45. Silverman, P. M., Wickersham, E., Rainwater, S., and Harris, R., 1991, Regulation of the F plasmid *traY* promoter in *Escherichia coli* K12 as a function of sequence context, *J. Mol. Biol.* **220:**271–279.

46. Smith, H. R., Humphreys, G. O., Grindley, N. D., Grindley, J. N., and Anderson, E. S, 1973, Molecular studies of an *fi+* plasmid from strains of *Salmonella typhimurium*, *Mol. Gen. Genet.* **126:**143–151.

47. Thompson, R., and Taylor, L., 1982, Promoter mapping and DNA sequencing of the F plasmid transfer genes traM and traJ, *Mol. Gen. Genet.* **188:**513–518.

48. Timmis, K. N., Andres, I., and Achtman, M., 1978, Fertility repression of F-like conjugative plasmids: physical mapping of the R6-5 *finO* and *finP* cistrons and identification of the FinO protein, *Proc. Natl. Acad. Sci. USA* **75:**5836–5840.

49. Tomizawa, J., 1986, Control of ColE1 plasmid replication: binding of RNA I to RNA II and inhibition of primer formation, *Cell* **47:**89–97.

50. Tsai, M-M., Fu, F., and Deonier, R. C., 1990, Intrinsic bends and integration host factor binding at F plasmid *oriT*, *J. Bacteriol.* **172:**4603–4609.

51. Turner, D. H., Sugimoto, N., and Freier, S. M., 1988, RNA structure prediction, *Ann. Rev. Biophys. Biophys. Chem.* **17:**167–192.

52. Turner, D. H., Sugimoto, N., Jaeger, J. A., Longfellow, C. E., Freier, S. M., and Kierzek, R., 1987, Improved parameters for prediction of RNA structure, *Cold Spring Harbor Symp. Quant. Biol.* **52:**123–133.

53. Watanabe, T., 1963, Infective heredity of multiple drug resistance in bacteria, *Bacteriol. Rev.* **27:**87–115.

54. Watanabe, T., and Fukasawa, T., 1961, Episome-mediated transfer of drug resistance in *Enterobacteriaceae*. I. Transfer of resistance factors by conjugation, *J. Bacteriol.* **81:**669–678.

55. Willetts, N., 1977, The transcriptional control of fertility of F-like plasmids, *J. Mol. Biol.* **112:**141–148.
56. Willetts, N., and Skurray, R., 1987, Structure and function of the F factor and mechanism of conjugation, in: *Escherichia coli and Salmonella typhimurium: cell and molecular biology* (F. C. Neidhardt), American Society for Microbiology, Washington, DC, pp. 1110–1133.
57. Womble, D. D., and Rownd, R. H., 1986, Regulation of IncFII plasmid DNA replication, *J. Mol. Biol.* **192:**529–548.
58. Yang, C-C., and Nash, H. A., 1989, The interaction of *E. coli* IHF protein with its specific binding sites, *Cell* **57:**869–880.
59. Yoshioka, Y., Fujita, Y., and Ohtsubo, E., 1990, Nucleotide sequence of the promoter-distal region of the *tra* operon of plasmid R100, including *traI* (DNA Helicase I) and *traD* genes, *J. Mol. Biol.* **214:**39–53.
60. Yoshioka, Y., Ohtsubo, H., and Ohtsubo, E., 1987, Repressor gene *finO* in plasmids R100 and F: constitutive transfer of plasmid F is caused by insertion of IS*3* into F *finO*, *J. Bacteriol.* **169:**619–623.
61. Zuker, M., 1989, Computer prediction of RNA structure, in: *Methods in Enzymology*, Vol. 180A (J. E. Dahlberg and J. N. Abelson), Academic Press, New York, pp. 262–288.

Chapter 4

Broad Host Range Conjugative and Mobilizable Plasmids in Gram-Negative Bacteria

DONALD G. GUINEY

1. Introduction

Bacterial conjugation mediates genetic exchange not only between cells of the same species but also between members of distantly related or even unrelated genera. These transfer events have been demonstrated among diverse members within both the gram-positive and gram-negative groups of organisms. Recently, experiments using natural conjugation systems have demonstrated gene transfer between gram-positive and gram-negative organisms, and even from bacteria to the lower eukaryote *Saccharomyces cerevisiae* (59, 124, 125). The significance of this promiscuous gene transfer is that it provides a mechanism for the availability of a huge pool of genes for bacterial evolution. A dramatic example of the ability of individual bacteria to acquire genes of selective value is the widespread development of resistance to antibiotics used in clinical medicine and agriculture.

Most conjugation systems in gram-negative bacteria are encoded by plasmids. More than 25 different groups of plasmids have been defined on the basis of incompatibility (Inc) properties, and many of these specify distinct transfer systems (30). The concept of plasmid host range arose as a result of attempts to transfer members of the Inc groups into various gram-negative genera. Many self-transmissible plasmids originally isolated in a species of *Enterobacteriaceae* could be transferred to other members of this family. However, only certain plasmids could be transferred to and maintained in *Pseudomonas* species. These were termed "broad host range plasmids," and those confined to the *Enterobacteriaceae*

DONALD G. GUINEY • Department of Medicine and Center for Molecular Genetics, University of California, San Diego, San Diego, California 92103.
Bacterial Conjugation, edited by Don B. Clewell. Plenum Press, New York, 1993.

were "narrow host range plasmids." Likewise, plasmids isolated from *Pseudomonas* could be divided into broad and narrow host range, depending on their ability to be transferred to and maintained in *Escherichia coli* (31, 63, 64).

A careful analysis of the host range of self-transmissible plasmids revealed that the property depended on many traits, including the conjugation system, the replication and maintenance functions, and the ability of plasmid-encoded selective markers to be expressed in the new host. For example, a *Pseudomonas aeruginosa* variant with enhanced recipient ability was isolated (119). Transfer of plasmid markers representing a number of Inc groups from *E. coli* to *P. aeruginosa* was detected, including several plasmids characterized as having a narrow host range. However, analysis of plasmid DNA in the recipients showed that most had undergone structural changes or were not present in the extrachromosomal state. These results suggested that a number of conjugation systems could mediate transfer from *E. coli* to *Pseudomonas* but that many plasmids could not be maintained in the latter host.

The advent of molecular cloning techniques facilitated direct testing of the contribution of various plasmid traits to the observed host range. The general approach has been to separate the conjugation system from the vegetative replication and maintenance region. The host range of each of these systems was tested for the *E. coli* fertility plasmid F (IncF1) and compared to the promiscuous IncP plasmid RK2 (48). A direct comparison of the F and RK2 transfer systems was accomplished by cloning the *oriT* regions of each plasmid together on an RK2 replicon that could be maintained in both *E. coli* and *P. aeruginosa*. The results showed that RK2 was 10^4 times more efficient than F at mobilizing the test plasmid from *E. coli* to *P. aeruginosa*, while both systems were equally effective in transfers between *E. coli* strains. However, a low but significant transfer between *E. coli* and *Pseudomonas* could be detected using F. An inability of F to replicate in *Pseudomonas* was demonstrated by cloning the F replicon together with the RK2 transfer system: This chimeric plasmid could not be maintained in *Pseudomonas* (48). This study demonstrated that plasmid conjugation systems have inherent differences in transfer efficiency between various bacteria. In addition, the results showed that F conjugation has a broader host range than the replication system.

The nonspecificity of plasmid conjugation, in contrast to replication, has been demonstrated by many investigators. A number of plasmids naturally found in either the *Enterobacteriaceae* or *Pseudomonas* can mediate transfer to but cannot replicate in the heterologous host (15, 41, 48, 68, 119). The promiscuous IncP plasmids can replicate in many species of the α, β, and γ subgroups of the purple bacteria, but not in the δ subgroup (*Myxococcus*), nor in cyanobacteria or the gram-negative anaerobe *Bacteroides fragilis* (23, 51, 120, 139). However, transfer of plasmid DNA mediated by RK2 conjugation occurs readily to all these organisms, as detected by chromosome integration (*Myxococcus*) or specially constructed shuttle vectors carrying an *oriT* region (cyanobacteria and *Bacteroides*). Recently, the shuttle vector approach has also been used to demonstrate IncP-mediated plasmid transfer from *E. coli* to gram-positive organisms and to yeast (59, 125). The nonspecificity of bacteria to yeast conjugation is underscored by the observation that the F transfer system is also able to mediate the process.

These studies on the host range of plasmid transfer indicate that most conjugation systems are capable of mediating promiscuous gene exchange, and the designation of these systems as "narrow" and "broad" host range is arbitrary. However, this chapter will focus on the clearly promiscuous self-transmissible plasmids that have been studied in detail, belonging to the P, N, and W incompatibility groups. These three systems share common phenotypic properties. In addition, the mobilization system of the broad host range IncQ plasmids will be reviewed.

2. IncP Plasmid Transfer

2.1. Classification of IncP Plasmids

A considerable number of IncP plasmids have been isolated from a wide variety of gram-negative hosts. Based on hybridization studies using the transfer origin (*oriT*) of RK2 as a probe, Yakobson and Guiney (141) found that the P plasmids could be divided into two groups, α and β. Subsequent analysis of several other loci, including *traC* (primase), has substantiated this classification (72). The IncPα plasmids are all closely related and differ primarily by insertions and deletions in a common backbone (128). Within IncPα, a group of plasmids isolated from a Birmingham, England, hospital, including RP1, RP4, RK2, R18, and R68, appear to be identical (26). The IncPβ plasmids have not been as well characterized. The best studied β plasmid, R751, has the same general genetic organization as RK2/RP4 but exhibits 20 to 40% sequence divergence at specific loci (86, 114, 148).

2.2. Phenotype of Conjugation Mediated by P Plasmids

Several important phenotypic properties have been defined for P plasmid conjugation. Transfer frequencies are considerably higher for mating performed on solid media rather than in the liquid phase (20). On agar plates or nitrocellulose filters, transfer efficiencies approach one per recipient cell for many matings. In the electron microscope, IncP plasmid-containing cells produce rigid pili that are thinner than N or W pili (18). P pili are antigenically distinct. Male-specific phage that infect IncP plasmid-containing cells include PRR1, PRD1, PR3, PR4, GU5, and PF3 (17, 19, 92, 93, 116). Based on the high transfer frequency, electron microscopy studies, and phage sensitivity, IncP conjugation systems are expressed constitutively. A recent study has shown a relatively high frequency of spontaneous variation to phage resistance accompanied by a change in the transfer phenotype of RK2/RP4, but the genetic basis for this phenomenon is unknown (L. Kornstein and V. Waters, submitted).

As noted in the Introduction, the host range of the IncP transfer system has been studied extensively. Efficient transfer to a wide variety of gram-negative bacteria has been documented (54). IncP conjugation has become a major tool for genetic manipulation of bacteria. For organisms in which P plasmids cannot replicate, shuttle vectors have been constructed using endogenous plasmids from the foreign host together with the *oriT* of RK2/RP4 or a mobilization region recognized by the IncP conjugation system. Using this approach, gene transfer has been detected to a variety of gram-positive organisms, *Mycobacteria*, cyanobacteria, *Bacteroides*, and yeast (51, 59, 74, 125, 139). The diversity of these recipients underscores the nonspecificity of the cell-to-cell interactions that characterize IncP-mediated conjugation, since it is difficult to postulate a common cell surface structure that is shared by all these organisms. Instead, P plasmid transfer appears to be a general system of conveying DNA across biological membranes. Significantly, the transferred DNA can be recovered as a double-stranded plasmid in most of the recipient organisms, indicating that the conjugal DNA processing system also has a very broad host range.

2.3. Genetic and Biochemical Analysis of IncP Conjugation

2.3.1. Genetic Organization of RK2/RP4 Transfer Regions

The initial deletion derivatives of RK2 constructed by Figurski et al (39) showed that loss of the region downstream from the kanamycin resistance (Km^R) locus resulted in defective transfer. In a series of studies by Barth and coworkers (7–9) Tn7 and Tn76 insertions were used to map the regions of RK2/RP4 required for transfer. The initial insertions (8, 9) identified two conjugation regions of RK2/RP4 that were separated by the IS8 (IS21) element and the Km^R locus: Tra1 extending toward the EcoRI site and Tra2 located toward the trfA locus (see Figure 4.1). Both regions appeared to be polycistronic and to contain loci required for sensitivity to donor-specific phage. Later, Barth described a single Tn7 insertion located about 3 kb from the KpnI site at coordinant 24 kb (7). This mutant, which was deficient in both surface exclusion (sex) and transfer, defined a new transfer region, Tra3. Later work, described below, suggests that Tra2 and Tra3 are contiguous and should be designated the Tra2 region (95). The origin of transfer (oriT) has been located by genetic techniques and shown to coincide with the relaxation complex (relaxosome) nick site (49, 53, 55, 98). oriT maps within the Tra1 region, with transfer proceeding as indicated by the arrow in Figure 4.1, so that a small portion of Tra1 is contained in the leading strand region, while most of Tra1 enters the recipient last (2, 47). The separation of the conjugation genes into two regions is unusual for self-transmissible plasmids. Subsequent analysis of the intervening DNA indicates that this region contains the aphA (Km^R) gene, IS8 (IS21); a partitioning site, parD; and a multimer resolution function, parCBA (32, 45, 97, 104, 107). These elements may have been inserted into this region during P plasmid evolution, resulting in the separation of Tra1 from Tra2. Tra1 also appears to differ in function from Tra2. To date, all genes shown to be active in conjugal DNA processing map in Tra1, and all mutations isolated in Tra2 affect surface exclusion and/or pilus biosynthesis as defined by sensitivity to donor-specific phage (50, 54, 95). However, certain insertions in Tra1 also result in phage resistance, indicating a role for the Tra1 region in pilus synthesis as well (see Section 2.3.3.).

2.3.2. Tra1 Core of RK2/RP4

The best understood portion of the RK2/RP4 conjugation system is the region including oriT and the adjacent genes. Major progress in elucidating the biochemical events that initiate transfer has been made by defining the DNA sequence requirements for oriT function and by purifying the plasmid-encoded proteins that interact with oriT. These results are described in detail in Chapter 5 and will be summarized here as they relate to the overall understanding of IncP conjugation.

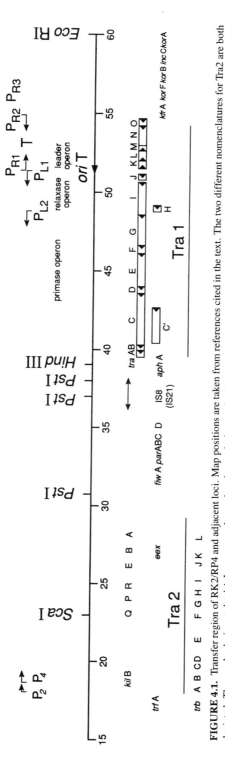

FIGURE 4.1. Transfer region of RK2/RP4 and adjacent loci. Map positions are taken from references cited in the text. The two different nomenclatures for Tra2 are both depicted: The newly designated *trbA-L* genes are located underneath the previously named *kilB*, *traQ*, *P*, *R*, *E*, *B*, *A*, and *eex* loci.

The transfer origin is defined genetically as the site required in *cis* for DNA transfer. Complete *oriT* function is present in a 250-bp fragment of RK2/RP4 (49, 132), and several functional domains have been identified in this *oriT* region, as shown in Chapter 5. A minimal sequence of 112 bp retains diminished but significant *oriT* activity (55). The single-stranded cleavage induced in vitro by the relaxation complex isolated from RK2/RP4-containing cells or by the relaxosome reconstituted from purified components is located within this minimal *oriT*, supporting the concept that this nick represents the physical origin of single-stranded DNA transfer (49, 96, 98). The nick occurs 8 bp from the foot of an imperfect inverted repeat containing 19-bp arms (55, 98). Deletions from either end of the extended 250-bp *oriT* sequence substantially decrease transfer, while loss of the nick site eliminates mobilization of *oriT*-containing plasmids (44, 49, 132).

A key to the definition of specific RK2/RP4 transfer proteins that interact with *oriT* was the discovery that R751 and other IncPβ plasmids cannot mobilize the minimal RK2/RP4 (IncPα) *oriT* sequence (141). Instead, an extended 2.2-kb region surrounding *oriT* is required for transfer by R751 (50). This segment, designated TraI core, was shown by Tn5 insertion mutagenesis and DNA sequencing to contain at least two genes required for the sequence-specific transfer of the IncPα *oriT* (44, 50, 147, 148). These genes, *traJ* and *traK*, are transcribed from divergent promoters, PR1 and PL1, located in the *oriT* region (Figure 4.1). A third genetic function, located downstream from *traJ*, appears to enhance R751 mobilization of plasmids containing intact *traJ* and *traK* loci (50). Sequence analysis showed that this region contains the N-terminal half of the *traI* gene as well as an 18-kDa ORF (99, 148). The role of this 18-kDa ORF in transfer has not been defined, since a protein product has not been detected and specific mutations in the ORF have not been isolated.

The biochemical properties of the TraI, TraJ, and TraK proteins have been elucidated by Erich Lanka and coworkers and are described in detail in Chapter 5. Both TraI and TraJ are required for nicking at *oriT* as evidenced by the formation of relaxosomes in vitro (96). Purified TraJ protein binds to the nick-proximal arm of the 19-bp inverted repeat as shown by DNA footprinting studies (147). TraI appears to be the nicking enzyme and is found covalently attached to the 5' end of the nicked strand following relaxation in vitro (99). The isolation of transfer-deficient base transition mutations within the nick region of *oriT* confirms this two-component model for nicking and transfer (132). Mutations in the arm of the inverted repeat decrease TraJ binding, while changes in the 8 bp between the repeat and the nick site interfere with TraI-inducing nicking. The TraH protein, encoded by a different reading frame in the C-terminal half of TraI (Figure 4.1), stabilizes the TraI/TraJ relaxosome complex in vitro (96). The role of TraH in vivo is unknown, since specific mutations have not been isolated. The TraK protein, although not required for nicking, is an essential protein for RK2/RP4 specific *oriT* transfer (50). Purified TraK binds at several sites to the right of the nick, in the leading region of DNA transfer (see Chapter 5).

2.3.3. Extended TraI Region of RK2/RP4

Lanka and coworkers have determined the sequence of the TraI region extending from the *aphA* (Km^R) gene through the *kfrA* locus located downstream of the *trfB* operon (54, 86, 121, 148). Computer analysis of this sequence using a codon preference approach together with subcloning and identification of protein products allowed this group to identify at least 14 ORFs as proven structural genes, as shown in Figure 4.1. For several of these, specific mutations have defined a role in the transfer process. As described in Section 2.3.2., the *traI*, *traJ*, and *traK* genes are all essential for transfer, and the gene products are involved in specific DNA processing events at *oriT*. Insertion mutations in *traF*, *traG*, and *traL*, verified by sequence analysis, also markedly decrease transfer, indicating that these loci are also essential conjugation genes (V. Waters, E. Lanka, and D. Guiney, submitted for publication). The DNA primases of RK2 are encoded by the *traC* locus, described in detail in Chapter 5. Two proteins, 118 kDa and 80 kDa, are generated from in-phase translation from two start sites (73). Both proteins have identical DNA priming activity and are synthesized in abundance. However, only the 118-kDa protein is transferred with the DNA strand to the recipient during mating, implying that this component functions in complementary strand synthesis in the recipient (100). Sites for efficient priming by the RK2-RP4 primase were found on both strands of the *oriT* region, suggesting that primase may function in conjugative DNA synthesis in both donor and recipient (140).

Despite the biochemical evidence suggesting that the DNA primase functions in complementary strand synthesis during mating, the specific role of the DNA primase in conjugation remains unclear. Mutations deficient in primase activity have been constructed by Tn7 insertion in RP4 and R18 (69, 71). RP4 primase mutants in *E. coli* exhibit a decreased transfer into several species, particularly *Salmonella typhimurium* and *Protens mirabilis*. Mobilization of the IncQ plasmid R300B was also diminished (71). Back-transfer experiments showed that the defect was specific for the recipient. Primase mutants of R18 in *P. aeruginosa* show markedly lower transfer frequency to *Pseudomonas stutzeri* (69). Provision of the cloned primase gene in the donor or the recipient complemented the primase mutation (89). Interestingly, primase-deficient R18 transferred at a frequency comparable to wild-type in matings between *P. stutzeri* strains and from *P. stutzeri* back to *P. aeruginosa*. These results raise the possibility that the primase has a role in inhibiting a host-specific DNA restriction system operating in *P. stutzeri*.

As noted previously, physical transfer of the 118-kDa primase protein, but not the smaller 80-kDa polypeptide, has been detected in *E. coli* by Rees and Wilkins (100). Earlier studies from this group presented evidence that the RP4 primase was involved in complementary strand synthesis in the recipient (84). However, definite proof is lacking that the recipient-specific transfer defect of primase mutants is due to a defect in priming complementary strand synthesis during or after transfer. A problem in interpreting the primase effect on transfer is the lack of well-defined and specific primase mutants, since the Tn7 inserts were not located precisely by DNA sequencing, and polar effects on the two downstream ORFs (*traA* and *traB*) of TraI were not excluded.

A second region of TraI, mapping to the right of *oriT*, has also been implicated in host range effects. Krishnapillai and coworkers isolated Tn7 mutants of R18 in *P. aeruginosa* and screened for mutations with decreased transfer into other hosts (69). A cluster of host range mutations was found to be located in the leading strand region between *oriT* and *korB* (109). These workers found two phenotypes exhibited by these mutants: three inserts (pM0510, pM0511, and pM0512) showed reduced transfer from *P. aeruginosa* to *E. coli*, *S. ty-phimurium*, and *P. maltophilia*, while a single mutant, pM01183, was defective in transfer only to *P. stutzeri*. A preliminary report located pM01183 in the N-terminal region of a 52-kDa ORF, while pM0511 is inserted within a smaller, overlapping ORF in the C-terminal region (68). The mechanism for the host range effects of mutations in this ORF is unknown, since the biochemical function of this protein is not understood.

The other known transfer genes of Tra1 are *traF*, *traG*, and *traL*. Tn*1725* insertions in these genes, followed by excision of the transposon by *Eco*RI deletion, has generated mutations in each locus consisting of a 35-bp insert containing the new *Eco*RI site. DNA sequencing has confirmed the mutations, which lead in each case to a marked decrease in transfer frequency of the RK2 derivative, pRK231 (V. Waters, E. Lanka, and D. Guiney, unpublished). The functions of these gene products in conjugation are not known. The TraF protein is required for sensitivity to donor-specific phage and therefore is likely to be involved in the assembly or function of the surface mating apparatus. The presence of a signal peptide sequence at the N-terminus of TraF is consistent with this proposal. *traG* mutants remain sensitive to phage. However, the TraG protein of RK2 shares striking amino acid homology with the VirD4 protein of the *Agrobacterium tumefaciens* Ti plasmids (148). *virD4* is an essential gene for the transfer of the T-DNA strand from *Agrobacterium* to susceptible plant cells during crown gall tumor formation and is located in the *virD* operon known to be involved in generating and processing the transferred DNA (146). There is an overall similarity in the organization of the *virD* operon and the *IncP* transfer region encoding *traJ*, *traI*, *traH*, and *traG* (see Section 2.6.).

In addition to the loci described above, the Tra1 region contains at least six additional ORFs that appear to encode protein products. However, the roles of these genes in transfer remain unclear. The Tra1 region appears to extend from the *aphA* gene clockwise to the *kfrA* locus near *korB* (54, 121). However, the exact extent of Tra1 has not been defined at either end, and roles in transfer for additional genes in the *trfB korA korB* region have not been excluded.

The organization of the Tra1 region of R751 (IncPβ) appears to be closely related to that of RK2/RP4. R751 genes corresponding to *tra G, H, I, J*, and *K* flank the *oriT* site, and the region shares 74% overall sequence identity with RK2/RP4 (148). The R751 primase locus (*traC*) also exhibits DNA homology with RK2/RP4, and the gene products show immunological cross-reactivity (72). R751 produces four primase gene products of 192, 152, 135, and 83 kDa on protein gels (86). Sequence analysis shows that the smaller three proteins are produced by in-phase translational starts within the *traC* open reading frame. The 135- and 83-kDa proteins are analogous to the two RK2/RP4 primase gene products. The 152-kDa primase protein is generated from a unique sequence fused to the *traC* gene of R751 but absent in RK2/RP4 (86). The largest protein, 192 kDa, may be a fusion between TraD and TraC resulting from translational readthrough. The functional significance of these larger primase polypeptides is unknown. In R751, the Tra2 region is separated from Tra1 by Tn*402* containing the dihydrofolate reductase gene.

2.3.4. Transcription of the Tra1 Region

In contrast to the F plasmid *tra* genes, the Tra1 region is not organized as a single operon. Promoters identified on the basis of *E. coli* RNA polymerase binding studies are shown in Figure 4.1 (54). Consensus *E. coli* promoters were confirmed at these sites by the subsequent sequence analysis (148). The Tra1 core is organized into two divergent operons beginning within the intergenic *oriT* region. PL1 defines the "relaxase operon," consisting of *traJ*, *traI*, and *traH*. PR1 initiates the "leader operon" containing *traK*, *traL*, and *traM*. A rho-independent transcription terminator is located at the end of *traM* and appears to function in vivo (148). The site of termination of the PL1 transcript is not known, but transcription of the genes distal to *traI* is likely to initiate at PL2, a promoter identified by sequence, RNA polymerase binding, and in vivo activity in a promoter probe vector (148, A. Greener, personal communication). On the distal right end of TraI, the *traN* gene appears to be transcribed from two clustered promoters, PR2 and PR3. Although the available evidence is consistent with this transcriptional organization, direct proof by Northern blot and primer extension analysis is lacking. In addition, no information is available on promoters used in heterologous hosts, although PL2 is active in a variety of gram-negative bacteria (A. Greener, personal communication).

2.3.5. Location and Phenotype of the Tra2 Region

Early mapping of the *tra* genes identified a second transfer region, Tra2, located between the *Pst*I site (30.8 kb) and the *Kpn*I site (24 kb) and separated from TraI by the *Pst*IC fragment, IS*8* (IS*21*) and the *aphA* (Km^R) gene (9) (Figure 4.1). Early deletion experiments showed that no essential *tra* genes were located on *Pst*IC (39). Later, a single transfer-deficient Tn*7* insertion was isolated in the *Sph*I J fragment at about 20 kb, and this region was designated Tra3 (7). Palombo et al (95) used a combination of deletions, insertions, and complementation analysis to define six transfer cistrons in the Tra2/Tra3 region. These loci were designated *A*, *B*, *E*, *R*, *P*, and *Q* in their map order counterclockwise (leftward) from the *Pst*I site at 30.8 kb (Figure 4.1). The phenotype of all the mutations in these loci is Tra$^-$ (transfer deficient) and Dps$^-$ (resistant to donor specific phage). Therefore, it is likely that all six genes are required for sex pilus biosynthesis or assembly. Certain *B*, *D*, and *R* mutants also exhibit decreased surface exclusion (Sfx$^-$), although this is not a consistent phenotype for all mutations at these loci. The single Tn*7* insertion defining Tra3 was also Sfx$^-$, but whether this mutation is in a structural gene or exhibits a polar effect has not been determined. Because of the proximity of *Q* to the Tra3 locus defined by this insertion, there is no evidence to justify the separate Tra3 designation, and the entire transfer region is now designated Tra2 (95). Recently, Lyras et al (80) have located a determinant (*S*) mapping between *B* and *E* that results in small colony formation in *P. aeruginosa* (Slo phenotype) when cloned on RSF1010 as part of the *B* to *E* region. In certain constructs, this region also expresses an Sfx$^+$ phenotype.

The *P* and *R* gene products have been identified as 26-kDa and 36-kDa proteins respectively, and the nucleotide sequence of the *P* region between the *Sca*I and *Kpn*I sites has been determined recently (94). The predicted protein has a molecular weight of 28 kDa and is generally hydrophilic with a hydrophobic N-terminus. Lessl et al (78) have identified and sequenced two entry exclusion genes (*eexA* and *eexB*) located in the region of the *S* and *B* loci of Palombo et al (95) and Lyras et al (80). *eexA* encodes a 28-kDa protein with a putative N-terminal signal sequence, and *eexB* encodes a predicted 7.3-kDa protein. Both genes are required for surface exclusion.

2.3.6. Reconstitution of the RK2/RP4 Conjugation System Using Cloned Tra1 and Tra2 Regions

Recently, the Tra1 and Tra2 regions have been cloned on separate, compatible plasmids (77). These constructions have been used to confirm that the Tra1 and Tra2 regions are sufficient for transfer, as implied by earlier results with pRK2013 (38). In addition, this approach has facilitated the detailed analysis of the sequence requirements for Tra2 function. The TraI region, composed of RK2/RP4 DNA extending from the *Pst*I site just upstream of *aphA* to the *Eco*RI site, was cloned on pBR329 to yield pVWDG23110. The Tra2 region was cloned as an *Nde*I/*Sal*I fragment (coordinates 16.83 to 34.4 kb) on a ColD replicon to yield pDB1726, which contains the *kilB* locus and is dependent on the *korB* region supplied *in trans* by pVWDG23110. The presence of both plasmids in the cell leads to high-frequency transfer of pVWDG23110 and can also mobilize RSF1010. Both RSF1010 mobilization and sensitivity to donor-specific phage require functions encoded by both TraI and Tra2.

Deletion analysis of pDB1726 showed that the minimal contiguous region required for Tra2 function extends from coordinate 18.03 kb to 29.26 kb, making it 11.23 kb in length (77). The *trfA* proximal end of Tra2 maps just upstream of the p116 protein, the first gene in the *trfA* operon (Figure 4.1). The *trfA* promoter maps in this region, as well as the start of a divergent gene (*trbA*). The designation *trb* is now given to transfer genes of the Tra2 region to differentiate these from the TraI loci (designated *tra*). Deletion of *trbA* results in a transfer-deficient phenotype that can be complemented by the cloned region *in trans*; therefore, the *trbA* gene appears to define one end of the Tra2 region. The distal end of Tra2 is located within a 543-bp *Not*I fragment between coordinates 28.72 and 29.26 kb.

2.3.7. Organization of Tra2

Lessl et al (77) have identified a putative promoter (P_2) upstream from *trbA* (Figure 4.1), very close to but divergent from the *trfA* promoter. All genes so far described in Tra2 are oriented in a clockwise manner. However, P_2 is unlikely to be solely responsible for Tra2 transcription. Previous studies on *kil* and *kor* genes mapped the *kilB* locus, defined as a region that could not be cloned without *korB* (40, 88), clockwise from the *trfA* operon. Recent evidence indicates that the *kilBI* gene, located downstream from *trbA*, and the intergenic region between *trbA* and *kilBI* are involved in the KilB phenotype (88). *kilBI* encodes a 35-kDa protein that, by itself, is not lethal to *E. coli*. Mutation and complementation of *kilBI* indicates that it is required for transfer, and sequence analysis reveals significant homology with the VirB11 protein of the *Agrobacterium tumefaciens* Ti plasmids (88). VirB11 is required for T-DNA transfer to plants and is known to possess ATPase and autophosphorylation activities. Because the *kilBI* gene is essential for conjugal transfer, it has also been designated *trbB*.

The precise transcriptional organization of Tra2 and the basis for the kilB phenotype remain unclear. In addition to the putative *trbA* promoter, Lessl et al (77) used RNA polymerase binding to locate a second probable Tra2 promoter (P_4) upstream of *trbB* (*kilBI*). The exact site of this promoter is in question, since earlier work suggested that the *kilB* promoter was approximately 250 bp upstream from the *kilBI* start codon on the basis of consensus sequence analysis (88). However, P_4 appears to be located only about 50 bp upstream of *trbB* (*kilBI*). The KilB phenotype requires a leader region upstream of *kilBI* as well as a portion of the gene. It is clear that cloning the functional Tra2 region requires the *korB* gene, yet the KorB binding site identified in the *kilB* region is well upstream of P_4. Additional KorA and KorB binding sites are located at the *trfA* promoter and could affect transcription of *trbA*. However, the actual location of promoters as determined by transcript mapping, and the effect of KorB on Tra2 transcription, remain to be determined. The possibility that KorB regulates both *trfA* and Tra2 expression has led to the proposal that conjugation and vegetative replication are coregulated in RK2/RP4 (88). However, the role of the level of *trfA* expression on *oriV* activity remains controversial, and recent evidence indicates that TrfA concentration is not the primary control element in RK2 replication (36, 67). Overexpression of the *trfA* promoter appears deleterious to the cells (3). Therefore, the Kor function may have evolved primarily as a mechanism to control expression of certain RK2/RP4 promoters whose overactivity is harmful to the host.

Recently, Lanka and colleagues have obtained the nucleotide sequence for the Tra2 region shown to be essential for transfer in conjunction with Tra1 (E. Lanka, personal communication). Analysis of the sequence reveals 12 open reading frames likely to be transfer genes, including *trbA* and *trbB* (*kilBI*), described previously. These loci are now designated *trbA-L* (Figure 4.1). Most striking is the similarity between Tra2 and the *virB* operon of the *Agrobacterium* Ti plasmids. In addition to the previously described relationship between *trbB* and *virBII*, five other tra2 gene products show similarities to VirB proteins: TrbC and VirB2, TrbD and VirB3, TrbE and VirB4, TrbF and VirB5, and TrbI and VirB10. The proposed functions of Tra2 and VirB are also analogous: Both are thought to be involved in the synthesis and assembly of the cell surface structures that facilitate the close interaction between donor and recipient required for DNA transfer.

2.4. Phenotypes of RK2/RP4 Related to Transfer: Fertility Inhibition and Retrotransfer

RP1 (probably identical to RK2/RP4) profoundly inhibits the transfer of the IncW plasmid R388 when both are coresident in the donor (145). R388 has no effect on RP1 transfer. Two loci on RP1 have independent fertility inhibiting activities on the IncW plasmid: *fiwA* and *fiwB*. *fiwA* is located in the *Pst*IC fragment between 31.7 kb and 32.8 kb (Figure 4.1), at the end closest to tra2 (42). Cloned on pBR322, this locus gives nearly the same inhibition of IncW transfer as the native RP1 but has no effect on sensitivity to the donor-specific phage PR4. The *fiwB* locus maps in the region of the *Eco*RI site, from 59.8 kb to 0.8 kb, a region also implicated in tellurite resistance (42). Whether *fiwB* and the tellurite resistance genes are identical has not been established, but expression of both phenotypes appears to be linked (46, 129). *fiwB* in the absence of *fiwA* specifies a level of fertility inhibition about 10^2-fold lower than the native RP1. *fiwB* appears to decrease the R388-mediated sensitivity to donor-specific phage, but whether this effect also requires *fiwA* has not been determined.

RP4 is the target of fertility inhibition by the F plasmid of *E. coli* and the IncN plasmid pKM101. The IncN locus is discussed in Section 3.1. Like the IncW inhibition just described, the F effect is also unidirectional: F decreases RP4 transfer by 10^2- to 10^3-fold, but RP4 has no effect on F transfer (117, 118). The F gene responsible for this activity has been identified as *pifC*, also designated F3 protein, located adjacent to *oriVI* (87, 118). *pifC* is involved in replication and also regulated expression of the *pif* operon. RP4 appears to have a binding site for PifC, since RP4 increases expression of *pif* genes analogous to the effect of the endogenous *pifO* regulatory sequence in F. However, the target site for *pifC* activity on RP4 has not been mapped.

A second phenotype of IncP plasmids related to transfer is the ability to mobilize both plasmids and chromosomal markers from the recipient to the donor, a phenomenon designated "retrotransfer" (82). In gene mapping experiments using pULB113 (RP4::mini-Mu, described in Section 2.5.2.) to mobilize the chromosomal genes of several gram-negative bacteria, results consistent with the transfer of recipient loci to the donor were noted. Careful study has documented that chromosomal genes from the recipient strain may frequently be transferred to the donor and recombined into the genome in homologous matings between members of the same species. These results obviously complicate the interpretation of mapping experiments. In heterologous matings between different species, R′ plasmids containing segments of the recipient chromosome can be detected in the donor after mating, implying that the transmissible plasmid must have recombined with the recipient chromosome prior to transfer back into the donor. Retrotransfer is also observed in the mobilization of non-self-transmissible plasmids present in the recipient back to the donor (83, 123). The mechanism of the retrotransfer process is not known, but the self-transmissible plasmid does not need to be stably maintained in the recipient, and kinetic data indicate that retrotransfer occurs early in the mating process (E. Top, personal communication). Retrotransfer could involve transfer of the conjugative plasmid into an intermediary recipient, with retransfer into another donor (triparental mating), or could result from a two-way exchange of DNA across a single mating bridge. A requirement for transfer of the conjugative plasmid DNA into the recipient has not been established, although the formation of R′ containing recipient DNA in the donor strongly suggests that plasmid transfer is occurring in this situation. Retrotransfer has been detected for IncP plasmids as well as IncM (R69.2) and IncN (pULG14) plasmids, but not for another IncN plasmid (pIP113) nor the IncW plasmids (pSa, R339, R388), indicating that the process is plasmid specific and not a general property of conjugation systems.

2.5. Use of IncP Conjugation for Genetic Manipulations in Bacteria

A fundamental property of conjugation systems is that the *oriT* sequence can be mobilized by the transfer genes *in cis* or *in trans*. In principle, any DNA molecule containing the *oriT* sequence can be transferred by the RK2/RP4 conjugation system into practically any bacterial recipient, gram-negative or gram-positive. This system provides an exceedingly powerful tool for the genetic analysis and manipulation of a wide variety of bacteria. In practice, the IncP conjugation system has been widely used to transfer both plasmids and chromosomal loci. This section will not attempt an exhaustive review of the applications of IncP transfer but will present the general principles involved in these genetic techniques.

2.5.1. Plasmid Mobilization and Shuttle Vectors

IncP plasmids are able to transfer a number of naturally occurring plasmids that contain a system for their mobilization (Mob⁺). These mobilization systems consist of a plasmid-specific *oriT* sequence and plasmid-encoded proteins that produce a site-specific nick at *oriT* and allow transfer by the IncP conjugation apparatus (see Chapter 5 for a discussion of the biochemical aspects of mobilization). These Mob⁺ plasmids include ColE1, IncQ plasmids, pSC101, the gonococcal penicillinase plasmid pWD2, and others

(54). However, a number of naturally occurring plasmids lack efficient Mob systems, and many cloning vectors constructed in vitro have lost their Mob regions. For these plasmids, efficient transfer by RK2/RP4 can be accomplished by insertion of the RK2/RP4 *oriT* sequence by in vitro construction or by transposition in vivo. For small target plasmids, the *oriT* region can usually be cloned into a nonessential region using convenient restriction sites. However, for large target plasmids, in vitro cloning is not practical. A convenient method of introducing the *oriT* sequence into large replicons uses transposon derivatives containing *oriT*, such as Tn5-*oriT* (Tn5-Mob) (113, 142). This system has the additional advantage that transposon mutagenesis can be carried out at the same time as plasmid transfer, so the phenotype of mutations can be tested by transfer into a plasmid-free derivative of the natural host (27).

 oriT-mediated plasmid mobilization also plays a central role in the construction of shuttle vectors used to transfer genes between distantly related bacteria. Many of the common cloning vectors used in *E. coli* molecular genetics cannot replicate in distantly related bacteria. Even the IncP and IncQ replicons cannot be maintained in certain gram-negative bacteria. To develop systems for genetic manipulation in these foreign hosts, shuttle vectors have been constructed that allow genes from these organisms to be cloned and manipulated in *E. coli*, then returned to the native host by IncP-mediated transfer. The requirements for such shuttle vectors consist of a replication region and selective marker that function in the desired foreign host. The shuttle vector then consists of a chimeric plasmid containing the foreign replicon and selective marker cloned on an *E. coli* vector containing *oriT* (54). If a suicide shuttle is desired, then the foreign replicon is omitted. The *oriT* need not necessarily be from RK2/RP4; the Mob regions of CoE1 or the IncQ plasmids will also be transferred by a helper IncP plasmid. Although other plasmid transfer system/*oriT* combinations can also be used, the IncP systems have proved to be extremely versatile, possibly because P plasmids transfer proteins during the conjugation event, such as primase, that are involved in DNA processing in the recipient (84, 100). This shuttle vector scheme was originally developed for the obligate anaerobe *B. fragilis*, which does not support replication of any known plasmid from facultative gram-negative organisms (51, 112). This approach has subsequently been employed for a number of other bacteria, including gram-positive cocci and bacilli, *Streptomyces*, *Clostridia*, and *Mycobacteria* (74, 81, 125, 135).

2.5.2. Transfer of Chromosomal Genes

 Because of their broad host range, IncP plasmids are potentially useful tools for genetic studies of chromosomal loci in a wide variety of gram-negative bacteria. Because the well-documented ability of the F plasmid to form Hfr strains was instrumental in mapping the *E. coli* chromosome, considerable interest has focused on the ability of IncP plasmids to mobilize chromosomal genes in a number of gram-negative hosts. For a detailed discussion of this topic, the reader is referred to a recent review (56).

 From analogy with the F plasmid of *E. coli*, it is generally assumed that the IncP plasmids must be integrated, at least transiently, into the host chromosome in order to transfer these loci to a recipient. Early experiments showed that certain native IncP plasmids, including RK2/RP4/R68, could mobilize chromosomal genes of some gram-negative hosts, including *P. aeruginosa*. In general, the transfer frequency for chromosomal loci is low, often below the range that is useful for genetic mapping. One mechanism for RK2/RP4/R68-mediated chromosomal transfer is replicative transposition of the Tn*1* (Tn*A*) element. The mechanism of Tn*1* transposition involves cointegrate formation with two copies of Tn*1* joining RK2/RP4 with the target replicon. The cointegrates are normally resolved by the *tnpR* (resolvase) gene product. Because Tn*1* transposes to chromosome sites only rarely, and because cointegrate formation is transient, chromosome mobilization by RK2/RP4/R68 in the cointegrate state occurs rarely, if at all. Inactivation of the *tnpR* gene stabilizes the cointegrate structure and can lead to detectable chromosome integration and transfer (101).

A second mechanism for chromosome mobilization by RK2/RP4 involves the IS*8* (designated IS*21* on R68) sequence. As discussed above, R68 (RK2/RP4) exhibits either very low or absent chromosome mobilization, but a spontaneous mutant was isolated, R68.45, that gave significantly higher frequencies of chromosomal gene transfer. Analysis of R68.45 showed that the plasmid contained a tandem duplication of the IS*21* element (75, 102, 105, 134). This arrangement enhances the transpositional activity of IS*21* by creating a new promoter reading from one IS*21* sequence into the second (56, 102, 104, 111). The tandem IS*21* repeat acts as a composite transposon so that the entire intervening R68 plasmid is inserted into the target, either another plasmid or the chromosome (102, 103, 106, 134). R68.45 is able to mobilize chromosomal genes in a wide variety of gram-negative bacteria. The IS*21*-mediated cointegrate is usually unstable, and variable lengths of the chromosome are transferred depending on the host (56). Stable Hfr strains are generally not found, limiting the usefulness of this approach.

In some bacterial hosts, the activity of a transposable element in the chromosome promotes integration of the P plasmid and chromosome transfer. This mechanism has been detected in *Pseudomonas* and *E. coli* but is obviously limited to certain isolates containing insertion sequences capable of this activity (5, 90, 91). Temperature-sensitive (TS) mutations in the *trfA* genes of IncP plasmids have been used to select for Hfr-like strains containing an integrated plasmid (56, 57). This approach is limited to those host species that can grow at the elevated temperature necessary to select for integration.

Because of the limitations on the chromosome mobilizing ability of native IncP plasmids, several artificial strategies have been used to increase the transfer frequencies or generate systems that yield Hfr-like polarized transfer. Three general approaches will be discussed here: (a) insertion of additional transposable elements on IncP plasmids, (b) generation of homology between the P plasmid and the chromosome, and (c) insertion of the *oriT* sequence into the chromosome.

The ability of RK2/RP4 to insert into the chromosome is enhanced by the presence of Mu or mini-Mu phage on the plasmid. An RP4::mini-Mu insertion, designated pULB113, has been used to transfer chromosomal genes in a large number of gram-negative bacteria that can maintain IncP plasmids (127). However, stable, integrated Hfr strains are not generally isolated using this plasmid. One mechanism to stabilize IncP chromosome cointegrates is the use of Tn*813* (a Tn*21* derivative deleted for the *tnpR* resolvase gene) on R751, or pME134, an RP1 derivative with a trimethoprim resistance gene cloned into the *tnpR* gene (16, 101). As noted previously, the ability of these plasmids to transfer chromosomal loci is limited by the preference of Tn*1*/Tn*3*-type transposons for plasmid rather than chromosomal insertion sites.

A second method of enhancing integration of the P plasmid is to insert a segment of the chromosome in order to facilitate homologous recombination (6, 10, 65). Large segments of the chromosome can be isolated in IncP plasmids during the formation of R′ plasmids. Construction of R′ plasmids in vivo results from mating a donor containing the chromosome mobilizing IncP plasmid with a recipient in which homologous recombination of the incoming DNA is not possible, due to either a *recA* mutation or lack of homology between the organisms (56). R′ plasmids involving mini-Mu contain phage DNA flanking the insert, while R68.45 primes have direct repeats of IS*21* at the junctions (56, 76). R′ plasmids can also be constructed in vitro by cloning restriction fragments of chromosomal DNA into the plasmid (6, 10, 65). While this method has the disadvantage of the large vector plasmid size, RK2/RP4 and derivatives have a number of single restriction sites that are convenient for these insertions. Chromosome mobilization using R′ plasmids results from single crossover recombination events that integrate the R′ at the site of DNA homology in the chromosome.

The third method of chromosome mobilization was designed to create random but stable sites of insertion from which polarized, Hfr-like transfer of the chromosome in individual clones could be detected. In this scheme, the *oriT* sequence of RK2/RP4 was inserted into a nonessential site of Tn*5*. The resulting transposons, designated Tn*5-oriT* and Tn*5*-Mob, were used to insert the *oriT* region into various sites of the *E. coli* and *Rhizobium meliloti* chromosomes (113, 142). In the presence of a helper RK2/RP4 plasmid containing the intact conjugation system, polarized transfer of chromosomal markers was detected. The advantage of this system for chromosomal transfer is that a library of different, reasonably stable *oriT* inserts can be constructed. Specific insertion sites, such as in auxotrophic markers, can also be selected. Mapping of genes in relation to the insertion can be achieved by introducing the helper plasmid and performing matings with the appropriate recipients.

In addition to the chromosomal mapping, the Tn*5-oriT* (or Tn*5*-Mob) system is useful in the genetic manipulation of larger plasmids in various gram-negative hosts. Many plasmids are not self-transmissible, and Tn*5-oriT* represents a convenient vehicle for introducing the *oriT* sequence into large plasmids for which in vitro cloning techniques are impractical. The Tn*5-oriT*-labeled plasmid can be transferred into various hosts using the RK2 transfer system cloned on ColE1 (pRK2073). This approach was used to map the virulence region on the 80-kb nonconjugative plasmid of *Salmonella dublin* (11, 27).

2.6. Relationship between IncP Conjugation and T-DNA Transfer to Plants Mediated by *Agrobacterium tumefaciens* Ti Plasmids

The bacteria-to-plant DNA transfer system of *A. tumefaciens* is a highly adapted mechanism for tumor induction and opine synthesis in the plant tissue (13, 146). This process can also be considered a special instance of very broad host range gene transfer. The biochemical mechanism of T-DNA transfer suggested a relationship to bacterial conjugation (115, 146), since transfer is initiated by nicking at specific border sequences, with covalent attachment of the VirD2 protein to the 5′ end of the nicked strand (1, 37, 60, 61, 130, 144). In addition, the *vir* gene products responsible for T-DNA transfer were also found to mobilize RSF1010 into plant cells (25). This process, analogous to transfer of RSF1010 between bacteria by IncP plasmids, depended on the RSF1010 *mob* genes.

Definitive evidence for a genetic relationship between IncP conjugation and T-DNA transfer has recently been obtained for three loci in RK2/RP4:*oriT*, *traG*, and the Tra2 region (132, 148, E. Lanka, personal communication). Comparison of the RK2 *oriT* sequence with the T-DNA borders reveals sequence identity in 8 bp at the nick sites and a consensus sequence extending for 12 bp (Figure 4.2). Furthermore, the location of the nick in RK2/RP4 and R751 is the same as in both the left and right nopaline Ti plasmid borders. Point mutations that interfere with RK2 *oriT* activity cluster within this nick region, confirming the importance of the consensus sequence for transfer function (132). The T-DNA borders of the Ti plasmids can be regarded as directly repeated *oriT* sequences that ensure the transfer of the intervening T-DNA to plants but not the entire Ti plasmid molecule.

Significant amino acid homology has been detected between the TraG protein of RK2/RP4 and the VirD4 protein of the Ti plasmids (148). R751 also contains a *traG* locus very similar to that of RK2/RP4 (83% identity). *traG* has recently been shown to be essential for RK2 conjugation, and *virD4* is required for crown gall tumor formation by *Agrobacterium*. However, the biochemical functions of TraG and VirD4 remain unknown. The *virD* operon begins with two genes, *virD1* and *virD2*, required for nicking at the border sequences to generate the transferred DNA (115, 143). There is an overall similarity in the genetic organization of the *virD* operon and the IncP transfer region encoding *traJ*, *traI*, *traH*, and *traG* (148). TraJ and TraI are responsible for site-specific nicking at *oriT* in analogy to VirD1 and VirD2. In each case, the nicking enzymes, TraI and VirD2, attach covalently to the 5′ end of the nicked strand, require the activity of a smaller accessory protein (TraJ and VirD1, respectively), and are transferred into the new host with the DNA. Furthermore, the nicking activities of both TraI and VirD2 are contained in the N-terminal portions of the proteins (50, 143). A limited consensus sequence within this region has been proposed. TraH and VirD3 are both encoded "out of frame" with respect to the corresponding nicking enzyme genes (*traI* and *virD2*). The homologous *traG* and *virD4* are located in corresponding positions immediately downstream of *traH* and *virD3*, respectively.

Recently, a striking similarity between the Tra2 region of RK2/RP4 and the *virB* operon of the Ti plasmid has been described (88, E. Lanka, personal communication). The *virB* operon appears to specify envelope proteins that have a role in the cell surface interactions required for DNA transport (70, 122, 131). Likewise, most or all of Tra2 is involved in the synthesis and/or assembly of the surface mating apparatus, including P pili, since the phenotypes of all Tra2 loci examined concern either surface exclusion or donor-specific phage sensitivity. As described in Section 2.3.7., six Tra2 products are similar to

RK2/RP4 *ori* T	A C C T A T C C T G C C
R751/*ori* T	A C A C A T C C T G C C
Ti LB (nopaline)	A T A T A T C C T G C C
Ti RB (nopaline)	A T A T A T C C T G T C
Consensus	A C C T/T A C A T C C T G C/T C

FIGURE 4.2. DNA sequence comparison of the IncP plasmid nick regions and *Agrobacterium* T-DNA border sequences, adapted from Waters et al (132). The locations of the nick sites are identical, as shown by wedges. The arrows indicate the leading strands and the direction of transfer.

VirB proteins, and the overall organization of genes in Tra2 and VirB is similar, except that *trbB* (*kilBI*) is located close to the start of Tra2, whereas its homologue, *virB11*, is at the end of the *virB* operon. For TrbB and VirB11, a functional relationship has also been established: Both are ATPases with autophosphorylating activity (28).

These results suggest a remarkable similarity between the broad host range conjugation systems of IncP plasmids and the *Agrobacterium* virulence system that infects the plant. Clearly, significant differences also exist, including the regulation of *vir* expression by a two-component sensing system involving *virA* and *virG* and the need for nuclear targeting and chromosomal integration of the T-DNA within the plant cell (see Chapter 9). However, the initiation events and DNA processing through the membranes during transfer may involve a similar mechanism used by both IncP and Ti plasmids to overcome barriers to genetic exchange between very dissimilar organisms.

3. IncN Plasmid Transfer

The IncN plasmids have been isolated from many species of *Enterobacteriaceae* as well as *Aeromonas* and *Vibrio* spp. (62, 63). IncN plasmids have been transferred into a number of gram-negative bacteria, including *Pseudomonas* and *Rhizobium* spp. The IncN conjugation system has been used to introduce suicide plasmid vectors into a wide variety of gram-negative genera, including *Agrobacterium*, *Azospirillum*, *Azotobacter*, *Bordetella*, *Pseudomonas*, and *Rhizobium*. IncN plasmid biology has been the subject of a recent review (62).

3.1. Mapping of the IncN Transfer Genes

Restriction site mapping and deletion analysis allowed Brown and Willetts (24) to construct a physical and genetic map of the IncN plasmid R46. Antibiotic and heavy metal resistance determinants were clustered; this region is deleted in the spontaneous derivative pKM101, which retains transfer proficiency. The conjugation system of pKM101 was characterized further by Winans and Walker (136) using deletion analysis, subcloning, and Tn5 mutagenesis. The transfer region comprises about 16 kb of DNA and is divided into three groups of genes, TRAI, TRAII, and TRAIII, on the basis of small intervening regions that are not essential for transfer. All Tn5 insertions in the TRAI/TRAII region that interrupted transfer also rendered the cells resistant to the donor-specific phage RPD1. A map of the pKM101 transfer region is shown in Figure 4.3. Three complementation groups, *traA*, *B*, and *C*, were identified in TRAI, and four groups, *traD*, *E*, *F*, and *G*, mapped in TRAII. Because most Tn5 insertions are polar, these are minimal numbers for actual genes. An entry exclusion locus, designated *eex*, is located between TRAI and TRAII. The presence of *eex* in the recipient decreases the transfer efficiency for an incoming PKM101 derivative by about 10^3-fold (137). *eex* is not required for donor-specific phage sensitivity or for transfer. However, because *eex* encodes a function related to transfer, TRAI and TRAII could be considered a single transfer region.

Although none of the individual genes in TRAI/TRAII has been characterized, two

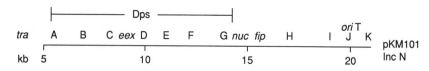

FIGURE 4.3. Transfer regions of the IncN plasmid pKM101 (136) and the IncW plasmid pSU1087 (14). The maps have been aligned to illustrate the similarity between the two systems. The kilobase scale for pKM101 reflects total plasmid coordinates, and the scale for pSU1087 refers to transfer region coordinates defined by Bolland et al (14). DpS and Pil$_W$ indicate regions involved in sensitivity to donor-specific phage. The *oriT* sites for each plasmid map at the ends of the transfer regions.

interesting phenotypes are associated with loci in this regions. Circumstantial evidence suggests that a product of the *traC* locus is transferred to the recipient, in analogy to the IncP plasmids that are known to transfer proteins during mating (136). The *kilB* function of pKM101 maps within the *traE* locus and is lethal to cells in the absence of the *korA* and *korB* functions. Such *kil* and *kor* loci are also prominent genetic features of the IncP plasmids.

The TraIII region contains transfer functions not required for donor-specific phage sensitivity (136). This region is presumed to be involved in conjugative DNA metabolism, although no specific biochemical activities have been characterized for the gene products of TraIII. Four complementation groups, *traH, I, J,* and *K*, were identified by Tn5 insertions (Figure 4.3). TraIII is separated from TRAII by approximately 2.5 kb. Two genetic functions map in this intervening region: *fip*, a locus that inhibits IncP plasmid transfer, and *nuc*, encoding a periplasmic endonuclease. *fip* decreases transfer of a coresident P plasmid by a factor of 10^4 or greater but does not affect donor-specific phage sensitivity encoded by the P plasmid (138). This phenotype resembles the *pifC* inhibition of IncP fertility: Neither locus appears to affect sex pilus synthesis as is seen in the classic fertility inhibition of F (*fi+* phenotype).

3.2. The IncN Transfer Origin

The transfer origin of R46 has been cloned and sequenced (29). Full transfer origin activity is encoded by a 600-bp fragment. Notable structural features include a set of 13 direct repeats of an 11-bp consensus sequence. Deletion of some or all of these repeats reduces *oriT* function. Inverted repeats are also present, including a pair located adjacent to each other. No homology was detected between the R46 *oriT* and *oriT* sequences of F, RK2/ RP4, ColE1, RSF1010, and CloDF13. However, the sequence adjacent to the nick site in pSC101 shared 23 of 31 bases with a region of the R46 *oriT*. Recently, the *oriT* of the IncW plasmid R388 has been shown to be related to the R46 *oriT* (see Section 4.2.).

The transfer system of a second IncN plasmid, pCU1, has been characterized by Iyer and coworkers (personal communication). The pCU1 Tra system can also be separated into regions encoding donor-specific phage sensitivity and DNA processing functions. The *oriT* of pCU1 is identical to R46. However, deletion of the 13 directly repeated 11-bp sequences had no effect on pCU1 *oriT* function. In vivo mapping of the nick site by cointegrate formation and resolution has located *nic* in the region of an inverted repeat.

4. The IncW Transfer System

IncW plasmids have been found in a variety of gram-negative bacteria including *Aeromonas liquefaciens* and many species of *Enterobacteriaceae*. pSa, a well-characterized W plasmid, has been transferred to many gram-negative genera, including *Acinetobacter*, *Agrobacterium*, *Alcaligenes*, *Legionella*, *Methylophilus*, *Myxococcus*, *Pseudomonas*, *Rhizobium*, *Vibrio*, *Xanthomonas*, and *Zymomonas*, as well as members of *Enterobacteriaceae*. In addition to the usefulness of IncW plasmids for genetic manipulation, considerable interest has focused on the ability of certain W plasmids such as pSa to suppress the ability of *Agrobacterium tumefaciens* to induce plant tumors. The biology of IncW plasmids has been reviewed recently (126).

4.1. Mapping the IncW Transfer Region

The transfer regions of the IncW plasmids pSa and R388 have been located on the physical maps of these plasmids, and both regions occupy approximately 12 kb (126). The conjugation system of R388 has been characterized in more detail using deletion analysis, subcloning, and insertion mutagenesis (14). The maximum size of the transfer region is 14.9 kb (see Figure 4.3). Like the IncN transfer system, the R388 transfer genes can be divided into two clusters with different phenotypes: One group contains genes required for pilus biosynthesis (designated Pil_W), and the other region is presumed to be involved in DNA processing (Mob_W). The Mob_W region can be mobilized *in trans* by the Pil_W functions cloned on a separate plasmid. Interestingly, Mob_W could also be transferred by providing the IncN plasmid pilus synthesis genes *in trans* but was not mobilized by the IncP conjugation system, underscoring the similarity between the N and W transfer systems. Mob_W is at least 4 kb in size, but no direct evident for DNA processing activities has been presented. Pil_W is located close or adjacent to Mob_W and comprises at least 7 kb. All transfer-deficient mutations in Pil_W are also resistant to donor-specific phage, whereas none of the inserts in Mob_W is phage resistant. An entry exclusion determinant mapped in one-half of Pil_W. The number of essential genes in the R388 conjugation system remains unknown.

4.2. The Transfer Origin of R388

Initial deletion analysis located *oriT* within a 2.4-kb segment at the end of Mob_W, analogous to the *oriT* of the IncN plasmids (14). Recently, the sequence of a 402-bp fragment containing the R388 *oriT* has been reported, and 330 bp was shown to be sufficient for efficient transfer origin function (79). Extensive sequence homology was found between the R388 sequence and the *oriT* of R46, the IncN plasmid discussed in Section 3.2. In addition, several structural motifs were similar. Both *oriT*s have a series of tandemly repeated 11-bp sequences located at the same end. In R388, there were 5 repeats, while R46 contains 13, and there are some similarities in the sequences of the repeats between the plasmids. In the region of an inverted repeat shared by the two *oriT*s, there is 83% sequence homology, and a 60-bp fragment containing this R388 sequence expressed diminished but detectable *oriT* activity. Overall, the sequence homology with R46 was 52%. As anticipated, the R388 *oriT* sequence could not be mobilized by the IncN conjugation system in the absence of the specific functions encoded by Mob$_W$.

In analogy with the IncP *oriT*, deletions from either end of the 330-bp R388 sequence produced successive loss of *oriT* function. The minimal 60-bp *oriT* segment is presumed to contain the nick site, but direct evidence for this location has not been obtained. This sequence also contains a consensus IHF binding site, but genetic evidence indicates that IHF is not essential for IncW transfer in *E. coli*.

5. Mobilization of IncQ Plasmids

The Q incompatibility group consists of mobilizable broad host range plasmids. Q plasmids are relatively small (8 to 14 kb) and can be transferred by certain conjugative plasmids into a variety of gram-negative bacteria (43, 58). Q plasmid mobilization frequency varies with the self-transmissible plasmid employed and is particularly efficient using IncP, IncIα, IncM, and IncX plasmids (133). A number of Q plasmids from natural isolates appear to be closely related; these plasmids are approximately 8.7 kb and encode resistance to streptomycin and sulfonamides. The three best studied Q plasmids, RSF1010, R300B, and R1162, were isolated from different sources but appear very similar or identical. IncQ plasmids have been extensively employed as cloning vectors in gram-negative bacteria. Of further significance is the finding that the *Agrobacterium* Ti plasmid *vir* function can mobilize an IncQ plasmid from the bacterial host into the plant cell, where it is integrated into the genome (25).

5.1. Organization of the Mobilization Region

The mobilization region of the IncQ plasmid RSF1010, shown in Figure 4.4, consists of the *cis*-acting *oriT* and three genes, *mobA*, *mobB*, and *mobC* (33, 34). The products of these genes have been identified as the B (78-kDa), D (15-kDa), and K (11-kDa) proteins, respectively, of RSF1010 (110). *mobA* is located adjacent to *oriT* and is identical to the N-terminal portion of the *repB* gene. This locus encodes two related proteins, RepB (MobA) and RepB′, a 36-kDa protein generated from a second in-frame translational start in the *repB* gene (58). The RepB/RepB′ proteins are plasmid-specific DNA primases that synthesize DNA primers for both strands at the origin of vegetative replication. Both proteins encode primase activity, indicating that this enzymatic function resides in the C-terminal domain of RepB. Mutations in the N-terminal domain of RepB affect mobilization, while the C-terminal portion can be deleted without interrupting mobilization (34, 58). Based on these results, the RepB/MobA protein appears to be bifunctional: The

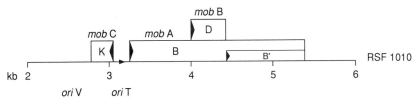

FIGURE 4.4. Mobilization region of RSF1010, taken from Scholz et al (110). *mobA*, *B*, and *C* encode the B, D, and K proteins, respectively. The B′ protein is generated from a second in-frame translational start within the gene, also designated *repB*. *mobB* overlaps *mobA* (*repB*) but is translated in a different frame.

C-terminal region is required for primer formation at *oriV*, and the N-terminus is involved with nicking and transfer of *oriT* (see Section 5.2.). The *mobB* gene overlaps *mobA* and is translated in a different frame. *mobC* is located on the other side of *oriT* and is transcribed/translated in the opposite direction from *mobA* and all the other identified genes of RSF1010. The biochemical functions of the *mobB* and *mobC* gene products have not been established, but together with MobA, they form the *oriT* nicking (relaxation) complex reconstituted in vitro. The *mob* genes are transcribed from divergent promoters located in the *oriT* region. Two promoters, P_1 and P_3, are located upstream of *mobA*/*mobB*, and a single promoter is oriented in the opposite direction to initiate transcription of *mobC* (58). Production of the *mob* gene products appears to be constitutive.

5.2. The *oriT* Region

The transfer origins of RSF1010 and R1162 have been mapped and subcloned. DNA sequence analysis indicates that these transfer origins are identical except for one possible base difference (21, 33). The *oriT* function of R1162 was further localized to a 38-bp sequence, containing an inverted repeat with 10-bp arms (22). Analysis of mutations and recombination at *oriT* located the proposed in vivo nick site 7 to 8 bp from the inverted repeat. RSF1010 has been isolated as a DNA-protein relaxation complex that exhibits site-specific nicking at *oriT* in vitro. The nick site was shown to be located after the seventh base from the foot of the inverted repeat (35). Deletion of the outer arm of this repeat (distal to the nick site) decreases mobilization by a factor of 100, while a specific deletion of the inner arm (proximal to the nick) completely abolishes *oriT* activity (22). Certain base mutations immediately surrounding the nick site also significantly decrease transfer. However, other base-pair changes close to the nick have no effect on mobilization. These studies indicate that both the inverted repeat and the specific bases at the nick site are required for optimal *oriT* activity (see Section 5.3.). The structural organization of the RSF1010/R1162 *oriT* is similar to that of RK2/RP4, in which the inverted repeat with 19-bp arms is situated 8 bp from the nick, and both the repeat and the specific nick site sequence are required for TraJ/TraI-induced nicking and transfer origin function (see Section 2.3.2.). However, no significant sequence homologies exist between the IncQ and IncP transfer origins.

Recently, *oriT*-specific nicking of RSF1010 has been achieved in vitro using the purified MobA, MobB, and MobC proteins (108). These proteins form a Mg^{+2}-dependent complex with *oriT* DNA. Similar to relaxation complexes isolated from cell lysates, nicking of the in vitro reconstituted complex occurs with addition of SDS. The nicking reaction requires MobA and MobC and is enhanced by MobB. The MobA protein appears to attach covalently to the 5′ end of the nicked strand at the site corresponding to the in vivo cleavage position. The MobA protein was also shown to cleave and ligate single-stranded DNA (see also Section 5.3.).

5.3. Site-Specific Recombination at *oriT*

Meyer and colleagues (4, 85) developed two assays for recombination at *oriT* that model events in DNA processing during the mobilization of R1162. One system employs plasmids containing two directly repeated *oriT* sequences flanking a kanamycin resistance (Km^r) marker. Nicking within both *oriT* sequences during transfer followed by ligation in the recipient yields a plasmid deleted for the Km^r gene. Analysis of *oriT* mutations suggests that the nicking reactions are not the same at each *oriT*. Mutation of bases at the nick site appears to affect the initial nicking reaction, while deletion of the outer arm of the inverted repeat inhibits the second nick and/or ligation step in the recipient (4, 66). A second *oriT*-specific recombination process was observed between two direct repeats of *oriT* cloned into M13. This recombination was not dependent on mobilization but does require the N-terminal portion of a gene analogous to *repB* (*mobA*) of RSF1010 (85). As noted in Section 5.1., the N-terminal domain of *repB* is required for efficient mobilization of both RSF1010 and R1162. The data on phage recombination suggest that this reaction reflects single-stranded cleavage and ligation of the transferred strand at *oriT*, consistent with a rolling-circle model for DNA transfer. A LacZ fusion with the N-terminal region of the MobA-analogous protein of R1162 was shown to possess *oriT*-specific cleavage and ligation activities on single-stranded DNA templates (12). This reaction appears to be responsible for the recombination between directly repeated copies of *oriT* in M13 phage. According to the rolling-circle model, this reaction represents the second cleavage at *oriT* and the subsequent ligation of the transferred single strand.

6. Conclusions

Broad host range plasmid conjugation has fundamental biological significance as a means of genetic exchange between diverse organisms. These processes are likely to have a major role in bacterial evolution, since such transfer greatly expands the gene pool available to individual species. Plasmid conjugation systems also facilitate the genetic manipulation of a wide variety of organisms whose endogenous genetic systems are poorly characterized.

Detailed genetic and biochemical studies of broad host range transfer systems have characterized events occurring at the transfer origin in considerable detail, particularly for the IncP and IncQ systems. A major advance has been the in vitro reconstitution of the RK2/RP4 *oriT*-specific relaxosome from purified components, as well as the analogous nicking complex from RSF1010. Fundamental features of the nicking reaction are site specificity and covalent attachment of the nicking enzyme to the 5′ end of the DNA strand, reactions predicted from earlier work on ColE1 (52) and a property of the *Agrobacterium* Ti plasmid VirD2 protein as well. The IncP conjugation system shares several important features with the *Agrobacterium tumefaciens* Ti plasmid *vir* system that transfers T-DNA into plants. These similarities may reflect the common requirement for DNA transfer between unrelated organisms that is exhibited by both systems.

In addition to elucidating the fundamental mechanisms of DNA transfer, the study of broad host range conjugation seeks to identify specific adaptive features of these systems that facilitate their promiscuous nature. These specific broad host range properties remain poorly characterized at present but will constitute a particularly fruitful aspect of the future work on conjugation systems.

ACKNOWLEDGMENTS. The willingness of colleagues to provide preprints and unpublished results is greatly appreciated. In particular, the close collaboration of Erich Lanka was instrumental in preparing the review of IncP plasmid conjugation. Work in the author's laboratory was supported by Public Health Service grants GM28924 and AI16463 from the National Institutes of Health.

References

1. Albright, L.M., Yanofsky, M.F., Leroux, B. Ma, D., and Nester, E.W., 1987, Processing of the T-DNA of *Agrobacterium tumefaciens* generates border nicks and linear, single-stranded T-DNA, *J. Bacteriol.* 169:1046–1055.
2. Al-Doori, Z., Watson, M., and Scaife, J., 1982, The orientation of transfer of the plasmid RP4, *Genet. Res. Camb.* 39:99–103.
3. Ayres, E., Saadi, S., Schreiner, H.C., Thomson, V.J., and Figurski, D.H., 1991, Differentiation of lethal and nonlethal, *kor*-regulated functions in the *kilB* region of broad host-range plasmid RK2, *Plasmid* 25: 53–63.
4. Barlett, M., Rickson, M.J., and Meyer, R.J., 1990, Recombination between directly repeated origins of conjugative transfer cloned in M13 bacteriophage DNA models ligation of the transferred plasmid strand, *Nucl. Acids Res.* 18:3579–3586.
5. Barsomian, G., and Lessie, T.G., 1986, Replicon fusions promoted by insertion sequences on *Pseudomonas cepacia* plasmid pTGL6, *Mol. Gen. Genet.* 204:273–280.
6. Barth, P.T., 1979, Plasmid RP4, with *Escherichia coli* DNA inserted in vitro, mediates chromosomal transfer, *Plasmid* 2:130–136.
7. Barth, P.T., 1979, RP4 and R300B as wide-host-range plasmid cloning vehicles, in: *Plasmids of Medical, Environmental and Commercial Importance* (K.N. Timmis and A. Pühler, eds.), Elsevier/North-Holland Biomedical Press, pp. 399–410.
8. Barth, P.T., and Grinter, N.J., 1977, Map of plasmid RP4 derived by insertion of transposon C, *J. Mol. Biol.* 113:455–474.
9. Barth, P.T., Grinter, N.J., and Bradley, D.E., 1978, Conjugal transfer system of plasmid RP4: analysis by transposon 7 insertion, *J. Bacteriol.* 133:43–52.
10. Beck, Y., Coetzee, W.F., and Coetzee, J.N., 1982, In vitro constructed RP4-prime plasmids mediate oriented mobilization of the *Proteus morganii* chromosome, *J. Gen. Microbiol.* 128:1163–1169.
11. Beninger, P.R., Chikami, G., Tanabe, K., Roudier, C., Fierer, J., and Guiney, D.G., 1988, Physical and genetic mapping of the *Salmonella dublin* virulence plasmid pSDL2, *J. Clin. Invest.* 81:1341–1347.
12. Bhattacharjee, M.K., and Meyer, R.J., 1991, A segment of a plasmid gene required for conjugal transfer encodes a site-specific, single-strand DNA endonuclease and ligase, *Nucl. Acids Res.* 19:1129–1137.
13. Binns, A. N., and Tomashow, M.F., 1988, Cell biology of *Agrobacterium* infection and transformation of plants, *Ann. Rev. Microbiol.* 42:575–606.
14. Bolland, S., Llosa, M., Avila, P., and de la Cruz, F., 1990, General organization of the conjugal transfer genes of the IncW plasmid R388 and interactions between R388 and IncN and IncP plasmids, *J. Bacteriol.* 172:5795–5802.

15. Boulnois, G.J., Varley, J.M., Sharpe, G.S., and Franklin, F.C.H., 1985, Transposon donor plasmids, based on CO1IB-p9, for use in *Pseudomonas putida* and a variety of other gram negative bacteria, *Mol. Gen. Genet.* 200:65–67.
16. Bowen, A.R.S.G., and Pemberton, J.M., 1985, Mercury resistance transposon Tn*813* mediates chromosome transfer in *Rhodopseudomonas sphaeroides* and intergeneric transfer of pBR322, in: *Plasmids in Bacteria* (D.R. Helinski, S.N. Cohen, D.B. Clewell, D.B. Jackson, and A. Hollaender, eds), Plenum Press, New York, pp. 105–115.
17. Bradley, D.E., 1974, Adsorption of bacteriophages specific for *Pseudomonas aeruginosa* R factors RP1 and R1822, *Biochim. Biophys. Res. Commun.* 57:893–900.
18. Bradley, D.E., 1980, Morphological and serological relationships of conjugative pili, *Plasmid* 4:155–169.
19. Bradley, D.E., and Rutherford, E.L., 1975, Basic characterization of a lipid-containing bacteriophage specific for plasmids of the P, N and W compatibility groups, *Can. J. Microbiol.* 21:152–163.
20. Bradley, D.E., Taylor, D.E., and Cohen, D.R., 1980, Specification of surface mating systems among conjugative drug resistance plasmids in *Escherichia coli* K-12, *J. Bacteriol.* 143:1466–1470.
21. Brasch, M.A., and Meyer, R.J., 1986, Genetic organization of plasmid R1162 DNA involved in conjugative mobilization, *J. Bacteriol.* 167:703–710.
22. Brasch, M.A., and Meyer, R. J., 1987, A 38 base-pair segment of DNA is required in *cis* for conjugative mobilization of broad host-range plasmid R1162, *J. Mol. Biol.* 198:361–369.
23. Breton, A.M., Jaona, S., and Guespin-Michel, J., 1985, Transfer of plasmid RP4 to *Myxococcus xanthus* and evidence for its integration into the chromosome, *J. Bacteriol.* 161:523–528.
24. Brown, A.C., and Willetts, N.S., 1981, A physical and genetic map of the IncN plasmid R46, *Plasmid* 5:188–201.
25. Buchanan-Wollaston, V., Passiatore, J.E., and Cannon, F., 1987, The *mob* and *oriT* mobilization functions of a bacterial plasmid promote its transfer to plants, *Nature*, London, 328:172–175.
26. Burkardt, H.J., Riess, G., and Pühler, A., 1979, Relationships of group P1 plasmids revealed by heteroduplex experiments: RP1, RP4, R68 and RK2 are identical, *J. Gen. Microbiol.* 114:341–348.
27. Chikami, G.K., Fierer, J., Guiney, D.G., 1985, Plasmid-mediated virulence in *Salmonella dublin* demonstrated by use of a Tn*5-oriT* construct, *Infect. Immun.* 50:420–422.
28. Christie, P.J., Ward, J.E. Jr., Gordon, M.P., and Nester, E.W., 1989, A gene required for transfer of T-DNA to plants encodes an ATPase with autophosphorylating activity, *Proc. Natl. Acad. Sci. USA* 86:9677–9681.
29. Coupland, G.M., Brown, A.M.C., and Willetts, N.S., 1987, The origin of transfer (*oriT*) of the conjugative plasmid R46: characterization of deletion analysis and DNA sequencing, *Mol. Gen. Genet.* 208:219–225.
30. Courturier, M., Bex, F., Bergquist, P.L., and Maas, W.K., 1988, Identification and classification of bacterial plasmids, *Microbiol. Rev.* 52:375–395.
31. Datta, N., and Hedges, R.W., 1972, Host ranges of R factor, *J. Gen. Microbiol.* 70:453–460.
32. Depicker, A., de Block, M., Inze, D., van Montagu, M., and Schell, J., 1980, IS-like element IS8 in RP4 plasmid and its involvement in cointegration, *Gene* 10:329–338.
33. Derbyshire, K.M., Hatfull, G., and Willetts, N., 1987, Mobilization of the non-conjugative plasmid RSF1010: a genetic and DNA sequence analysis of the mobilization region, *Mol. Gen. Genet.* 206:161–168.
34. Derbyshire, K.M., and Willetts, N.S., 1987, Mobilization of the nonconjugative plasmid RSF1010: a genetic analysis of its origin of transfer, *Mol. Gen. Genet.* 206:154–160.
35. Drolet, M., Zanga, P., and Lau, P.C.K., 1990, The mobilization and origin of transfer regions of a *thiobacillus ferrooxidans* plasmid: relatedness to plasmids RSF1010 and pSC101, *Mol. Microbiol.* 4:1381–1391.
36. Durland, R. H., and Helinski, D.R., 1990, Replication of the broad-host-range plasmid RK2: direct measurement of intracellular concentrations of essential TrfA proteins and their effect on plasmid copy number, *J. Bacteriol.* 172:3849–3858.
37. Durrenberger, F., Crameri, A., Hohn, B., and Koukolilova-Nocola, Z., 1989, Covalently bound VirD2 protein of *Agrobacterium tumefaciens* protects the T-DNA from exonucleolytic degradation, *Proc. Natl. Acad. Sci. USA* 886:9154–9158.
38. Figurski, D.H., and Helinski, D., 1979, Replication of an origin containing derivative of plasmid RK2 dependent on a plasmid function in *trans*, *Proc. Natl. Acad. Sci. USA* 76:1648–1652.
39. Figurski, D., Meyer, R., Miller, D.S., Helinski, D.R., 1976, Generation in vitro of deletions in the broad-host-range plasmid RK2 using phage Mu insertions and a restriction endonuclease, *Gene* 1:107–119.
40. Figurski, D.H., Pohlman, R.F., Bechhofer, D. H., Prince, A.S., and Kelton, C.A., 1982, Broad host range

plasmid RK2 encodes multiple *kil* genes potentially lethal to *Escherichia coli* host cells, *Proc. Natl. Acad. Sci. USA* 79:1935–1939.

41. Finger, J., and Krishnapillai, V., 1980, Host range, entry exclusion, and incompatibility of *Pseudomonas* FP plasmids, *Plasmid* 3:332–342.
42. Fong, S.T., and Stanisich, V.A., 1989, Location and characterization of two functions on RP1 that inhibit the fertility of the IncW plasmid R388, *J. Gen. Microbiol.* 135:499–502.
43. Frey, J., and Bagdasarian, M., 1989, The molecular biology of IncQ plasmids, in: *Promiscuous Plasmids of Gram-Negative Bacteria* (C.M. Thomas, ed.), Academic Press, London, pp. 79–94.
44. Fürste, J.P., Pansegrau, W., Ziegelin, G., Kröger, M., and Lanka, E., 1989, Conjugative transfer of promiscuous IncP plasmids: interaction of plasmid-encoded products with the transfer origin, *Proc. Natl. Acad. Sci. USA* 86:1771–1775.
45. Gerlitz, M., Hrabak, O., and Schwab, H., 1990, Partitioning of broad host-range plasmid RP4 is a complex system involving site-specific recombination, *J. Bacteriol.* 172:6194–6203.
46. Goncharoff, P., Saadi, S., Chang, C., Saltman, L.H., and Figurski, D.H., 1991, Structural, molecular, and genetic analysis of the *kilA* operon of broad-host-range plasmid RK2, *J. Bacteriol.* 173:3463–3477.
47. Grinter, N., 1981, Analysis of chromosome mobilization using hybrids between plasmid RP4 and a fragment of bacteriophage carrying IS1, *Plasmid* 5:267–276.
48. Guiney, D.G., 1982, Host range of conjugation and replication functions of *Escherichia coli* sex plasmid F *lac*: comparison with the broad host range plasmid RK2, *J. Mol. Biol.* 162:699–703.
49. Guiney, D.G., Deiss, C., and Simnad, V., 1988, Location of the relaxation complex nick site within the minimal origin of transfer of RK2, *Plasmid* 20:259–265.
50. Guiney, D.G., Deiss, C. Simnad, V., Yee, L., Pansegrau, W., and Lanka, E., 1989, Mutagenesis of the TraI core region of RK2 by using Tn5: identification of plasmid-specific transfer genes, *J. Bacteriol.* 171:4100–4103.
51. Guiney, D.G., Hasegawa, P., and Davis, C.E., 1984, Plasmid transfer from *Escherichia coli* to *Bacteroides fragilis*: differential expression of antibiotic resistance phenotypes, *Proc. Natl. Acad. Sci. USA* 81:7203–7206.
52. Guiney, D.G., and Helinski, D.R., 1975, Relaxation complexes of plasmid DNA and protein. III. Association of protein with the 5′ terminus of the broken DNA strand in the relaxed complex of plasmid ColE1, *J. Biol. Chem.* 250:8796–8803.
53. Guiney, D.G., and Helinski, D.R., 1979, The DNA-protein relaxation complex of the plasmid RK2: location of the site-specific nick in the region of the proposed origin of transfer. *Mol. Gen. Genet.* 176:183–189.
54. Guiney, D.G., and Lanka, E., 1989, Conjugative transfer of IncP plasmids, in: *Promiscuous Plasmids of Gram-negative Bacteria* (C.M. Thomas, ed.), Academic Press, London, pp. 27–56.
55. Guiney, D.G., and Yakobson, E., 1983, Location and nucleotide sequence of the transfer origin of the broad host range plasmid RK2, *Proc. Natl. Acad. Sci. USA* 80:3595–3598.
56. Haas, D., and Reimmann, C., 1989. Use of IncP plasmids in chromosomal genetics of gram-negative bacteria, in: *Promiscuous Plasmids of Gram-Negative Bacteria* (C.M. Thomas, ed.), Academic Press, London, pp. 185–206.
57. Haas, D., Watson, J., Krieg, R., and Leisinger, T., 1981, Isolation of an Hfr donor of *Pseudomonas aeruginosa* PAO by insertion of the plasmid RP1 into the tryptophan synthase gene, *Mol. Gen. Genet.* 182:240–244.
58. Haring, V., and Scherzinger, E., 1989, Replication proteins of the IncQ plasmid RSF1010, in: *Promiscuous Plasmids of Gram-Negative Bacteria* (C.M. Thomas, ed.), Academic Press, London, pp. 95–124.
59. Heinemann, J.A., and Sprague, G.F., Jr., 1989, Bacterial conjugative plasmids mobilize DNA transfer between bacteria and yeast, *Nature* 340, 205–209.
60. Herrera-Estrella, A., Chen, Z., Van Montagu, M., and Wang, K., 1988, VirD proteins of *Agrobacterium tumefaciens* are required for the formation of a covalent DNA-protein complex at the 5′ terminus of T-strand molecules, *EMBO J.* 7:4055–4062.
61. Howard, E.A., Winsor, B.A., DeVos, G., and Zambryski, P., 1989, Activation of the T-DNA transfer process in *Agrobacterium* results in the generation of a T-strand-protein complex: tight association of VirD2 with the 5′ ends of T-strand, *Proc. Natl. Acad. Sci. USA* 86:4017–4021.
62. Iyer, V.N., 1989, IncN group plasmids and their genetic systems, in: *Promiscuous Plasmids of Gram-Negative Bacteria* (C.M. Thomas, ed.), Academic Press, London, pp. 165–183.
63. Jacob, A.E., Shapiro, J.A., Yamamoto, L., Smith, D.L.., Cohen, S.N., and Berg, D., 1977, Plasmids

studied in *Escherichia coli* and other enteric bacteria, in: *DNA Insertion Elements, Plasmids and Episomes* (A.I. Bukhari, J.A. Shapiro, and S.L. Adhya, eds.), Cold Spring Harbor Laboratory, New York, pp. 607–638.

64. Jacoby, G.A., and Shapiro, J.A., 1977, Plasmids studies in *Pseudomonas aeruginosa* and other Pseudomonads, in: *DNA Insertion Elements, Plasmids and Episomes* (A.I. Bukhari, J.A. Shapiro, and S.L. Adhya, eds.), Cold Spring Harbor Laboratory, New York, pp. 639–656.

65. Julliot, J.S., and Boistard, P., 1979, Use of RP4-prime plasmids constructed in vitro to promote a polarized transfer of the chromosome in *Escherichia coli* and *Rhizobium meliloti*, *Mol. Gen. Genet.* 173:289–298.

66. Kim, K., and Meyer, R.J., 1989, Unidirectional transfer of broad host-range plasmid R1162 during conjugative mobilization. Evidence for genetically distinct events at *oriT*, *J. Mol. Biol.* 122:287–300.

67. Kittell, B.L., and Helinski, D.R., 1991, Iteron inhibition of plasmid RK2 replication in vitro: evidence for intermolecular coupling of replication origins as a mechanism for RK2 replication control, *Proc. Natl. Acad. Sci. USA* 88:1389–1393.

68. Krishnapillai, V., 1988, Molecular genetic analysis of bacterial plasmid promiscuity, *FEMS Microbiol. Rev.* 54:223–238.

69. Krishnapallai, V., Nash, J., and Lanka, E., 1984, Insertion mutations in the promiscuous IncP-1 plasmid R18 which affect its host range between *Pseudomonas* species, *Plasmid* 12:170–180.

70. Kuldau, G.A., De Vos, G., Owen, J., McCaffrey, G., and Zambryski, P., 1990, The *virB* operon of *Agrobacterium tubefaciens* pTiC58 encodes 11 open reading frames, *Mol. Gen. Genet.* 221:256–266.

71. Lanka, E., and Barth, P.T., 1981, Plasmid RP4 specifies a deoxyribonucleic acid primase involved in its conjugal transfer and maintenance, *J. Bacteriol.* 148:769–781.

72. Lanka, E., Fürste, J.P., Yakobson, E., and Guiney, D.G., 1985, Conserved regions at the DNA primase locus of the IncPα and IncPβ plasmids, *Plasmid* 14:217–223.

73. Lanka, E., Luz, R. Kröger, M., and Fürste, J.P., 1984, Plasmid RP4 encodes two forms of a DNA primase, *Mol. Gen. Genet.* 194:65–72.

74. Lazraq, R., Clavel-Seres, S., David, H.L., and Roulland-Dussoix, D., 1990, Conjugative transfer of a shuttle plasmid for *Escherichia coli* to *Mycobacterium smegmatis*, *FEMS Microbiol. Lett.* 69:135–138.

75. Leemans, J., Villarroel, R., Silva, B., Van Montagu, M., and Schell, J., 1980, Direct repetition of a 1.2 Md DNA sequence is involved in site-specific recombination by the P1 plasmid R68, *Gene* 10:319–328.

76. Lejeune, P., Mergeay, M., Van Gijsegem, F., Faelen, M., Geritis, J., and Toussaint, A., 1983, Chromosome transfer and R-prime plasmid formation mediated by plasmid pULB113 (RP4::mini-Mu) in *Alcaligenes eutrophus* CH34 and *Pseudomonas fluorescens* 6.2, *J. Bacteriol.* 155:1015–1026.

77. Lessl, M., Balzer, D., Lurz, R., Waters, V., Guiney, D.G., and Lanka, E., 1992, Dissection of IncP conjugation plasmid transfer: definition of the transfer region Tra2 by mobilization of the Tra2 region in *trans*, *J. Bacteriol.* 174:2493–2500.

78. Lessl, M., Krishnapillai, V., and Schilf, W., 1991, Identification and characterization of two entry exclusion genes of the promiscuous IncP plasmid R18, *Mol. Gen. Genet.* 227:120–126.

79. Llosa, M., Bolland, S., and de la Cruz, F., 1990, Structural and functional analysis of the origin of conjugal transfer of the broad-host-range IncW plasmid R388 and comparison to the related IncN plasmid R46, *Mol. Gen. Genet.*, in press.

80. Lyras, D., Palombo, E.A., and Stanisich, V.A., 1992, Characterization of a Tra2 function of RP1 that affects growth of *Pseudomonas aeruginosa* PAO and surface exclusion in *Escherichia coli* K12, *Plasmid* 27: 105–108.

81. Mazodier, P., Petter, R., and Thompson, C., 1989, Intergeneric conjugation between *Escherichia coli* and *Streptomyces* species, *J. Bacteriol.* 171:3583–3585.

82. Mergeay, M., Lejeune, P., Sadouk, A., Gerits, J., and Fabry, L., 1987, Shuttle transfer (or retrotransfer) of chromosomal markers mediated by plasmid pULB113, *Mol. Gen. Genet.* 209:61–70.

83. Mergeay, M., Springael, D., and Top, E., 1990, Gene transfer in polluted soils, in: *Bacterial Genetics in Natural Environments* (J.C. Fry and M.J. Day, ed.), Chapman & Hall, London, New York, pp. 152–171.

84. Merryweather, A., Barth, P.T., and Wilkins, B.M., 1986, Role and specificity of plasmid RP4-encoded DNA primase in bacterial conjugation, *J. Bacteriol.* 167:12–17.

85. Meyer, R., 1989, Site-specific recombination at *oriT* of plasmid R1162 in the absence of conjugative transfer, *J. Bacteriol.* 171:799–806.

86. Miele, L., Strack, B., Kruft, V., and Lanka, E., 1991, Gene organization and nucleotide sequence of the primase region of IncP plasmids RP4 and R751, *DNA Sequence* 2:145–162.

87. Miller, J., Lanka, E., Malamy, M., 1985, F-factor inhibition of conjugal transfer of broad-host-range plasmid RP4: requirement for the protein product of *pif* operon regulatory gene *pifC*, *J. Bacteriol.* 163:1067–1073.

88. Motallebi-Veshareh, M., Balzer, D., Lanka, E., Jagura-Burdzy, G., and Thomas, C.M., 1992, Conjugative transfer functions of broad host range plasmid RK2 are coregulated with vegetative replication, *Mol. Microbiol.* 6:907–920.

89. Nash, J., and Krishnapillai, V., 1988, Role of IncP-1 plasmid primase in conjugation between *Pseudomonas* species, *FEMS Microbiol. Lett.* 49:257–260.

90. Nayudu, M., and Holloway, B.W., 1981, Isolation and characterization of R-plasmid variants with enhanced chromosomal mobilization ability in *Escherichia coli* K12, *Plasmid* 6:53–66.

91. O'Hoy, K., and Krishnapillai, V., 1985, Transposon mutagenesis of the *Pseudomonas aeruginosa* PAO chromosome and the isolation of high frequency of recombination donors, *FEMS Microbiol. Lett.* 29:299–303.

92. Olsen, R.H., Siak, J., and Gray, R.H., 1974, Characteristics of PRD1, a plasmid-dependent broad host range DNA bacteriophage, *J. Virol.* 14:689–699.

93. Olsen, R.H., and Thomas, D.D., 1973, Characteristics and purification of PRR1, an RNA phage specific for the broad host range *Pseudomonas* R1822 drug resistance plasmid, *J. Virol.* 12:1560–1567.

94. Palombo, E.A., 1990, Genetic and molecular analysis of the Tra2 and Tra2-Tra3 regions of the plasmid RP1, Ph.D. dissertation, LaTrobe Univeristy, Bundoora, Australia, pp. 78–95.

95. Palombo, E.A., Yusoff, K., Stanisich, V.A., Krishnapillai, V., and Willetts, N.S., 1989, Cloning and genetic analysis of *tra* cistrons of the Tra2/Tra3 region of plasmid RP1, *Plasmid* 22:59–69.

96. Pansegrau, W., Balzer, D., Kruft, V., Lurz, R., and Lanka, E., 1990, In vitro assembly of relaxosomes at the transfer origin of plasmid RP4, *Proc. Natl. Acad. Sci. USA* 87:6555–6559.

97. Pansegrau, W., Miele, L., Lurz, R., and Lanka, E., 1987, Nucleotide sequence of the kanamycin resistance determinant of plasmid RP4: homology of other aminoglycoside 3'-phosphotransferase, *Plasmid* 18:193–204.

98. Pansegrau, W., Ziegelin, G., and Lanka, E., 1988, The origin of conjugative IncP plasmid transfer: interaction with plasmid-encoded products and the nucleotide sequence at the relaxation site, *Biochim. Biophys. Acta* 951:365–374.

99. Pansegrau, W., Ziegelin, G., and Lanka, E., 1990, Covalent association of the *traI* gene product of plasmid RP4 with the 5'-terminal nucleotide at the relaxation nick site, *J. Biol. Chem.* 265:10637–10644.

100. Rees, C.E.D., and Wilkins, B.M., 1990, Protein transfer into the recipient cell during bacterial conjugation: studies with F and RP4, *Mol. Microbiol.* 4:1199–1205.

101. Reimmann, C., and Haas, D., 1986, IS21 insertion in the *trfA* replication control gene of chromosomally integrated plasmid RP1: a property of stable *Pseudomonas aeruginosa* Hfr strains, *Mol. Gen. Genet.* 203:511–519.

102. Reimmann, C., and Haas, D., 1987, Mode of replicon fusion mediated by the duplicated insertion sequence IS21, in *Escherichia coli*, *Genetics* 115:619–625.

103. Reimann, C., and Haas, D., 1990, The *ist*A gene of insertion sequence IS21 is essential for cleavage at the inner 3' ends of tandemly repeated IS21 elements in vitro, *EMBO J.* 9:4055–4063.

104. Reimmann, C., Moore, R., Little, S., Savioz, A., Willetts, N.S., and Haas, D., 1989, Genetic structure, function and regulation of the transposable element IS21, *Mol. Gen. Genet.* 215:416–424.

105. Riess, G., Holloway, B.W., and Phler, A., 1980, R68.45, a plasmid with chromosome mobilizing ability (Cma) carries a tandem duplication, *Genet. Res.* 36:99–109.

106. Riess, G., Masepohl, B., and Phler, A., 1983, Analysis of IS21-mediated mobilization of plasmid pACYC184 by R68.45 in *Escherichia coli*, *Plasmid* 10:111–118.

107. Roberts, R.C., Burioni, R., and Helinski, D.R., 1990, Genetic characterization of the stabilizing functions of a region of broad-host-range plasmid RK2, *J. Bacteriol.* 172, 6204–6216.

108. Scherzinger, E., Lurz, R., Otto, S., and Dobrinski, R., 1992, In vitro cleavage of double- and single-stranded DNA by plasmid RSF1010-encoded mobilization proteins, *Nucl. Acids Res.* 20:41–48.

109. Schilf, W., and Krishnapillai, V., 1986, Genetic analysis of insertion mutations of the promiscuous IncP-1 plasmid R18 mapping *oriT* which affect its host range, *Plasmid* 15:48–56.

110. Scholz, P., Haring, V., Wittmann-Liebold, B., Ashman, K., Bagdasarian, M., and Scherzinger, E., 1989, Complete nucleotide sequence and gene organization of the broad-host-range plasmid RSF1010, *Gene* 75:271–288.

111. Schurter, W., and Holloway, B.W., 1987, Interactions between the transposable element IS21 on R68.45 and Tn7 in *Pseudomonas aeruginosa* PAO, *Plasmid* 17:61–64.

112. Shoemaker, N., Guthrie, E., Salyers, A., and Gardener, J., 1985, Evidence that the clindamycin-erythromycin resistance gene of *Bacteroides* plasmid pBF4 is on a transposable element, *J. Bacteriol.* 162:626–632.

113. Simon, R., 1984, High frequency mobilization of Gram-negative bacterial replicons by the in vitro constructed Tn5-Mob transposon, *Mol. Gen. Genet.* 196:413–420.

114. Smith, C.A., and Thomas, C.M., 1985, Comparison of the nucleotide sequences of the vegetative replication origins of broad host range *IncP* plasmids R751 and RK2 reveals conserved features of probably functional significance, *Nucl. Acids Res.* 13:557–572.

115. Stachel, S.E., and Zambryski, P., 1986, *Agrobacterium tumefaciens* and the susceptible plant cell: a novel adaptation of extracellular recognition and DNA conjugation, *Cell* 47:155–157.

116. Stanisich, V.A., 1974, The properties and host range of male-specific bacteriophages of *Pseudomonas aeruginosa*, *J. Gen. Microbiol.* 84:332–342.

117. Tanimoto, K., and Iino, T., 1983, Transfer inhibition of RP4 by F-factor, *Mol. Gen. Genet.* 192:104–109.

118. Tanimoto, K., Iino, T., Ohtsubo, H., and Ohtsubo, E., 1985, Identification of a gene, *tir*, of R100, functionally homologous to the F3 gene of F in the inhibition of RP4 transfer, *Mol. Gen. Genet.* 198: 356–357.

119. Tardiff, G., and Grant, R.B., 1983, Transfer of plasmids from *Escherichia coli* to *Pseudomonas aeruginosa*: characterization of a *Pseudomonas aeruginosa* mutant with enhanced recipient ability for enterobacterial plasmids, *Antimicrob. Ag. Chemother.* 24:201–208.

120. Thomas, C.M., and Helinski, D.R., 1989, Vegetative replication and stable inheritance of IncP plasmids, in: *Promiscuous Plasmids of Gram-Negative Bacteria* (C.M. Thomas, ed.), Academic Press, London, pp. 1–25.

121. Thomas, C.M., Theophilus, B.D., Johnston, L., Jagura-Burdzy, G. Schilf, W., Lurz, R., and Lanka, E., 1990, Identification of a seventh operon on plasmid RK2 regulated by the *korA* gene product, *Gene* 89: 29–35.

122. Thompson, D.V., Melchers, L.S., Idler, K.B., Schilperoort, R.A., and Hooykaas, P.J.J., 1988, Analysis of the complete nucleotide sequence of the *Agrobacterium tumefaciens virB* operon, *Nucl. Acids Res.* 16:4621–4636.

123. Top, E., Mergeay, M., Springael, D., and Verstraete, W., 1990, Gene escape model: transfer of heavy metal resistance genes from *Escherichia coli* to *Alcaligenes eutrophus* on agar plates and in soil samples, *Appl. Environ. Microbiol.* 56:2471–2479.

124. Trieu-Cuot, P., Carlier, C., and Courvalin, P., 1988, Conjugative plasmid transfer from *Enterococcus faecalis* to *Escherichia coli*, *J. Bacteriol.* 170:4388–4391.

125. Trieu-Cuot, P., Carlier, C., Martin, P., and Courvalin, P., 1987, Plasmid transfer by conjugation from *Escherichia coli* to gram-positive bacteria, *FEMS Microbiol. Lett.* 48:289–294.

126. Valentine, C.R., and Kado, C.I., 1989, Molecular genetics of IncW plasmids, in: *Promiscuous Plasmids of Gram-Negative Bacteria* (C.M. Thomas, ed.), Academic Press, London, pp. 125–163.

127. Van Gijsegem, F., Toussaint, A., 1982, Chromosome transfer and R-prime formation by an RP4::mini-Mu derivative in *Escherichia coli*, *Salmonella typhimurium*, *Klebsiella pneumoniae* and *Proteus mirabilis*, *Plasmid* 7:30–44.

128. Villarroel, R., Hedges, R.W., Maenhaut, R., Leemans, J., Engler, G., Van Montagu, M., and Schell, J., 1983, Heteroduplex analysis of P-plasmid evolution: the role of insertion and deletion of transposable elements, *Mol. Gen. Genet.* 189:390–399.

129. Walter, E.G., Thomas, C.M., Ibbotson, J.P., and Taylor, D.E., 1991, Transcriptional analysis, translational analysis, and sequence of the *kil*A-tellurite resistance region of plasmid RK25e^r, *J. Bacteriol.* 173:1111–1119.

130. Wang, K., Stachel, S.E., Timmerman, B., Van Montagu, M., and Zambryski, P.C., 1987, Site-specific nick in the T-DNA border sequence as a result of *Agrobacterium vir* gene expression, *Science* 235:587–591.

131. Ward, J.E., Akiyoshi, D., Regier, D., Datta, A., Gordon, M.P., and Nester, E.W., 1988, Characterization of the *virB* operon from an *Agrobacterium tumefaciens* Ti plasmid, *J. Biol. Chem.* 263:5804–5814.

132. Waters, V.L., Hirata, K., Pansegrau, W., Lanka, E., and Guiney, D.G., 1991, Sequence identity in the nick regions of IncP plasmid transfer origins and T-DNA borders of *Agrobacterium* Ti plasmids, *Proc. Natl. Acad. Sci. USA* 88:1456–1460.

133. Willetts, N.S., and Crowther, C., 1981, Mobilization of the nonconjugative IncQ plasmid RSF1010, *Genet. Res.* 37:311–316.

134. Willetts, N.S., Crowther, C., and Holloway, B.W., 1981, The insertion sequence IS21 of R68.45 and the molecular basis for mobilization of the bacterial chromosome, *Plasmid* 6:30–52.

135. Williams, D.R., Young, D.I., and Young, M., 1990, Conjugative plasmid transfer form *Escherichia coli* to *Clostridium acetobutylicum*, *J. Gen. Microbiol.* 136:819–826.

136. Winans, S.C., and Walker, G.C., 1985a, Conjugal transfer system of the IncN plasmid pKM101, *J. Bacteriol.* 161:402–410.

137. Winans, S.C., and Walker, G.C., 1985b, Entry exclusion determinants of IncN plasmid pKM101, *J. Bacteriol.* 161:411–416.

138. Winans, S.C., and Walker, G.C., 1985c, Fertility inhibition of RP1 by IncN plasmid pKM101, *J. Bacteriol.* 161:425–427.

139. Wolk, C.P., Vonshak, A., Kehoe, P., and Elhai, J., 1984, Construction of shuttle vectors capable of conjugative transfer from *Escherichia coli* to nitrogen-fixing filamentous cyanobacteria, *Proc. Natl. Acad. Sci. USA* 81:1561–1565.

140. Yakobson, E., Deiss, C., Hirata, K., and Guiney, D.G., 1990, Initiation of DNA synthesis in the transfer origin region of RK2 by the plasmid-encoded primase: detection using defective M13 phage, *Plasmid* 23:80–84.

141. Yakobson, E., and Guiney, D., 1983, Homology in the transfer origins of broad-host-range IncP plasmids: definition of two subgroups of P-plasmids, *Mol. Gen. Genet.* 192:436–438.

142. Yakobson, E., and Guiney, D.G., 1984, Conjugal transfer of bacterial chromosomes mediated by the RK2 plasmid transfer origin cloned into transposon Tn5, *J. Bacteriol.* 160:451–453.

143. Yanofsky, M.F., Porter, S.G., Young, C., Albright, L.M., Gordon, M.P., and Nester, E.W., 1986, The *virD* operon of *Agrobacterium tumefaciens* encode a site-specific endonuclease, *Cell* 47:471–477.

144. Young, C., and Nester, E.W., 1988, Association of the *virD2* protein with the 5′ end of T-strands in *Agrobacterium tumefaciens*, *J. Bacteriol.* 170:3367–3374.

145. Yusoff, K., and Stanisich, V., 1984, Location of a function on RP1 that fertility inhibits IncW plasmids, *Plasmid* 11:178–181.

146. Zambryski, P., 1988, Basic processes underlying *Agrobacterium*-mediated DNA transfer to plant cells, *Annu. Rev. Genet.* 22:1–30.

147. Ziegelin, G., Fürste, J.P., and Lanka, E., 1989, TraJ protein of plasmid RP4 binds to a 19-base pair invert sequence repetition within the transfer origin, *J. Biol. Chem.* 264:11989–11994.

148. Ziegelin, G., Pansegrau, W., Strack, B., Balzer, D., Kröger, M., Kruft, V., and Lanka, E., 1990, Nucleotide sequence and organization of genes flanking the transfer origin of promiscuous plasmid RP4, *DNA Sequence* 1:303–327.

Chapter 5

DNA Processing and Replication during Plasmid Transfer between Gram-Negative Bacteria

BRIAN WILKINS and ERICH LANKA

1. Introduction

This chapter concerns the processing and synthesis of plasmid DNA during its transmission between conjugating gram-negative bacteria, focusing on conjugation systems specified by plasmids isolated in or transferred experimentally to enterobacteria. This extensive collection of plasmids can be classified into more than 25 different incompatibility (Inc) groups (34), each of which is generally associated with a distinct conjugation system (158). Only a few of these systems have been investigated at the biochemical and molecular levels, but studies have identified unifying themes as well as an interesting diversity of enzymatic strategies for the conjugative processing of plasmid DNA.

Gram-negative conjugation can be viewed as a specialized replicative event that increases the population size of the plasmid during its horizontal transfer between organisms. The process can be divided operationally into two stages. The first involves the formation of a specific bridge between the plasmid-containing donor bacterium and the recipient cell, which are brought into contact by the extracellular conjugative pilus. The second stage concerns the transfer and processing of DNA. This stage is initiated by nicking of the plasmid at the specific origin of transfer (*oriT*) site and is followed by unwinding of the duplex by one or more DNA helicases and the transfer of the open DNA strand to the recipient cell. The transferred DNA is then circularized by a mechanism involving

BRIAN WILKINS • Department of Genetics, University of Leicester, Leicester LE1 4RH, United Kingdom.   ERICH LANKA • Max-Planck-Institut für Molekulare Genetik, Abteilung Schuster, Ihnestrasse 73, D-1000 Berlin 33, Federal Republic of Germany.

Bacterial Conjugation, edited by Don B. Clewell. Plenum Press, New York, 1993.

interactions at the *oriT* sequence. Such single-stranded (ss) DNA transfer is normally associated with conjugative DNA synthesis, generating a replacement strand in the donor cell and a complementary strand in the recipient.

This review is based on information from the study of the genetically and mor-phologically distinct conjugation systems determined by plasmids belonging to the F and I complexes of incompatibility groups and to the IncP group. Exemplars are plasmids F, R1, R6-5, and R100 (NR1) of the F complex (76, 159, 163) and IncI1 plasmids ColIb-P9 and R64 of the I complex (84, 128). Pertinent IncP plasmids are RP1, RP4, RK2, and R18, considered to be a single biological entity belonging to the IncPα subgroup, and R751 of the IncPβ subgroup (66). These are all large (>50 kb) conjugative plasmids that carry a complement of *tra* genes necessary to promote both the cellular interactions and the DNA processing reactions basic to conjugation.

The *oriT* site on many plasmids is located at or close to one end of the segment that includes the *tra* genes, and it is oriented such that the *tra* genes are transmitted late to the recipient cell. The DNA segment transferred early from the *oriT* site is termed the leading region. This portion of some F and I complex plasmids is known to include some conserved genes that promote installation of the transferred plasmid in the newly infected recipient cell. These genes are considered in the final section of the chapter.

Important insights of the DNA transfer process have also come from the study of small, naturally occurring, mobilizable plasmids. These plasmids transfer autonomously from donor cells that also harbor a conjugative plasmid and are exemplified by ColE1 (21) and the similar, if not identical, IncQ plasmids, R300B, R1162, and RSF1010 (11, 41, 135). Such mobilizable plasmids (<10 kb) carry their *oriT* site and a few adjacent *mob* genes inferred to mediate such events as *oriT* nicking and DNA circularization. The conjugative plasmid contributes to mobilization by providing at least the functions necessary for the bacterial interactions specific to conjugation.

The aim of this chapter is to provide a general perspective of the subject area and information on recent advances. Further documentation is given in the reviews on the conjugative processing of plasmid DNA by Willetts and Wilkins (160) and on the conjuga-tion system of F-like plasmids by Ippen-Ihler and Minkley (76) and Willetts and Skurray (159). Conjugative transfer of IncP plasmids has been reviewed recently by Guiney and Lanka (66).

2. Processing of the Transferred DNA Strand

2.1. Sequence of Gene Transfer

The *oriT* site maps on several different conjugative plasmids at or near one end of the DNA segment carrying the defined *tra* genes. This arrangement is found on F-like plasmids (76, 159) and on representative IncI1 (128), IncN (162), IncP (66, 91), and IncW plasmids (16). All of those conjugative plasmids examined so far transfer in a preferred direction from *oriT*, with most, if not all, of the *tra* genes entering the recipient last (5, 32, 64, 72, 159). The same generalization applies to ColE1, where the *oriT* region, termed *bom* (*basis of mobility*), is situated about 400 bp from the cluster of mobility genes that are transferred late during mobilization (20, 21). IncQ plasmids also transfer unidirectionally, but on these elements the *oriT* site lies within the mobility region (41, 79).

This pattern showing late transfer of *tra* genes may be more than fortuitous, indicating that the order of gene transmission is functionally important for a conjugative plasmid. An appealing speculation is that the arrangement has evolved to allow early transfer of genes for plasmid replication and maintenance. However, this hypothesis is scarcely consistent with the location of the replication region on CoIIb, R100, and RP4, since these regions map at least 20 kb from *oriT* and are transferred immediately before the *tra* genes (66, 128, 163). Alternatively, late transfer of the *tra* genes per se may confer some advantage, possibly to delay their expression in the newly infected recipient cell until completion of DNA transport and circularization. It is known that expression of plasmid genes in the recipient actively terminates the conjugation cycle, but the timing of this termination event relative to a cycle of DNA transfer has not been established (3, 17, 101).

2.2. Single-Stranded Transfer

The general model of bacterial conjugation proposes that a specific plasmid strand is transmitted in the 5' to 3' orientation to the recipient cell and that the complementary strand is retained in the donor. It is stressed that these features have been demonstrated directly only for the F transfer system, although there is molecular evidence for transmission of a unique DNA strand in conjugation directed by IncFII and IncI1 plasmids (116, 132, 147, 148). Despite the limited range of plasmids that have been analyzed in this respect, parallels between different plasmids at the level of protein-*oriT* interactions provide compelling indications that ssDNA transport in the 5' to 3' direction is a unifying property of the several transfer systems considered in this chapter. It is unknown whether the 5' terminus of the DNA enters the recipient cell or is retained in the donor in the vicinity of the transport pore. In the latter case, the transferring strand would loop into the recipient bacterium.

A particularly elusive aspect of conjugation is the nature of the bridge or pore supporting DNA transfer. The conjugative pilus is necessary for the formation of intercellular contact, but there is no conclusive evidence that DNA undergoing transfer is associated with an extended pilus. Most models postulate that DNA transfer occurs between cells that are brought into surface-surface contact by pilus retraction and subsequently stabilized as aggregates (76, 101). Irrespective of its structure, the pore for DNA transport allows no general mixing of the cytoplasmic contents of conjugating bacteria (126, 127, 137).

Erection of the F pilus requires 15 or more *tra* gene products, and some of these may contribute to a membrane-spanning complex at the base of the pilus that also supports DNA transport (76, 159). Of the F *tra* products known to function in the conjugative processing of DNA, TraD protein is implicated in the DNA transport process. This protein (calculated molecular mass of 81,400 da [77]) is associated with the inner membrane of F+ cells (118a). Properties of cells carrying mutant plasmids indicate that TraD protein is required at a stage following effective cell aggregation, *oriT* nicking, and the initiation of DNA unwinding (47, 80, 118). This protein was therefore postulated to form, or be part of, the pore for DNA transfer. Purified TraD protein binds to ss and double-stranded (ds) DNA and apparently contains DNA-dependent ATPase activity, consistent with a model in which TraD actively facilitates transport of DNA out of the donor cell (E. G. Minkley, Jr., personal communication).

The transferred F strand enters the cytoplasm of the recipient cell without detectable association with any *tra* gene product (127). However, the transferred DNA of some other plasmids, such as ColIb and RP4, is accompanied into the recipient by specific *tra* products called DNA escort proteins. Such proteins, discussed more extensively in Section 3.2.2., are transmitted in multiple copies per DNA strand, presumably as DNA binding proteins. Apart from promoting conjugative events in the recipient cell, some DNA escort proteins might facilitate DNA transport across cell envelopes, as evidenced by the properties of ColIb mutants defective in the 3′ region of *sog* (109, 126).

2.3. Events at the Origin of Transfer

2.3.1. Nature of *oriT* Regions

In the general model of conjugation, *oriT* represents the site on a conjugative or mobilizable plasmid where the transfer process is initiated by a specific single-stranded cleavage event. Interaction of the *oriT* region with plasmid-specified transfer gene products is required to mediate this cleavage process. These components belong to the Dtr (DNA transfer and replication) system of a conjugative plasmid (156). Transfer origins function in *cis* and can be readily isolated by molecular cloning through their ability to convert a nontransmissible vector into a mobilizable plasmid. Deletion analyses and nucleotide sequencing have demonstrated that transfer origins comprise intergenic regions of up to about 500 bp, which are flanked by *tra* genes on either one or both sides.

The nucleotide sequence of the *oriT* region of numerous plasmids has been determined. These include the following conjugative plasmids: F (IncFI) (142), P307 (IncFI) (60), R1 (IncFII) (117), R100 (IncFII) (49,106), ColB4-K98 (IncFII) (49), pED208 (IncFV) (42), R64 (IncI1) (83), R46 (IncN) (33), pCU1 (IncN) (123), RP4/RK2 (IncPα), (53, 67, 171), R751 (IncPβ) (53, 171), R388 (IncW) (96) and pTiC58 (31a) of *Agrobacterium tumefaciens*. The nucleotide sequence of the *oriT* region on the following mobilizable plasmids is known: ColE1 (26), pSC101 (13), RSF1010/R1162 (IncQ) (23, 40, 135), Col DF13 (139), pACYC184 (131) (nucleotide positions 515-653; D. Haas, personal communication), ColK (6), ColA (111), and also pTF1 (45) and pTF-FC2 (130) isolated from *Thiobacillus ferrooxidans*. Extensive nucleotide sequence similarities are found only between the *oriT* regions of related plasmids, such as those within the F complex, the IncP group, or the ColE1-like family. Lack of a general similarity implies that transfer systems are quite diverse.

The organization of the *oriT* regions of the two conjugative plasmids, F and RP4/RK2, and of the two mobilizable plasmids, ColE1 and R1162/RSF1010, is summarized in Figure 5.1 and discussed in detail because genetic and biochemical studies of these systems are the most advanced. Common features are inverted and direct sequence repetitions, thought to function as targets for protein-DNA recognition. The AT content of *oriT* regions is usually higher than that of the flanking regions, and this is particularly obvious in the case of IncP and IncQ plasmids (135, 171). High AT content is also characteristic of vegetative replication origins (85). Experiments involving replacement of the primary origin for

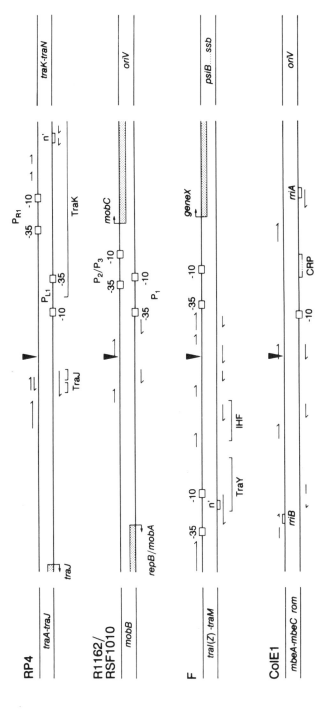

FIGURE 5.1. Comparative maps of *oriT* regions. The central part of the figure corresponds to 300 bp of each *oriT* region. A single scale is used. In all cases the upper line corresponds to the strand known or predicted to be transferred, and it contains the relaxation nick site indicated by the solid arrowhead. The leading region is to the right of the nick site. Genes and proteins are discussed in the text. Sequence repetitions are shown (→). Horizontal brackets designate protein binding sites. Hatched boxes depict genes. Open squares indicate known or putative promoters, and open rectangles labeled n′ and *rri* are presumptive and known PriA protein recognition sites, respectively (PriA = protein n′ = primosomal protein replication factor Y) (85, 168). CRP is a cyclic AMP-dependent promoter.

complementary strand synthesis in phage M13 DNA by the RK2 *oriT* region suggest that the transfer origin might act as an origin for synthesis of replicative form from ssDNA in the presence of certain *tra* gene products such as the IncP plasmid DNA primase (164). Computer searches and functional tests suggest that *oriT* regions may contain sequences recognized by host-specified proteins involved in initiation of DNA synthesis. These will be considered in Section 3.2.1. In addition, *oriT* regions include promoters responsible for transcription of *tra* and *mob* genes. Essential *tra* and *mob* genes of IncP (53, 171), IncQ (135), and pTF1 (44) plasmids are transcribed divergently from these promoters in the *oriT* region, indicating a similar gene organization around the IncP and IncQ transfer origins (Figure 5.1). Products of these genes interact specifically with the corresponding *oriT*-DNA (Section 2.3.3.).

Deletion analysis has shown that the F plasmid *oriT* locus is extensive, with domains that individually contribute to nicking, transfer, and overall structure (51). Major nicking determinants are inferred to lie in the region from about 22 bp to the right to 70 bp to the left of the nick site (Figure 5.1). This domain includes an IHF binding site. Further to the left is the transfer domain of ~125 bp, which includes the inverted repeat shown at the left end of the *oriT* region in Figure 5.1. This repeat is implicated as an important sequence element for transfer, and it requires proper phasing with respect to the nick site for full *oriT* function.

IHF (integration host factor) (165) binding sites have been detected in the AT-rich regions of F-type plasmids, and it has been reported that IHF is required for efficient expression of the R100 conjugative transfer system (39, 56). IHF binding sites (consensus sequence YAANNNNTTGATa/t) (35) enhance intrinsic bending in the F *oriT* DNA and may function in forming higher order DNA-protein structures (146). IHF-dependent plasmid transfer has not been demonstrated for other than F-type plasmids.

Part of the *oriT* region of IncP plasmids is bent (122). Dissection of fully functional IncP transfer origins revealed that removal of DNA on one side resulted in reduced transfer frequencies (53, 119). Frequencies dropped about 200-fold when DNA to the right of the nick site was absent (Figure 5.1). A special conformational feature of this DNA is a static bend, recognized by classical analyses for curved DNA (122). A similar intrinsic bend at a corresponding location has been predicted for R751 *oriT* by computer analysis. Therefore, it appears likely that structural requirements for efficient transfer are the curvature of DNA, including binding to this region by the TraK protein (170). The IncP TraK protein might have an analogous function to IHF in the F system.

2.3.2. Nicking at *oriT*

The *oriT* region contains the relaxation nick site needed to provide the substrate for generation of the single strand to be transported to the recipient cell. Structural studies of nick sites were made possible by the fundamental observation of Clewell and Helinski (31) that conjugative and mobilizable plasmids can be isolated from cells as protein-plasmid DNA relaxation complexes. Such complexes are referred to as relaxosomes (52). Treatment of relaxosomes with protein-denaturing agents releases an open circular DNA that is an ideal substrate for sequence analysis. The methodology described by the two authors to prepare relaxed DNA has been applied to determine the exact location of the nick site in ColE1 DNA (12) and, more recently, on F (143), R100 (75), R64 (54), RSF1010 (133), and IncP plasmids (119, 122). A common conclusion of these studies is that one specific phosphodiester bond is cleaved between two different nucleotides 5'R/Y or 5'Y/R. This holds true for other replicative systems, like those of ssDNA phages and certain gram-positive bacterial plasmids that replicate via asymmetric rolling circles (65, 141). Similarities have been detected between nick regions of closely related plasmids, but there are some instructive exceptions (Section 2.3.4.). The nick regions of IncP plasmids and IncQ plasmids (RSF1010 and R1162) show some sequence identity. However, this comprises only four nucleotides, G▼CCC,* and its occurrence might therefore be a matter of chance.

What is the chemical nature of the terminal nucleotides at a nick site? In accordance with the mechanistic model for initiation of transfer DNA replication by a rolling circle, F- and P-type plasmids as well as RSF1010 offer an unmodified 3' hydroxyl terminus at the relaxation nick site, which is susceptible in vitro to elongation by DNA polymerase I of *Escherichia coli* (119, 122, 133, 143). Mutational analyses of the RK2/RP4 nick site demonstrate that the 3' terminal nucleotide position appears to be more important than the 5' position. A change from G▼C to AC prevents nicking, whereas G▼C to G▼T mutation only slightly reduces transfer frequency and in vitro relaxation (152).

The first evidence of the nature of the 5' terminus of the nick site came again from a thorough study of ColE1 relaxosomes in the Helinski laboratory (100). This study showed that a 60-kDa polypeptide is associated with the 5' terminal nucleotide at the nick. The polypeptide is probably identical to the *mob3* gene product of 57,895 Da predicted from the nucleotide sequence of ColE1 (26). Mob3 has been identified by minicell analyses and renamed MbeA (21). In addition, it has been suggested that the 5' terminal nucleotide at the R100 nick site is modified by covalent association with a protein (75). Modification of the 5' terminus of the cleaved strand has also been demonstrated for the systems of RP4, RSF1010, and F, where TraI, MobA (RepB), and TraI is the covalently attached protein, respectively (105, 110, 133).

These studies of nick regions of three different transfer systems support the model that the strand to be transferred is covalently linked to a plasmid-encoded protein at its 5' terminal nucleotide. It is proposed that the linkage consists of a phosphodiester with the side chain hydroxyl of a seryl, threonyl, or tyrosyl residue. For RP4 TraI, the tyrosine residue Y22 is indeed bound to the 5' phosphate of the deoxycytidylic residue of the nick site. A mutation at Y22 to phenylalanine yields a cleavage-defective TraI protein (W. Pansegrau and E. Lanka, unpublished data). Furthermore, it is envisaged that this covalent protein-DNA complex is essential for circularizing the transferred strand (Figure 5.2).

2.3.3. Relaxosomes

It has been demonstrated genetically that more than one plasmid gene is necessary for formation of the relaxosome. This means the relaxosome must be formed by a specific protein-DNA interaction and additional protein-protein interactions at the transfer origin. Genetic study of the nicking process indicated that the F system needs at least the *traY* gene and the region containing *traI* (47). These two genes are at opposite ends of the *tra* operon, indicating that relaxation does not necessarily involve closely linked loci, as is the case for IncP and Q plasmids (40, 53, 119). The *traI* region of F and R100 is complex, and it is not clear which part is needed for relaxation (22, 75, 166). It was proposed that F *oriT* nicking

---

*The symbol ▼ indicates the relaxation nick site.

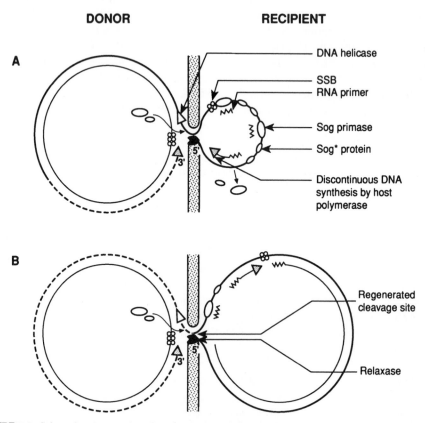

FIGURE 5.2. Schematic representation of two stages in the processing of DNA during the transfer of IncI1 and IncP plasmids. Panel A shows transfer of the preexisting DNA strand (heavy continuous line) with a 5′ to 3′ polarity. The relaxase protein (ColIb NikB; RP4 TraI) is covalently linked to the unique 5′ terminus at the nicked *oriT* site, and the complex is located close to the DNA transport pore. The transferring strand is escorted into the recipient cell by plasmid DNA primase (ColIb Sog; RP4 Pri = TraC) and the sequence-related Sog* protein. RNA primers are elongated by the replicative DNA polymerase of the host (*E. coli* DNA polymerase III holoenzyme). In the donor cell, DNA unwinding is mediated by a plasmid or host-encoded DNA helicase yet to be identified; here the helicase is translocating processively in the 5′ to 3′ direction on the bound DNA strand. Synthesis of the replacement strand (broken line) by a rolling-circle mode of DNA replication has reconstituted an *oriT* nick region in the donor. Small circles indicate single-stranded DNA binding (SSB) protein. No attempt has been made to indicate the complexity of the cell envelopes (hatched barrier between cells) or the DNA transport pore. Panel B shows the stage immediately prior to circularization of the transferred DNA strand by the second cleavage mechanism. DNA transfer has brought the reconstituted nick region in single-stranded form into close proximity with the relaxase protein linked to the 5′ terminus of the transferring strand (Figure 5.3). Relaxase cuts the reconstituted nick region and ligates the 5′ and 3′ termini to give a monomeric circle of transferred DNA. Circularization is occurring before the completion of complementary strand synthesis in the recipient cell.

requires the activities of *traY* and *traZ* (48), but it is now thought that *traZ* is part of *traI* (144). F *traI* specifies a 180-kDa polypeptide (TraI, predicted molecular mass of 192 kDa) and a second 94-kDa polypeptide (TraI*, predicted molecular mass of 88 kDa) generated from an internal in-phase translational start (22, 144).

TraY products of F and R100 are plasmid-specific *oriT* binding proteins (74, 90) that are structurally similar to the Arc and Mnt repressors of phage P22 (19). An in vitro system of purified TraY and TraI proteins in the presence of superhelical F *oriT* plasmid DNA was inert for nicking, suggesting that one or more components are missing (90). However, a crude in vitro R100 system containing *oriT*-DNA, a TraY fusion polypeptide, and an extract of cells carrying a TraI overproducing plasmid appears to be active in complex formation as well as in the specific nicking reaction (75). The overproducing plasmid contained the complete *traI* gene and, in addition, ~ 3 kb of DNA downstream of *traI* extending beyond *finO* (166). The success of the crude system might depend on a product(s) specified by the region downstream of *traI* or on host factors. Participation of host proteins cannot yet be ruled out.

F TraI is a multifunctional protein also known as DNA helicase I (Section 2.4.). Recently it has been found that purified helicase I alone can mediate specific nicking at F *oriT* in a process requiring a superhelical DNA substrate and Mg^{2+} ions. The in vitro reaction is protein concentration dependent, but only 50 to 70% of the input DNA substrate is converted to the nicked species. The phosphodiester bond interrupted in vitro is identical to the site nicked in vivo, suggesting that helicase I is the DNA relaxase initiating conjugative transfer of F. The 3' terminus at the nick contains a free hydroxyl, but the 5' end is blocked, most likely by covalent linkage to helicase I. Interestingly, the cleavage reaction is independent of TraY and occurs in the absence of ionic detergents or proteases (104, 129, S. M. Matson, personal communication). This observation questions the trigger hypothesis, discussed later in this section, and opens the possibility of studying rolling-circle replication in vitro, including the helicase I-catalyzed unwinding reaction. The intriguing question of the role of TraI* remains to be solved.

The ColE1 relaxosome has been shown to contain three proteins of 60, 16, and 11 kDa (100). Mobilization of ColE1 by R64*drdll* is dependent on four genes: *mbeA, B, C*, and *D*. The predicted sizes for MbeA (57,744 Da), MbeB (19,525 Da), and MbeC (12,883 Da) agree quite well with the sizes of proteins found in the complex (21).

Mobilization of RSF1010 by IncP plasmids requires a ~1.8-kb region of RSF1010 DNA that includes the transfer origin (40, 41). This region extends on both sides of *oriT* and contains *mobA, B*, and *C*, specifying the proteins B, D, and K, respectively (135). MobA, also called RepB, possesses RSF1010 replicon-specific priming activity. Purified Mob proteins have been used to demonstrate in vitro strand-specific nicking at RSF1010 *oriT*. In the presence of Mg^{2+} ions, MobA, MobB, and MobC form large amounts of a cleavage complex with supercoiled or linear duplex *oriT* DNA. On addition of sodium dodecyl sulfate (SDS), a single-stranded break is generated in the DNA, and MobA is found linked to the 5' terminus, presumably covalently. The double-stranded nicking activity of MobA requires Mg^{2+} and MobC, and it is stimulated by MobB. The position of the nick site generated in vitro and in vivo is identical (nucleotide positions 3138/9, r-strand [135]). In the absence of SDS, MobA also cleaves ssDNA lacking *oriT* sequences. Reclosure occurs after prolonged incubation, as shown by the formation of circular DNA molecules of different sizes (133).

The key to understanding protein interactions of *oriT* of IncP plasmids came from studies of heterologous mobilization of RP4 *oriT*-containing plasmids by R751 and vice versa (52, 53, 67, 119). This approach was applied successfully to the F complex of plasmids to define specificity of *tra* genes for the cognate *oriT* region (157) (see "*oriT* specificity" [160]). Judged by such tests, the IncP genes *traJ* and *traK* are specificity determinants for *oriT* function. These two genes are located on opposite sides of *oriT* in the TraI core region (~2.2 kb). TraI core is defined as the minimal region needed for heterologous mobilization, and it also contains the plasmid genes necessary for formation of the relaxosome. Capturing of specifically relaxed DNA from cells requires part of the relaxase operon, namely the genes *traJ* and the 5' three-fifths of *traI*. This operon extends to the left of the RP4 *oriT* region (Figure 5.1) and contains a third gene, *traH*, which is within *traI* occupying a different reading frame. The presence of *traK* increases the yield of relaxed molecules. This gene is located in the leading region at the 5' end of an operon, called the leader operon.

These findings were essentially confirmed by in vitro reconstitution of the RP4 relaxosome from purified components. Use of expression vectors allowed overproduction of the proteins encoded by the TraI core region in amounts that facilitated their identification and purification. TraI and TraJ proteins in the presence of Mg^{2+} ions were sufficient for in vitro relaxation. Addition of TraH, an acidic oligomeric protein (pI = 4.2), stabilized the protein-*oriT* DNA complex, allowing its detection on agarose gels and visualization on electron micrographs (120). The presence of TraK in the *in vitro* reactions increased the yield of relaxed *oriT* plasmids as it does *in vivo* (53, 152, W. Pansegrau and E. Lanka, unpublished data).

Negatively supercoiled *oriT* DNA is apparently the preferred substrate for the formation of relaxosomes in vitro. This is in agreement with the basic observation that relaxosomes prepared from cells contain superhelical plasmid DNA. TraJ protein binds specifically to *oriT*, recognizing the right arm of the imperfect 19-bp inverted sequence repetition (Figure 5.1). The TraJ binding site and the nick site are located on the same side of the DNA helix. Relaxosome assembly is proposed to involve a cascadelike mechanism. The initial step is binding of TraJ to *oriT*, giving a complex detectable by gel electrophoresis (169). Next this complex is recognized by TraI, which is unable to bind *oriT* DNA directly. Relaxation products are identical, irrespective of whether relaxosomes were isolated from cells or reconstituted in vitro. The in vitro cleavage reaction at RP4 *oriT* is independent of host-encoded components, which might be a contributory property of a broad host range transfer system (120).

A general feature of three of the systems discussed here is their genetic complexity; they contain overlapping genes in different reading frames. Such genes have not been reported for the regions of F-like plasmids that contribute to relaxosomes, but F *traI* contains an in-frame translational start site for TraI* expression (22, 144). A similar overlapping gene arrangement for R100 *traI* has not been described (166). All the proteins discussed here normally occur in very small amounts, making their study difficult without the application of in vitro recombinant DNA techniques to allow overexpression. Again there is one exception: F TraI has a copy number of about 500 to 700 molecules per F+ cell (81).

Conjugative DNA synthesis and DNA transport are initiated in response to an as yet unidentified signal generated by mating-pair formation (160). It is also unclear whether *oriT* nicking is a conjugative event triggered by mating-pair formation or is a reversible process occurring in the absence of recipient cells (47). The in vitro IncP relaxation system releases only part of the *oriT* plasmid DNA as a complex of Tra proteins and open circular DNA. This could mean that the method of "inducing" the nicking reaction in fact freezes a thermodynamic equilibrium between noncovalent complexed supercoiled plasmid DNA and nicked plasmid DNA, the latter kept in the superhelical state by covalent and noncovalent interactions with *tra* gene products (120). Action of the DNA relaxase (Tra proteins involved in specific cleavage at *oriT*) (Figure 5.2) therefore could resemble a type I topoisomerase, conserving energy of the cleaved phosphoester bond via formation of the covalent protein-*oriT* intermediate. This hypothesis is supported by the observation that under certain conditions a variety of topoisomers are released from relaxosomes. In addition, a striking increase in the number of topoisomers is seen during the relaxation reaction with TraI mutant proteins containing serine to alanine substitutions at two different positions (S74A and S14A). This is interpreted as a decrease of binding strength of TraI in the complex to the nick region upstream of the nick site. It also indicates that the religation reaction is a property of the relaxosomal proteins.

The TraI protein alone cleaves short, single-stranded deoxyribooligonucleotides containing the nick region of the transferred strand. In this reaction TraI becomes covalently attached to the same nucleotide as in the cleavage reaction carried out with relaxosomes. A hydrolysis equilibrium is reached at about 30% of the input substrate. In this TraI-mediated in vitro reaction, the use of two oligonucleotide species different in length yields hybrid oligonucleotides. Therefore, this TraI-catalyzed reaction mimics a site-specific recombination at the nick region, and it provides experimental indication that second cleavage by transesterification can occur (W. Pansegrau and E. Lanka, unpublished data) (Figure 5.3). For RP4 TraI, the tyrosine residue Y22 is indeed bound to the 5′ phosphate of the

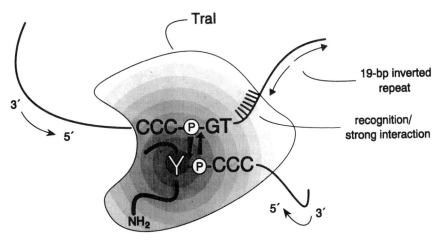

FIGURE 5.3. Model of relaxase action for circularization of transferred DNA by the second cleavage mechanism. The 5′ phosphate of the nucleotide at the nick site is covalently attached to the tyrosyl residue 22 (Y) of RP4 TraI protein. The strand undergoing transfer is constantly scanned for the reconstituted nick region sequence. A transesterification occurs when this sequence is recognized by the TraI protein to yield a closed circular DNA molecule. Single-stranded DNA is drawn as a thin line and the N-terminal portion of the polypeptide chain as a bold line.

deoxycytidylic residue of the nick site. A mutation at Y22 to phenylalanine yields a cleavage-defective TraI protein (W. Pansegrau and E. Lanka, unpublished data).

Thus, the relaxosome possesses the intrinsic capacity to break and reseal its target DNA strand in a reversible reaction without additional cofactors or a conjugative trigger. Other plasmid maintenance functions like vegetative DNA replication and transcription may therefore be undisturbed by the presence of relaxosomal proteins. However, the mechanism by which the relaxosome in vivo is used by the DNA transfer machinery during the conjugative process remains to be elucidated.

Transfer gene products have been described that interact specifically with nucleotide sequences within *oriT* regions, although the proteins are inessential for the cleavage reaction. One of these proteins, TraM of F, is thought to process the signal that a competent mating pair has formed and hence initiate the transfer process (159). The *traM* gene maps adjacent to *oriT*, and the presence of a functional *traM* is essential for the conjugative process (4, 80). TraM protein was found in the cytoplasm and inner membrane, which is consistent with its proposed role as a signaling protein in the initiation of transfer (42). *oriT* binding studies with a crude extract of cells overexpressing TraM protein of $F_o lac$ (IncFV) demonstrated recognition of the cognate *oriT* because the protein interacted with the R1 *oriT* region weakly and not at all with an equivalent fragment of F. TraM protein bound to three large motifs in the *oriT* region containing one inverted repeat, two direct repeats, and the *traM* promoter region. The latter suggests that the protein may have an autoregulatory role in its own expression. Similar results have been obtained from binding studies with purified R1 and R100 TraM protein (2, 136). The size of the protected DNA regions comprises up to 110 nucleotides, indicating that several TraM molecules must participate in binding. The region that binds TraM is the last portion of the plasmid to enter the recipient cell. The TraM binding region does not overlap the TraY binding site(s), although they lie next to each other, separated by only a few base pairs. However, additive influence of both binding regions on the conformation of the *oriT* DNA might be possible.

The TraK product of IncP plasmids is essential to the transfer machinery because *traK* mutants are phenotypically Tra⁻ and the gene belongs to the TraI core, defined as the minimal region of TraI that still allows heterologous mobilization by IncPβ helper plasmids and vice versa. The assumption that the TraK protein binds to *oriT* and recognizes only cognate sequences was verified (170). However, the binding properties turned out to be unusual because high protein-to-DNA ratios were necessary to observe *oriT* fragment-TraK complexes with altered electrophoretic mobilities in a fragment retention assay, and the reduction in mobility was greater than would be expected from other known binding studies. TraK binds cooperatively to DNA in the leading region of RP4, and DNase I and hydroxyl radical-protection studies revealed that a rather large segment of about 150 bp interacts with TraK protein. This contains protected and nonprotected DNA stretches of 10 to 12 bp arranged in an alternating manner. Static bending of this *oriT* region is dramatically increased on TraK interaction, and the complexes can be visualized by electron microscopy, demonstrating that more than one TraK molecule (molecular mass of 14.6 kDa) must be associated with the DNA. Interestingly, complexing with TraK reduced the apparent length of *oriT*-containing DNA fragments by a consistent value of about 180 bp. This may indicate that the DNA is wound around a protein core of several TraK molecules (170). Despite these remarkable DNA-binding properties, the mechanistic role of TraK in the transfer process is unknown. Because the protein is not essential for the relaxation reaction in vivo and in vitro (53, 120, 152), one might speculate that TraK is involved in the circularization reaction.

2.3.4. Sequence Similarities between Different Systems

Some intriguing similarities have been detected recently between the DNA transfer systems of quite different plasmids. One unexpected relationship has been described between the *oriT* region of IncP plasmids and the border sequences of *Agrobacterium* Ti and Ri plasmids (152). The latter are thought to transfer DNA to plant cells by a process resembling bacterial conjugation, the border sequences being direct repeats delimiting the transferred DNA segment (161, 167). Comparison of the IncP nick regions and the Ti/Ri border sequences leads to the extraction of the following 12-nucleotide consensus sequence (121, 152):

$$A\ {{CCT}\atop{TAC}}\ ATCCTG\ {{C}\atop{T}}\ C$$

Resemblances are also found at the amino acid level between Tra and Vir proteins specified by the IncP and Ti plasmid transfer systems, respectively. The RP4 Tra2 region (now defined as the segment between kb coordinates 18.03 to 29.26 on the standard map of RP4) encodes at least 12 proteins (95). One of these, TrbB (KilB), shows 50% positional amino acid identity with VirB11 (112). Five other Tra2 products share sequences with VirB2, B3, B4, B5, and B10. Moreover, relative gene orders are identical (95a). Similarities are also apparent between RP4 TraI products and Vir proteins. Examples are the following pairs: TraI and VirD2, both involved in the specific cleavage process; TraL and VirC1; TraG and VirD4 with 45% identity (171). The TraI Y22 residue forms the phosphodiester with the deoxycytidylic residue at the nick site. The corresponding conserved amino acid position in VirD2 is Y29. A mutation in this position to phenylalanine appears to cause a cleavage defect, as is the case for TraI Y22F (149, W. Pansegrau and E. Lanka, unpublished data). These various similarities at the amino acid level, together with partially identical gene orders and the consensus sequence at nicked regions, provide compelling evidence that parts of the IncP and Ti transfer systems are related by divergent evolution.

Similarities between RP4 Tra2 and Ti VirB extend to the pilus-determining Pil$_W$ region of plasmid R388 (IncW). All three regions constitute one class of transfer operon as judged by (a) extensive sequence similarity of several of the proposed proteins; (b) hydrophobicity, basic character, and signal sequences of predicted products; (c) relative order of similar open reading frames; (d) proposed roles of products in preparing the donor cell envelope for exit of DNA to the recipient cell. The DNA processing genes on R388 include *trwA* and *trwC*. TrwA shares some similarity to RP4 TraJ. However, TrwC is related to F TraI. Presence of a nucleotide-binding motif similar with that of the F TraI helicase domain suggests that TrwC might catalyze duplex unwinding in conjugation as well as nicking (F. de la Cruz, personal communication).

IncI1 plasmids specify an unusually elaborate *tra* system, apparently composed of components from two ancestrally distinct systems. One component confers phenotypes similar to those endowed by IncP and IncM plasmids, whereas the second shows resemblances to F-like plasmids (128, Sections 3.2.2. and 4.). Formation of the R64 (IncI1) relaxosome requires *oriT* and the adjacent *trans*-acting *nikA* and *nikB* genes. These genes are arranged in an operon having the same orientation relative to *oriT* as the RP4 relaxase operon. More significant, NikA protein shares 30% sequence similarity with RP4 TraJ, and only the N-terminal portion of NikB is required for relaxation, as is the case for RP4 TraI protein (55). Relatedness is further indicated by the detection of common sequence motifs in the N-terminal portions of NikB and TraI and by the finding that R64 *oriT* contains a consensus sequence present at the nick sites of IncP plasmids and in the border sequences of Ti/Ri plasmids (121).

Relationship has been detected between RSF1010 and pSC101, which are different and compatible replicons. The *oriT* nick site on these plasmids is within an identical dodeca-nucleotide, and there is 47% sequence identity between the N-terminal regions of the RSF1010 MobA protein and a presumptive pSC101 mobility protein (45).

Another apparently hybrid plasmid is pTF-FC2, isolated from the biomining bacterium *Thiobacillus ferrooxidans*. This mobilizable plasmid has a broad host range and shows interesting relationships to IncQ and IncP plasmids. The minimal pTF-FC2 replicon, consisting of *oriV*, *repA*, and *repC*, shows a high degree of sequence similarity to an IncQ-type replicon (43). However, similarity does not extend to the *mob* regions of these plasmids. Instead, pTF-FC2 *mob* is almost identical to that of the relaxase and leader operons of IncP plasmids. The five *mob* genes (*mobA–E*) correspond to the IncP transfer genes *traI, J, K, L*, and *M* in size and with respect to map locations relative to the adjacent *oriT* locus (130). Common sequence motifs in the corresponding proteins point to an evolutionary relationship between the two transfer systems, each of which is coupled to a different type of replicon. Clearly there is no obligatory coupling between a DNA transfer system and replicon type, and some conjugation systems have apparently evolved by recombinational rearrangement of components from different systems.

Ti plasmids specify, in addition to the Vir transfer system for transmitting DNA from agrobacteria to plant cells, a conjugative Tra system allowing transmission of DNA between agrobacteria. The Vir system is extensively similar to both Tra regions of RP4 (95a, 121, 152, 171), whereas the Ti Tra system shares similarity with RSF1010 *oriT* (31a), and at least one of the Ti *tra* gene products is sequence-related to one of the RSF1010 proteins, namely MobA (S. K. Farrand, personal communication). Despite these relationships, only the Vir system mediates RSF1010 mobilization between agrobacteria; Ti Tra does not (12a, 31a). Lack of mobilization probably reflects a functional incompatibility between systems at the level of processing of the mobilizable plasmid by the conjugative machinery of the self-transmissible plasmid. IncP or Vir-mediated RSF1010 mobilization requires the analogous proteins TraG and VirD4, respectively (12a). Neither TraG nor VirD4 is directly involved in the cleavage process at *oriT*/T-border sequences or in pilus production. TraG is probably located at or in the inner membrane of the donor cell and is thought to act in RP4-directed conjugation to facilitate interaction of the relaxosome with the system directing productive contact between mating cells. Thus, the specificity observed between many mobilization and conjugative systems (160) may reside in the ability of TraG-like proteins to recognize the relaxosome of the mobilizable plasmid as a prelude to DNA export.

2.4. DNA Unwinding

Following nicking at *oriT*, the plasmid must be unwound by one or more DNA helicases as a prerequisite for ssDNA transfer (Figure 5.2). Conjugative plasmids may encode their own helicase for this purpose, as illustrated by the TraI protein of F. This 180-kDa protein, first purified and partially characterized as DNA helicase I by Hoffmann-Berling and colleagues (57, 103), possesses interrelated ssDNA-dependent ATPase and helicase activities. In vitro studies with heterologous DNAs have shown that the enzyme is highly processive and proceeds unidirectionally in the 5' to 3' direction relative to the single strand on which the protein is bound (88, 89). The protein readily aggregates, and it was originally proposed to unwind DNA as a multimer following the binding of some 80 molecules to a single-stranded region adjacent to duplex DNA (88). However, cooperative binding of helicase I is not a prerequisite for the unwinding function (153). Possibly the enzyme is active as a monomer, since substantial unwinding was detected in a reaction involving a 1:1 ratio of helicase I molecules to DNA molecules (89).

Identification of DNA helicase I as TraI protein allows the biochemical properties described previously to be applied to the conjugative process (1). The enzyme is thought to be the primary helicase involved in unwinding the F plasmid in the donor cell, and, in this role, it would migrate on the strand destined for transfer, since this has the necessary 5' to 3' polarity. The rate of unwinding in vitro is in the range of 1 kbp per sec at 37°C (57), which is somewhat in excess of the rate of transfer of chromosomal DNA from Hfr donor cells. The unwinding reaction could provide the motive force for DNA transport, provided that the enzyme is anchored with respect to the cell envelope to allow displacement of the unwound strand relative to the cell boundary. TraI protein is located in the cytoplasmic fraction of the cell (159), so presumably one or more other *tra* gene products provide the hypothetical helicase I anchor. TraD or TraM protein has been suggested for this role (138).

Another conceptual challenge stems from the requirement for a ssDNA region of about 200 nucleotides to initiate helicase I activity on nonspecific substrates in vitro (88). Possibly this region is generated during conjugation by another DNA helicase, such as bacterial Rep protein migrating 3' to 5' on the nontransferred plasmid strand, or the requirement is bypassed in vivo by a coupling of DNA unwinding to the *oriT*- nicking reaction (76, 145). This coupling might be achieved directly by TraI protein, since it is both a DNA helicase and a component of the *oriT*-nicking mechanism (Section 2.3.3.). Appreciation of these dual activities of TraI rationalizes the paradox that, whereas helicase I is active on heterologous DNA substrates in vitro, the *traI* genes of F and the closely related R100 plasmid show system specificity in vitro (157).

Apart from F-like plasmids, no other group of conjugative plasmids has been shown yet to specify a DNA helicase. ColE1 mobilization from F-containing cells is independent of *traI* (157), suggesting that a host DNA unwinding enzyme may function in this process or that ColE1 encodes its own helicase. The *repA* gene of IncQ plasmids encodes a protein with ssDNA-stimulated ATPase and DNA helicase activities, but this gene functions in vegetative DNA replication and is inessential for RSF1010 mobilization in conjugative transfer (41, 68).

Depending on the topology of a plasmid undergoing transfer, efficient DNA unwinding may require the ancillary activity of a DNA topoisomerase. Previous models (159, 160) anticipated that *oriT* nicking relaxes the plasmid entirely. However, this concept may be an oversimplification, since plasmid F retains domains of negative supercoiling following random nicking by irradiation (124). If the conjugatively nicked DNA strand is restrained from free rotation, DNA helicase activity would overwind the duplex ahead of the enzyme, providing a progressive impediment to further unwinding. The problem, analogous to that created by replication fork movement in circular DNA, would be overcome by one or more topoisomerases acting to prevent accumulation of positive torsional stress. Enzymes catalyzing negative supercoiling, such as DNA gyrase, would have this effect (58).

DNA gyrase activity is implicated in conjugation through numerous experiments with nalidixic acid, an inhibitor of the A subunit of the enzyme. The drug is inferred to affect initiation of DNA transfer and also to curtail ongoing DNA transfer (160). While the latter effect is consistent with a role for gyrase in facilitating DNA unwinding, it might also be a consequence of gyrase-inhibitor-DNA ternary complexes known physically to block other aspects of DNA metabolism (58). Such secondary effects of drug treatment are obviated by the use of temperature-sensitive *gyr* mutants. R64 transfer from a *gyrB* (Ts) donor strain was shown to be about 10-fold deficient at 40°C, but it was not possible to identify the stage affected in the process (70). Coumermycin A1, an inhibitor of the B subunit of DNA gyrase, also reduced the efficiency of plasmid transfer from sensitive donor cells, but, in contrast to nalidixic acid, it appeared to function as a reversible inhibitor of Hfr DNA transfer, allowing the process to resume from the point of interruption following removal of the drug (70). This is the response predicted for an inhibitor that impedes conjugative unwinding of DNA.

2.5. Circularization of Transferred DNA

Circularization of the transferred plasmid strand is independent of RecA protein and de novo synthesis of Tra proteins in the recipient cell (160). Current models were stimulated by structural parallels between relaxed plasmid DNA (Section 2.3.2.) and the replication intermediate involved in the formation of ssDNA circles during bacteriophage φX174 multiplication. In this phage system, circularization of ssDNA is mediated by the viral gene A protein linked covalently to the 5' phosphoryl group of the unwound DNA strand (85). Circularization of transferred DNA is therefore envisaged to involve a ligation event that is formally the reverse of *oriT* nicking and mediated by at least the Tra or Mob protein linked to the 5' end of the transferring strand (47, 122, 151). Participating proteins are expected to include the MbeA protein of ColE1, the TraI protein of RP4, and the TraI protein of F (Section 2.3.2.). It has yet to be established whether these DNA-associated proteins are transmitted into the recipient cell or are retained in the donor in the vicinity of the conjugative bridge. The second terminus used for ligation might be the 3'OH group at the trailing end of a monomeric DNA strand, providing a simple circularization mechanism. Alternatively, if this terminus created by the original nick at *oriT* is extended by a rolling-circle mode of DNA synthesis, a second nick at the reconstituted *oriT* sequence would be required to generate a unit length of DNA for circularization (Figures 5.2 and 5.3). Such a mechanism is described here as the second cleavage model.

Cis and *trans* recombination of homologous *oriT* sequences during conjugation provides evidence for such rejoining mechanisms, but the data do not distinguish unequivocally between the two models. *Trans* recombination of *oriT* sequences is manifest by the cointegration of physically independent plasmids during their transfer from the same donor cell. This would occur if the 5' end of one linear transfer intermediate were joined to the 3' end of a second molecule and vice versa (25, 40, 150). *Cis* recombination at *oriT* sites has been studied more extensively and is demonstrated by the production of monomeric plasmids in recipient cells following conjugation with donors carrying a multimeric form of the plasmid. Such conjugative monomerization of plasmids was less than 100% efficient, consistent with an inefficient second cleavage mechanism as well as with a process in which two contiguous *oriT* sites are nicked prior to DNA transfer to provide the substrate for simple circularization (40, 48, 150). The second cleavage model would also explain formation of F' plasmids deleted of the *tra* operon. Such plasmids could be generated during Hfr conjugation if the F nicking-ligation complex recognizes chromosomal sites possessing a similar DNA sequence motif to *oriT* (71).

More elaborate evidence for a second cleavage mechanism is provided by the patterns of transfer-dependent recombination at two copies of the 38-bp R1162 (IncQ) *oriT* sequence cloned as direct repeats in a vector replicon. A large proportion of the plasmid molecules recovered in recipient cells following mobilization contained only one *oriT* region and had lost the intervening DNA between the *oriT* copies. Combination of a normal and a mutant *oriT* region in this system indicated that the initial cleavage reaction and the subsequent rejoining step necessary for recombinant production are partially distinct mechanisms with different sequence requirements at *oriT*. Maybe *oriT* nicking requires supercoiled DNA, whereas the second cleavage occurs in ssDNA. Such a mechanism for breaking and rejoining transferred ssDNA raises the possibility that transfer of IncQ plasmids normally occurs by a rolling-circle mechanism (24, 79).

In accord with this interpretation, site-specific recombination also occurs between two direct repeats of the R1162 *oriT* sequence cloned in a phage M13 vector. This recombination during phage multiplication is strand specific and dependent on an R1162 Mob protein supplied *in trans*. The probable explanation is that a phage replication intermediate provides the substrate for a plasmid-encoded, breaking-resealing system operating at successive *oriT* sites on ssDNA, and that the recombination models circularization of the transferred plasmid strand (10, 110). Such in vivo recombination of directly repeated copies of *oriT* is mediated by a fusion protein containing the N-terminal region of a large Mob protein. This is identifiable as RepB (MobA) (135). The hybrid protein exerts a ssDNA endonuclease activity that cleaves a specific *oriT* strand at the nick site, and it is apparently retained by a fraction of the cut molecules through covalent attachment to the 5' terminus at the cleavage site. After prolonged incubation, some of the cleaved molecules are religated (14). An implication of these findings is that the 3' hydroxyl terminus of the transferring DNA strand is extended by replication or repair synthesis to regenerate a continuous *oriT* sequence. Secondary structure created by the inverted repeats in the extension might allow recognition of the reconstituted *oriT* by MobA protein linked to the leading end of the DNA strand. Inverted repeats in the trailing portion of the *oriT* region of other plasmids, such as RP4 (Figure 5.1), might serve the same function.

While these experiments on *oriT* recombination favor the second cleavage model, there is no unambiguous evidence that circularization involves the processing of a transferring DNA intermediate that is detectably greater than unit length (160). We consider in the next section whether or not DNA transfer is associated with continuous extension of the 3' terminus of the transferring plasmid strand.

3. Conjugative DNA Synthesis

3.1. Replacement Strand Synthesis

The replicative process generating a replacement for the transferred plasmid strand is termed DCDS (donor conjugative DNA synthesis). Current understanding of the process is fragmentary and relies on data from in vivo investigations using temperature-sensitive *dna* mutations to inactivate defined *E. coli* replication proteins and rifampin to inhibit RNA polymerase. The functions of these proteins in DNA replication are reviewed elsewhere (85, 168).

DNA polymerase III holoenzyme is implicated as the enzyme responsible for the elongation reaction in DCDS by the finding that *E. coli dnaE* (Ts) mutants were proficient donors of the F plasmid at high temperature but were defective in supporting DCDS (80). Besides indicating that the replicative DNA polymerase of the donor cell synthesizes the replacement strand, these results show that there is no obligatory coupling between the DNA transport process and DCDS on the template of the retained plasmid strand. In the normal situation, the retained strand is unlikely to accumulate in single-stranded form because the rate of DNA synthesis by DNA polymerase III holoenzyme is about 700 nucleotides per second at 37°C (85), a value similar to the rate of DNA transport.

The nature of the primer(s) initiating DCDS is more obscure. In principle, the 3'-hydroxyl terminus of the strand undergoing transfer might be extended, as predicted by the classic rolling-circle model (59). Such a 3'-hydroxyl terminus flanks the relaxation nick site in F and RP4 DNA isolated by procedures involving SDS and proteinase K (119, 122, 143), but it is unknown whether this terminus is accessible *in vivo*. Indeed, there is indirect evidence that de novo synthesis of an RNA primer is necessary for DCDS. Such a requirement was inferred from experiments based on *dnaB* (Ts) donor strains. These are blocked for vegetative DNA replication but support apparently normal conjugative DNA metabolism, thus providing a system allowing direct measurement of DCDS by thymine incorporation (160).

Requirement of RNA polymerase to synthesize a primer for DCDS was indicated by experiments involving rifampin treatment of *dnaB* donors of plasmid F (80). A putative promoter oriented such that it could provide a primer was identified near F *oriT*, but there is no evidence that this sequence is functional (142) (Figure 5.1). Similar experiments with rifampin-treated *dnaB* donors of an IncI1 plasmid suggested that an untranslated RNA species is needed not only for DCDS of such plasmids but also for their transfer (37). However, these results may reflect indirect effects of rifampin on DNA metabolism in IncI1 plasmid-containing cells (160). A more recent suggestion is that the DNA primase encoded by IncI1 plasmids generates primers for DCDS in a process that is inessential for transfer of the preexisting plasmid strand (30). IncP plasmids also determine a DNA primase, and the RK2 strand retained in the donor cell has sites in the *oriT* region for RK2 primase-directed initiation of DNA synthesis (164). Possibly one or more of these sites normally serve as the origin of replacement strand synthesis in conjunction with the plasmid enzyme (Section 3.2.2.).

These in vivo studies of DCDS involved *dnaB* mutants and therefore precluded activity of a DnaB-dependent priming mechanism involving *E. coli* primase (DnaG protein). One such mechanism is provided by the primosome, which is a complex including DnaB and DnaG proteins and four other gene products (PriA, PriB, PriC, and DnaT). The primosome is assembled at a specific site on ssDNA, which is recognized by PriA, and the protein complex can then translocate in the 5' to 3' direction to allow primase to generate primers at downstream locations. The sequence characteristics of a primosome assembly site (PAS) are unclear. Considerable secondary structure in a region of about 100 nucleotides is important, but the hexanucleotide -AAGCGG-, originally considered to be a consensus sequence, is not a standard feature of the numerous PAS sites now known (85, 102, 115, 168).

The strand of ColE1 retained in the donor cell contains a defined PAS sequence termed *rriA* (Figure 5.1). This site is likely to serve as the origin of lagging strand synthesis during vegetative DNA replication, as discussed by Marians (102). However, it might also function as the origin for DCDS, allowing formation of a primer for the continuous synthesis of the replacement strand in the rightward direction according to Figure 5.1 (160).

3.2. DNA Synthesis on the Transferred Strand

DNA synthesis on the transferred plasmid strand is termed RCDS (recipient conjugative DNA synthesis). Information on the process stems from in vivo investigations, many involving inhibitors of transcription and use of replication-defective mutants of *E. coli*. These studies have identified some of the host proteins active in RCDS, as discussed in the following section. The process does not require expression of plasmid *tra* genes in the newly infected recipient cell (17, 69), but, in some conjugation systems, specific Tra proteins are transported from the donor to the recipient cell to facilitate the priming of this replicative event (Section 3.2.2.). Primer removal and ligation of nascent DNA as a prelude to the formation of negatively supercoiled DNA are presumably mediated by replicative enzymes of the host cell.

3.2.1. Host Enzymes

The conjugative properties of *E. coli dnaE* recipients indicate that the replicative DNA polymerase of the recipient cell synthesizes DNA on the transferred plasmid strand (160). An appealing model is that this DNA synthesis occurs concurrently with DNA transfer, thereby preventing accumulation of ssDNA in the recipient cell. Concurrent DNA synthesis on a strand transferred in the 5' and 3' direction would involve de novo synthesis of multiple primers, and, by analogy with the size distribution of Okazaki fragments formed in DNA replication, these de novo starts would occur at 1- to 2-kb intervals (85).

The effects of rifampicin and DnaB inactivation on RCDS of plasmid F suggested that primers are generated by RNA polymerase or some DnaB-dependent process acting as alternative mechanisms (160). Likewise, more than one primer-generating mechanism can act on the transferred strand of IncI1 and IncP plasmids. The DNA primases encoded by these plasmids are clearly implicated in RCDS, but because primase-defective mutants were found to retain a nearly normal conjugative potential in *E. coli* matings, some bacterial primer-generating mechanism can substitute for the plasmid enzyme in RCDS (30, 91).

Activity of a primosome is an attractive hypothesis for the initiation of RCDS by host enzymes. ColE1 contains a functional PAS sequence (PAS-CH or *rriB*) on the strand assumed to be transferred during conjugation (15, 114, 172). The site has a role in vegetative replication, and it might also function in RCDS (102). While PAS-CH is close to the relaxation nick site, it is in the region transferred late to the recipient cell (Figure 5.1). Thus it could only function in RCDS following transfer of a monomeric length of ColE1, possibly acting after the circularization of this incoming DNA strand.

A computer-aided search for putative PAS sites in the leading region of the transferred strand of F has identified a segment with secondary structure and three sequences showing 5/6 matches to -AAGCGG- (see Section 3.1.). This segment is about 2.5 kb from *oriT* (98). Functional tests are needed to determine whether the DNA segment has any significance in primosome activity. Such tests have recently shown that the transferable strand of F does contain a single-stranded initiation sequence, designated *ssiE* (F·f6). The sequence, requiring about 200 nucleotides for full activity, is located in the leading region on the *oriT* distal side of ORF169 (gene *X*) (Figure 5.1). The *ssiE* signal is thought to allow primase (DnaG) to synthesize a unique primer by a primosome-independent, phage G4-type mechanism (115). If *ssiE* acts in RCDS, the primer will initiate DNA synthesis toward the nearby *oriT* site.

3.2.2. Plasmid DNA Primases

Some conjugative plasmids encode their own DNA primase for the initiation of RCDS. These enzymes were recognized by their ability to synthesize functional primers on ssDNA from small bacteriophages and to substitute for bacterial primase during DNA replication in *dnaG* (Ts) mutants of *E. coli* (93, 154, 155). By these criteria, about one-third of the Inc groups studied in *E. coli* were found to specify a DNA primase (160). The set includes representatives of the I complex of incompatibility groups (B, I1, I2, and K) and members of the IncC, J, M, P, and U groups. Other plasmids may encode a DNA primase with greater template specificity. The RepB protein of IncQ plasmids provides such an example, but there is no evidence that the primase activity of this protein functions in conjugation (68). There is no clear-cut correlation between carriage of a DNA primase gene and any other plasmid phenotype, including host range. However, most of the identified primase-determining plasmids specify a rigid rather than a flexible conjugative pilus.

The best characterized plasmid DNA primases are those determined by the *sog* gene of ColIb and by the *pri* (*traC*) gene of RP4. Each gene determines two polypeptides as separate, in-frame, translation products differing at the N-terminus. Molecular mass values are 210 kDa and 160 kDa for Sog proteins (109, 155) and 118 kDa and 80 kDa for Pri proteins (91, 94). Judged from the properties of mutants, DNA primase activity is determined by a domain in the unique N-terminal region of the larger Sog polypeptide (18) but by a region common to the C-terminal portion of both Pri polypeptides (94, 140). Nucleotide sequence comparisons indicate that *pri*, the 5′ region of *sog*, and the *E. coli* *dnaG* gene are apparently unrelated (140).

However, the primase domains of RP4 and R751 Pri proteins (107) and of the larger Sog polypeptide contain the common motif of -EGYATA-. Interestingly, the same motif is present in the primase domain of the α protein of the *E. coli* satellite phage P4 (50), and also in a rudimentary form in all known sequences of prokaryotic DNA primases including *E. coli* DnaG protein. Depending on the residue altered, mutation of this motif in the smaller RP4 Pri primase polypeptide (TraC2) and the P4 α protein can increase, decrease, or eliminate enzymatic activity, indicating that amino acids E, Y, and T are probably part of the catalytic center (140).

The biochemical properties of Pri and Sog DNA primases were recently reviewed in detail (66). Both enzymes can use a similar diversity of DNA templates in vitro, and they synthesize oligoribonucleotide primers ranging in size from 2 to 10 nucleotides. These primers contain cytidine or cytidine 5′ monophosphate at the 5′ terminus with AMP as the second nucleotide (92). This distinguishes them from primers made by DnaG protein and by the DNA primases specified by bacteriophages T4 and T7, which contain a purine triphosphate as the initiating nucleotide.

The *pri* and *sog* genes map in the Tra1 region of RP4 and in the Tra2 region of ColIb, respectively (91, 128). Despite these locations, primase-negative mutants of both plasmids exhibit no consistently strong defects in conjugation. While *pri* mutants of RP4 and of R18 showed normal conjugative proficiency in matings of *E. coli* strains and of *Pseudomonas aeruginosa*, respectively, they were conjugation defective in some interspecies matings, depending on the organism used as the recipient. Back transfers to the donor host strain were unaffected by *pri* mutation (87, 91, 113). These properties suggest that IncP plasmid primase functions in RCDS but that the primer-generating enzymes native to some bacteria can substitute effectively. A complementary approach to studying the role of plasmid DNA primases has involved measurement of RCDS in different genetic backgrounds. This procedure has shown that Sog primase of ColIb initiates DNA synthesis on the transferred ColIb strand but that primase of the recipient cell can substitute in this process (30).

Such conjugative deficiencies of RP4- and ColIb-primase mutants were complemented when the recipient or, more significantly, the donor cell carried the cognate *pri*$^+$ or *sog*$^+$ gene on a nontransmissible cloning vector (30, 108, 113). The implication is that plasmid DNA primases or their RNA products are provided by the donor cell and are transferred to the recipient, where they initiate RCDS.

There is direct evidence that both Sog proteins are transmitted into the recipient cell without apparent processing (109, 126) (Figure 5.2). Transmission of the proteins required an active DNA transfer system and occurred in the same direction as DNA transport with an estimated stoichiometry of 250 molecules of each Sog polypeptide per transferred plasmid strand. Transferred Sog polypeptides were found in the cytoplasmic fraction of the recipient cells. This is consistent with the proposed role of plasmid primases in RCDS and with a functional test showing that Sog primase is transferred from the donor cell to a location in the recipient where it can act in bacterial DNA replication (29). Likewise, it was demonstrated that the larger Pri protein is transferred to the cytoplasm of the recipient cell in RP4-mediated matings. No transfer of the 80-kDa Pri polypeptide was observed, and it might be retained in the donor cell (127). Possibly this smaller Pri protein functions in DCDS (Section 3.1.).

The amino acid sequences of Pri proteins and of the N-terminal region of the larger Sog protein lack a typical signal sequence, indicating that their transfer occurs by a process other than the classical protein export pathway (140). Plasmid DNA primases bind to ssDNA (91, 155), and they are presumably transmitted in a complex with the transferred plasmid strand. If the proteins are distributed on this DNA, the polypeptides would provide an efficient mechanism for initiating complementary strand synthesis as conjugative transfer proceeds. Such a priming system would render initiation of RCDS independent of host enzymes, and this may represent the contribution of *pri* to the conjugative promiscuity of IncP plasmids (86, 91).

Despite enzymatic similarities, Pri and Sog polypeptides are functionally distinguishable. Each is specific for the cognate conjugation system (108). More significant, the properties of transposon-insertion mutants indicate that, whereas Pri proteins are unnecessary for RP4 DNA transfer itself, the C-terminal region of Sog polypeptides contributes to IncI1 plasmid transport at some stage following cell aggregation and *oriT* nicking (91, 109).

Genetic evidence implies that this second function of Sog polypeptides in promoting DNA transfer is more important in conjugation than Sog primase activity, which is partially substitutable by host DNA replication enzymes (30, 109).

Plasmid F and its relatives fail to specify a DNA primase detectable in assays involving nonhomologous DNA templates, but these plasmids might encode a replicon-specific enzyme (160). The F TraI protein was hypothesized to determine such a DNA primase activity on the basis of the similar genetic organization of *traI*, *sog*, and *pri*, and the functional link between replicative DNA helicases and primases (144). However, no priming has been detected using purified F TraI protein and single-stranded DNA templates (S. W. Matson, personal communication). Furthermore, no transfer of any TraI-related protein was detected in F-mediated conjugation, and no F-specified protein was observed to escort the transferred DNA from the donor to the cytoplasm of the recipient cell (127). The implication is that F lacks a DNA primase gene, and this is consistent with the conclusion that RCDS of the plasmid is mediated by host enzymes (Section 3.2.1.).

These comparative studies indicate that diverse strategies are exploited to promote initiation of DNA synthesis on the transferred plasmid strand. The DNA primase of RP4 clearly contributes to plasmid promiscuity at the level of conjugation. However, the carriage of a DNA primase gene is not diagnostic of a broad host range plasmid, since IncI1 plasmids are apparently maintained as autonomous entities in only a limited set of enterobacteria. For these plasmids, DNA primase activity appears to be an ancillary function of a large polypeptide that contributes to the DNA transport process. Further analysis of these transferred primase proteins is required to identify domains that facilitate their transport across the cell envelopes of conjugating cells and their interaction with DNA. It is also possible that these proteins have other functions in the recipient cell, such as directing the incoming DNA to an assembly of host replication enzymes, as in the pilot protein concept (85), and protecting the immigrant plasmid from nucleases. This might explain the findings that R18 *pri*::Tn7 mutants were severely defective in *P. aeruginosa* to *Pseudomonas stutzeri* conjugation yet showed normal conjugative proficiency in homologous matings of *P. stutzeri* strains (113). These results are difficult to reconcile if *pri* contributes solely to the priming of RCDS.

4. Leading Region Functions

The leading region of a plasmid is the first segment to enter the recipient cell during conjugation and it may carry determinants called installation genes that promote establishment of the immigrant plasmid in the new host cell. The leading region of the F plasmid is defined as the 13 kb of DNA between *oriT* at 66.7F and the primary replication region, RepFIA. The majority of the transcriptional activity in this segment of F occurs within the sequences 59.4 to 66.7F, which encode six known polypeptides (36, 97). Genes identified in this portion of the leading region include *flm* of the *hok* gene family, which specifies a maintenance function; *psiB*, which determines plasmid SOS inhibition; and *ssb*, which encodes a single-stranded DNA binding protein (SSB) of the type that binds cooperatively to DNA with no sequence specificity (7, 82, 97).

Although the leading region of F is dispensable for bacterial conjugation (98), *ssb* and *psiB* are thought to play ancillary roles in DNA metabolism in conjugating cells. Circumstantial evidence is the presence of sequences homologous to *ssb* and *psiB* on conjugative plasmids from nine different Inc groups (61, 63). While most of these Inc groups belong to the F complex and the Il-B-K subset of the I complex, it is striking that the *ssb* and *psiB* homologues map in the same relative orientation within the leading region on representatives of both complexes of plasmids (46, 73, 78, 163). A more significant pointer to the role of this conserved region is the increase of plasmid SSB activity detected when the Tra systems of IncFII and IncI1 plasmids are derepressed (62, 73). Use of a promoter probe has confirmed that expression of ColIb *ssb* is regulated by the fertility inhibition system and, more intriguingly, by a second mechanism allowing zygotic induction of the gene in the recipient cell following conjugation (78).

Plasmid *ssb* genes are homologous to *E. coli ssb*, and they complement known defects of bacteria carrying an *ssb* (Ts) mutation, indicating shared functions (61, 62). *E. coli* SSB has a central role in DNA replication (85). In vitro, the protein prevents reassociation of unwound DNA, protects DNA from nucleolytic cleavage, confers specificity on priming mechanisms, and stimulates DNA polymerase III holoenzyme (85). The SSBs determined by *E. coli*, plasmid F, and ColIb have a similar molecular mass, and the N-terminal two-thirds of the proteins are strikingly similar (28, 73). This portion of *E. coli* SSB contains the DNA binding domain and residues important for monomer: monomer interactions (27).

The biological significance of plasmid SSBs remains unclear. The regulatory patterns already described imply that the proteins function in the conjugative metabolism of DNA in the recipient cell. However, *ssb* mutants of F-like and IncI1 plasmids showed no obvious defect in conjugation, and tests with *ssb* (Ts) strains of *E. coli* have so far proved to be uninstructive (62, 73). Furthermore, there is no evidence that SSBs coat the transferred DNA during its passage to the recipient cell (127). Possibly plasmid *ssb* genes function to prevent depletion of SSB reserves in conjugating cells, and there may be a more stringent requirement for their products in bacteria other than *E. coli* K-12 or in conditions differing from those routinely used in the laboratory.

The *psiB* locus specifies a polypeptide of ~ 12 kDa that inhibits the bacterial SOS response as a function of its intracellular concentration (7, 9, 46). SOS induction requires activation of the coprotease function of RecA protein by a signal inferred to be ssDNA. PsiB protein is thought to exert its inhibitory effect by direct interaction with activated RecA protein (9, A. Bailone and R. Devoret, personal communication). A gene designated *psiA* is located immediately downstream of *psiB*; *psiA* is inessential for the Psi function, and its role has yet to be determined (7, 99).

Manifestation of the Psi phenotype by established plasmid-containing strains requires overexpression of *psiB*. This state is achieved when *psiB* is cloned on a multicopy plasmid or there is an insertion upstream of *psiB*, as found on natural plasmids such as R6-5 and on artificially constructed derivatives of ColIb (46, 63, 73, 78). In conjugation, expression of *psiB* on ColIb and F is strongly but transiently induced in the infected recipient cell, giving levels of expression sufficient to cause the Psi phenotype. The implication is that *psiB* genes facilitate installation of the plasmid in the new host by preventing expression of the SOS response. Consistent with this hypothesis is a sixfold increase in activity of an SOS reporter gene following conjugation mediated by a ColIb*drd psiB* mutant. Carriage of *psiB*+ on the plasmid prevents such SOS induction. An attractive possibility is that the burst of PsiB synthesis in the conjugatively infected cell prevents SOS induction by progressive transfer of ssDNA. However, plasmids carrying a *psiB* gene generally possess a replicon of the

RepFIC (or RepFIIA) family. This correlation could reflect a role for PsiB in the initial replication of the immigrant plasmid (8, 78).

The ColIb leading region also contains an antirestriction gene designated *pra* or *ard* (alleviation of restriction of DNA) (38, 125). The gene is located between *oriT* and *psiB* and, like other known genes in the leading region, its activity is increased by derepression of the I1 transfer system. Ard alleviates restriction of DNA by the three known families of type I restriction-modification system, including *Eco*K. Transmission of unmodified ColIb*drd* by conjugation is remarkably resistant to *Eco*K restriction in the recipient cell, in contrast to the sensitivity of plasmid transfer by transformation. Genetic studies have indicated that *ard* contributes to conjugative evasion of *Eco*K restriction and that the process requires Ard product expressed in the recipient by the transferred plasmid. This raises the timing problem of how Ard accumulates before the immigrant plasmid is destroyed by the restriction enzyme. Possibly the route of DNA entry during conjugation allows Ard production before the plasmid becomes accessible to a type I restriction enzyme. These enzymes have been found only in *E. coli* and relatives, which include the organisms known to maintain ColIb. Thus *ard* may be an adaptation that facilitates transfer of the plasmid between its different natural hosts (125). Sequences homologous to ColIb *ard* are detectable on plasmids belonging to several different incompatibility groups (P. Chilley and B. M. Wilkins, unpublished data). One such plasmid is pKM101 (IncN), which carries a functional *ard* gene in the leading region (12b).

Transcription of *ssb*, *psiB*, *psiA*, and *ard* occurs on the same DNA strand as that transferred during conjugation, and each gene has a nearby promoter. The possibility that transferred leading region genes can be transcribed before they are converted into duplex form by complementary strand synthesis warrants consideration.

The leading region of RP4 has no known homology to the equivalent part of plasmids in the set carrying related *ssb* and *psiB* genes. Moreover, unlike plasmid F, there are some essential *tra* genes in the leading region of RP4. Three of these genes (*traK*, *traL*, and *traM*) comprise the leader operon expressed from the promoter in the *oriT* region (P_{R1}) (Figure 5.1). Downstream of this operon is *traN*, which is close to one terminus of the TraI region (66). The biochemical function of TraN protein is unknown, but it appears to be another accessory protein that facilitates conjugative processing of DNA in certain recipient cell backgrounds (86, 134).

5. Conclusions

Extension of molecular studies to plasmids outside the F complex has proved to be fruitful in elucidating basic principles and mechanistic diversity. The recent development of simple assays of *oriT* cleavage has made it feasible to identify and purify some of the proteins involved. This has led to reconstitution of transfer initiation complexes that will form the core of in vitro systems mimicking the conjugative processing of DNA. There remain some challenging questions.

- What is the physical relationship between the relaxosome and the basal structure of the conjugative pilus?
- What is the molecular signal that triggers *oriT* cleavage and DNA export?

- What is the nature of the DNA transport bridge?
- Where is the 5' terminus of the transferring DNA strand located?
- Is DNA transport associated with a rolling-circle mode of DNA synthesis in the donor cell? Resolution of this question is important for elucidating the process that circularizes the transferred DNA.
- What is the precise function of leading-region genes that appear to play an accessory role in installing the plasmid in the new cell?

To address these problems, genetic and biochemical studies with organisms other than *E. coli* K-12 might be required.

ACKNOWLEDGMENTS. We thank numerous colleagues for supplying personal communications and preprints. We are grateful for the secretarial assistance of Joan Warriner and Renate Spann. E. Lanka thanks Heinz Schuster for generous support and stimulating discussions. E. Lanka's work was supported by Sonderforschungsbereich grant 344/B2 of the Deutsche Forschungsgemeinschaft.

References

1. Abdel-Monem, M., Taucher-Scholz, G., and Klinkert, M.-Q., 1983, Identification of *Escherichia coli* DNA helicase I as the *traI* gene product of the F sex factor, *Proc. Natl. Acad. Sci. USA* **80**:4659–4663.
2. Abo, T., Inamoto, S., and Ohtsubo, E., 1991, Specific DNA binding of the TraM protein to the *oriT* region of plasmid R100, *J. Bacteriol.* **173**:6347–6354.
3. Achtman, M., Morelli, G., and Schwuchow, S., 1978, Cell-cell interactions in conjugating *Escherichia coli*: role of F pili and fate of mating aggregates, *J. Bacteriol.* **135**:1053–1061.
4. Achtman, M., Willetts, N., and Clark, A. J., 1972, Conjugational complementation analysis of transfer-deficient mutants of F*lac* in *Escherichia coli*, *J. Bacteriol.* **110**:831–842.
5. Al-Doori, Z., Watson, M., and Scaife, J., 1982, The orientation of transfer of the plasmid RP4, *Genet. Res. Camb.* **39**:99–103.
6. Archer, J. A. K., 1985, Sequence analysis of plasmid ColK, Ph.D. thesis, University of Glasgow, United Kingdom.
7. Bagdasarian, M., Bailone, A., Bagdasarian, M. M., Manning, P. A., Lurz, R., Timmis, K. N., and Devoret, R., 1986, An inhibitor of SOS induction specified by a plasmid locus in *Escherichia coli*, *Proc. Natl. Acad. Sci. USA* **83**:5723–5726.
8. Bagdasarian, M., Bailone, A., Angulo, J. F., Scholz, P., Bagdasarian, M., and Devoret, R., 1992, PsiB, an anti-SOS protein, is transiently expressed by the F sex factor during its transmission to an *Escherichia coli* K-12 recipient, *Mol. Microbiol.* **6**:885–893.
9. Bailone, A., Bäckman, A., Sommer, S., Célérier, J., Bagdasarian, M. M., Bagdasarian, M., and Devoret, R., 1988, PsiB polypeptide prevents activation of RecA protein in *Escherichia coli*, *Mol. Gen. Genet.* **214**:389–395.
10. Barlett, M. M., Erickson, M. J., and Meyer, R. J., 1990, Recombination between directly repeated origins of conjugative transfer cloned in M13 bacteriophage DNA models ligation of the transferred plasmid strand, *Nucl. Acids Res.* **18**:3579–3586.
11. Barth, P. T., Tobin, L., and Sharpe, G. S., 1981, Development of broad host-range plasmid vectors, in: *Molecular Biology, Pathogenicity, and Ecology of Bacterial Plasmids* (S. B. Levy, R. C. Clowes, and E. L. Koenig, eds.), Plenum Press, New York,. pp. 439–448.
12. Bastia, D., 1978, Determination of restriction sites and the nucleotide sequence surrounding the relaxation site of ColE1, *J. Mol. Biol.* **124**:601–639.
12a. Beijersbergen, A., Den Dulk-Ras, A., Schilperoort, R. A., and Hooykaas, P. J. J., 1992, Conjugative transfer by the virulence system of *Agrobacterium tumefaciens*, *Science* **256**:1324–1327.

12b. Belogurov, A. A., Delver, E. P., and Rodzevich, O. V., 1992, IncN plasmid pKM101 and IncI1 plasmid ColIb-P9 encode homologous antirestriction proteins in their leading regions, *J. Bacteriol.* **174:**5079–5085.

13. Bernardi, A., and Bernardi, F., 1984, Complete sequence of pSC101, *Nucl. Acids Res.* **12:**9415–9426.

14. Bhattacharjee, M. K., and Meyer, R. J., 1991, A segment of a plasmid gene required for conjugal transfer encodes a site-specific, single-strand DNA endonuclease and ligase, *Nucl. Acids Res.* **19:**1129–1137.

15. Böldicke, T. W., Hillenbrand, G., Lanka, E., and Staudenbauer, W. L., 1981, Rifampicin-resistant initiation of DNA synthesis on the isolated strands of ColE plasmid DNA, *Nucl. Acids Res.* **9:**5215–5231.

16. Bolland, S., Llosa, M., Avila, P., and de la Cruz, F., 1990, General organization of the conjugal transfer genes of the IncW plasmid R388 and interactions between R388 and IncN and IncP plasmids, *J. Bacteriol.* **172:**5795–5802.

17. Boulnois, G. J., and Wilkins, B. M., 1978, A ColI-specified product, synthesized in newly infected recipients, limits the amount of DNA transferred during conjugation of *Escherichia coli* K-12, *J. Bacteriol.* **133:**1–9.

18. Boulnois, G. J., Wilkins, B. M., and Lanka, E., 1982, Overlapping genes at the DNA primase locus of the large plasmid ColI, *Nucl. Acids Res.* **10:**855–869.

19. Bowie, J. U., and Sauer, R. T., 1990, TraY proteins of F and related episomes are members of the Arc and Mnt repressor family, *J. Mol. Biol.* **211:**5–6.

20. Boyd, A. C., and Sherratt, D. J., 1986, Polar mobilization of the *Escherichia coli* chromosome by the ColE1 transfer origin, *Mol. Gen. Genet.* **203:**496–504.

21. Boyd, A. C., Archer, J. A. K., and Sherratt, D. J., 1989, Characterization of the ColE1 mobilization region and its protein products, *Mol. Gen. Genet.* **217:**488–498.

22. Bradshaw, H. D., Jr., Traxler, B. A., Minkley, E. G., Jr., Nester, E. W., and Gordon, M. P., 1990, Nucleotide sequence of the *traI* (helicase I) gene from the sex factor F, *J. Bacteriol.* **172:**4127–4131.

23. Brasch, M. A., and Meyer, R. J., 1986, Genetic organization of plasmid R1162 DNA involved in conjugative mobilization, *J Bacteriol.* **167:**703–710.

24. Brasch, M. A., and Meyer, R. J., 1987, A 38 base-pair segment of DNA is required in *cis* for conjugative mobilization of broad-host-range plasmid R1162, *J. Mol. Biol.* **198:**361–369.

25. Broome-Smith, J., 1980, *RecA* independent, site-specific recombination between ColE1 or ColK and a miniplasmid they complement: implications for the mechanism of DNA transfer during mobilization, *Plasmid* **4:**51–63.

26. Chan, P. T., Ohmori, H., Tomizawa, J., and Lebowitz, J., 1985, Nucleotide sequence and gene organization of ColE1 DNA, *J. Biol. Chem.* **260:**8925–8935.

27. Chase, J. W., and Williams, K. R., 1986, Single-stranded DNA binding proteins required for DNA replication, *Annu. Rev. Biochem.* **55:**103–136.

28. Chase, J. W., Merrill, B. M., and Williams, K. R., 1983, F sex factor encodes a single-stranded DNA binding protein (SSB) with extensive sequence homology to *Escherichia coli* SSB, *Proc. Natl. Acad. Sci. USA* **80:**5480–5484.

29. Chatfield, L. K., and Wilkins, B. M., 1984, Conjugative transfer of IncI₁ plasmid DNA primase, *Mol. Gen. Genet.* **197:**461–466.

30. Chatfield, L. K., Orr, E., Boulnois, G. J., and Wilkins, B. M., 1982, DNA primase of plasmid ColIb is involved in conjugal DNA synthesis in donor and recipient bacteria, *J. Bacteriol.* **152:**1188–1195.

31. Clewell, D. B., and Helinski, D. R., 1969, Supercoiled circular DNA-protein complex in *Escherichia coli*: purification and induced conversion to an open circular DNA form, *Proc. Natl. Acad. Sci. USA* **62:**1159–1166.

31a. Cook, D. M., and Farrand, S. K., 1992, The *oriT* region of *Agrobacterium tumefaciens* Ti plasmid pTiC58 shares DNA sequence identity with the transfer origins of RSF1010 and RK2/RP4 and with T-region borders, *J. Bacteriol.* **174:**6238–6246.

32. Coupland, G. M., 1984, The conjugation system and insertion sequences of the IncN plasmid R46, Ph.D. thesis, University of Edinburgh, United Kingdom.

33. Coupland, G. M., Brown, A. M. C., and Willetts, N. S., 1987, The origin of transfer (*oriT*) of the conjugative plasmid R46: characterization by deletion analysis and DNA sequencing, *Mol. Gen. Genet.* **208:**219–225.

34. Couturier, M., Bex, F., Bergquist, P. L., and Maas, W. K., 1988, Identification and classification of bacterial plasmids, *Microbiol. Rev.* **52:**375–395.

35. Craig, N. L., and Nash, H. A., 1984, *E. coli* integration host factor binds to specific sites in DNA, *Cell* **39:**707–716.

36. Cram, D., Ray, A., O'Gorman, L., and Skurray, R., 1984, Transcriptional analysis of the leading region in F plasmid DNA transfer, *Plasmid* **11:**221–233.

37. Curtiss R., III, and Fenwick R. G., Jr., 1975, Mechanism of conjugal plasmid transfer, in: *Microbiology– 1974* (D. Schlessinger, ed.), American Society for Microbiology, Washington DC, pp. 156–165.

38. Delver, E. P., Kotova, V. U., Zavilgelsky, G. B., and Belogurov, A. A., 1991, Nucleotide sequence of the gene (*ard*) encoding the antirestriction protein of plasmid ColIb-P9, *J. Bacteriol.* **173:**5887–5892.

39. Dempsey, W. B., 1987, Integration host factor and conjugative transfer of the antibiotic resistance plasmid R100, *J. Bacteriol.* **169:**4391–4392.

40. Derbyshire, K. M., and Willetts, N. S., 1987, Mobilization of the non-conjugative plasmid RSF1010: a genetic analysis of its origin of transfer, *Mol. Gen. Genet.* **206:**154–160.

41. Derbyshire, K. M., Hatfull, G., and Willetts, N., 1987, Mobilization of the non-conjugative plasmid RSF1010: a genetic and DNA sequence analysis of the mobilization region, *Mol. Gen. Genet.* **206:**161–168.

42. Di Laurenzio, L., Frost, L. S., Finlay, B. B., and Paranchych, W., 1991, Characterization of the *oriT* region of the IncFV plasmid pED208, *Mol. Microbiol.* **5:**1779–1790.

43. Dorrington, R. A., and Rawlings, D. E., 1990, Characterization of the minimum replicon of the broad-host-range plasmid pTF-FC2 and similarity between pTF-FC2 and the IncQ plasmids, *J. Bacteriol.* **172:**5697–5705.

44. Drolet, M., and Lau, P. C. K., 1992, Mobilization protein-DNA binding and divergent transcription at the transfer origin of the *Thiobacillus ferrooxidans* pTF1 plasmid, *Mol. Microbiol.* **6:**1061–1071.

45. Drolet, M., Zanga, P., and Lau, P. C. K., 1990, The mobilization and origin of transfer regions of a *Thiobacillus ferrooxidans* plasmid: relatedness to plasmids RSF1010 and pSC101, *Mol. Microbiol.* **4:**1381–1391.

46. Dutreix, M., Bäckman, A., Célérier, J., Bagdasarian, M. M., Sommer, S., Bailone, A., Devoret, R., and Bagdasarian, M., 1988, Identification of *psiB* genes on plasmids F and R6-5. Molecular basis for *psiB* enhanced expression in plasmid R6-5, *Nucl. Acids Res.* **16:**10669–10679.

47. Everett, R., and Willetts, N., 1980, Characterization of an in vivo system for nicking at the origin of conjugal DNA transfer of the sex factor F, *J. Mol. Biol.* **136:**129–150.

48. Everett, R., and Willetts, N., 1982, Cloning, mutation and location of the F origin of conjugal transfer, *EMBO J.* **1:**747–753.

49. Finlay, B. B., Frost, L. S., and Paranchych, W., 1986, Origin of transfer of IncF plasmids and nucleotide sequences of the type II *oriT*, *traM*, and *traY* alleles from ColB4-K98 and the type IV *traY* allele from R100-1, *J. Bacteriol.* **168:**132–139.

50. Flensburg, J., and Calendar, R., 1987, Bacteriophage P4 DNA replication, nucleotide sequence of the P4 replication gene and the *cis* replication region, *J Mol. Biol.* **195:**439–445.

51. Fu, Y-H. F., Tsai, M.-M., Luo, Y., and Deonier, R. C., 1991, Deletion analysis of the F plasmid *oriT* locus, *J. Bacteriol.* **173:**1012–1020.

52. Fürste, J. P., Ziegelin, G., Pansegrau, W., and Lanka, E., 1987, Conjugative transfer of promiscuous plasmid RP4: plasmid-specified functions essential for formation of relaxosomes, in: *DNA Replication and Recombination* (R. McMacken and T. J. Kelly, eds.), UCLA Symposia on Molecular Cell Biology, New Series Vol. 47, Alan R. Liss, New York, pp. 553–564.

53. Fürste, J. P., Pansegrau, W., Ziegelin, G., Kröger, M., and Lanka, E., 1989, Conjugative transfer of promiscuous IncP plasmids: interaction of plasmid-encoded products with the transfer origin, *Proc. Natl. Acad. Sci USA* **86:**1771–1775.

54. Furuya, N., and Komano, T., 1991, Determination of the nick site at *oriT* of IncI1 plasmid R64: global similarity of *oriT* structures of IncI1 and IncP plasmids, *J. Bacteriol.* **173:**6612–6617.

55. Furuya, N., Nisioka, T., and Komano, T., 1991, Nucleotide sequence and functions of the *oriT* operon in IncI1 plasmid R64, *J. Bacteriol.* **173:**2231–2237.

56. Gamas, P., Caro, L., Galas, D., and Chandler, M., 1987, Expression of F transfer functions depends on the *Escherichia coli* integration host factor, *Mol. Gen. Genet.* **207:**302–305.

57. Geider, K., and Hoffmann-Berling, H., 1981, Proteins controlling the helical structure of DNA, *Annu. Rev. Biochem.* **50:**233–260.

58. Gellert, M., 1981, DNA topoisomerases, *Annu. Rev. Biochem.* **50:**879–910.

59. Gilbert, W., and Dressler, D., 1968, DNA replication: the rolling circle model, *Cold Spring Harbor Symp. Quant. Biol.* **33:**473–484.
60. Göldner, A., Graus, H., and Högenauer, G., 1987, The origin of transfer of P307, *Plasmid* **18:**76–83.
61. Golub, E. I., and Low, K. B., 1985, Conjugative plasmids of enteric bacteria from many different incompatibility groups have similar genes for single-stranded DNA-binding proteins, *J. Bacteriol.* **162:**235–241.
62. Golub, E. I., and Low, K. B., 1986, Derepression of single-stranded DNA-binding protein genes on plasmids derepressed for conjugation, and complementation of an *E. coli ssb⁻* mutation by these genes, *Mol. Gen. Genet.* **204:**410–416.
63. Golub, E., Bailone, A., and Devoret, R., 1988, A gene encoding an SOS inhibitor is present in different conjugative plasmids, *J. Bacteriol.* **170:**4392–4394.
64. Grinter, N., 1981, Analysis of chromosome mobilization using hybrids between plasmid RP4 and a fragment of bacteriophage λ carrying IS*1*, *Plasmid* **5:**267–276.
65. Gruss, A., and Ehrlich, S. D., 1989, The family of highly interrelated single-stranded deoxyribonucleic acid plasmids, *Microbiol. Rev.* **53:**231–241.
66. Guiney, D. G., and Lanka, E., 1989, Conjugative transfer of IncP plasmids, in: *Promiscuous Plasmids of Gram-Negative Bacteria* (C. M. Thomas, ed.), Academic Press, London, pp. 27–56.
67. Guiney, D. G., and Yakobson, E., 1983, Location and nucleotide sequence of the transfer origin of the broad host range plasmid RK2, *Proc. Natl. Acad. Sci. USA* **80:**3595–3598.
68. Haring, V., and Scherzinger, E., 1989, Replication proteins of the IncQ plasmid RSF1010, in: *Promiscuous Plasmids of Gram-Negative Bacteria* (C. M. Thomas, ed.), Academic Press, London, pp. 95–124.
69. Hiraga, S., and Saitoh, T., 1975, F deoxyribonucleic acid transferred to recipient cells in the presence of rifampin, *J. Bacteriol.* **121:**1000–1006.
70. Hooper, D. C., Wolfson, J. S., Tung, C., Souza, K. S., and Swartz, M. N., 1989, Effects of inhibition of the B subunit of DNA gyrase on conjugation in *Escherichia coli*, *J. Bacteriol.* **171:**2235–2237.
71. Horowitz, B., and Deonier, R. C., 1985, Formation of Δ *tra* F' plasmids: specific recombination at *oriT*, *J. Mol. Biol.* **108:**267–274.
72. Howland, C. J., and Wilkins, B. M., 1988, Direction of conjugative transfer of IncI1 plasmid ColIb-P9, *J. Bacteriol.* **170:**4958–4959.
73. Howland, C. J., Rees, C. E. D., Barth, P. T., and Wilkins, B. M., 1989, The *ssb* gene of plasmid ColIb-P9, *J. Bacteriol.* **171:**2466–2473.
74. Inamoto, S., and Ohtsubo, E., 1990, Specific binding of the TraY protein to *oriT* and the promoter region for the *traY* gene of plasmid R100, *J. Biol. Chem.* **265:**6461–6466.
75. Inamoto, S., Yoshioka, Y., and Ohtsubo, E., 1991, Site- and strand-specific nicking in vitro at *oriT* by the TraY-TraI endonuclease of plasmid R100, *J. Biol. Chem.* **266:**10086–10092.
76. Ippen-Ihler, K. A., and Minkley, E. G., Jr., 1986, The conjugation system of F, the fertility factor of *Escherichia coli*, *Annu. Rev. Genet.* **20:**593–624.
77. Jalajakumari, M. B., and Manning, P. A., 1989, Nucleotide sequence of the *traD* region in the *Escherichia coli* F sex factor, *Gene* **81:**195–202.
78. Jones, A. L., Barth, P. T., and Wilkins, B. M., 1992, Zygotic induction of plasmid *ssb* and *psiB* genes following conjugative transfer of IncI1 plasmid ColIb-P9, *Mol. Microbiol.* **6:**605–613.
79. Kim, K., and Meyer, R. J., 1989, Unidirectional transfer of broad-host-range plasmid R1162 during conjugative mobilization. Evidence for genetically distinct events at *oriT*, *J. Mol. Biol.* **208:**501–505.
80. Kingsman, A., and Willetts, N., 1978, The requirements for conjugal DNA synthesis in the donor strain during F *lac* transfer, *J. Mol. Biol.* **122:**287–300.
81. Klinkert, M.-Q., Klein, A., and Abdel-Monem, M., 1980, Studies on the functions of DNA helicase I and DNA helicase II of *Escherichia coli*, *J. Biol. Chem.* **255:**9746–9752.
82. Kolodkin, A. L., Capage, M. A., Golub, E. I., and Low, K. B., 1983, F sex factor of *Escherichia coli* K-12 codes for a single-stranded DNA binding protein, *Proc. Natl. Acad. Sci. USA* **80:**4422–4426.
83. Komano, T., Toyoshima, A., Morita, K., and Nisioka, T., 1988, Cloning and nucleotide sequence of the *oriT* region of the IncI1 plasmid R64, *J. Bacteriol.* **170:**4385–4387.
84. Komano, T., Funayama, N., Kim, S.-R., and Nisioka, T., 1990, Transfer region of IncI1 plasmid R64 and role of shufflon in R64 transfer, *J. Bacteriol.* **172:**2230–2235.
85. Kornberg, A., and Baker, T. A., 1992, *DNA Replication* (2nd ed.). W. H. Freeman, New York.

86. Krishnapillai, V., 1988, Molecular genetic analysis of bacterial plasmid promiscuity, *FEMS Microbiol. Rev.* **54:**223–238.

87. Krishnapillai, V., Nash, J., and Lanka, E., 1984, Insertion mutations in the promiscuous IncP-1 plasmid R18 which affect its host range between *Psuedomonas* species, *Plasmid* **12:**170–180.

88. Kuhn, B., Abdel-Monem, H., Krell, H., and Hoffmann-Berling, H., 1979, Evidence for two mechanisms for DNA unwinding catalyzed by DNA helicases, *J. Biol. Chem.* **254:**11343–11350.

89. Lahue, E. E., and Matson, S. W., 1988, *Escherichia coli* DNA helicase I catalyzes a unidirectional and highly processive unwinding reaction, *J. Biol. Chem.* **263:**3208–3215.

90. Lahue, E. E., and Matson, S. W., 1990, Purified *Escherichia coli* F-factor TraY protein binds *oriT*, *J. Bacteriol.* **172:**1385–1391.

91. Lanka, E., and Barth, P. T., 1981, Plasmid RP4 specifies a deoxyribonucleic acid primase involved in its conjugal transfer and maintenance, *J. Bacteriol.* **148:**769–781.

92. Lanka, E., and Fürste, J. P., 1984, Function and properties of RP4 DNA primase, in: *Proteins Involved in DNA Replication* (U. Hübscher and S. Spadari, eds.), Plenum Press, New York, pp. 265–280.

93. Lanka, E., Scherzinger, E., Günther, E., and Schuster, H., 1979, A DNA primase specified by I-like plasmids, *Proc. Natl. Acad. Sci. USA* **76:**3632–3636.

94. Lanka, E., Lurz, R., Kröger, M., and Fürste, J. P., 1984, Plasmid RP4 encodes two forms of a DNA primase, *Mol. Gen. Genet.* **194:**65–72.

95. Lessl, M., Balzer, D., Lurz, R., Waters, V. L., Guiney, D. G., and Lanka, E., 1992, Dissection of IncP conjugative plasmid transfer: definition of the transfer region Tra2 by mobilization of the Tra1 region in *trans*, *J. Bacteriol.* **174:**2493–2500.

95a. Lessl, M., Balzer, D., Pansegrau, W., and Lanka, E., 1992, Sequence similarities between the RP4 Tra2 and the Ti VirB region strongly support the conjugation model for T-DNA transfer, *J. Biol. Chem.* **267:**20,471–20,480.

96. Llosa, M., Bolland, S., and de la Cruz, F., 1991, Structural and functional analysis of the origin of conjugal transfer of the broad-host-range IncW plasmid R388 and comparison to the related IncN plasmid R46, *Mol. Gen. Genet.* **226:**473–483.

97. Loh, S. M., Cram, D. S., and Skurray, R. A., 1988, Nucleotide sequence and transcriptional analysis of a third function (Flm) involved in F-plasmid maintenance, *Gene* **66:**259–268.

98. Loh, S., Cram, D., and Skurray, R., 1989, Nucleotide sequence of the leading region adjacent to the origin of transfer on plasmid F and its conservation among conjugative plasmids, *Mol. Gen. Genet.* **219:**177–186.

99. Loh, S., Skurray, R., Célérier, J., Bagdasarian, M., Bailone, A., and Devoret, R., 1990, Nucleotide sequence of the *psiA* (plasmid SOS inhibition) gene located on the leading region of plasmids F and R6-5, *Nucl. Acids Res.* **18:**4597.

100. Lovett, M. A., and Helinski, D. R., 1975, Relaxation complexes of plasmid DNA and protein. II. Characterization of the proteins associated with the unrelaxed and relaxed complexes of plasmid ColE1, *J. Biol. Chem.* **250:**8790–8795.

101. Manning, P. A., and Achtman, M., 1979, Cell-to-cell interactions in conjugating *Escherichia coli*: the involvement of the cell envelope, in: *Bacterial Outer Membranes: Biogenesis and Functions* (M. Inouye, ed.), John Wiley & Sons, New York, pp. 409–447.

102. Marians, K. J., 1984, Enzymology of DNA in replication in prokaryotes, *Crit. Rev. Biochem.* **17:**153–215.

103. Matson, S. W., and Kaiser-Rogers, K. A., 1990, DNA helicases, *Annu. Rev. Biochem.* **59:**289–329.

104. Matson, S. W., and Morton, B. S., 1991, *Escherichia coli* DNA helicase I catalyzes a site- and strand-specific nicking reaction at the F plasmid *oriT*, *J. Biol. Chem.* **266:**16232–16237.

105. Matson, S. W., and Morton, B. S., 1993, *Escherichia coli* DNA helicase I is covalently bound to the 5′ side of the F plasmid *oriT* nick site, *EMBO J.*, in press.

106. McIntire, S. A., and Dempsey, W. B., 1987, *oriT* sequence of the antibiotic resistance plasmid R100, *J. Bacteriol.* **169:**3829–3832.

107. Miele, L., Strack, B., Kruft, V., and Lanka, E., 1991, Gene organization and nucleotide sequence of the primase region of IncP plasmids RP4 and R751, *DNA Sequence* **2:**145–162.

108. Merryweather, A., Barth, P. T., and Wilkins, B. M., 1986, Role and specificity of plasmid RP4-encoded DNA primase in bacterial conjugation, *J. Bacteriol.* **167:**12–17.

109. Merryweather, A., Rees, C. E. D., Smith, N. M., and Wilkins, B. M., 1986, Role of *sog* polypeptides specified by plasmid ColIb-P9 and their transfer between conjugating bacteria, *EMBO J.* **5:**3007–3012.

110. Meyer, R., 1989, Site-specific recombination at *oriT* of plasmid R1162 in the absence of conjugative transfer, *J. Bacteriol.* **171**:799–806.
111. Morlon, J., Chartier, M., Bidaud, M., and Lasdunski, C., 1988, The complete nucleotide sequence of the colicinogenic plasmid ColA. High extent of homology with ColE1, *Mol. Gen. Genet.* **211**:231–243.
112. Motallebi-Veshareh, M., Balzer, D., Lanka, E., Jagura-Burdzy, G., and Thomas, C. M., 1992, Conjugative transfer functions of broad host range plasmid RK2 are coregulated with vegetative replication, *Mol. Microbiol.* **6**:907–920.
113. Nash, J., and Krishnapillai, V., 1988, Role of IncP-1 plasmid primase in conjugation between *Pseudomonas* species, *FEMS Microbiol. Lett.* **49**:257–260.
114. Nomura, N., Low, R. L., and Ray, D. S., 1982, Identification of ColE1 DNA sequences that direct single strand-to-double strand conversion by a φX174 type mechanism, *Proc. Natl. Acad. Sci. USA* **79**:3153–3157.
115. Nomura, N., Masai, H., Inuzuka, M., Miyazaki, C., Ohtsubo, E., Itoh, T., Sasamoto, S., Matsui, M., Ishizaki, R., and Arai, K., 1991, Identification of eleven single-strand initiation sequences (*ssi*) for priming of DNA replication in the F, R6K, R100 and ColE2 plasmids, *Gene* **108**:15–22.
116. Ohki, M., and Tomizawa, J., 1968, Asymmetric transfer of DNA strands in bacterial conjugation, *Cold Spring Harbor Symp. Quant. Biol.* **33**:651–657.
117. Ostermann, E., Kricek, F., and Högenauer, G., 1984, Cloning the origin of transfer region of the resistance plasmid R1, *EMBO J.* **3**:1731–1735.
118. Panicker, M. M., and Minkley, E. G., Jr., 1985, DNA transfer occurs during a cell surface contact stage of F sex factor-mediated bacterial conjugation, *J. Bacteriol.* **162**:584–590.
118a. Panicker, M. M., and Minkley, Jr., E. G., 1992, Purification and properties of the F sex factor TraD protein, an inner membrane conjugal transfer protein, *J. Biol. Chem.* **267**:12761–12766.
119. Pansegrau, W., Ziegelin, G., and Lanka, E., 1988, The origin of conjugative IncP plasmid transfer: interaction with plasmid-encoded products and the nucleotide sequence at the relaxation site, *Biochim. Biophys. Acta* **951**:365–374.
120. Pansegrau, W., Balzer, D., Kruft, V., Lurz, R., and Lanka, E., 1990, In vitro assembly of relaxosomes at the transfer origin of plasmid RP4, *Proc. Natl. Acad. Sci. USA* **87**:6555–6559.
121. Pansegrau, W., and Lanka, E., 1991, Common sequence motifs in DNA relaxases and nick regions from a variety of DNA transfer systems, *Nucl. Acids Res.* **19**:3455.
121a. Pansegrau, W., and Lanka, E., 1992, A common sequence motif among prokaryotic DNA primases, *Nucl. Acids Res.* **20**:4931.
122. Pansegrau, W., Ziegelin, G., and Lanka, E., 1990, Covalent association of the *traI* gene product of plasmid RP4 with the 5′-terminal nucleotide at the relaxation nick site, *J. Biol. Chem.* **265**:10637–10644.
123. Paterson, E. S., and Iyer, V. N., 1992, The *oriT* region of the conjugative transfer system of plasmid pCU1 and specificity between it and the *mob* region of other N *tra* plasmids, *J. Bacteriol.* **174**:499–507.
124. Pettijohn, D. E., and Pfenninger, O., 1980, Supercoils in prokaryotic DNA restrained in vivo, *Proc. Natl. Acad. Sci. USA* **77**:1331–1335.
125. Read, T. D., Thomas, A. T., and Wilkins, B. M., 1992, Evasion of type I and type II restriction systems by IncI1 plasmid ColIb-P9 during transfer by bacterial conjugation, *Mol. Microbiol.* **6**:1933–1941.
126. Rees, C. E. D., and Wilkins, B. M., 1989, Transfer of *tra* proteins into the recipient cell during conjugation mediated by plasmid ColIb-P9, *J. Bacteriol.* **171**:3152–3157.
127. Rees, C. E. D., and Wilkins, B. M., 1990, Protein transfer into the recipient cell during bacterial conjugation: studies with F and RP4, *Mol. Microbiol.* **4**:1199–1205.
128. Rees, C. E. D., Bradley, D. E., and Wilkins, B. M., 1987, Organization and regulation of the conjugation genes of IncI1 plasmid ColIb-P9, *Plasmid* **18**:223–236.
129. Reygers, U., Wessel, R., Müller, H., and Hoffmann-Berling, H., 1991, Endonuclease activity of *Escherichia coli* DNA helicase I directed against the transfer origin of the F factor, *EMBO J.* **10**:2689–2694.
130. Rohrer, J., and Rawlings, D. E., 1992, Nucleotide sequence and functional analysis of the mobilization region of the broad-host-range plasmid pTF-FC2, *J. Bacteriol.* **174**:6230–6237.
131. Rose, R. E., 1988, The nucleotide sequence of pACYC184, *Nucl. Acids Res.* **16**:355.
132. Rupp, W. D., and Ihler, G., 1968, Strand selection during bacterial mating, *Cold Spring Harbor Symp. Quant. Biol.* **33**:647–650.
133. Scherzinger, E., Lurz, R., Otto, S., and Dobrinski, B., 1992, In vitro cleavage of double- and single-stranded DNA by plasmid RSF1010-encoded mobilization proteins, *Nucl. Acids Res.* **20**:41–48.

134. Schilf, W., and Krishnapillai, V., 1986, Genetic analysis of insertion mutations of the promiscuous IncP-1 plasmid R18 mapping near *oriT* which affect its host range, *Plasmid* **15**:48–56.

135. Scholz, P., Haring, V., Wittmann-Liebold, B., Ashman, K., Bagdasarian, M., and Scherzinger, E., 1989, Complete nucleotide sequence and gene organization of the broad-host-range plasmid RSF1010, *Gene* **75**:271–288.

136. Schwab, M., Gruber, H., and Högenauer, G., 1991, The TraM protein of plasmid R1 is a DNA-binding protein, *Mol. Microbiol.* **5**:439–446.

137. Silver, S. D., Moody, E. E. M., and Clowes, R. C., 1965, Limits on material transfer during $F^+ \times F^-$ matings in *Escherichia coli* K12, *J. Mol. Biol.* **12**:283–286.

138. Silverman, P. M., 1987, The structural basis of prokaryotic DNA transfer, in: *Bacterial Outer Membranes as Model Systems* (M. Inouye, ed.), John Wiley & Sons, New York, pp. 277–309.

139. Snijders, A., van Putten, A. J., Veltkamp, E., and Nijkamp, H. J. J., 1983, Localization and nucleotide sequence of the *bom* region of Clo DF13, *Mol. Gen. Genet.* **192**:444–451.

140. Strack, B., Lessl, M., Calendar, R., and Lanka, E., 1992, A common sequence motif, -E-G-Y-A-T-A-, identified within the primase domains of plasmid-encoded I- and P-type DNA primases and the α protein of the *Escherichia coli* satellite phage P4, *J. Biol. Chem.* **267**:13062–13072.

141. Thomas, C. D., Balson, D. F., and Shaw, W. V., 1990, In vitro studies of the initiation of staphylococcal plasmid replication, *J. Biol. Chem.* **265**:5519–5530.

142. Thompson, R., Taylor, L., Kelly, K., Everett, R., and Willetts, N., 1984, The F plasmid origin of transfer: DNA sequence of wild-type and mutant origins and location of origin-specific nicks, *EMBO J.* **3**:1175–1180.

143. Thompson, T. L., Centola, M. B., and Deonier, R. C., 1989, Location of the nick at *oriT* of the F plasmid, *J. Mol. Biol.* **207**:505–512.

144. Traxler, B. A., and Minkley, E. G., Jr., 1987, Revised genetic map of the distal end of the F transfer operon: implications for DNA helicase I, nicking at *oriT*, and conjugal DNA transport, *J. Bacteriol.* **169**:3251–3259.

145. Traxler, B. A., and Minkley, E. G., Jr., 1988, Evidence that DNA helicase I and *oriT* site-specific nicking are both functions of the F TraI protein, *J. Mol. Biol.* **204**:205–209.

146. Tsai, M-M., Fu, Y.-H. F., and Deonier, R. C., 1990, Intrinsic bends and integration host factor binding at F plasmid *oriT*, *J. Bacteriol.* **172**:4603–4609.

147. Vapnek, D., and Rupp, W. D., 1970, Asymmetric segregation of the complementary sex-factor DNA strands during conjugation in *Escherichia coli*, *J. Mol. Biol.* **53**:287–303.

148. Vapnek, D., Lipman, M. B., and Rupp, W. D., 1971, Physical properties and mechanism of transfer of R factors in *Escherichia coli*, *J. Bacteriol.* **108**:508–514.

149. Vogel, A. M., and Das, A., 1992, Mutational analysis of *Agrobacterium tumefaciens virD2*: tyrosine 29 is essential for endonuclease activity, *J. Bacteriol.* **174**:303–308.

150. Warren, G. J., and Clark, A. J., 1980, Sequence-specific recombination of plasmid ColE1, *Proc. Natl. Acad. Sci. USA* **77**:6724–6728.

151. Warren, G. J., Twigg, A. J., and Sherratt, D. J., 1978, ColE1 plasmid mobility and relaxation complex, *Nature* (London) **274**:259–261.

152. Waters, V. L., Hirata, K. H., Pansegrau, W., Lanka, E., and Guiney, D. G., 1991, Sequence identity in the nick regions of IncP plasmid transfer origins and T-DNA borders of *Agrobacterium* Ti plasmids, *Proc. Natl. Acad. Sci. USA* **88**:1456–1460.

153. Wessel, R., Müller, H., and Hoffmann-Berling, H., 1990, Electron microscopy of DNA · helicase-I complexes in the act of strand separation, *Eur. J. Biochem.* **189**:277–285.

154. Wilkins, B. M., 1975, Partial suppression of the phenotype of *Escherichia coli* K-12 *dnaG* mutants by some I-like conjugative plasmids, *J. Bacteriol.* **122**:899–904.

155. Wilkins, B. M., Boulnois, G. J., and Lanka, E., 1981, A plasmid DNA primase active in discontinuous bacterial DNA replication, *Nature* (London) **290**:217–221.

156. Willetts, N., 1981, Sites and systems for conjugal DNA transfer in bacteria, in: *Molecular Biology, Pathogenicity, and Ecology of Bacterial Plasmids* (S. B. Levy, R. C. Clowes, and E. L. Koenig, eds.), Plenum Press, New York, pp. 207–215.

157. Willetts, N., and Maule, J., 1979, Investigations of the F conjugation gene *traI*: *traI* mutants and λ*traI* transducing phages, *Mol. Gen. Genet.* **169**:325–336.

158. Willetts, N., and Maule, J., 1985, Specificities of IncF plasmid conjugation genes, *Genet. Res. Camb.* **47**: 1–11.

159. Willetts, N., and Skurray, R., 1987, Structure and function of the F factor and mechanism of conjugation, in: *Escherichia coli and Salmonella typhimurium: Cellular and Molecular Biology* (F. C. Neidhardt, J. L. Ingraham, K. B. Low, B. Magasanik, and H. E. Umbarger, eds.), American Society for Microbiology, Washington DC, pp. 1110–1133.

160. Willetts, N., and Wilkins, B., 1984, Processing of plasmid DNA during bacterial conjugation, *Microbiol. Rev.* **48:**24–41.

161. Winans, S. C., 1992, Two-way chemical signaling in *Agrobacterium*-plant interactions, *Microbiol. Rev.* **56:**12–31.

162. Winans, S. C., and Walker, G. C., 1985, Conjugal transfer system of the IncN plasmid pKM101, *J. Bacteriol.* **161:**402–410.

163. Womble, D. D., and Rownd, R. H., 1988, Genetic and physical map of plasmid NR1: comparison with other IncFII antibiotic resistance plasmids, *Microbiol. Rev.* **52:**433–451.

164. Yakobson, E., Deiss, C., Hirata, K., and Guiney, D. G., 1990, Initiation of DNA synthesis in the transfer origin region of RK2 by the plasmid-encoded primase: detection using defective M13 phage, *Plasmid* **23:**80–84.

165. Yang, C.-C., and Nash, H. A., 1989, The interaction of *E. coli* IHF protein with its specific binding sites, *Cell* **57:**869–880.

166. Yoshioka, Y., Fujita, Y., and Ohtsubo, E., 1990, Nucleotide sequence of the promoter-distal region of the *tra* operon of plasmid R100, including *traI* (DNA helicase I) and *traD* genes, *J. Mol. Biol.* **214:**39–53.

167. Zambryski, P., 1988, Basic processes underlying *Agrobacterium*-mediated DNA transfer to plant cells, *Annu. Rev. Genet.* **22:**1–30.

168. Zavitz, K. H., and Marians, K. J., 1991, Dissecting the functional role of PriA protein-catalyzed primosome assembly in *Escherichia coli* DNA replication, *Mol. Microbiol.* **5:**2869–2873.

169. Ziegelin, G., Fürste, J. P., and Lanka, E., 1989, TraJ protein of plasmid RP4 binds to a 19-base pair invert sequence repetition within the transfer origin, *J. Biol. Chem.* **264:**11989–11994.

170. Ziegelin, G., Pansegrau, W., Lurz, R., and Lanka, E., 1992, TraK protein of conjugative plasmid RP4 forms a specialized nucleoprotein complex with the transfer origin, *J. Biol. Chem.* **267:**17279–17286.

171. Ziegelin, G., Pansegrau, W., Strack, B., Balzer, D., Kröger, M., Kruft, V., and Lanka, E., 1991, Nucleotide sequence and organization of genes flanking the transfer origin of promiscuous plasmid RP4, *DNA Sequence* **1:**303–327.

172. Zipursky, S. L., and Marians, K. J., 1981, *Escherichia coli* factor Y sites of plasmid pBR322 can function as origins of DNA replication, *Proc. Natl. Acad. Sci. USA* **78:**6111–6115.

Chapter 6

Mobilization of Chromosomes and Nonconjugative Plasmids by Cointegrative Mechanisms

CORNELIA REIMMANN and DIETER HAAS

1. Introduction

Conjugative plasmids in bacteria bring about the transfer of DNA by cell contact between a donor and a recipient (250). Naturally occurring conjugative plasmids transfer themselves. Furthermore, in many cases they can also transfer (mobilize) chromosomal DNA or nonconjugative plasmids. Thus, mobilization, as we define it, is the process by which a conjugative plasmid accomplishes the transfer of DNA that is not self-transmissible.

Basically, mobilization can happen in two ways. First, the conjugative plasmid and the mobilized DNA remain physically separated throughout the mobilization process. This type of mobilization is referred to as mobilization *in trans* (188); some authors have used the term *donation* to describe the same process (61, 313).

Second, mobilization proceeds via the formation of a cointegrate between the conjugative plasmid and the replicon to be mobilized. The essential feature of this type of mobilization *in cis* is that the transfer-proficient plasmid recombines with the other replicon, resulting in a covalent link between the two replicons. The cointegrates thus formed may be quite stable and amenable to physical analysis. In some instances, however, the cointegrates dissociate rapidly into their components after conjugative transfer. Because of this instability, it can be difficult to prove that a cointegrative mechanism occurs in a particular mobilization experiment. Therefore, we will use the term *mobilization* irrespective of the actual transfer mechanism, and we will not adopt the term *conduction*, which has been

CORNELIA REIMMANN and DIETER HAAS • Mikrobiologisches Institut, Eidgenössische Technische Hochschule, CH-8092 Zürich, Switzerland.
Bacterial Conjugation, edited by Don B. Clewell. Plenum Press, New York, 1993.

coined to describe specifically the mobilization of a replicon by physical association with a conjugative plasmid (61, 313).

Mobilization experiments play an important role in bacterial genetics. Historically, the discovery of chromosome mobilization by the F plasmid in *Escherichia coli* opened up the genetic analysis of the bacterial chromosome, and F has become the paradigm of conjugation and chromosome mobilization (139, 206). A crucial finding was that stable integration of a conjugative plasmid into the chromosome produces high frequency of recombination (Hfr) donor strains, which give oriented chromosome transfer from the site of insertion (139, 206). Subsequently, conjugative chromosome transfer systems were established in many bacterial genera (*Pseudomonas, Streptomyces, Rhizobium*, etc.) and permitted the construction of chromosome maps. Mobilization experiments have also contributed importantly to the characterization of large bacterial plasmids such as those encoding toluene degradation, symbiotic nitrogen fixation, or induction of plant tumors. Today, mobilization techniques continue to be important in genetic analysis and are routinely used to transfer recombinant plasmids between bacterial species. An overview of some important mobilization systems, with special emphasis on gram-negative bacteria, is given in Section 3.

A distinction between mobilization *in trans* and mobilization *in cis* can be made in a number of well-defined systems. A nonconjugative replicon can be mobilized *in trans* when it carries an origin of transfer (*oriT*), which is recognized by the transfer (*tra*) functions of a conjugative plasmid present in the same cell (105, 323, 341, 347, 349, 350). Certain nonconjugative plasmids such as ColE1 of *E. coli* or the broad host range plasmid RSF1010 are naturally mobilizable because they contain an *oriT* and, in addition, specific mobilization (*mob*) genes. The *mob* products interact with the *oriT* and allow the *tra* functions of a conjugative, mobilizing plasmid (e.g., F or RK2) to bring about the conjugative transfer of the plasmid to be mobilized (36, 104). Typically, in this type of mobilization the transfer frequency of the nonconjugative plasmid may be as high as that of the conjugative plasmid. Mobilization *in trans* can be achieved quite simply by genetic manipulation. For example, when the *oriT* region of the broad host range plasmid RK2 (=RP1=RP4≅R68) is inserted into any replicon, the latter becomes mobilizable *in trans* by the transfer functions of RK2. The broad host range properties of the RK2 conjugation system enable conjugative plasmid transfer between *E. coli* and a variety of gram-negative and even gram-positive bacteria (121, 217, 218, 286, 329). Tn5 and mini-Mu derivatives carrying $oriT_{RK2}$ have been constructed; they greatly facilitate the random insertion of *oriT* into any replicon to be mobilized in Gram-negative bacteria (120, 192, 297–299, 354). A replication-deficient RP1 derivative, pME28, which carries *oriT* and transposes effectively because of an IS*21* tandem duplication (see Sections 2.2.2 and 4.2), can be used for the same purpose (245, 276). The mechanisms governing mobilization *in trans* are discussed in Chapters 3 and 4.

Although the strategy of *oriT* cloning and introduction of a corresponding transfer-proficient helper plasmid are widely applicable and have distinct advantages for plasmid or chromosome mobilization (192, 245, 276, 297, 300, 354), most examples of mobilization reported in the literature are not of this type. Rather, mobilization is commonly mediated by cointegrate formation; the replicon to be mobilized is fused to a conjugative plasmid, which thus supplies both the *oriT* and the *tra* functions *in cis*. Following conjugative transfer, the cointegrates may persist or be resolved. Typically, in this type of mobilization the mobilizing plasmid alone shows a higher transfer frequency than does the mobilized nonconjugative plasmid, because as a rule only a fraction of the latter undergoes cointegrate formation. The relevant questions in this context are: What are the structures of the cointegrates? Which mechanisms are involved in replicon fusion? How frequently do cointegrates arise? Why are some cointegrates structurally unstable? In Section 2, we discuss various mechanisms that allow recombination between replicons in bacteria and give specific examples.

It should be mentioned here that mobilization *in cis* does not always depend on a conjugative plasmid but can also be brought about by some conjugative transposons. An artificial conjugative transposon for use in enteric bacteria was constructed by Johnson and Willetts (173); these authors inserted the entire *oriT/tra* region of the F plasmid into transposon Tn*1*. Transposition of this Tn*1* derivative into a nonconjugative plasmid renders the plasmid self-transmissible (173). Similarly, the *oriT/tra* functions of RP4 have been introduced into Tn*1* and Tn*5*. The recombinant transposons promote chromosome mobilization in *E. coli* and *Rhizobium* sp. (172). Interestingly, a *natural* conjugative transposon, Tn*4399*, which mobilizes nonconjugative plasmids *in cis*, has been identified in *Bacteroides fragilis*. Insertion of Tn*4399* into the Tra⁻ shuttle vector pGAT400 confers transferability on this vector in *B. fragilis* × *B. fragilis* and *B. fragilis* × *E. coli* crosses (140, 141). There is evidence that the conjugative transposon Tn*925* from *Enterococcus faecalis* can mobilize chromosomal genes in *E. faecalis* and *Bacillus subtilis* (325). The related conjugative transposon Tn*916* of *E. faecalis* (Chapter 14), in contrast, does not mobilize nonconjugative replicons into which it is inserted (D. B. Clewell, personal communication).

In Section 4, we focus on the molecular mechanism of cointegrate formation by the broad host range plasmid R68.45, which has been an important tool in chromosomal genetics of gram-negative bacteria (125, 126, 152), and we point out some analogies between retrovirus integration and R68.45-mediated replicon fusion.

2. Mechanisms of Cointegrate Formation

Many chromosome mobilization experiments were done at a time when little was known about the physical interactions between conjugative plasmids with chromosomes and other replicons. In the period from 1950 to about 1980, plasmids having chromosome mobilizing ability (Cma) were called "sex factors" or "fertility factors" and they were used primarily as tools for chromosome mapping. As we now realize, a crucial step in many mobilization experiments is the recombination of a conjugative plasmid with the DNA to be mobilized. During the past 10 to 15 years, several mechanisms that govern recombination in bacteria have been studied in molecular detail. In the following section we briefly discuss these mechanisms and their relevance to cointegrate formation. We classify them into four different types: homologous, transpositional, site-specific, and illegitimate recombination. We adopt this classification scheme to explain (or to speculate about) the formation of various plasmid cointegrates, both old and new ones. The sophistication with which the different types of recombination have been studied varies considerably and, to some extent, this is reflected in the terminology. "Illegitimate recombination" was originally defined as being "rare, haphazard and not obviously dependent upon genetic (i.e., DNA sequence) homology" (103). A more recent view of illegitimate recombination (93) excludes recombination events resulting from normal or "legitimate" activities of transposition or specialized recombination. Essentially, the terms "legitimate" and "illegitimate" are just synonyms for "better known" and "lesser known," respectively. Thus, cointegrate formation due to transposition or site-specific recombination will be considered to be legitimate events in this review.

2.1. Homologous Recombination

In *E. coli*, homologous recombination requires the RecA protein and at least 40 to 50 base pairs (bp) of DNA sequence homology between recombining DNA molecules (305). Longer stretches of sequence similarity improve the recombination frequencies (293). In *E. coli* K-12, recombination between circular double-stranded (ds) DNA molecules—plasmid × plasmid or plasmid × chromosome—proceeds primarily via the RecE and RecF pathways and is independent of the RecBCD enzyme (67, 305). The molecular mechanisms and the components of the RecE and RecF pathways are poorly understood. It is known, however, that *recA* and *recF* mutations each lower the frequency of interplasmidic recombination about 100-fold, indicating important roles for the RecA and RecF proteins (67, 197). According to current views (305, 345), homologous recombination is initiated when the 3'-OH end of a single-stranded (ss) DNA molecule invades and pairs with a homologous dsDNA molecule. This step is catalyzed by RecA protein and is aided by ss-binding (SSB) protein, the product of the *ssb* gene (195, 305). RecF protein also binds to ssDNA and may interact with RecA (113). If the RecA protein has indeed this function in the RecE and RecF pathways, then it remains to be seen whether and how ssDNA is produced from circular dsDNA.

The situation is different in conjugation and transduction. Here, the incoming DNA is at least in part linear. The ends of dsDNA are suitable for binding of the RecBCD enzyme (305, 306). This enzyme unwinds and occasionally nicks the dsDNA, producing long ssDNA tails that extend from the nicked dsDNA molecule. Nicking takes place 4–6 nucleotides to the 3' side of a DNA sequence (5'-GCTGGTGG-3') known as Chi site. In enteric bacteria, Chi sites are recombinant hotspots (305, 345). The 3' ends of the tails formed by the RecBCD enzyme may be used to initiate RecA-dependent recombination with homologous dsDNA (305, 306, 345). This model is supported by an *in vitro* system containing the RecA, RecBCD, and SSB proteins and a Chi site in the linear donor DNA (89). Furthermore, *recBC* mutants show a ~100-fold reduction in conjugational and transductional recombination relative to the wild type (67, 305, 360). The RecBCD pathway is strongly dependent on RecA but independent of RecF in a wild-type background (207). Mutations in *recA* reduce conjugational and transductional recombination >1000-fold (67, 305, 360).

Thus, the RecA protein is essential for homologous recombination in *E. coli* irrespective of the recombination pathway. The reactions following initiation of recombination (RecA-promoted strand exchange, formation of a Holliday junction, translocation and resolution of this junction by the RuvABC and RecG proteins) have been described and discussed in recent overviews (239, 305, 321, 345, 346). RecA-like proteins are present in a wide range of gram-negative and gram-positive bacteria. In *B. subtilis* the RecE protein is analogous to RecA. The properties of RecA-like proteins are very similar among different bacteria, suggesting that the mechanisms of recombination are highly conserved (230, 244, 305, 314).

Let us illustrate the importance of *recA*-dependent cointegrate formation (Figure 6.1A) with two examples. Peterson et al (263) examined the recombinants between the conjugative R plasmid NR1 and a ColE1 derivative, RSF2124, in *E. coli*. Cointegrates consisting of the two plasmids were selected and analyzed after conjugation with a *polA* recipient, in which RSF2124 cannot replicate. When a 20.5-kilobase (kb) fragment of NR1 was cloned into RSF2124, about 40% of the NR1 molecules formed a cointegrate with the engineered RSF2124 derivative. When a smaller (4.8-kb) fragment of NR1 was inserted into RSF2124 giving pRR134 (Figure 6.1A), the cointegrate frequency dropped to about 4% (263). It is known from other work that the recombination frequencies are lowered further to 10^{-3} or 10^{-4} when the two plasmids carry homologous sequences of only several hundred nucleotides (67, 197). These results show that plasmids that share extensive homology form cointegrates readily in *recA*$^+$ host cells.

Trieu-Cuot et al (330) constructed a cointegrate between the broad host range streptococcal plasmid pAMβ1 (26.5 kb) and a pBR322 derivative, pAT190 (6 kb), carrying a 0.35-kb fragment from pAMβ1 and a kanamycin-resistance determinant. The cointegrate (32.5 kb) was obtained *in vivo* by the following procedure. *Streptococcus sanguis* harboring pAMβ1 was transformed with pAT190 DNA. The *E. coli* plasmid pAT190 cannot replicate in *Streptococcus*, but following transformation the plasmid could be rescued efficiently by homologous recombination with the resident plasmid pAMβ1; recombination took place in the 0.35-kb region of homology (330). Natural transformation in gram-positive bacteria involves the linearization of donor plasmid molecules at the surface of competent host cells and entry of ssDNA into the cells. The ssDNA then recombines with the homologus dsDNA of the recipient genome (227, 314, 330). Dimeric forms of the plasmid are much more efficient in recombination than monomeric forms, at least in *B. subtilis* (227). The construction of the pAMβ1-pAT190 cointegrate illustrates the ease with which two plasmids can be fused *in vivo*, provided the plasmids carry a region of sequence homology. Since pAMβ1 is a broad host range conjugative plasmid (see Chapter 2), the pAMβ1-pAT190 cointegrate can be transferred by conjugation to other gram-positive bacteria such as *E. faecalis* and interestingly also to the gram-negative bacterium *E. coli* (330).

The formation of Hfr strains in *E. coli* probably represents the most prominent example of mobilization by a cointegrative mechanism that depends essentially on homologous recombination. We will discuss the chromosomal integration of the F plasmid in Section 3.1.1.

2.2. Transposition

2.2.1. Replicative Transposition

The ampicillin-resistance transposon Tn*3* and related transposable elements such as Tn*21* promote replicon fusion; cointegrates are formed as intermediates during *replicative transposition* (4, 115, 165, 295) (Figure 6.1B). Transposons of the Tn*3* family are bordered by inverted repeats of 35 to 38 bp (occasionally 48 bp). These transposons code for transposase (the product of the *tnpA* gene) and resolvase (the *tnpR* gene product) and they contain a 120-bp sequence that constitutes an internal recombination site (*res* or IRS). In the first step of transposition, the donor replicon is joined to the target replicon by two copies of the transposable element in direct orientation (Figure 6.1B). This step requires transposase

A) Homologous recombination

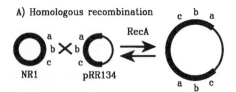

B) Replicative transposition

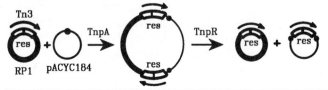

C) Nonreplicative transposition combined with homologous recombination: transposition first

D) Nonreplicative transposition combined with homologous recombination: recombination (dimerization) first

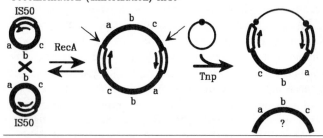

E) Inverse transposition (nonreplicative)

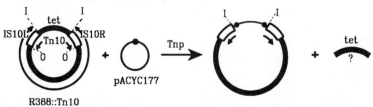

FIGURE 6.1. Mechanisms of cointegrate formation. Heavy lines denote the donor plasmids, thin lines the recipient (target) replicons. Target duplications, which are formed as a consequence of transpositional recombination, are designed by ● (B, C, D, E, F, H). Crossovers in homologous recombination are indicated by ✗ (A, C, D). The inner ends of IS*10* are abbreviated by I (E). The bar (Ɩ) signals a short sequence having similarity with an active transposon end (◁) (H). Specific plasmids and their functions are described and referenced in the text. (Figure continued on next page.)

F) Nonreplicative transposition depending on the reactive IS21 tandem

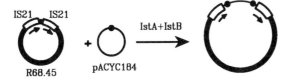

G) Site-specific recombination

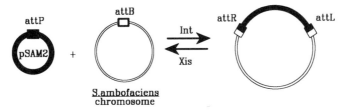

H) One-ended transposition

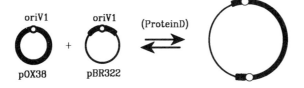

I) Recombination at oriV

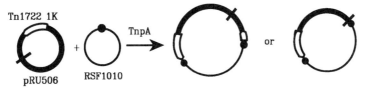

J) Recombination at oriT

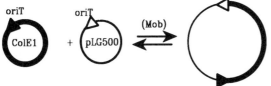

FIGURE 6.1. (*continued*)

and, as the donor element is duplicated, also host replication functions (165). The cointegrates formed are short-lived because in a second step resolvase promotes highly efficient site-specific recombination at the *res* sites. The products of the resolution process are shown in Figure 6.1B. The target replicon now carries one copy of the transposable element, and the donor replicon is restored (295). The copy in the target replicon is flanked by 5-bp direct repeats (i.e., by a target duplication). For unknown reasons, Tn*3* inserts into plasmids much more readily than it does into the chromosome (295). When the resolvase gene is inactivated by mutation, cointegrates accumulate and can be detected. They are stable in a *recA* host but are resolved slowly in a *recA*+ *background. The tnpA* promoter of Tn*3* lies within the *res* site; therefore, when resolvase binds to *res*, this enzyme also acts as a repressor of transposase. As a consequence, *tnpR* mutants show an increase in transposition frequencies (211, 219, 295).

Tn*3*-promoted mobilization can be demonstrated as shown by the following example. The conjugative IncP plasmid pUB307 (RP1 deleted for the Tn*3*-like element Tn*1*) does not mobilize the *E. coli* vector pACYC184 at detectable frequencies ($< 10^{-7}$). When pACYC184 carries Tn*3*, this plasmid can form a transient cointegrate with pUB307; the cointegrate is conjugatively transferred to a recipient. Alternatively, Tn*3* can be carried by the conjugative plasmid (pUB307::Tn*3*), which is now able to mobilize pACYC184 via the same type of transient cointegrate as above. Typical transposition frequencies range from 10^{-5} to 10^{-7} per generation (17, 19, 295). In either case, resolution in the recipient produces two separate plasmids, both of which carry a Tn*3* element (295) (Figure 6.1B).

Tn*1000* (= γδ), a member of the Tn*3* family without antibiotic-resistance marker, is found on the *E. coli* F plasmid. F mobilizes nonconjugative plasmids such as pBR322 (which is a Mob⁻ derivative of ColE1) via Tn*1000*-mediated cointegrate formation. After conjugative transfer and resolution, pBR322 contains a Tn*1000* copy (122). Therefore, this procedure is frequently used for transposon insertion mutagenesis of pBR322-derived recombinant plasmids (22).

Members of the Tn*3* family also occur in gram-positive bacteria. The *Staphylococcus* transposon Tn*551* and the *Streptococcus* (*Enterococcus*) transposon Tn*917* resemble Tn*3* in that they contain similar inverted repeats at their ends, create 5-bp target duplications, and transpose via cointegrate formation and resolution (265, 295). In *E. faecalis*, the conjugative plasmid pAD1 mobilizes the nonconjugative plasmid pAD2, which carries the erythromycin-resistance transposon Tn*917*. After mobilization, some transconjugants contain pAD1::Tn*917* and pAD2. It is therefore postulated that mobilization of pAD2 can occur by a Tn*917*-mediated cointegrative mechanism, analogous to the mechanism already described for Tn*3* (62).

When a mobilized or mobilizing plasmid acquires a new insertion of a transposable element during mobilization, this can be taken as evidence for a cointegrate intermediate formed by transposition. Does the absence of a new transposon insertion in the mobilized or mobilizing plasmid exclude a cointegrate pathway of transposition? The answer is no. For instance, F-mediated transfer of pBR322 produces some transconjugants that carry pBR322 without Tn*1000* insertion (203). In *recA*+ cells, pBR322 forms dimers and multimers. Hence, Tn*1000* transposition can fuse F to a pBR322 dimer. Resolution in the recipient produces F and a pBR322 dimer with a Tn*1000* insertion. If the recipient is *recA*+, the latter plasmid can dissociate into monomeric pBR322 and pBR322::Tn*1000* via intramolecular recombination. As a consequence, there will be daughter cells of the recipient that contain pBR322 without Tn*1000* insertion. It is possible to avoid this complication by using a monomeric pBR322 plasmid in the F donor and a *recA* recipient (203).

2.2.2. Nonreplicative Transposition

Whereas Tn*3* uses predominantly, if not exclusively, a replicative pathway of transposition, many insertion sequences (IS elements) and transposons transpose by a nonreplicative mode. Tn*5* and Tn*10* are well-documented examples in this category of transposable elements (16, 23, 24, 114, 190). There is good evidence that they insert into target DNA by a "cut-and-paste" mechanism and give simple insertions (Figure 6.1C, left) Transposition does not produce cointegrate intermediates (134, 149). The donor replicon may be lost (or repaired) following transposition, but this has not been determined conclusively. Tn*5* and Tn*10* both produce target duplications of 9 bp (24, 190).

Tn*5* and Tn*10* are composite transposons. Their ends are formed by large inverted repeats that are transposable elements themselves: IS*50* and IS*10*, respectively. The central parts of Tn*5* and Tn*10* determine resistance to kanamycin/bleomycin/streptomycin and tetracycline, respectively. In both Tn*5* and Tn*10* the right copy (IS*50*R, IS*10*R) codes for a functional transposase, whereas the left copy (IS*50*L, IS*10*L) specifies a defective transposase (24, 190). The terms IS*50* and IS*10* will be used as generic designations for IS*50*R, IS*10*R, and their synthetic derivatives (131, 324).

Although cointegrates are not intermediates in nonreplicative transposition, IS*50* and IS*10* do promote cointegrate formation in *recA*$^+$ donor cells (24, 134, 149, 169, 190). Cointegrate formation can be studied, for instance, by a mating-out assay in which pBR322 is mobilized by a conjugative mini-F plasmid without Tn*1000* but with Tn*5*. In such a system, mobilization depends on Tn*5* (149). Historically, it was important to show that RecA is not required for transposition of Tn*5* and Tn*10*. However, when RecA is functional in donor cells, both Tn*5* and Tn*10* can give rise to the formation of cointegrates. Due to the absence of a site-specific resolvase in Tn*5* and Tn*10*, the cointegrates formed are stable in *recA*$^-$ recipients (24, 190). In the context of this review, we will limit the discussion of cointegrate formation to the models shown in Figure 6.1C and D. IS*50*-mediated replicon fusion was studied in detail by Lichens-Park and Syvanen (202). Cointegrate formation was found to require multiple copies of the IS*50* donor molecule and a *recA*$^+$ background. In Figure 6.1C, it is assumed that IS*50* first transposes in a nonreplicative way to the target replicon. Then, IS*50* in the target molecule recombines with IS*50* in another donor molecule to form a cointegrate. In the scheme shown in Figure 6.1D, which is preferred by several authors (24, 149, 202), the donor plasmid is thought to dimerize first. From this dimeric plasmid, a transposable element consisting of a plasmid monomer and flanking IS*50* copies (Figure 6.1D) transposes into the target replicon. Note that the final product is the same as in Figure 6.1C. In *recA*$^-$ donor strains, cointegrate formation is strongly reduced but not completely absent, perhaps because some dimerization of the donor replicon still occurs (24).

Inverse transposition is a nonreplicative transposition process associated with composite transposons such as Tn*10*, Tn*5*, or Tn*9* (51, 134, 240, 281). In normal transposition, transposase acts on the outer ends (O) of the insertion sequences bordering the resistance determinant of the transposon (Figure 6.1E). However, the inner ends (I) of the insertion sequences may also be recognized by transposase. In this event, the donor replicon flanked by two IS elements transposes into the target replicon, and the central part carrying the resistance determinant of the transposon is lost (Figure 6.1E). For instance, the conjugative plasmid R388::Tn*10* mobilizes pACYC177 at a frequency of 10^{-4}, in a *recA*⁻ donor. R388 alone does not mobilize pACYC177. Most of the transconjugants obtained with an R388::Tn*10* + pACYC177 donor contain the trimethoprim-resistance marker of R388 but not the tetracycline-resistance determinant of Tn*10*, and in the cointegrates formed R388 and pACYC177 are linked by two IS*10* elements (134).

Reactive tandems of the insertion sequence IS*21* provide the basis for yet another mode of cointegrate formation that appears to be nonreplicative. Plasmids carrying an IS*21* tandem duplication [= (IS*21*)$_2$] form cointegrates with other replicons at frequencies of 10^{-2} . . . 10^{-5} in *E. coli* (274, 275, 279). The archetype (IS*21*)$_2$ plasmid is R68.45, which is a derivative of the broad host range IncP plasmid R68 (124). R68 contains a single IS*21* copy and promotes replicon fusion only rarely (279), indicating the essential role of (IS*21*)$_2$ in cointegrate formation. R68.45-mediated mobilization of nonconjugative plasmids (e.g., pACYC184) results in cointegrates in which the two plasmids are fused by single IS*21* copies in direct orientation (279) (Figure 6.1F). In Section 4, we will discuss the molecular mechanism of transposition via the reactive tandem in more detail. We do not know how widespread this transposition process is in bacteria, but at least one other IS element, IS*30*, is known to form tandem repeats (76), and hence might use the same mechanism to generate cointegrates (252).

To conclude this section, we would like to point out that some insertion sequences and transposons, e.g., IS*1* and Tn*9*, generate both simple insertions and cointegrates. It has been proposed that these transposable elements may alternatively use nonreplicative or replicative pathways of transposition (29, 114).

2.3. Site-Specific Recombination

Site-specific recombination, like transposition, is mediated by system-specific proteins and does not rely on extended sequence homology and on the RecA protein (70). The sites of specific recombination usually share a short region of sequence homology. The recombination event is reciprocal and conservative; no new nucleotides are synthesized. Integration and excision of bacteriophage λ in *E. coli* are the first and best-characterized systems of site-specific recombination (46, 70). *E. coli* can be lysogenized by phage λ in the following way. After infection, λ integrates as a circular dsDNA molecule into the bacterial chromosome, by recombination between the phage *attP* site and the chromosomal *attB* site. The integrated prophage is flanked by *attP/attB* hybrids, which are called *attL* and *attR*. Integration requires a phage enzyme, the Int protein, and integration host factor (IHF), an *E. coli* protein. During prophage induction, the phage Xis protein promotes λ excision (70).

There is a family of conjugative plasmids that mobilize chromosomal markers and elicit lethal zygosis in *Streptomyces* (see Chapter 11). A member of this family, the *S. ambofaciens* plasmid pSAM2, integrates into the chromosome of a *S. ambofaciens* strain that is naturally devoid of pSAM2. The integration event is promoted by an integrase that has structural similarities with the λ Int protein. Moreover, pSAM2 has an open reading frame (*xis*) whose predicted protein product resembles the λ Xis protein (34). Integration occurs by site-specific recombination between a 58-bp sequence present in the plasmid (*attP*) and in the chromosome (*attB*). The *xis* and *int* genes are closely linked to *attP* (30). Other plasmids in *Actinomycetes*—SLP1, pSE211, and pIJ408—appear to integrate into the chromosome by a similar integration mechanism and at similar *attB* sites (53, 184, 309).

Site-specific insertion of antibiotic resistance genes into *integrons* plays an important role in the evolution of multiple-drug-resistance plasmids and transposons (130). Integrons code for a site-specific recombination enzyme, integrase, and contain a conserved sequence where recombination takes place (31, 130, 214). The incoming resistance gene is associated with a characteristic 59-bp element and thought to occur as a circular molecule. The recombination crossover between the integron and the incoming gene is catalyzed by integrase and has been localized to a GTT triplet that is conserved in the integron and at the 3' end of the 59-bp element (130). Integron-dependent recombination was first discovered in Tn*21* and related elements (214). Under experimental conditions, integrons can promote replicon fusion at frequencies of $\sim 10^{-4}$ (130). For example, plasmid R388, which carries an integron, forms cointegrates with pACYC184 into which a 59-bp element has been inserted. Integrase not only performs cointegrate formation, but also resolves cointegrates, with excision of the insert as a covalently closed circle (69, 130).

The mobilization of the small, nonconjugative, broad host range plasmid pMV158 (originally isolated from *Streptococcus agalactiae*) by the conjugative plasmids pAMβ1 or pIP501 in lactococci might involve site-specific recombination. Plasmid pMV158 was found to encode a protein that resembles the Pre (plasmid recombination enzyme) protein of the staphylococcal plasmid pT181 (108, 338). Moreover, pMV158, pT181, and further small plasmids from gram-positive bacteria contain similar 24-bp palindromic sequences, designated RS_A sites, upstream of the open reading frame for the Pre protein (337). It is known that the plasmids pT181 and pE194 of *Staphylococcus aureus* undergo site-specific recombination at RS_A sites. This recombination produces cointegrates, is mediated by Pre protein, and does not involve RecA (108). Since mobilization of pMV158 requires its RS_A site and Pre-like protein (269, 337), it is possible that mobilization proceeds via site-specific recombination. However, no cointegrates between pMV158 and pAMβ1 (or pIP501) could be detected in *Lactococcus lactis* recipients. This suggests that either cointegrates escaped detection because of very rapid resolution or that, in fact, cointegrate formation is not necessary for pMV158 mobilization (269, 338). If the latter possibility is true, then it would be more appropriate to call the RS_A site *oriT* and the Pre-like protein Mob (269). Further experimental work is needed to clarify the situation. At any rate, this example illustrates how difficult it can be in practice to distinguish between mobilization *in trans* and mobilization *in cis*.

2.4. Illegitimate Recombination

Illegitimate recombination events, as they are currently defined (93), can arise when enzymes that break and join DNA make "errors." As a result, dsDNA molecules can be fused at sites with little or short (3 to 20 bp) sequence homology. Several types of illegitimate recombination that lead to cointegrate formation are discussed below. For a comprehensive review of illegitimate recombination, the recent paper by Ehrlich (93) should be consulted.

2.4.1. One-Ended Transposition

Members of the Tn*3* family and bacteriophage Mu can promote replicon fusion even when only one end of the transposable element is present. This process has been called *one-ended transposition*. Its mechanism was mainly studied with artificial constructs of transposons Tn*3*, Tn*21*, and Tn*1721* lacking one terminal inverted repeat but retaining the transposase gene (5, 6, 18, 115, 234, 235), and with plasmids carrying one end of Mu plus the Mu transposition functions A and B (119). In one-ended transposition, cointegrates are generated at a lower frequency than in normal replicative transposition with wild-type transposable elements. The cointegrates formed during one-ended transposition are composed of the entire target replicon and variable lengths of the donor replicon. At one junction the truncated transposable element is fused via its single "correct" end to the target. Many different target sites can be used, as expected for transpositional recombination. The second, "illegitimate" junction can be formed at different sites in the donor plasmid. Quite often the entire donor plasmid is joined to the target, and in this case the second junction is located at or very near the duplicated single active end of the transposon (Figure 6.1H) (234, 287). One-ended transposition of Tn*1721* and Tn*21* generates target duplications as in normal transposition (235), suggesting that the mechanisms of both forms of transposition are essentially the same; a model of asymmetric replicative transposition has been proposed (287).

One-ended transposition is not limited to laboratory constructs. IS*102*, which naturally occurs on pSC101, and IS*911*, an insertion sequence recently identified in *Shigella dysenteriae*, both mediate cointegration according to the scheme shown in Figure 6.1H. Cointegrates contain one copy of the IS element with an adjacent donor replicon fragment. The "illegitimate" junctions are formed by donor sequences that resemble the IS ends (209, 268). It therefore appears that in the donor replicon the IS-encoded transposase recognizes one IS end (indicated by ◁ in Figure 6.1H) and, at a certain distance from the IS element, a short similar sequence present in inverted orientation (indicated by I in Figure 6.1H). The IS end and the inverted similar sequence thus form the ends of a new transposable element. In the cointegrates, this element is flanked by target duplications (209, 268). It seems reasonable to assume that IS*911*-mediated cointegrate formation occurs by a nonreplicative mechanism (268), akin to the mechanism described for IS*21* tandems (Figure 6.1F).

Other naturally occurring plasmids also form cointegrates by mechanisms that may be analogous to one-ended transposition. The ColE1-like plasmid pUB2380 transposes into the conjugative plasmid R388 in a way that leads to cointegrates of highly variable sizes. However, one junction in the cointegrates always involves a particular site in the pUB2380 donor replicon. This common end might come from a transposable element. The other junction between pUB2380 and R388 appears to be variable (18).

The broad host range conjugative plasmid R772 forms cointegrates with the octopine Ti plasmid of *Agrobacterium tumefaciens*. An insertion sequence carried by R772, IS*70*, was found to be involved in cointegrate formation. One intact IS*70* copy and a short segment of IS*70*, respectively, were present at the junctions between R772 and Ti (148). The junction containing the truncated IS*70* element might result from one-ended IS*70* transposition or from a secondary deletion in IS*70*. In conclusion, transpositional replicon fusion does not always produce cointegrates with intact copies of the transposable element at both junctions (as in Figure 6.1B, C, D, E, and F).

2.4.2. Recombination at *oriV*

The replication origin *oriV1* of the *E. coli* F plasmid can undergo *recA*-independent site-specific recombination with another *oriV1* sequence or with a sequence having some similarity to *oriV1* (188, 253, 255). This recombination can be demonstrated in a mobilization experiment as follows. The F derivative pOX38 (55 kb), which carries *oriV1* and the *tra* functions, mobilizes a recombinant plasmid consisting of pBR322 and an *oriV1* fragment from F at high frequencies, 10^{-3} to 10^{-4} per pOX38 transferred (Figure 6.1I). Mobilization depends on an 8-bp spacer sequence located between the 10-bp inverted repeats within *oriV1*. The spacer sequence has been designated *rfsF* (replicon fusion site of F) (255). Although pOX38-pBR322*oriV1* cointegrates could not be isolated from the recipient cells (even if these were *recA*), it has been postulated that these cointegrates are intermediates in mobilization and that they are very rapidly resolved by site-specific recombination at *oriV1*, i.e., by a reversal of the process of cointegrate formation (Figure 6.1I). Protein D encoded by F and a region near *oriV1* are necessary for resolution, which is assumed to be much faster than cointegrate formation (255). Site-specific recombination at *oriV1* may be important for the resolution of F multimers and hence may improve maintenance of this low-copy-number plasmid (349).

The mechanism of *oriV1*-dependent replicon fusion could be studied by an analysis of stable cointegrates between pBR322 (without *oriV1*) and pOX38. These arise rarely, at 10^{-8} per pOX38 transfer. Fusions occur by site-specific recombination between *oriV1* on pOX38 and sequences on pBR322 that bear some resemblance to *oriV1*. As a result, the junction sequences in the pOX38-pBR322 cointegrates are similar but not identical and this may impede cointegrate resolution. O'Connor et al (255) have shown that site-specific recombination between *oriV1* and related sequences takes place within the 8-bp spacer; there are no added, deleted, or changed bases at the junctions. In this respect, site-specific recombination at *oriV1* is analogous to λ integration (255).

2.4.3. Recombination at *oriT*

RecA-independent site-specific recombination can also occur at *oriT*. Broome-Smith (41) demonstrated that, during mobilization of ColE1 by R64*drd*-11 *in trans*, a small cryptic plasmid (pLG500) present in the donor was comobilized. In the recipient, ColE1 and pLG500 were found to be linked together; replicon fusion occurred at (or very close to) the *oriT* (*nic*) of ColE1 and a specific site of pLG500, presumably also representing an origin of transfer (Figure 6.1J). Plasmid pLG500 alone is not mobilized by R64*drd*-11; it appears that pLG500 does not code for mobilization (*mob*) functions. The following model has been proposed to account for the formation of pLG500-ColE1 cointegrates (41). Both pLG500 and ColE1 are mobilized *in trans* by R64*drd*-11. After transfer to the recipient, the leading 5' end of the ColE1 single strand (perhaps with a Mob protein of ColE1 attached to it) would occasionally fail to recognize its own 3' end but might ligate to the 3' terminus of pLG500. In turn, the 5' end of pLG500 would be joined to the 3' end of ColE1. Thus, a composite pLG500-ColE1 plasmid can be generated and circularized in the recipient. This composite plasmid does not always dissociate and therefore can be isolated from the recipient. In the donor, pLG500-ColE1 cointegrates have not been detected (41). Site-specific illegitimate recombination at *oriT* is also involved in the formation of a certain type of Δ*tra* F' plasmids

(116, 149) (see Section 3.1.3). Illegitimate, *oriT*-dependent recombination is stimulated by the *traY* and *traI* products of F (47).

Recent experiments (10, 28, 226) provide direct evidence for site-specific recombination at *oriT*. When two directly repeated copies of *oriT* from the mobilizable broad host range plasmid R1162 (= RSF1010) are cloned into bacteriophage M13mp19, they undergo recombination; a phage molecule with a single *oriT* is formed and the segment lying between the two *oriT*s is deleted. This recombination requires mobilization functions of R1162, but can be observed in the absence of conjugation. Recombination takes place at the nick sites in the *oriT*s (28). It has been proposed that the M13 system models the ligation of the incoming, ssDNA molecules in the recipient. In this ligation step, the 3′ end generated at one *oriT* is joined to the 5′ end at the cleavage site of the second *oriT* (10, 28, 226).

2.4.4. Recombination at Short Homologous Sequences

Various illegitimate genome rearrangements (deletions, duplications, cointegrate formation) can occur by recombination at short (3- to 20-bp) homologous sequences (3, 13, 93, 215). In some of these recombination events, RecA (or RecE in *Bacillus*) play a role, whereas in others, RecA is not required (93). There is evidence that DNA gyrase (topoisomerase II) can promote *recA*-independent inter- and intramolecular recombination by a mechanism that is believed to involve double strand breakage and reunion of DNA (168, 232, 241). Gyrase-mediated recombination has been demonstrated *in vitro* in cell extracts; it is possible that gyrase alone does not carry out the entire recombination process but is assisted by other proteins in the cell extract (167). Gyrase is known to cleave ds DNA *in vivo* and *in vitro* with some specificity, and a fairly degenerate consensus sequence has been proposed for the cleavage sites (204). Some, but by no means all, sequences where illegitimate recombination has taken place show resemblance to the consensus sequence for gyrase cleavage sites (168, 215). Thus, gyrase-promoted recombination may account for some events of illegitimate recombination, but gyrase cannot be the sole factor determining the site of strand exchange. This conclusion is also supported by the analysis of a rare class of pOX38-pBR322 cointegrates that were generated outside the *oriV1* region (254).

Our knowledge of the enzymatic systems and reactions that perform illegitimate recombination is rudimentary at present. Frequencies of illegitimate recombination vary widely, ranging from 10^{-3} to $< 10^{-8}$. Short homologous sequences are found often, but not always at the junctions of rearranged DNA (3, 13, 93). A number of different mechanisms of illegitimate recombination remains to be discovered.

3. Use of Mobilization Involving Cointegrative Mechanisms

3.1. Chromosome Mobilization and Mapping

In this section we will focus on chromosome mobilization in Gram-negative bacteria. With the exception of the *Actinomycetes* (see Chapter 11), much less is known about chromosome transfer in gram-positive microorganisms, although there can be little doubt that the basic principles described here also apply to these organisms.

3.1.1. Hfr Strains in Gram-Negative Bacteria

Stable integration of a conjugative plasmid into the bacterial chromosome gives rise to Hfr (high frequency of recombination) donor strains. Chromosome transfer proceeds unidirectionally from the site of plasmid integration (126, 139, 206), and proximal chromosomal markers are transferred at frequencies of 10^{-1} to about 10^{-4} per donor, depending on the efficiency of conjugation, i.e., on the type of conjugative plasmid integrated. Hfr strains are used to map chromosomal markers according to time of entry by the classical technique of interrupted mating, and the data thus obtained can be extended and corroborated by linkage analysis (139, 206). The classical circular chromosome maps of *E. coli*, *S. typhimurium*, *Pseudomonas aeruginosa*, and *P. putida* have all been calibrated in time units as determined by Hfr crosses (7, 37, 156, 258, 282, 316). Time-of-entry data have also helped to order markers in other gram-negative bacteria, e.g., in *Citrobacter freundii* (81), *Erwinia amylovora* (55), *P. syringae* (249), and *Methylobacillus flagellatum* (331).

The F plasmid of *E. coli*, the archetypal episome, integrates spontaneously into the chromosome and, in many cases, remains there stably (74, 139). The frequency of spontaneous, stable integration of F is sufficiently high so that Hfr strains can be isolated from F^+ cultures by screening a large number of individual F^+ clones (74, 170). Such behavior is not commonly observed for other conjugative plasmids in gram-negative bacteria (Table 6.1). Usually some kind of selective pressure has to be applied to isolate strains carrying an integrated plasmid. The following procedures have proved useful.

1. Mutagenic treatment of plasmid carrier strains with ultraviolet (UV), alkylating agents, or acridine orange (21) may enhance the frequency of plasmid integration and/or stabilize an integrated plasmid. For instance, Hfr strains could be isolated from *E. coli* strains harboring the colicinogenic plasmid ColV only after mutagenesis with nitrosoguanidine (181), and a similar treatment was necessary to obtain stable chromosomal integration of the sex factor FP2 in *P. aeruginosa* (145). Integration of the F plasmid in *E. coli* can also be favored by mutagenic treatment (21, 206). In these cases it is not clear what kind of mutagenic event was responsible for the isolation of Hfr strains. In contrast, there is plausible explanation for the mode of action of acridine orange during Hfr formation; this curing agent interferes with the autonomous replication of F and F primes (206). For example, an Hfr donor was obtained from an *Erwinia* strain carrying $F' lac^+$ after treatment with acridine orange and subsequent selection for Lac^+ (55).

2. Plasmids that are temperature-sensitive (ts) for replication but are maintained normally at low temperatures (usually 30°C) have been used successfully for the isolation of Hfr donors in *E. coli*, *Salmonella*, *Citrobacter*, *Rhodobacter capsulatus*, and *P. aeruginosa* (Table 6.1). When plasmid markers are selected at nonpermissive temperature (usually 40–43°C), it is often possible to recover surviving bacteria in which the ts plasmid has recombined with the chromosome. The plasmid is then passively replicated along with the chromosome. This principle has been exploited with F primes, R*ts*1, pSC101, and IncP broad host range plasmids of the RP1 type (Table 6.1). In most instances, the ts character was introduced by mutation (75, 78, 128, 133, 257); R*ts*1 is an exception in that it is naturally ts (322). Since the ts defects of these plasmids are usually manifested at temperatures above 37°C, the method of ts plasmid integration is obviously limited to host

TABLE 6.1   Hfr Donor Strains in Gram-Negative Bacteria

Bacteria	Conjugative plasmid or *oriT* element integrated[a]	Methods of Hfr isolation[b]	References
Citrobacter freundii	F'*ts*	c	81
Erwinia amylovora	F'	b	55
Erwinia chrysanthemi	F'	a	56
Escherichia coli	F	a, b, d	21, 72, 139, 206, 247, 333
	F',F'*ts*	c, d, f	75, 82, 153, 166, 205, 236
	F'*ts*::Tn*10*[c]	c	191
	ColB	d	236
	ColV	b, d	181, 236, 248
	R1	a, d	142, 236
	R6K	d	310
	R100	d	248, 355
	R483, R144	d	79
	Rts1	c	322
	pVF9 (= pSC101*tra*⁺)	c	101
	RP1	d, f	213, 298
	RP4'	a	117, 343
	pTH10 (= RP4*ts*)	c	133
	pRP19.6 (= R68.45*ts*)	c	78
Methylobacillus flagellatum	pAS8-121 (= RP4::ColE1)	e	291, 331
Paracoccus denitrificans	Tn*1* (*oriT*/*tra*$_{RP4}$)	g	110
Pseudomonas aeruginosa	FP2	b, d	145, 146
	pME319 (= RP1*ts*)	c	128
	pMO190 (= R68*ts*)	c	154
	pME134 (= RP1*ts tnpR*)	c	273
	pME487 (= R68.45*ts* Ap^s)	c	276
	pMO514 (= R68*ts*::Tn*2521*)	c	257, 258
	pMT1000 (= R68*ts*::Tn*501*)	c	136
Pseudomonas putida	pMO22 (= R91-5::Tn*501*)	e	83
	pMO75 (= R91-5::Tn*5*)	e	316
Pseudomonas syringae	R91-5	e	249
	pMO22	e	249
	pMO75	e	249
Rhizobium meliloti	Tn5-Mob/pRK600	g	192
Rhodobacter capsulatus	pJMP7 (= RP4*ts*)	c	68, 210
Rhodobacter sphaeroides	R751::Tn*813*	a	35
	pAS8-121	e	182´
	Tn*5*-*oriT*$_s$/Factor S	g	319
Salmonella abony	F	a	282, 283
Salmonella typhimurium	F	a, d	7, 282, 283
	F'*ts*::Tn*10*[c]	c	60

[a]To keep the size of this table manageable, we use F', F'*ts*, ColV, ColB, RP1 (= RP4), and RP4' as generic designations for groups of related plasmids in Enterobacteriaceae, and we do not indicate individual plasmid designations.

[b]Chromosomal integration of conjugative plasmids was obtained as follows: a, spontaneous; b, after mutagenic treatment; c, plasmid with temperature-sensitive replication, selected at nonpermissive temperature; d, integrative suppression; e, no autonomous plasmid replication due to narrow host range of replication; f, incompatibility; g, in some instances, an *oriT* was chromosomally inserted on a transposable element. For details, refer to text.

[c]The simultaneous presence of a chromosomal Tn*10* insertion provides a region of "portable homology" for plasmid integration.

strains that can grow at these high temperatures. In strains that do not tolerate high temperatures, such as *P. putida*, other methods must be tried.

3. Integrative suppression of a chromosomal *dnaA*(ts) mutation provides a powerful selection for isolating Hfr cells (247). In *dnaA*(ts) mutants the initiation of chromosome replication is blocked at nonpermissive temperature. After integration of a plasmid, the initiation of replication, at nonpermissive temperature, is directed by the plasmid's origin of replication. At permissive temperature the normal chromosome origin is used (50, 247, 310). Many conjugative plasmids (F, ColV, ColB, R1, R6K, R100, R483, R144, RP1) (Table 6.1) have been inserted into the *E. coli* chromosome by integrative suppression and they always conferred Hfr properties on the host (50, 79, 166, 213, 236, 247, 248, 310, 355). In *S. typhimurium*, a *dnaA*(ts) mutation is also available and has been exploited to construct Hfr strains by integrative suppression (7). In *P. aeruginosa*, the *dna-11* mutation has *dnaA*(ts) characteristics and proved helpful in the stabilization of the Hfr-donor properties of a strain carrying a chromosomally integrated FP2 plasmid (146).

4. Plasmids having a narrow host range of replication but a wide host range of conjugative transfer can be used to construct Hfr donors in host strains that do not support autonomous replication of the plasmids. For instance, the conjugative IncP-10 plasmid R91-5 is able to replicate autonomously in *P. aeruginosa* but not in *P. putida* or *P. syringae*. R91-5 derivatives have been transferred by conjugation to *P. putida* and *P. syringae*; selection for the plasmid-encoded antibiotic resistance has resulted in Hfr strains carrying a chromosomally integrated plasmid (83, 249, 316). By genetic manipulation, the broad host range replication machinery of plasmid RP4 has been replaced by the narrow host range replication system of ColE1. In the RP4-ColE1 cointegrate pAS8-121, the replication control gene *trfA* of RP4 is inactivated through a Tn7 insertion and replication is directed by ColE1. Thus, pAS8-121 replicates in *E. coli*, but fails to do so in nonenteric bacteria. This construct has been used to generate Hfr strains in *Rhodobacter sphaeroides* (182) and in *Methylobacillus flagellatum* (291, 331).

5. Incompatibility properties can also be utilized to select for chromosomal integration of a plasmid. When an *E. coli* recipient harbors an integrated F plasmid, an incoming F' plasmid fails to replicate because it is incompatible with the chromosomal F copy; however, the F' plasmid can be rescued by insertion into the chromosome. Thus, a "double male," i.e., an Hfr strain with two transfer origins, can be created (82). A similar approach allowed the construction of the *E. coli* Hfr donor strain S17-1, which is widely used for mobilization *in trans* of plasmids to other bacteria (298). An RP4 derivative (RP4-2 Tc::Mu Km::Tn7) was forced to integrate into the chromosome of a *recA E. coli* strain by the introduction of an incompatible plasmid and selection for the continued presence of both plasmids. The autonomous incoming plasmid was then cured by several rounds of treatment with acridine orange (298). Parenthetically, an F-derived Hfr strain does not tolerate an additional autonomous F plasmid, whereas an RP4-derived Hfr strain does allow autonomous replication of RP4 (117, 344). This difference in incompatibility behavior is due to different copy numbers. F has a single copy per chromosome, and hence one chromosomal copy of F expresses strong incompatibility toward an additional F replicon (183). In contrast, RP4 has six to eight copies per *E. coli* chromosome. A chromosomal RP4 slightly lowers the copy number of autonomously replicating RP4 plasmids but does not eliminate them (117). This phenomenon, which has been termed integrative compatibility (344), is consistent with a negative control of plasmid replication and incompatibility (117).

As mentioned in Section 1, chromosomal integration of a transposable element containing an *oriT* can provide Hfr origins. The corresponding *tra* functions may be supplied by the transposable element itself (172) or by a conjugative plasmid *in trans* (192, 276, 319). Hfr strains of *Paracoccus*, *Rhodobacter*, and *Rhizobium* sp. have thus been created (Table 6.1).

We now turn to the structures of chromosomally integrated plasmids. This topic has been studied most thoroughly in the case of F with essential contributions by the groups of N. Davidson, E. Ohtsubo, and R. Deonier (80, 84–86, 129, 216, 333). Campbell's model of λ integration in *E. coli* (46) stimulated the hypothesis that circular forms of F (and F primes) insert themselves into the circular chromosome by homologous recombination [reviewed by Hayes (139)]. Support for the involvement of homologous recombination came from the observation that in *recA* mutants, F and F primes are normally maintained but promote chromosome transfer at much reduced frequencies (63, 139) and, in parallel, generate Hfr strains about 100 times less frequently compared to a *rec*⁺ host (72). In addition to RecA, the RecF pathway also seems to be important for chromosomal integration of F, as demonstrated by integrative suppression experiments (39). It is now quite clear that the insertion sequences carried by F and by the *E. coli* chromosome are the sites where F integration and excision occur via homologous recombination (80, 84, 85, 333). F carries one copy of IS2, two copies of IS3, and one copy of Tn*1000* (= γδ) (349). When F lacks these insertion sequences, it is unable to mobilize chromosomal markers (348). The chromosomal IS2 and IS3 copies have recently been mapped; *E. coli* W3110 carries 12 to 14 IS2 copies and six IS3 copies. The origins of chromosomal transfer in many Hfr strains correlate well with the locations of the IS2 and IS3 elements (30, 333), suggesting that in these strains the IS elements have determined the site of F integration (Figure 6.2, p. 162). A detailed analysis has revealed that IS2 and IS3 elements in the *lac-purE* region of the chromosome account for the chromosomal integration of F in 16 of about 19 Hfr strains having origins in that region (85).

The numbers and sites of IS2 and IS3 insertions vary between individual *E. coli* K-12 strains (30, 107, 333). It is therefore possible that some Hfr strains, whose transfer origins do not correlate with the known chromosomal IS2 and IS3 elements of strains W3110, contain *new* chromosomal insertions of IS2 or IS3, permitting the integration of F. Formally, this situation corresponds to the scheme depicted in Figure 6.1C. Whether IS2 and IS3 can also promote F integration via replicative transposition (Figure 6.1B) is not known. It is known, however, that Tn*1000*, which promotes replicative transposition of the Tn3 type (Figure 6.1B), accounts for some Hfr strains. In fact, a chromosomal Tn*1000* copy can act as a "sex factor affinity site" (123). Because Tn*1000* codes for an active resolvase, one can expect that the Tn*1000*-dependent F–chromosome cointegrates are unstable unless some secondary rearrangements take place at the junctions between F and the chromosome. The paper by Umeda and Ohtsubo (333) nicely recapitulates the experimental evidence that has led to our current understanding of F–chromosome interaction (Figure 6.2).

We have already hinted at the *recA*-dependent insertion of F′ plasmids. The structures of F′ plasmids will be discussed in Section 3.1.3. Chromosomal segments carried by F′ plasmids effectively recombine with the homologous chromosome regions (285), and thus provide specific sites for F integration. Hence, both the origin and the orientation of chromosome transfer can be predetermined. The use of F′ ts derivatives facilitates the selection of Hfr strains (75, 205). The same principles also apply to R′ plasmids.

Plasmid R100.1, which carries the tetracycline-resistance (*tet*) transposon Tn*10*, has been inserted into the *E. coli* chromosome by the technique of integrative suppression. This insertion can result in Hfr strains that have lost the *tet* determinant (51). In one such Hfr strain, which was analyzed in detail, the R plasmid was found to have integrated by inverse transposition (Figure 6.1E). The inner ends of the IS*10* elements, i.e., the ends opposite to those normally serving in transposition of Tn*10*, were used to transpose the entire R100.1 plasmid but without the *tet* region (51).

When the *E. coli* chromosome contains a single Tn*10* element, IS*10*-mediated inverse transposition can fuse any plasmid (not carrying Tn*10*) to the chromosome. As described above for transposition of R100.1, the inner ends of IS*10* are cut and ligated to the target DNA (i.e., the plasmid), with concomitant loss of the *tet* determinant of Tn*10*. This event is rare (10^{-8} per cell and generation) but can clearly be observed when appropriate selection for integration is available (135). The strategy of inverse transposition has been used to generate new *E. coli* Hfr strains. A ts derivative of pSC101 carrying the entire *oriT/tra* region of F (pVF9) (Table 6.1) was integrated into the chromosome at the sites of Tn*10* insertions. Selection for the pVF9 chloramphenicol-resistance marker at nonpermissive temperature was combined with an enrichment for tetracycline-sensitive clones (101). Since Tn*10* insertions are available at many different locations of the *E. coli* chromosome (22), Hfr strains can be created at virtually any predetermined site.

In another strategy for Hfr isolation, chromosomal Tn*10* insertions serve as regions of "portable homology" with a Tn*10* element carried by a conjugative plasmid, e.g., F'*ts* (60, 191). Thus, integration of F'*ts*::Tn*10* occurs by homologous recombination, and the integrated plasmid will be flanked by Tn*10* copies in the same orientation. This procedure, which has been used in *S. typhimurium*, has the advantage that both the origin and the direction of chromosome transfer can be predetermined (60). A collection of Hfr donors in *P. putida* has been obtained by a variation of this method (Table 6.1). Chromosomal Tn*5* insertions were used to direct the integration of the narrow host range plasmid pMO75 (= R91-5::Tn5), which does not replicate in *P. putida* (316). The majority (~ 90%) of the Hfr donors isolated from any particular Tn*5* mutant transferred the chromosome in one direction, whereas the Hfr strains of the minority class gave chromosome transfer in the opposite direction (316). It is not clear how the minority class Hfrs were generated as no physical analysis of integrated plasmids was carried out. A possible explanation might be provided by the structure of Tn*5*, which is bordered by large (1.5-kb) inverted repeats consisting of IS*50* elements in opposite orientation (24). The minority class Hfrs might arise after recombination between, say, IS*50*L (left) at one Tn*5* element and IS*50*R (right) of the other Tn*5* element, with the internal sequences of the transposons in opposite alignment (316). Although the formation of a similar minority class of Hfr would be expected in the Tn*10*-dependent plasmid integration events, in fact no such class has been found (60).

The λ attachment site *attP*, when inserted into the IncP plasmid RP4 with the λ *int* gene, allows site-specific recombination between RP4 λ*att* and the chromosomal *attB* site in *E. coli*. As a result, an Hfr strain has been isolated that has the plasmid integrated at *attB* (343, 344).

In *P. aeruginosa*, chromosomal integration of pME487, a ts derivative of R68.45, occurs via nonreplicative transposition mediated by (IS*21*)$_2$, as outlined in Figure 6.1F. The integrated plasmid is flanked by single IS*21* copies in the same orientation, and excision can be prevented by a *recA* mutation in the host (276). Another ts IncP plasmid, the RP1 derivative pME134, has been inserted into the *P. aeruginosa* chromosome via replicative transposition of Tn*801*, which is a member of the Tn*3* family (273). Both pME487 and pME134 carry a ts mutation in their replication control gene *trfA*, enabling selection for chromosomal integration at 42°C, and both plasmids give rise to Hfr donors (273, 276) (Table 6.1). In pME134, the resolvase (*tnpR*) gene of Tn*801* was inactivated by the insertion of a trimethoprim (Tp) resistance determinant. After integration, the pME134–chromosome junctions consist of directly repeated Tn*801*Tp elements, as predicted by the scheme of Figure 6.1B. Plasmid excision is blocked by the *tnpR* mutation and, in addition, by a *recA* mutation in the host (273). Without the *tnpR* and *recA* mutations, an integrated RP1*ts* plasmid (pME319) is highly unstable in the *P. aeruginosa* chromosome, although one stable Hfr strain has been obtained after repeated subculturing at permissive and nonpermissive temperatures (128). A ts derivative of R68 loaded with Tn*2521*, pMO514, has been used to construct another set of Hfr strains in *P. aeruginosa* (257, 258). The integrated pMO514 is flanked by an intact Tn*2521* copy at one border and a truncated Tn*2521* copy at the other border, suggesting that transposition of Tn*2521* is involved in the insertion process. However, the mechanism of Tn*2521* transposition and the reason(s) for truncation of one copy are unknown (256).

In *Myxococcus xanthus*, RP4 cannot replicate autonomously and integrates into the chromosome. Chromosome transfer is promoted by the integrated plasmid at low frequencies and, therefore, we have not included these donors in Table 6.1. The IS*21-par* region of RP4 (Figure 6.4) is important for integration, but the precise mechanism has not been elucidated (40, 109, 171).

Some Hfr strains as first isolated are unstable. This problem occurs especially when the plasmid–chromosome cointegrates can be resolved by RecA, resolvase, or a Xis function (Figure 6.1). For instance, *S. typhimurium* Hfr strains obtained by homologous recombination of F′ *ts*::Tn*10* with chromosomal Tn*10*s are quite unstable and should be used immediately after isolation of temperature-resistant clones at 42°C (60, 282). Similarly, most Hfr strains of *E. coli* and *P. aeruginosa* created by integration of pTH10 or pME319 (ts derivatives of RP4 and RP1, respectively) are highly unstable, presumably because the Tn*1*/Tn*801*-promoted cointegrates are rapidly resolved by resolvase (128, 133). Plasmid copy numbers > 1 probably aggravate the instability problem. In *E. coli*, RP4 has six to eight copies per chromosome in the autonomous state. RP4 plasmids recovered after stable chromosomal integration are defective for replication (118). When the RP1 (RP4) origin of replication remains functional after integration, the plasmid has a marked tendency to excise and the Hfr donor properties are highly unstable. This was demonstrated in *E. coli* when a *dnaA*(ts) mutation was suppressed by RP1 integration; as soon as selective pressure (growth at 41°C) was relieved, autonomous RP1 molecules reappeared and donor ability was lost (213). In stable *P. aeruginosa* Hfr strains, a significant proportion of chromosomally integrated pME134 and pME487 molecules has acquired an insertion element inactivating the *trfA* replication control gene. It should be emphasized that these Hfrs were selected after repeated purification at both permissive and nonpermissive temperatures, leaving time for insertional inactivation of *trfA* (273, 276). Thus, it appears that multiple rounds of replication initiation from the origin of an integrated plasmid are deleterious to the host, at least when normal chromosome replication from *oriC* remains possible. A similar situation occurs in *E. coli* Hfr strains created by the integration of the ts pSC101 derivative pVF9. In the autonomous state, pSC101 has about five copies per chromosome. Strains carrying an integrated pVF9 can be easily selected at nonpermissive temperature, but they grow poorly and show a reduced efficiency of plating when shifted to permissive temperature (30°C). This deleterious effect on the host is linked to the RepA function of pSC101 (101).

When the integrated plasmid is a low-copy-number replicon, like F, its own replication origin may be partially switched off in exponentially growing cells because under these conditions its copy number may already exceed the normal copy number of the autonomous plasmid, and this may inhibit further initiation of replication at the plasmid origin. Thus, in classical Hfr strains of *E. coli*, replication appears to initiate mostly at the chromosomal origin and the integrated F plasmid seems to be replicated passively (49). However, the site of F integration may influence the replication pattern and evidence for the functioning of the F replication origin has also been obtained in an *E. coli* Hfr strain during exponential growth (208). In conclusion, the chromosomal replication machinery of *E. coli* tolerates an integrated F plasmid quite well, whereas integrated intermediate- or high-copy-number plasmids tend to be deleterious to the host.

Chromosomal integration of a plasmid with an intermediate- or high-copy number can lead to a further complication. Structural analysis of *P. aeruginosa* Hfr strains derived from the IncP plasmids pME134 or pME487 indicates that two (or more) plasmid copies have been integrated into the chromosome in tandem. Perhaps the conditions of antibiotic selection have favored the integration of multiple plasmid copies (273, 276). A tandem configuration does not affect Hfr donor properties; unidirectional chromosome transfer still occurs. However, bidirectional chromosome transfer has also been observed in some R plasmid-derived Hfr strains. Such cases are probably due to the joint insertion of (at least) two plasmid copies in opposite orientation (133, 210, 316). Double-male strains can be obtained with F as well, although special conditions of selection must be employed (82, 206). The caveat is that it should not be assumed automatically that any newly isolated Hfr strain transfers the chromosome unidirectionally from a single origin; this has to be demonstrated by experimental evidence (time-of-entry data, linkage analysis).

3.1.2. Chromosome Transfer without Stable Integration of a Conjugative Plasmid

Escherichia coli strains carrying an autonomous F plasmid are chromosome donors although markers are transferred at much (about 10^{-5}-fold) lower frequencies than those observed in Hfr crosses. Some recombinants in $F^+ \times F^-$ crosses are due to the presence of stable Hfr donors in the F^+ population. Eighty to eighty-five percent of the recombinants, however, arise by another mechanism (74), either by transient and unstable integration of F into the chromosome or by some kind of mobilization *in trans*. Although there is no direct evidence for either mechanism, the data of Willetts and Johnson (348) are in favor of a physical interaction between the mobilizing F plasmid and the chromosome. To our knowledge, the natural occurrence of an *oriT* in the chromosomes of gram-negative bacteria has not been demonstrated. (However, as stated in Section 1, an *oriT* can be introduced into the chromosome by artificial transposons such as Tn5-Mob.)

It is probably true that any conjugative plasmid can transfer chromosomal markers by mobilization *in cis* if three conditions are fulfilled. First, conjugation should be efficient; ideally, the frequency of plasmid transfer per donor cell should be 10^{-4} or better. Second, the plasmid should be capable of effective homologous, transpositional, site-specific, or illegitimate recombination with the chromosome. Third, in the recipient the chromosomal segments transferred should recombine readily with the resident chromosome, as linear chromosome fragments are rapidly lost. The last condition is clearly satisfied in wild-type ($recA^+recBCD^+$) recipients of *E. coli*. As shown by the recovery of selected proximal markers in Hfr crosses, the frequency of recombination between the linear incoming chromosome fragment and the resident chromosome may approach or even exceed 50% in Rec[+] recipients (139, 305, 306). The principal reason for the high frequency of this conjugational recombination is that the incoming linear DNA is an excellent substrate for the RecBCD enzyme (305, 306). In *P. aeruginosa* the frequency of conjugational recombination in the recipient is also very high, at least 10% per conjugation event (273).

F and the IncP plasmids RP1, RP4, and R68 satisfy the first condition in that they are naturally derepressed for conjugative transfer (cf. Chapters 2 and 3). Several R plasmids whose conjugative functions are normally repressed have been mutated to give derepressed (*drd*) variants. Several of these, for instance, R1*drd*-19 in *E. coli* (260), R144*drd*-3 in *Klebsiella pneumoniae* (91), and R100*drd*-56 in *Erwinia amylovora* (57), have been shown to have chromosome-mobilizing ability (Cma). Recombinational proficiency, the second condition, is probably inherent in "good" sex factors. Indirect experimental evidence for the importance of recombination comes from conjugative plasmids that have been genetically manipulated to carry active transposable elements or DNA fragments having homology with chromosomal regions. For instance, transposons Tn*501* and Tn*813* (= Tn*21 tnpR*), which belong to the Tn*3* family, enhance the Cma of IncP plasmids in *R. sphaeroides* (35, 261). The concept of "portable homology" has been extended to conjugative plasmids that do not stably integrate into the chromosome. The same transposable elements (e.g., Tn*1* or Tn*5*) present in the chromosome and on conjugative plasmids give rise to "transposon-facilitated recombination" (Tfr) donors, which promote polar chromosome transfer. Tfr donors have been constructed in *P. aeruginosa* (196, 308), *Vibrio cholerae* (175), a marine *Vibrio* strain (164), *Agrobacterium tumefaciens* (162, 264), and recently also in a gram-positive bacterium, *Staphylococcus aureus* (315). *In vitro* or *in vivo* insertions of chromosomal fragments into IncP plasmids produce R' plasmids that, like F' plasmids, give oriented chromosome transfer in *E. coli*, *P. aeruginosa*, *Proteus morganii*, or *Rhizobium meliloti* (12, 14, 116, 150, 179, 265). In all cases mentioned, the modified IncP plasmids have a much better Cma than their parental natural plasmids, supporting the idea that enhanced tranpositional or homologous recombination stimulates Cma. Two additional IncP plasmids with much increased transpositional activity and Cma, pULB113 and R68.45, will be considered below (p. 159–161).

There is a long list of natural sex factors in gram-negative bacteria. Holloway (151) compiled data on some 30 Cma[+] plasmids in 1979. Many of these are R plasmids, with narrow or broad host ranges. The obvious speculation is that transposable elements are responsible, at least in part, for the Cma displayed by these R plasmids. Some R plasmids give oriented chromosome transfer; examples are provided by R68 in *P. aeruginosa* strain PAT (342), R40a, R391, Rip69, and R447b in *Proteus mirabilis* (65, 66), and R1*drd*-16 in *E. coli* (142). Other R plasmids show no preference for oriented transfer and presumably interact with a range of chromosomal sites. For instance, this kind of Cma is observed for RP1 (RP4, R68) in *Acinetobacter calcoaceticus* (327), *Caulobacter crescentus* (11), *Pseudomonas glycinea* (106), and *Rhizobium meliloti* (22).

Holloway's (151) list also contains several Cma[+] plasmids that do not code for antibiotic resistances, such as the sex plasmids FP2, FP5, FP39, and FP110 in *P.*

aeruginosa, the XYL-K plasmid in *P. putida*, or ColI plasmids in enteric bacteria. More recent work indicates that the Ti plasmid of *Agrobacterium* is also able to mediate chromosome transfer (88). These Cma^+ plasmids give either random or specific, oriented transfer. The FP2 plasmid transfers the chromosome of *P. aeruginosa* strain PAO from a major and a minor origin, in opposite direction. Both origins lie close to each other (308). It can be speculated that these origins are sites of preferential recombination with FP2; however, the physical interaction of FP2 with the *P. aeruginosa* chromosome has not been studied. Other FP plasmids of *P. aeruginosa* and the XYL-K plasmid of *P. putida* appear to have single origins of chromosome transfer, and this property has been exploited for genetic mapping by time-of-entry (151). RecA function in the donor is necessary for Cma of FP2 in *P. aeruginosa* (52) but dispensable for Cma of XYL-K and TOL plasmids in *P. putida* (48) and for Cma of ColI plasmids in *E. coli* (63).

Chromosome-mobilizing efficiencies (i.e., the frequency of chromosomal marker transfer per frequency of plasmid transfer) range from about 10^{-2} to 10^{-6} for various plasmids having Cma (65, 151). It appears likely that chromosome mobilizing efficiencies largely reflect the frequencies of plasmid–chromosome recombination in the donor cells, but for most Cma^+ plasmids other than F, little is known about the mechanisms that determine the efficiencies and origins of chromosome transfer.

However, for two widely used Cma^+ plasmids, R68.45 and pULB113 (Table 6.2), we do have fairly accurate ideas of how these plasmids interact with the chromosome and other replicons. Both plasmids are IncP derivatives and thus capable of replication in a wide range of gram-negative bacteria; these plasmids carry highly active transposable elements. R68.45 has an IS*21* tandem duplication [= $(IS21)_2$] that very efficiently promotes cointegrate formation by a "cut-and-paste" (nonreplicative) mode of transposition (Figure 6.1F). R68.45 was isolated from *P. aeruginosa* as a variant of R68, which carries a single IS*21* element and forms cointegrates infrequently (124, 127, 198, 278, 351). Mechanistic aspects of $(IS21)_2$-mediated cointegrate formation will be discussed in Section 4. The other frequently used plasmid, pULB113, is an RP4 derivative carrying mini-Mu (Mu3A). Mini-Mu (8 kb) has intact Mu ends and an active transposase (A) gene but lacks all the viral functions that are deleterious to the host (334). Infecting Mu integrates into target DNA by nonreplicative, simple transposition (Figure 6.1C), but during lytic growth Mu produces predominantly cointegrates by a replicative mechanism (Figure 6.1B). Mini-Mu plasmids generate both simple insertions and cointegrates *in vivo* and *in vitro*, with 5-bp target duplications (233). Both $(IS21)_2$- and Mu-promoted replicon fusions are *recA*-independent (14, 336, 351).

A survey (Table 6.2) reveals that R68.45 displays good Cma in *Pseudomonas*, *Rhizobium/Agrobacterium* and related bacteria, whereas pULB113 has the highest activity in enteric bacteria. Cma clearly depends on $(IS21)_2$ in R68.45 (73, 127) and on mini-Mu in pULB113 (334), respectively. Although in different bacteria the recovery of recombinants per donor varies considerably (Table 6.2), this is to some extent a consequence of different plasmid transfer frequencies. Chromosome-mobilizing efficiencies (calculated per plasmid transfer) generally range between 10^{-3} and 10^{-5}, indicating that $(IS21)_2$- and Mu-mediated transposition have extended host ranges. In a few bacteria, e.g., cowpea rhizobia and *Pseudomonas glycinea*, R68.45 and RP4 have equal Cma, and in other bacteria, e.g., *Proteus mirabilis*, some strains of *Erwinia carotovora* and *Acinetobacter calcoaceticus*, R68.45 is reported not to have Cma (58, 64, 106, 220, 328). It is not clear whether in these

TABLE 6.2   Chromosome Mobilization by the IncP Plasmids R68.45 and pULB113 in Gram-negative Bacteria

Bacteria	Demonstration of Cma (expressed as recombinants/donors) by		References
	R68.45	pULB113	
Aeromonas hydrophila		10^{-4}–10^{-5}	111
Agrobacterium tumefaciens	10^{-4}–10^{-6}		43, 162, 228
Alcaligenes eutrophus		10^{-4}–10^{-5}	200, 225
Ancylobacter aquaticus	10^{-6}–10^{-7}		32
Azospirillum brasilense	10^{-6}		100
Azotobacter vinelandii	10^{-4}–10^{-6}		326
Bordetella pertussis	10^{-6}–10^{-8}		303
Enterobacter cloacae		10^{-5}–10^{-7}	290
Erwinia amylovora		10^{-6}	58
Erwinia carotovora	10^{-5}–10^{-7}	10^{-6}–10^{-7a}	57, 58, 99
Erwinia chrysanthemi	10^{-6}–10^{-7}	10^{-3}	54, 56, 58, 288
Escherichia coli	10^{-6}–10^{-7}	10^{-4}	246, 334
Klebsiella pneumoniae	10^{-5}	10^{-6}–10^{-7}	201, 334
Legionella pneumophila	10^{-8}		92
Methylobacterium AM1	10^{-4}–10^{-5}		155, 320
Methylobacterium extorquens		ND	2
Methylophilus methylotrophus	10^{-6}–10^{-7b}		155, 237
Paracoccus denitrificans	10^{-8}		352
Proteus mirabilis		10^{-6}–10^{-7}	334
Pseudomonas aeruginosa PAO	10^{-3}–10^{-5}		124, 125
Pseudomonas aeruginosa PAT	10^{-3}–10^{-4}		342
Pseudomonas cepacia PCJ	10^{-5}–10^{-7}		H. Matsumoto
Pseudomonas fluorescens	10^{-5c}	10^{-4}–10^{-5}	199, 225
Pseudomonas glycinea	10^{-6}–10^{-8}		106
Pseudomonas putida PPN	10^{-4}–10^{-6d}		38
Rhizobium leguminosarum	10^{-6}		25
Rhizobium meliloti	10^{-3}–10^{-5}		194
Rhizobium phaseoli	ND		176
Rhizobium trifolii	10^{-4}–10^{-5}		222
Cowpea rhizobia	10^{-6}		220
Rhodobacter capsulatus	10^{-6e}		212, 356, 357
Rhodobacter sphaeroides	10^{-4}–10^{-8}		301, 332
Salmonella typhimurium		10^{-6}–10^{-7}	225, 334
Thiobacillus A2	10^{-5}–10^{-6}		266
Vibrio cholerae		10^{-5}–10^{-7}	312
Vibrio sp.	10^{-7}		164
Xanthomonas sp.	10^{-5}–10^{-6}		231
Zymomonas mobilis	10^{-4}–10^{-7}		152, 302

ND, not determined.
[a]including the pULB113-like plasmid R68::Mu (99).
[b]The R68.45-like plasmid pMO172 was used.
[c]The R68.45-like plasmid pMO47 was used.
[d]The R68.45-like plasmid pMO61 was used.
[e]Including the R68.45-like plasmid pBLM2 (212).

organisms (IS21)$_2$ is transpositionally inactive or whether during genetic manipulation one IS21 copy was lost from (IS21)$_2$, an event that occurs spontaneously (73, 127). Loss of one IS21 copy was observed, for instance, after transfer of an R68.45 derivative to *Brady-rhizobium japonicum* (292).

Genetic data indicate that both R68.45 and pULB113 interact with many different sites on plasmids and chromosomes and promote nonpolar chromosome transfer (125, 334). This is a consequence of low target specificity of the IS21 and Mu transposition functions (233, 277). The structures of the R68.45–chromosome and pULB113–chromosome cointegrates are not amenable to direct experimental examination because these plasmids apparently do not stably integrate into the chromosome and do not form spontaneous Hfr strains. However, as already mentioned, a ts derivative of R68.45—pME487—has been forced into the *P. aeruginosa* chromosome. The integrated pME487 is bordered by single IS21 copies in direct orientation (276). The structure of chromosomally integrated pULB113 can be deduced from pULB113′ plasmids; mini-Mu elements flank the integrated plasmid (334) (Figure 6.3).

There is circumstantial evidence that the physical association of R68.45 with the *P. aeruginosa* chromosome is short-lived. R68.45 donors transfer short chromosome segments (< 10% of the total chromosome), whereas pME487-derived Hfr strains donate much longer chromosome regions, as indicated by linkage analysis (124, 276). It can be estimated that exponentially growing *P. aeruginosa* cells do not tolerate replication-proficient R68.45 in the chromosome for more than 10–15 min (276). In rhizobia the behavior of R68.45 is different in that it transfers large chromosome segments (26, 194, 222).

The principal application of sex plasmids is classical chromosome mapping. In addition to the establishment of circular chromosome maps for *E. coli*, *S. typhimurium*, *P. aeruginosa*, and *P. putida* already mentioned, genetic circularity has also been demonstrated for the chromosomes of several gram-negative bacteria: *A. tumefaciens* (162, 228), *Rhodobacter capsulatus* (353), *Proteus mirabilis* (66), *Proteus morganii* (14), *Acinetobacter calcoaceticus* (340), *Rhizobium leguminosarum*, and *R. meliloti* (27). The chromosomes of streptomycetes have been mapped extensively by conjugation (53). In other gram-positive bacteria, chromosome transfer by plasmids has been demonstrated (102, 315) but not used for the construction of chromosome maps.

3.1.3. Formation of F and R Prime Plasmids

Excision of plasmids that are stably or transiently integrated in the chromosome can sometimes lead to the formation of prime plasmids. These carry chromosomal fragments that have bordered the integrated plasmid on either side or on both sides. The selection methods and properties of prime plasmids derived from F, ColV, and R plasmids have been reviewed (126, 153, 205). As shown in Figure 6.2, F can excise by homologous recombination at directly oriented, homologous sequences (153, 333). "Type I" excision by recombination between IS2 or IS3 in F and the same insertion sequence in chromosomal DNA gives rise to F′ plasmids with a chromosomal segment on only one side of F. "Type II" excision by recombination between flanking chromosomal regions of homology (IS2, IS3, IS5, IS30, or rRNA genes) produces F′ plasmids that carry chromosomal genes on both sides of F (Figure 6.2). Some F′ plasmids may be due to Tn1000-mediated transpositional recombination, since Tn1000 has been found to occur at some F–chromosome junctions

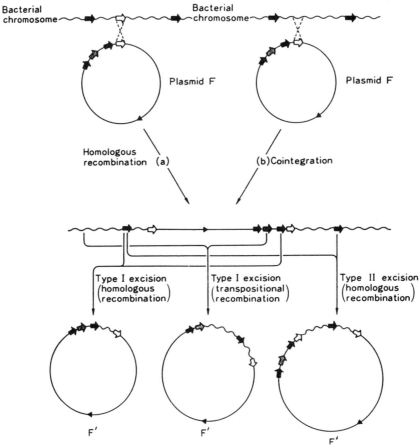

FIGURE 6.2. Models for chromosomal integration and excision of the F plasmid by homologous recombination and formation of F′ plasmids in *E. coli*. Hfr donors are created by F integration. IS2 (thick open arrows) and IS3 (thick filled arrows) are shown to provide homology and Tn*1000* (thick hatched arrows) to promote transpositional recombination. The *oriT* of F is indicated by ◄. Reproduced from reference 333, by permission of Academic Press.

(Figure 6.2) (86, 153, 333). Illegitimate recombination involving the *oriT* of F and distinct chromosomal sequences displaying partial homology with *oriT* can generate certain F primes that are deleted for *tra* functions (163). A set of F′ plasmids whose chromosomal inserts together cover the entire *E. coli* chromosome is available for complementation tests and for rapid, rough mapping of chromosomal mutations in *E. coli* and *S. typhimurium* (153, 205).

In prime plasmids formed by R68.45 or pULB113, the chromosomal sequences are bordered by directly oriented copies of IS*21* or mini-Mu, respectively (126, 153, 334, 359). Van Gijsegem and Toussaint (334) proposed that pULB113′ formation occurs by replicative transposition of mini-Mu (Figure 6.3A). Our model of R68.45′ formation (126) postulates that two copies of R68.45 in the same orientation transpose directly into nearby chromo-

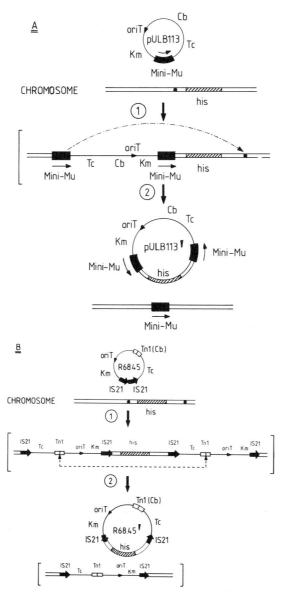

FIGURE 6.3. Models for R′ formation by pULB113 and R68.45. (A) Replicative transposition of mini-Mu mediates integration (①) and imprecise excision of pULB113 (②). (B) Nonreplicative transposition of (IS21)₂ accounts for integration of R68.45 (①). The model assumes that a second R68.45 copy is inserted near the first copy. Site-specific resolution at the *res* site of Tn*1* or homologous recombination then promotes excision of the R68.45′ plasmid (②). Reprinted from reference 126, by permission of Academic Press.

somal sites and then excise by Tn*I*-encoded resolvase or by homologous recombination (Figure 6.3B). Alternatively, one copy of R68.45 and an additional single copy of IS*21*, in the same orientation, could be inserted into nearby chromosomal locations. "Type I" excision of an R68.45′ plasmid could then occur as a consequence of homologous recombination between the single IS*21* element and the distal IS*21* element at the R68.45– chromosome border (not shown).

In a number of gram-negative bacteria, R′ plasmids have found application for the characterization of chromosomal genes by complementation. Typically, R′ plasmids can be isolated from crosses between a donor strain carrying an IncP plasmid (e.g., R68.45) and a recipient of another species (often *E. coli* or *P. aeruginosa*) that does not have a high degree of sequence homology with the donor. The gene(s) of interest thus can be cloned *in vivo* by selection, in the recipient, for expression of the gene(s) plus any R plasmid-encoded antibiotic-resistance marker(s). Heterologous gene expression is a prerequisite for this type of R′ isolation. This approach has been used to produce a variety of R′ plasmids from pseudomonads, enteric bacteria, methylotrophs, rhizobia, and other gram-negative bacteria (1, 15, 33, 58, 95, 143, 144, 155, 177, 189, 200, 237, 238, 331, 334). Alternatively, when *recA* mutants are available, R′ plasmids can be recovered from matings between R+ donors and *recA* recipients (150). Although in both R′ isolation methods the risk of homologous recombination between the chromosomal insert of the prime plasmid and the chromosome is minimized, *recA*-independent structural instability of the inserts is not uncommon (9, 126, 150, 200), and this might be a consequence of the intermediate copy number of IncP plasmids. Presumably for the same reason, some chromosome regions fail to be translocated to R′ plasmids (38), and it may be difficult to construct sets of R′ plasmids covering an entire bacterial chromosome.

Some R′ plasmids have been reported that carry large (> 100-kb) pieces of passenger DNA (9, 284, 296). Thus, R′ construction offers the possibility to clone extended genomic regions *in vivo* and to transfer them readily to heterologous hosts for functional analysis. Nevertheless, we would like to caution that physical analysis of R′ plasmids can be a time-consuming exercise because preparation of sufficient quantities of pure plasmid DNA is tedious and because, as mentioned above, some large R′ plasmids are difficult to maintain stably. Therefore, *in vitro* cloning methods (e.g., cosmid cloning) usually seem preferable whenever the genomic fragments to be cloned do not exceed about 30 to 40 kb.

3.2. Plasmid Mobilization

Plasmid cointegrates have been constructed *in vivo* for several reasons. First, a number of interesting phenotypic properties of gram-negative and gram-positive bacteria could be associated with particular plasmids after these plasmids had been transferred, as cointegrates, to other bacteria. Second, many commonly used cloning vectors have a narrow host range of replication; however, their host range can often be expanded *in vivo* by fusion with a promiscuous plasmid. Moreover, some bacteria are poorly transformable with plasmid DNA but act as good recipients in conjugation. In such cases it may be advantageous to introduce recombinant plasmids, again as cointegrates, by conjugation. In the following we give some examples for these applications.

3.2.1. Localization of Genes of Interest on Plasmids

Agrobacterium tumefaciens strains that carry a Ti plasmid induce crown gall tumors in susceptible plants. In early experiments, RP4-Ti cointegrates greatly helped to establish the role of the large Ti plasmids in tumor formation. The derepressed RP4 transfer functions facilitated the transfer of RP4-Ti plasmids to nontumorigenic *Agrobacterium* strains, *Rhizobium*, and *E. coli*, and the RP4-encoded resistance determinants provided selectable markers (59, 157–159, 335). RP4-Ti cointegrates can be maintained in *E. coli*, where the Ti plasmids normally cannot replicate. *Escherichia coli* transconjugants harboring an RP4-Ti hybrid, however, do not cause plant tumors because the *E. coli* chromosome lacks essential virulence functions (158). In the cointegrates, an IS8 (= IS21) element has been found at both RP4-Ti junctions, indicating that replicon fusion is due to IS8 transposition (87). Ti::Tn*1* plasmids are mobilized by RP4, which naturally carries Tn*1*, at a higher frequency than are the wild-type Ti plasmids (160). The increased transfer frequency is probably due to homologous recombination between the Tn*1* elements present on both plasmids. RP1::Tn*904* and RP4::Tn*1831* also show enhanced mobilization of Ti plasmids, presumably because of transpositional recombination (160). Cointegrates between R and Ti plasmids usually dissociate in *E. coli*, with loss of the Ti part (87, 160). However, in yet another type of cointegrate, the conjugative plasmid R772, which harbors the insertion sequence IS70, has been fused stably to an octopine Ti plasmid. As mentioned in Section 2.4.1., the junctions between these plasmids consist of one complete and one partially deleted copy of IS70. The partial deletion of one IS70 element might explain why such a cointegrate is stable in *E. coli* (148).

Using a similar approach of *in vivo* replicon fusion, Hooykaas et al (161) transferred the Ri plasmid, which causes hairy root disease in plants, from *Agrobacterium rhizogenes* to plasmid-free *Agrobacterium* strains. Cointegrates were generated with RP1::Tn*904*, which provided selectable markers and an effective conjugation system.

The TOL plasmid of *P. putida* codes for enzymes catabolizing xylenes and toluene (or toluate). Although TOL is a Tra+, a broad host range plasmid (20), it does not enable distantly related bacteria such as *E. coli* to grow on toluate (242), and hence toluate media cannot be used for selection of TOL transfer to *E. coli*. Nevertheless, TOL catabolic enzymes are expressed in *E. coli*. The isolation of RP4-TOL hybrid plasmids therefore facilitated the initial genetic manipulation of the TOL genes and their functional analysis in *E. coli* (242, 243). Similarly, R68.45 has been used to pick up the genes for the entire phenol (Phl) degradation pathway from the *P. putida* plasmid pPGH1 (147). Owing to the broad host range properties of R68.45, the *phl* genes can be transferred to *E. coli* and be expressed in this host. IS21 appears to be involved in the integration of pPHG1 DNA into R68.45 (147). Both RP4-TOL and R68.45-Phl contain only parts of the native catabolic plasmids, and thus can be considered to be R′ plasmids (147, 243).

In fast-growing rhizobia, the genes for symbiotic nitrogen fixation (*nif, fix*) occur on extremely large plasmids (pSym). R′ plasmids have been constructed *in vivo* that carry *nif* and *fix* genes on large segments originating from the pSym megaplasmid of *R. meliloti*. Clusters of *nif* and *fix* genes could then be localized conveniently on the R′ plasmids by transposon mutagenesis (9, 180).

Lactococci are used as starter cultures by the dairy industry. In *Lactococcus lactis* ssp.

lactis, the nonconjugative plasmid pSK08 determines lactose catabolism. The transfer factor pRS01 mobilizes the Lac plasmid pSK08 by forming cointegrates, and thus allows conjugative Lac⁺ transfer. Replicon fusions occur by *rec*-independent, replicative transposition (Figure 6.1B) of ISS*1*, an insertion sequence on pSK08 (98, 313). Proteinase activity (Prt), another property of lactococci that is essential for growth of these bacteria in milk, is associated with the nonconjugative plasmids pCI301 and pCI203 present in *L. lactis* ssp. *lactis* and *S. lactis* ssp. *cremoris*, respectively. These plasmids can be mobilized to Prt⁻ lactococci by the broad host range plasmid pAMβ1. Mobilization clearly involves replicon fusion; high-frequency, site-specific recombination (cf. Section 2.3) accounts for the formation of pAMβ1-pCI301 cointegrates (137, 138).

3.2.2. Expansion of Host Range

The host range of pBR322 and related cloning vectors is limited to enteric bacteria. The conjugative IncW broad host range plasmid R388 carrying Tn*10* forms cointegrates with other replicons by Tn*10*-mediated inverse transposition. These cointegrates can then be transferred to and replicated in a range of gram-negative bacteria. The usefulness of the method has been demonstrated with TOL genes, which were cloned in pBR322. Because of their poor expression in *E. coli*, they had to be transferred back to *P. putida* for fine-structure physical analysis. This was achieved by fusing pBR322-TOL recombinant plasmids with R388::Tn*10* (132).

Escherichia coli vectors that contain the *oriT* (*mob*) region of RP4 (pSUP vectors) can be fused with RP4 by a single crossover in the *mob* region. The formation of RP4-pSUP cointegrates occurs at a frequency of about 10^{-2} in *recA⁺ E. coli* hosts. The cointegrates can then be transferred to other gram-negative bacteria such as *Rhizobium* sp. (270, 299). Provided that these bireplicon constructs are sufficiently stable, they can be used as an alternative to recombinant plasmids based on vectors with broad host range replication properties.

Some lactic acid bacteria are difficult to transform with plasmid DNA by conventional procedures. This problem may be overcome by conjugative mobilization. For instance, the broad host range conjugative plasmid pVA797 mobilizes the partially homologous shuttle vector pVA838 (which replicates in *E. coli* and streptococci/lactococci) by cointegrate formation. After recombination at the homologous sequences in *Streptococcus sanguis*, the pVA797-pVA838 cointegrates can be introduced by conjugation into *Enterococcus faecalis* and various strains of lactic streptococci (280, 307). In the recipients, the cointegrates are resolved into two separate plasmids, which segregate because of mutual incompatibility (280, 307). The availability of pVA797, pAMβ1 (discussed in Sections 2.3 and 3.2.1) and similar plasmids for plasmid mobilization *in cis* has substantially contributed to the development of molecular genetics in lactic acid bacteria.

4. Replicon Fusion by R68.45

The mechanisms of homologous, transpositional, site-specific, and illegitimate recombination (summarized in Figure 6.1) have been reviewed extensively during the past five years (18, 22, 24, 53, 70, 71, 93, 107, 114, 190, 233, 295, 305), with the exception of the mechanism of IS*21* transposition (Figure 6.1F), which has been the subject of some recent studies (275, 277). IS*21*-mediated cointegrate formation is largely responsible for Cma of the widely used sex plasmid R68.45 (Table 6.2). Here, we would like to point out some particular features of IS*21* transposition.

4.1. Structure of IS*21*

The total nucleotide sequence of IS*21* (2131 bp) is known. The termini are formed by short, inverted repeats of 11 bp, with one mismatch (Figure 6.4). A few additional bases up to 30 bp "inland" are also repeated in inverted orientation (not shown) (107). As stated in Section 2.2.2, IS*21* forms tandem duplications [= (IS*21*)$_2$], which actively promote replicon fusions. In the prototype plasmid carrying (IS*21*)$_2$, R68.45, IS*21*L (left) and IS*21*R (right) are separated by 3 bp termed junction sequence (277). At the very end of the right IS*21*R terminus, a perfect "-35" sequence of the *E. coli* σ^{70} consensus promoter is found (Figure 6.4), and at an appropriate distance (18 bp) there is a "-10" sequence in the adjacent IS*21*L. Thus, a promoter (Pj) is formed at the IS*21*-IS*21* junction (Figure 6.4). This promoter drives the expression of two open reading frames (ORFs), the *istA* and *istB* genes, in IS*21*L (277). Similarly, the right IS*21*L terminus contains a "-35" sequence that is part of the *aphA* (aminoglycoside phosphotransferase) promoter on R68.45 (259) (Figure 6.4). The *istA* and *istB* gene products have been detected in maxicell experiments with (IS*21*)$_2$ plasmids. Plasmids having a single IS*21* element do not express the *istA* and *istB* gene products unless a strong external promoter allows transcription of *istA* and *istB* (277). This indicates that the transposition functions of IS*21* are normally expressed at low levels but turned on after tandem duplication of the IS element. Three IS*21*-encoded proteins have been detected; the *istA* gene has two translational starts, leading to polypeptides of 46 kDa and 45 kDa (S. Schmid, T. Seitz, and D. Haas, unpublished results). The *istB* gene codes for a 30-kDa protein (277). To date, however, there is no evidence for further IS*21* products, although several short ORFs in addition to *istA* and *istB* exist on both strands of IS*21* (107, 277). A translation stop in *istA* completely blocks the replicon fusion activity of (IS*21*)$_2$ (275), whereas *istB* mutations strongly reduce activity but do not abolish it entirely (S. Schmid and D. Haas, unpublished results). A transcriptional stop signal in *istA* eliminates the expression of IstB and both IstA proteins, indicating that the *istAB* genes form an operon that is transcribed from Pj (277). When the *istA* and *istB* genes are physically separated and expressed individually, the transpositional activities of IS*21* are not affected (S. Schmid and D. Haas, unpublished results). Thus, it seems unlikely that transposase activity should reside in a (hypothetical) IstAB fusion protein. In contrast, translational frameshifting in the *insA* gene of IS*1* produces an InsAB' fusion protein, the IS*1* transposase (267).

The *istA* gene has some homology with the *tnpA* (transposase) gene of Tn*501* and reveals a potential helix-turn-helix DNA binding motif in the N-terminal part of the IstA proteins (277). Significantly, in the IstA proteins there is another amino acid sequence motif that recurs in retroviral integrases (IN) and in the IS*3* family of insertion sequences (97, 174, 186). Integrase is essential for the integration of newly synthesized retroviral DNA into the host genome (339). These structural similarities are compatible with the idea that

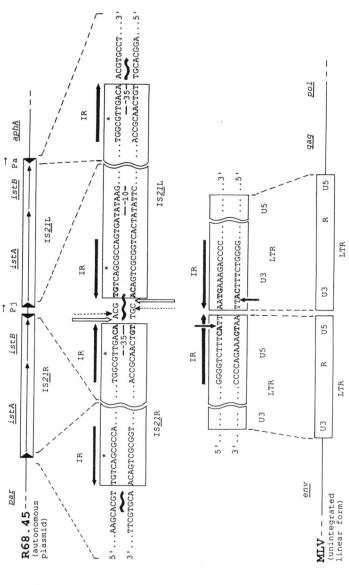

FIGURE 6.4. Processive cuts at the termini of (IS21)₂ from plasmid R68.45 and of long terminal repeats (LTRs) from Moloney murine leukemia virus (MLV). The right and left copy of IS21 from R68.45 are designated IS21R and IS21L, respectively, according to an earlier convention (274, 289). (IS21)₂ is flanked by ACGT target duplications (underlined with wavy lines) and the same sequence also occurs at the junction. Eleven base pair inverted repeats (IR) with one mismatch (*) form the IS21 termini. (IS21)₂ is located between the *par* (partitioning) region and the *aphA* (kanamycin-resistance) gene (109, 259, 277). Pj and Pa are promotors formed by "−35" sequences at the IS21 termini and "−10" sequences in IS21L and upstream of *aphA*, respectively. The junction promoter Pj drives the expression of the *istA* and *istB* genes in IS21L, and Pa is the promoter for the *aphA* gene (259). In an *istA*-dependent *in vitro* reaction, cuts (vertical open arrows) have been observed at the inner 3′ ends of IS21; additional cuts (vertical dashed arrows) may also occur at the neighboring bases in the ACG junction (275). Long terminal repeats (LTRs) are placed at the ends of MLV; U3, R, and U5 are defined regions in LTR. *gag* (group-specific antigen), *pol* (reverse transcriptase and integration protein), and *env* (viral structural protein) are viral ORFs (339). The integration protein removes the terminal dinucleotide (TT) from the 3′ ends by cuts as indicated with heavy vertical arrows.

the *istA* products have transposase functions. The IstB protein contains an ATP-binding motif and shows some similarity with the DnaA and DnaC replication proteins of *E. coli* (191). The MuB protein, which allosterically activates strand transfer by MuA transposase and helps capture the target DNA, shows the same likely ATP-binding site (233).

Recently, Menou et al (223) described a 2184-bp insertion sequence from the gram-positive bacterium *Bacillus thuringiensis*, IS*232*, which is very similar to IS*21*, except for the terminal inverted repeats. The amino acid sequence derived from ORF1 of IS*232* shows 30% similarity (identical residues plus conservative changes) with the 46-kDa IstA protein of IS*21*. The putative translation product of ORF2 of IS*232* has 55% similarity with IstB of IS*21*. An IS*232* derivative transposes in *E. coli* (223). A "−10" promoter region occurs at one end, a "−35" region at the other end of IS*232*. Although the existence of IS*232* tandems has not been demonstrated, it is nevertheless tempting to speculate that, by analogy with IS*21*, IS*232* could form a tandem repeat and this could create a junction promoter for the downstream *orf-1* and *orf-2*. Because of these similarities, it has been proposed that IS*21* and IS*232* have a common ancestor (223).

IS*640*, a 1.1-kb insertion sequence from *Shigella sonnei*, is almost identical to the first 1.1 kb of IS*21* but two (−1) frameshift mutations disrupt the homology between the (only) deduced protein of IS*640* and the IstA proteins after the first 280 amino acid residues. IS*640* does not show a second ORF (216) and apparently lacks an IS*21*-like terminal repeat distal to the ORF. Transpositional activity of IS*640* has not been demonstrated. Therefore, IS*640* might be a truncated form of IS*21*. Another sequence related to IS*21* has been found on the *P. fluorescens* plasmid pEG1. This sequence transposes in *Pseudomonas* and has two ORFs resembling *istA* and *istB*. However, the 12-bp terminal inverted repeats differ from those of IS*21* (358).

4.2. A Model for IS*21* Transposition and IS*21*-Mediated Cointegrate Formation

A single IS*21* element carrying a kanamycin-resistance gene insertion transposes to give simple insertions at about 10^{-4} when the *istAB* gene products *in trans* are expressed under the control of a good external promoter. It appears that the IstA (46 kDa) and the IstB proteins are important for catalyzing these simple insertions (T. Seitz, S. Schmid, and D. Haas, unpublished results). Moreover, it is known that a single IS*21* copy can promote replicon fusion at frequencies of 10^{-6} in *E. coli*; the cointegrates formed contain one IS*21* element at each junction (274). Since this process occurs in a *recA* donor, the pathways of Fig. 6.1C, D are unlikely. A replicative pathway of transposition (Figure 6.1B) without resolution is possible but at present an alternative explanation seems more plausible. Spontaneous duplication of IS*21* in the donor plasmid at ~10^{-4}, followed by (IS*21*)$_2$-promoted replicon fusion (Figure 6.1F) would also explain the cointegrate structure observed. Spontaneous formation of (IS*21*)$_2$ has been reported in *P. aeruginosa* (278, 351), *Rhodobacter capsulatus* (1, 212), *Proteus mirabilis* (77), and, after chromosomal integration of RP4, in *Myxococcus xanthus* (171).

How are IS*21* tandems formed? One model proposes that duplications occur by an event that joins one end of IS*21* to the opposite end of another IS*21* copy lying on the other branch of the replication fork (126). Alternatively, a single IS*21* copy might transpose next to an IS21 element carried by another plasmid copy. Both views imply that the sequences flanking an IS21 insertion are preferred target sites for a second IS21 insertion. This prediction has been verified in mobilization experiments. When a mobilizing conjugative plasmid and a mobilized nonconjugative plasmid each carry a single IS21 copy, cointegrates can be formed that contain (IS21)$_2$ at one junction between the two replicons (77, 274). The same mode of tandem formation has also been found for IS30 (76). (IS21)$_2$ formation has also been observed in pME28, an RP1::IS21 derivative in which a segment lying between two directly repeated copies of IS21 was deleted spontaneously (274). The junction sequence in pME28 comprises 2 bp, which are different from the 3-bp junction of R68.45 (Figure 6.4) (277). The R68.45-like plasmid pMO61, which is a good sex factor in P. putida (38), also has a 2-bp junction (E. Brugnera, S. Schmid, and D. Haas, unpublished results). Taken together, these data indicate that IS21 can, but need not, insert near the termini of another IS21 element. It appears likely that the transposition functions of IS21 are required for tandem formation. However, experimental evidence for this assumption is not yet available.

IS21 tandems are quite stable, even in a recA$^+$ background, and the (IS21)$_2$ plasmid R68.45 has been maintained in gram-negative bacteria for many years, although occasional loss of one IS21 copy has been observed (127). However, when R68.45 is conjugatively transferred to P. aeruginosa or Erwinia recipients, spontaneous plasmid deletions occur at high frequencies. These deletions eliminate one IS21 copy and a flanking region of R68 (aphA . . . tra-1 or par . . . tra-2) (73, 127). They are probably due to intramolecular transposition of IS21. The fact that they are so frequent after conjugative transfer is reminiscent of zygotic induction of prophages (139) and suggests that IS21 transposition functions may be negatively controlled.

Cointegrates formed by transpositional fusion of (IS21)$_2$ plasmids with other replicons contain one IS21 element at each junction (279). When either IS21 copy in (IS21)$_2$ is marked with an Ω insertion, the cointegrates have a single IS21::Ω element at one replicon junction and a single IS21 element at the other (274). This indicates that during replicon fusion no new IS21 copies are generated and is consistent with a nonreplicative pathway of transposition (Figure 6.1F). This model is supported by the finding that the inner IS21 ends in (IS21)$_2$ are sufficient for cointegrate formation, whereas the outer ends can be deleted without any effect (274). The IstA 45-kDa protein is highly active in cointegrate formation with (IS21)$_2$ plasmids but promotes simple IS21 insertions with poor efficiency (S. Schmid, T. Seitz, and D. Haas, unpublished results). Therefore, the IstA 45-kDa protein has been named cointegrase, whereas the term transposase is used for the IstA 46-kDa protein.

As a first step in (IS21)$_2$-dependent replicon fusion, it is proposed that cointegrase makes a double strand cut at the IS21-IS21 junction. The configuration of this cut is crucial for the polarity of strand transfer during replicon fusion. When an IS21-IS21 junction fragment, in supercoiled form, is incubated in a crude E. coli extract containing overproduced IstA proteins, a double strand cut is produced at the junction. Cleavage takes place precisely at the inner 3' termini of IS21 (indicated by open vertical arrows in Figure 6.4) and the resulting staggered cut generates 5' protrusions of 2 bp. [Because of technical problems, the 5' ends have not been determined unambiguously, and a 3-bp stagger also remains possible (Fig. 6.4).] The istA gene, but not the istB gene, is required for this in vitro cleavage of the junction (275). For three prokaryotic transposable elements, Mu, Tn7 and Tn10, the polarity of strand transfer has been determined (8, 71, 94, 190, 233). Transposition is initiated by cuts at the 3' ends of these elements in the donor. The 3' ends are then ligated to the 5' ends of cleaved target DNA (8, 71, 233). The recent development of an *in vitro* system has allowed us to demonstrate that (IS*21*)$_2$ replicon fusion requires cointegrase and IstB in a crude *E. coli* extract (S. Schmid and D. Haas, unpublished results).

Insertion of R68.45 of pME28 creates target duplication of 4 bp (Figure 6.1F). In this process, the 3-bp junction sequence of R68.45 or the 2-bp junction sequence of pME28 is lost (277). If the *in vitro* cuts observed at the IS*21*-IS*21* junction reflect what happens in the (IS*21*)$_2$ donor replicon at the onset of replicon fusion, then the target duplications observed *in vivo* can be explained by the following model. The target is opened by a staggered cut with 5' protruding ends of 4 or 5 bp (Figure 6.5, ①). After joining of the inner 3' ends of IS*21* to the 5' ends of the cleaved target, repair synthesis produces the target duplications (Figure 6.5, ②). This model also accounts for the loss of the junction sequence of the donor replicon during repair (Figure 6.5, ③).

Cointegrates are resolved in *recA*$^+$ hosts by recombination between the IS*21* elements; no IS*21*-encoded resolvase has been found (Figure 6.5, ③) (277, 351). IS*21* can be inserted into many different sites on plasmids or on the chromosome; however, the recovery of auxotrophic mutations induced by IS*21* is infrequent (276, 279). This indicates that IS*21* has a low, but not totally random specificity of insertion.

In conclusion, IS*21* tandems are highly reactive in replicon fusion for three reasons. First, at the IS*21*-IS*21* junction a promoter is formed, permitting the expression of the transposition proteins in the downstream IS*21* element. Second, there is a truncated form of transposase, cointegrase, which appears to have specialized in replicon fusion. Third, the two IS*21* termini that are used for cointegration are held together by the junction sequence. It is likely that this configuration greatly facilitates the productive interaction of cointegrase with the linked IS*21* ends. The "reactive tandem" concept was originally proposed for Mu and IS*1* transposition (96, 272). Although there is no compelling evidence for Mu or IS*1* tandem intermediates, the concept has proved fruitful to explain replicon fusions promoted by IS*21* and, moreover, by IS*30* (252, 277, W. Arber, personal communication). Interestingly, IS*3*-mediated cointegration is stimulated at least 100-fold when an IS*3* end is fused, *in vitro*, to an intact IS*3* element with 4-bp spacing. Tandems composed of two entire IS*3* copies cannot be constructed *in vitro*, perhaps because they would be highly unstable (311). Thus, replicon fusion via reactive tandems is not restricted to IS*21* and IS*30*.

There are striking analogies between IS*21*-mediated cointegration and retrovirus integration. Although it may ultimately turn out that some of these analogies are just formal and not due to similar transposition mechanisms, we would like to provide a brief comparison. Following reverse transcription of retroviral RNA, retrovirus DNA contains long terminal repeats (LTRs). This linear DNA molecule, when entering the nucleus, may circularize to give a covalently closed 2-LTR circle. In the 2-LTR circle of Moloney murine leukemia virus (MLV), the LTRs are blunt-end ligated (42, 339). Formally, the LTRs in the 2-LTR circle and (IS*21*)$_2$ in R68.45 are located at analogous positions. The 2-LTR circle is not the immediate precursor to the integrated provirus. Rather, linear viral DNA while still in the cytoplasm is processed at the 3' ends; this reaction removes two terminal bases (TT in MLV; Figure 6.4), and recessed 3' OH groups are formed, which determine the boundaries of the integrated provirus (42, 112, 185). IS*21* and the LTRs of MLV (and other retroviruses) have terminal inverted repeats of 11 bp and 13 bp, respectively. All these repeats have the dinucleotide CA at their 3' termini (Figure 6.4) (42, 277). The nucleotides that follow CA

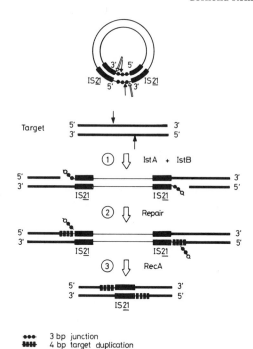

FIGURE 6.5. Model for transposition of (IS*21*)$_2$ plasmids. ① The donor plasmid, e.g., R68.45, is opened at the IS*21*-IS*21* junction by cuts (open arrows) producing 3' ends of IS*21*. The 5' ends of the 3-bp junction sequence are due to the same cuts or to additional nicks (solid arrows). The uncertainty concerning the 5' ends is indicated by open circles representing the 5'-terminal nucleotides in the junction sequence. Target cleavage is pictured to give 5' protruding ends of 4 bp. Joining of the IS*21* ends to the cleaved target is thought to involve the *istAB* gene products. ② Repair synthesis removes the junction sequences and results in 4-bp target duplications. *Mutatis mutandis*, the joining and repairing reactions appear to be analogous to retrovirus integration. ③ In a *recA*+ host, the integrated plasmid can be excised by homologous recombination, leaving behind a simple IS*21* insertion. Reproduced from reference 275, by permission of Oxford University Press.

at the 3' ends and form the junctions in 2-LTR circles and in (IS*21*)$_2$ plasmids show flexibility: their number and base composition can vary without affecting integration (42, 277).

The viral integration (IN) protein is necessary for the formation of recessed 3' ends at the LTRs (42, 178, 185, 294). The cuts at the 3' ends of IS*21* and LTR occur at analogous positions (Figure 6.4), and both the IN protein and the IS*21* transposase are active *in trans* (42, 275). Upon integration of linear, recessed retrovirus DNA mediated by the IN protein, a 4- to 6-bp target sequence of the host is duplicated, flanking the integrated provirus (42, 44). The model for these target duplications assumes that a staggered cut with 4- to 6-bp 5' protrusions is made in the target DNA and that the joining (45) and repairing reactions are similar to those shown for (IS*21*)$_2$ (Figure 6.5 ②). Thus, the retroviral junction nucleotides (e.g., the two AA dinucleotides of MLV; Figure 6.4) are lost upon virus integration (42, 44, 339).

In conclusion, substantial progress has recently been made in the characterization of the retroviral integration reactions *in vitro* (42, 44, 45, 112, 178, 185, 294). It will be interesting to see whether mechanistically similar steps occur in IS21-mediated replicon fusion.

5. Conclusions

Chromosome mobilization, together with generalized transduction and natural transformation, is fundamental to classical bacterial genetics and remains important for mapping of chromosomal genes. Separation of large DNA fragments by pulsed-field gel electrophoresis and ordering of genomic libraries provide straightforward, rapid methods for physical chromosome mapping. The first combined physical and classical genetic map was that of *E. coli* (229); subsequent combined maps have become available, e.g., for *P. aeruginosa* and *Streptomyces coelicolor* (187, 271). Current emphasis has shifted to physical mapping (90, 251, 304, 318). Nevertheless, classical chromosome mobilization techniques continue to be useful for mapping as illustrated by some recent studies (101, 192, 249, 316). For example, the existence of two circular chromosomes in *Rhodobacter sphaeroides* was first established by physical methods (318) and subsequently confirmed by Hfr mapping (319).

Plasmid mobilization techniques are routinely used for intra- and interspecies transfer of recombinant plasmids and their analysis, e.g., by complementation tests. In some situations, the need for conjugation and mobilization can now be bypassed by electroporation, which permits the introduction of plasmid DNA into a wide variety of bacterial species (224). It is even possible to "electrotransfer" the nonconjugative plasmid pBR322 from an *E. coli* donor to an *E. coli* recipient directly, without isolation of plasmid DNA (317). *In vivo* cloning methods are still quite useful in several bacterial species for the rapid and simple construction of prime plasmids. These methods can be used side-by-side with *in vitro* cloning procedures, which are of course more widely applicable.

The most important remaining problems in cointegrate formation concern the molecular mechanisms of DNA recombination. Partial reactions of homologous, transpositional, and illegitimate recombination have been carried out successfully *in vitro* but many aspects of these recombination pathways still need to be defined by appropriate *in vitro* systems.

ACKNOWLEDGMENTS. We cordially thank our many colleagues who provided us with reprints and reminded us of older publications. We are particularly grateful to R. C. Deonier and E. Ohtsubo for supplying bibliographies on the F plasmid and to A. Mercenier and D. B. Clewell for providing references concerning plasmid interactions in gram-positive bacteria. We are grateful to S. Patil and S. Schmid for carefully reading the manuscript and to M. T. Lecomte and E. Jäggi for providing expert secretarial assistance. Our work has been supported by the Swiss National Foundation for Scientific Research and by the Roche Research Foundation.

References

1. Allibert, P., Willison, J. C., and Vignais, P. M., 1987, Complementation of nitrogen-regulatory (*ntr*-like) mutations in *Rhodobacter capsulatus* by an *Escherichia coli* gene: Cloning and sequencing of the gene and characterization of the gene product, *J. Bacteriol.* **169**:260–271.

2. Al-Taho, N. M., and Warner, P. J., 1987, Restoration of phenotype in *Escherichia coli* auxotrophs by pULB113-mediated mobilisation from methylotrophic bacteria, *FEMS Microbiol. Lett.* **43:**235–239.

3. Anderson, P., 1987, Twenty years of illegitimate recombination, *Genetics* **115:**581–584.

4. Arthur, A., and Sherratt, D., 1979, Dissection of the transposition process: A transposon-encoded site-specific recombination system, *Mol. Gen. Genet.* **175:**267–274.

5. Arthur, A., Nimmo, E., Hettle, S., and Sherratt, D., 1984, Transposition and transposition immunity of transposon Tn*3* derivatives having different ends, *EMBO J.* **3:** 1723–1729.

6. Avila, P., de la Cruz, F., Ward, E., and Grinsted, J., 1984, Plasmids containing one inverted repeat of Tn*21* can fuse with other plasmids in the presence of Tn*21* transposase, *Mol. Gen. Genet.* **195:**288–293.

7. Bagdasarian, M., Hryniewicz, M., Zdzienicka, M., and Bagdasarian, M., 1975, Integrative suppression of a *dna* mutation in *Salmonella typhimurium*, *Mol. Gen. Genet.* **139:**213–231.

8. Bainton, R., Gamas, P., and Craig, N. L., 1991, Tn7 transposition in vitro proceeds through an excised transposon intermediate generated by staggered breaks in DNA, *Cell* **65:** 805–816.

9. Bánfalvi, Z., Randhawa, G. S., Kondorosi, E., Kiss, A., and Kondorosi, A., 1983, Construction and characterization of R-prime plasmids carrying symbiotic genes of *R. meliloti*, *Mol. Gen. Genet.* **189:** 129–135.

10. Barlett, M. M., Erickson, J. M., and Meyer, R. J., 1990, Recombination between directly repeated origins of conjugative transfer cloned in M13 bacteriophage DNA models ligation of the transferred plasmid strand, *Nucleic Acids Res.* **18:**3579–3586.

11. Barrett, J. T., Rhodes, C. S., Ferber, D. M., Jenkins, B., Kuhl, S. A., and Ely, B., 1982, Construction of a genetic map for *Caulobacter crescentus*, *J. Bacteriol.* **149:**889–896.

12. Barth, P. T., 1979, Plasmid RP4, with *Escherichia coli* DNA inserted *in vitro*, mediates chromosomal transfer, *Plasmid* **2:**130–136.

13. Bashkirov, V. I., Khasanov, F. K., and Prozorov, A. A., 1987, Illegitimate recombination in *Bacillus subtilis*: Nucleotide sequences at recombinant DNA junctions, *Mol. Gen. Genet.* **210:**578–580.

14. Beck, Y., Coetzee, W. F., and Coetzee, J. N., 1982, *In vitro*-constructed RP4-prime plasmids mediate orientated mobilization of the *Proteus morganii* chromosome, *J. Gen. Microbiol.* **128:**1163–1169.

15. Beeching, J. R., Weightman, A. J., and Slater, J. H., 1983, The formation of an R-prime carrying the fraction I dehalogenase gene from *Pseudomonas putida* PP3 using the IncP plasmic R68.44, *J. Gen. Microbiol.* **129:**2071–2078.

16. Benjamin, H. W., and Kleckner, N., 1992, Excision of Tn*10* from the donor site during transposition occurs by flush double strand cleavages at the transposon termini, *Proc. Natl. Acad. Sci. USA* **89:**4648–4652.

17. Bennett, P. M., de la Cruz, F., and Grinsted, J., 1983, Cointegrates are not obligatory intermediates in transposition of Tn*3* and Tn*21*, *Nature* **305:**743–744.

18. Bennett, P. M., Heritage, J., Comanducci, A., and Dodd, H. M., 1986, Evolution of R plasmids by replicon fusion, *J. Antimicrob. Chemother.* **18** (Suppl. C):103–111.

19. Bennett, P. M., Grinsted, J., and Foster, T. J., 1990, Detection and use of transposons, in: *Methods in Microbiology*, Vol. 21 (J. Grinsted and P. M. Bennett, eds.), Academic Press, London, pp. 205–231.

20. Benson, S., and Shapiro, J., 1978, TOL is a broad-host-range plasmid, *J. Bacteriol.* **135:**278–280.

21. Berg, C. M., and Curtiss, III, R., 1967, Transposition derivatives of an Hfr strain of *Escherichia coli* K-12, *Genetics* **56:**503–525.

22. Berg, C. M., Berg, D. E., and Groisman, E. A., 1989, Transposable elements and the genetic engineering of bacteria, in: *Mobile DNA* (D. E. Berg and M. M. Howe, eds.), American Society for Microbiology, Washington, D. C., pp. 879–925.

23. Berg, D. E., 1977, Insertion and excision of the transposable kanamycin resistance determinant Tn5, in: *DNA Insertion Elements, Plasmids, and Episomes* (A. I. Bukhari, J. A. Shapiro, and S. L. Adhya, eds.), Cold Spring Harbor Laboratory, Cold Spring Harbor, N. Y., pp. 205–212.

24. Berg, D. E., 1989, Transposon Tn5, in: *Mobile DNA* (D. E. Berg and M. M. Howe, eds.), American Society for Microbiology, Washington, D.C., pp. 185–210.

25. Beringer, J. E., and Hopwood, D. A., 1976, Chromosomal recombination and mapping in *Rhizobium leguminosarum*, *Nature* **264:**291–293.

26. Beringer J. E., Hoggan, S. A., and Johnston, A. W. B., 1978, Linkage mapping in *Rhizobium leguminosarum* by means of R plasmid-mediated recombination, *J. Gen. Microbiol.* **104:**201–207.

27. Beringer, J. E., Johnston, A. W. B., and Kondorosi, A., 1987, *Rhizobium meliloti* and *Rhizobium leguminosarum*, in: *Genetic Maps*, Vol. 4 (S. J. O'Brien, ed.), Cold Spring Harbor Laboratory, Cold Spring Harbor, N.Y., pp. 245–251.

28. Bhattacharjee, M. K., and Meyer, R. J., 1991, A segment of a plasmid gene required for conjugal transfer encodes a site-specific, single-strand DNA endonuclease and ligase, *Nucleic Acids Res.* **19:**1129–1137.

29. Biel, S. W., and Berg, D. E., 1984, Mechanism of IS*1* transposition in *E. coli*: Choice between simple insertion and cointegration, *Genetics* **108:**319–330.

30. Birkenbihl, R. P., and Vielmetter, W., 1989, Complete maps of IS*1*, IS*2*, IS*3*, IS*4*, IS*5*, IS*30* and IS*150* locations in *Escherichia coli* K12, *Mol. Gen. Genet.* **220:**147–153.

31. Bissonette, L., and Roy, P. H., 1992, Characterization of In0 of *Pseudomonas aeruginosa* plasmid pVS1, an ancestor of integrons of multiresistance plasmids and transposons of gram-negative bacteria, *J. Bacteriol.* **175:**1248–1257.

32. Bittle, C., and Konopka, A., 1990, IncP-mediated transfer of loci involved with gas vesicle production in *Ancylobacter aquaticus*, *J. Gen. Microbiol.* **136:**1259–1263.

33. Blanco, G., Gutierrez, J. C., Ramos, F., and Tortolero, M., 1991, Isolation and characterization of R-primes of *Azotobacter vinelandii*, *FEMS Microbiol. Lett.* **80:**213–216.

34. Boccard, F., Smokvina, T., Pernodet, J.-L., Friedmann, A., and Guérineau, M., 1989, The integrated conjugative plasmid pSAM2 of *Streptomyces ambofaciens* is related to temperate bacteriophages, *EMBO J.* **8:**973–980.

35. Bowen, A. R. S. G., and Pemberton, J. M., 1985, Mercury resistance transposon Tn*813* mediates chromosome transfer in *Rhodopseudomonas sphaeroides* and intergeneric transfer of pBR322, in: *Plasmids in Bacteria* (D. R. Helinski, S. N. Cohen, D. B. Clewell, D. A. Jackson, and A. Hollaender, eds.), Plenum Press, New York, pp. 105–115.

36. Boyd, A. C., Archer, J. A. K., and Sherratt, D. J., 1989, Characterization of the ColE1 mobilization region and its protein products, *Mol. Gen. Genet.* **217:**488–498.

37. Brandt, R., Günther, E., and Herrmann, H., 1984, Mapping of cysteine genes on the chromosome of *Pseudomonas aeruginosa* PAO, *Mol. Gen. Genet.* **197:**292–296.

38. Bray, R., Strom, D., Barton, J., Dean, H. F., and Morgan, A. F., 1987, Isolation and characterization of *Pseudomonas putida* R-prime plasmids, *J. Gen. Microbiol.* **133:**683–690.

39. Bresler, S. E., Krivonogov, S. V., and Lanzov, V. A., 1979, Genetic determination of the donor properties in *Escherichia coli* K-12, *Mol. Gen. Genet.* **177:**177–184.

40. Breton, A. M., Jaoua, S., and Guespin-Michel, J., 1985, Transfer of plasmid RP4 to *Myxococcus xanthus* and evidence for its integration into the chromosome, *J. Bacteriol.* **161:**523–528.

41. Broome-Smith, J., 1980, RecA independent, site-specific recombination between ColE1 or ColK and a miniplasmid they complement for mobilization and relaxation: Implications for the mechanism of DNA transfer during mobilization, *Plasmid* **4:**51–63.

42. Brown, P. O., 1990, Integration of retroviral DNA, in: *Retroviruses—Strategies of Replication* (R. Swanstrom and P. K. Vogt, eds.), Springer Verlag, Berlin, pp. 19–48.

43. Bryan, J., Saeed, N., Fox, D., and Sastry, G. R. K., 1982, R68.45 mediated chromosomal gene transfer in *Agrobacterium tumefaciens*, *Arch. Microbiol.* **131:**271–277.

44. Bushman, F. D., Fujiwara, T., and Craigie, R., 1990, Retroviral DNA integration directed by HIV integration protein in vitro, *Science* **249:**1555–1558.

45. Bushman, F. D., and Craigie, R., 1991, Activities of human immunodeficiency virus (HIV) integration protein *in vitro*: Specific cleavage and integration of HIV DNA, *Proc. Natl. Acad. Sci. USA* **88:**1339–1343.

46. Campbell, A., 1962, Episomes, *Adv. Genet.* **11:**101–145.

47. Carter, J. R., and Porter, R. D., 1991, *traY* and *traI* are required for *oriT*-dependent enhanced recombination between *lac*-containing plasmids and λ*plac5*, *J. Bacteriol.* **173:**1027–1034.

48. Chakrabarty, A. M., and Gunsalus, I. C., 1979, Chromosomal mobilization from a *recA* mutant of *Pseudomonas putida*, *Mol. Gen. Genet.* **176:**151–154.

49. Chandler, M., Silver, L., Roth, Y., and Caro, L., 1976, Chromosome replication in an Hfr strain of *Escherichia coli*, *J. Mol. Biol.* **104:**517–523.

50. Chandler, M., Silver, L., and Caro, L., 1977, Suppression of an *Escherichia coli dnaA* mutation by the integrated R factor R100.1: Origin of chromosome replication during exponential growth, *J. Bacteriol.* **131:**421–430.

51. Chandler, M., Roulet, E., Silver, L., Boy de la Tour, E., and Caro, L., 1979, Tn*10* mediated integration of the plasmid R100.1 into the bacterial chromosome: Inverse transposition, *Mol. Gen. Genet.* **173**:23–30.

52. Chandler, P. M., and Krishnapillai, V., 1974, Isolation and properties of recombination-deficient mutants of *Pseudomonas aeruginosa*, *Mutation Res.* **23**:15–23.

53. Chater, K. F., Henderson, D. J., Bibb, M. J., and Hopwood, D. A., 1988, Genome flux in *Streptomyces coelicolor* and other streptomycetes and its possible relevance to the evolution of mobile antibiotic resistance determinants, in: *Transposition* (A. J. Kingsman, K. F. Chater, and S. M. Kingsman, eds.), Cambridge University Press, Cambridge, England, pp. 7–42.

54. Chatterjee, A. K., 1980, Acceptance by *Erwinia* spp. of R plasmid R68.45 and its ability to mobilize the chromosome of *Erwinia chrysanthemi*, *J. Bacteriol.* **142**:111–119.

55. Chatterjee, A. K., and Starr, M. P., 1973, Gene transmission among strains of *Erwinia amylovora*, *J. Bacteriol.* **116**:1100–1106.

56. Chatterjee, A. K., and Starr, M. P., 1977, Donor strains of the soft-rot bacterium *Erwinia chrysanthemi* and conjugational transfer of the pectolytic capacity, *J. Bacteriol.* **132**:862–869.

57. Chatterjee, A. K., and Starr, M. P., 1980, Genetics of *Erwinia* species, *Annu. Rev. Microbiol.* **34**:645–676.

58. Chatterjee, A. K., Ross, L. M., McEvoy, J. L., and Thurn, K. K., 1985, pULB113, an RP4::mini-Mu plasmid, mediates chromosomal mobilization and R-prime formation in *Erwinia amylovora*, *Erwinia chrysanthemi*, and subspecies of *Erwinia carotovora*, *Appl. Environ. Microbiol.* **50**:1–9.

59. Chilton, M-D., Farrand, S. K., Levin, R., and Nester, E. W., 1976, RP4 promotion of transfer of a large *Agrobacterium* plasmid which confers virulence, *Genetics* **83**:609–618.

60. Chumley, F. G., Menzel, R., and Roth, J. R., 1979, Hfr formation directed by Tn*10*, *Genetics* **91**:639–655.

61. Clark, A. J., and Warren, G. J., 1979, Conjugal transmission of plasmids, *Annu. Rev. Genet.* **13**:99–125.

62. Clewell, D. B., Tomich, P. K., Gawron-Burke, M. C., Franke, A. E., Yagi, Y., and An, F. Y., 1982, Mapping of *Streptococcus faecalis* plasmids pAD1 and pAD2 and studies relating to transposition of Tn*917*, *J. Bacteriol.* **152**:1220–1230.

63. Clowes, R. C., and Moody, E. E. M., 1966, Chromosomal transfer from "recombination-deficient" strains of *Escherichia coli*, *Genetics* **53**:717–726.

64. Coetzee, J. N., 1978, Mobilization of the *Proteus mirabilis* chromosome by R plasmid R772, *J. Gen. Microbiol.* **108**:103–109.

65. Coetzee, J. N., 1979, Patterns of mobilization of the *Proteus mirabilis* chromosome by R plasmids, *J. Gen. Microbiol.* **111**:243–251.

66. Coetzee, J. N., van Dijken, M. C., and Coetzee, W. F., 1987, *Proteus mirabilis*, in: *Genetic Maps*, Vol. 4 (S. J. O'Brien, ed), Cold Spring Harbor Laboratory, Cold Spring Harbor, N. Y., pp. 224–229.

67. Cohen, A., Silberstein, Z., Broido, S., and Laban, A., 1985, General genetic recombination of bacterial plasmids, in: *Plasmids in Bacteria* (D. R. Helinski, S. N. Cohen, D. B. Clewell, D. A. Jackson, and A. Hollaender, eds.), Plenum Press, New York, pp. 505–519.

68. Colbeau, A., Magnin, J.-P., Cauvin, B., Champion, T., and Vignais, P. M., 1990, Genetic and physical mapping of an hydrogenase gene cluster from *Rhodobacter capsulatus*, *Mol. Gen. Genet.* **220**:393–399.

69. Collis, C. M., and Hall, R. M., 1992, Gene cassette from the insert region of integrons are excised as covalently closed circles, *Mol. Microbiol.* **6**:2875–2885.

70. Craig, N. L., 1988, The mechanism of conservative site-specific recombination, *Annu. Rev. Genet.* **22**:77–105.

71. Craig, N. L., 1991, Tn7: A target site-specific transposon, *Mol. Microbiol.* **5**:2569–2573.

72. Cullum, J., and Broda, P., 1979, Chromosome transfer and Hfr formation by F in *rec⁺* and *recA* strains of *Escherichia coli* K12, *Plasmid* **2**:358–365.

73. Currier, T. C., and Morgan, M. K., 1982, Direct DNA repeat in plasmid R68.45 is associated with deletion formation and concomitant loss of chromosome mobilization ability, *J. Bacteriol.* **150**:251–259.

74. Curtiss, III, R., and Stallions, D. R., 1969, Probability of F integration and frequency of stable Hfr donors in F⁺ populations of *Escherichia coli* K-12, *Genetics* **63**:27–38.

75. Cuzin, F., and Jacob, F., 1967, Mutations de l'épisome F d'*Escherichia coli* K12. 2. Mutants à réplication thermosensible, *Ann. Inst. Pasteur* **112**:397–418.

76. Dalrymple, B., 1987, Novel rearrangements of IS*30* carrying plasmids leading to the reactivation of gene expression, *Mol. Gen. Genet.* **207**:413–420.

77. Danilevich, V. N., and Kostyuchenko, D. A., 1986, Immunity to repeated insertion of IS*21* sequence, *Mol. Biol.* **19**:1016–1022 (English translation).

78. Danilevich, V. N., Kostyuchenko, D. A., and Negrii, N. V., 1986, Interaction (incorporation and excision) of the plasmid pRP19.6, a derivative of RP1 containing a duplicated sequence IS21, with the chromosome of *Escherichia coli* K12, *Mol. Biol.* **19:**858–867 (English translation).

79. Datta, N., and Barth, P. T., 1976, Hfr Formation by I pilus-determining plasmids in *Escherichia coli* K-12, *J. Bacteriol.* **125:**811–817.

80. Davidson, N., Deonier, R. C., Hu, S., and Ohtsubo, E., 1975, Electron microscope heteroduplex studies of sequence relations among plasmids of *Escherichia coli*. X. Deoxyribonucleic acid sequence organization of F and F-primes, and the sequences involved in Hfr formation, in: *Microbiology—1974* (D. Schlessinger, ed.), American Society for Microbiology, Washington, D. C., pp. 56–65.

81. De Graaff, J., Kreuning, P. C., and Stouthamer, A. H., 1974, Isolation and characterization of Hfr males in *Citrobacter freundii*, *Antonie van Leeuwenhoek J. Microbiol.* **40:**161–170.

82. DeVries, J. K., and Maas, W. K., 1971, Chromosomal integration of F' factors in recombination-deficient Hfr strains of *Escherichia coli*, *J. Bacteriol.* **106:**150–156.

83. Dean, H. F., and Morgan, A. F., 1983, Integration of R91-5::Tn*501* into the *Pseudomonas putida* PPN chromosome and genetic circularity of the chromosomal map, *J. Bacteriol.* **153:**485–497.

84. Deonier, R. C., and Davidson, N., 1976, The sequence organization of the integrated F plasmid in two Hfr strains of *Escherichia coli*, *J. Mol. Biol.* **107:**207–222.

85. Deonier, R. C., and Hadley, R. G., 1980, IS2-IS2 and IS3-IS3 relative recombination frequencies in F integration, *Plasmid* **3:**48–64.

86. Deonier, R. C., and Mirels, L., 1977, Excision of F plasmid sequences by recombination at directly repeated insertion sequence 2 elements: Involvement of recA, *Proc. Natl. Acad. Sci. USA* **74:**3965–3969.

87. Depicker, A., De Block, M., Inzé, D., Van Montagu, M., and Schell, J., 1980, IS-like element IS8 in RP4 plasmid and its involvement in cointegration, *Gene* **10:**329–338.

88. Dessaux, Y., Petit, A., Ellis, J. G., Legrain, C., Demarez, M., Wiame, J.-M., Popoff, M., and Tempé, J., 1989, Ti plasmid-controlled chromosome transfer in *Agrobacterium tumefaciens*, *J. Bacteriol.* **171:**6363–6366.

89. Dixon, D. A., and Kowalczykowski, S. C., 1991, Homologous pairing in vitro stimulated by the recombinant hotspot, Chi, *Cell* **66:**361–371.

90. Dixon, R., and Kinghorn, J. R., 1990, Separation of large DNA molecules by pulsed-field gel electrophoresis, *Soc. Gen. Microbiol. Q.* **17:**86–88.

91. Dixon, R., Cannon, F. C., and Postgate, J. R., 1975, Properties of the R-factor R144drd3 in *Klebsiella pneumoniae* strain M5a1, *Genet. Res.* **28:**327–338.

92. Dreyfus, L. A., and Iglewski, B. H., 1985, Conjugation-mediated genetic exchange in *Legionella pneumophila*, *J. Bacteriol.* **161:**80–84.

93. Ehrlich, S. D., 1989, Illegitimate recombination in bacteria, in: *Mobile DNA* (D. E. Berg and M. M. Howe, eds.), American Society for Microbiology, Washington, D. C., pp. 799–832.

94. Engelman, A., Mizuuchi, K., and Craigie, R., 1991, HIV-1 DNA integration: Mechanism of viral DNA cleavage and DNA strand transfer, *Cell* **67:**1211–1221.

95. Espin, G., Alvarez-Morales, A., and Merrick, M., 1981, Complementation analysis of glnA-linked mutations which affect nitrogen fixation in *Klebsiella pneumoniae*, *Mol. Gen. Genet.* **184:**213–217.

96. Faelen, M., Toussaint, A., and De Lafontaine, J., 1975, Model for the enhancement of λ-*gal* integration into partially induced Mu-1 lysogens, *J. Bacteriol.* **121:**873–882.

97. Fayet, O., Ramond, P., Polard, P., Prère, M. F., and Chandler, M., 1990, Functional similarities between retroviruses and the IS3 family of bacterial insertion sequences, *Mol. Microbiol.* **4:**1771–1777.

98. Fitzgerald, G. F., and Gasson, M. J., 1988, *In vivo* gene transfer systems and transposons, *Biochimie* **70:**489–502.

99. Forbes, K. J., and Pérombelon, M. C. M., 1985, Chromosomal mapping in *Erwinia carotovora* subsp. *carotovora* with the IncP plasmid R68::Mu, *J. Bacteriol.* **164:**1110–1116.

100. Franche, C., Canelo, E., Gauthier, D., and Elmerich, C., 1981, Mobilization of the chromosome of *Azospirillum brasilense* by plasmid R68-45, *FEMS Microbiol. Lett.* **10:**199–202.

101. François, V., Conter, A., and Louarn, J.-M., 1990, Properties of new *Escherichia coli* Hfr strains constructed by integration of pSC101-derived conjugative plasmids, *J. Bacteriol.* **172:**1436–1440.

102. Franke, A. E., Dunny, G. M., Brown, B. L., An, F., Oliver, D. R., Damle, S. P., and Clewell, D. B., 1978, Gene transfer in *Streptococcus faecalis*: Evidence for the mobilization of chromosomal determinants by

transmissible plasmids, in: *Microbiology—1978* (D. Schlessinger, ed.), American Society for Microbiology, Washington, D.C., pp. 45–47.

103. Franklin, N. C., 1971, Illegitimate recombination, in: *The Bacteriophage Lambda* (A. D. Hershey, ed.), Cold Spring Harbor Laboratory, Cold Spring Harbor, N. Y., pp. 175–194.

104. Frey, J., and Bagdasarian, M., 1989, The molecular biology of IncQ plasmids, in: *Promiscuous Plasmids of Gram-negative Bacteria* (C. M. Thomas, ed.), Academic Press, London, pp. 79–94.

105. Fürste, J. P., Pansegrau, W., Ziegelin, G., Kröger, M., and Lanka, E., 1989, Conjugative transfer of promiscuous IncP plasmids: Interaction of plasmid-encoded products with the transfer origin, *Proc. Natl. Acad. Sci. USA* **86:**1771–1775.

106. Fulbright, D. W., and Leary, J. V., 1978, Linkage analysis of *Pseudomonas glycinea*, *J. Bacteriol.* **136:** 497–500.

107. Galas, D. J., and Chandler, M., 1989, Bacterial insertion sequences, in: *Mobile DNA* (D. E. Berg and M. M. Howe, eds.), American Society for Microbiology, Washington, D.C., pp. 109–162.

108. Gennaro, M. L., Kornblum, J., and Novick, R. P., 1987, A site-specific recombination function in *Staphylococcus aureus* plasmids, *J. Bacteriol.* **169:**2601–2610.

109. Gerlitz, M., Hrabak, O., and Schwab, H., 1990, Partitioning of broad-host-range plasmid RP4 is a complex system involving site-specific recombination, *J. Bacteriol.* **172:**6194–6203.

110. Ghozlan, H. A., Ahmadian, R. M., Fröhlich, M., Sabry, S., and Kleiner, D., 1991, Genetic tools for *Paracoccus denitrificans*, *FEMS Microbiol. Lett.* **82:**303–306.

111. Gobius, K. S., and Pemberton, J. M., 1986, Use of plasmid pULB113 (RP4::mini-Mu) to construct a genomic map of *Aeromonas hydrophila*, *Curr. Microbiol.* **13:**111–115.

112. Grandgenett, D. P., and Mumm, S. R., 1990, Unraveling retrovirus integration, *Cell* **60:**3–4.

113. Griffin, IV, T. J., and Kolodner, R. D., 1990, Purification and preliminary characterization of the *Escherichia coli* K-12 RecF protein, *J. Bacteriol.* **172:**6291–6299.

114. Grindley, N. D. F., and Reed, R. R., 1985, Transpositional recombination in prokaryotes, *Annu. Rev. Biochem.* **54:**863–896.

115. Grinsted, J., De la Cruz, F., and Schmitt, R., 1990, The Tn*21* subgroup of bacterial transposable elements, *Plasmid* **24:**163–189.

116. Grinter, N. J., 1981, Analysis of chromosome mobilization using hybrids between plasmid RP4 and a fragment of bacteriophage λ carrying IS*1*, *Plasmid* **5:**267–276.

117. Grinter, N. J., 1984, Replication control of IncP plasmids, *Plasmid* **11:**74–81.

118. Grinter, N. J., 1984, Replication defective RP4 plasmids recovered after chromosomal integration, *Plasmid* **11:**65–73.

119. Groenen, M. A. M., Kokke, M., and van de Putte, P., 1986, Transposition of mini-Mu containing only one of the ends of bacteriophage Mu, *EMBO J.* **5:**3687–3690.

120. Groisman, E. A., and Casadaban, M. J., 1986, Mini-Mu bacteriophage with plasmid replicons for *in vivo* cloning and *lac* gene fusing, *J. Bacteriol.* **168:**357–364.

121. Guiney, D. G., and Lanka, E., 1989, Conjugative transfer of IncP plasmids, in: *Promiscuous Plasmids of Gram-negative Bacteria* (C. M. Thomas, ed.), Academic Press, London, pp. 27–56.

122. Guyer, M., 1978, The γδ sequence of F is an insertion sequence, *J. Mol. Biol.* **126:**347–365.

123. Guyer, M. S., Reed, R. R., Steitz, J. A., and Low, K. B., 1980, Identification of a sex-factor-affinity site in *E. coli* as γδ, *Cold Spring Harbor Symp. Quant. Biol.* **45:**135–140.

124. Haas, D., and Holloway, B. W., 1976, R factor variants with enhanced sex factor activity in *Pseudomonas aeruginosa*, *Mol. Gen. Genet.* **144:**243–251.

125. Haas, D., and Holloway, B. W., 1978, Chromosome mobilization by the R plasmid R68.45: A tool in *Pseudomonas* genetics, *Mol. Gen. Genet.* **158:**229–237.

126. Haas, D., and Reimmann, C., 1989, Use of IncP plasmids in chromosomal genetics of gram-negative bacteria, in: *Promiscuous Plasmids of Gram-negative Bacteria* (C. M. Thomas, ed.), Academic Press, London, pp. 185–206.

127. Haas, D., and Riess, G., 1983, Spontaneous deletions of the chromosome-mobilizing plasmid R68.45 in *Pseudomonas aeruginosa* PAO, *Plasmid* **9:**42–52.

128. Haas, D., Watson, J., Krieg, R., and Leisinger, T., 1981, Isolation of an Hfr donor of *Pseudomonas aeruginosa* PAO by insertion of the plasmid RP1 into the tryptophan synthase gene, *Mol. Gen. Genet.* **182:**240–244.

129. Hadley, R. G., and Deonier, C., 1980, Specificity in the formation of Δtra F-prime plasmids, *J. Bacteriol.* **143:**680–692.

130. Hall, R. M., Brookes, D. E., and Stokes, H. W., 1991, Site-specific insertion of genes into integrons: Role of the 59-base element and determination of the recombination crossover point, *Mol. Microbiol.* **5:**1941–1959.

131. Haniford, D. B., Chelouche, A. R., and Kleckner, N., 1989, A specific class of IS*10* transposase mutants are blocked for target site interactions and promote formation of an excised transposon fragment, *Cell* **59:** 385–394.

132. Harayama, S., and Rekik, M., 1989, A simple procedure for transferring genes cloned in *Escherichia coli* vectors into other gram-negative bacteria: Phenotypic analysis and mapping of TOL plasmid gene *xylK*, *Gene* **78:**19–27.

133. Harayama, S., Tsuda, M., and Iino, T., 1980, High frequency mobilization of the chromosome of *Escherichia coli* by a mutant of plasmid RP4 temperature-sensitive for maintenance, *Mol. Gen. Genet.* **180:**47–56.

134. Harayama, S., Oguchi, T., and Iino, T., 1984, Does Tn*10* transpose via the cointegrate molecule? *Mol. Gen. Genet.* **194:**444–450.

135. Harayama, S., Oguchi, T., and Iino, T., 1984, The *E. coli* K-12 chromosome flanked by two IS*10* sequences transposes, *Mol. Gen. Genet.* **197:**62–66.

136. Harayama, S., Lehrbach, P. R., Tsuda, M., Leppik, R., Iino, T., Reineke, W., Knackmuss, H. J., and Timmis, K. N., 1984, Genetic engineering systems for *Pseudomonas* and their use in the analysis and manipulation of metabolic pathways, in: *Transferable Antibiotic Resistance. Plasmids and Gene Manipulation* (S. Mitsuhashi and V. Krcméry, eds.), Avicenum Czechoslovak Medical Press, Prague, pp. 361–372.

137. Hayes, F., Caplice, E., McSweeny, A., Fitzgerald, G. F., and Daly, C., 1990, pAMβ1-associated mobilization of proteinase plasmids from *Lactococcus lactis* subsp. *lactis* UC317 and *L. lactis* subsp. *cremoris* UC205, *Appl. Environ. Microbiol.* **56:**195–201.

138. Hayes, F., Daly, C., and Fitzgerald, G. F., 1990, High-frequency, site-specific recombination between lactococcal and pAMβ1 plasmid DNAs, *J. Bacteriol.* **172:**3485–3489.

139. Hayes, W., 1968, *The genetics of bacteria and their viruses*, Blackwell Scientific Publications, Oxford.

140. Hecht, D. W., and Malamy, M. H., 1989, Tn*4399*, a conjugal mobilizing transposon of *Bacteroides fragilis*, *J. Bacteriol.* **171:**3603–3608.

141. Hecht, D. W., Thompson, J. S., and Malamy, M. H., 1989, Characterization of the termini and transposition products of Tn*4399*, a conjugal mobilizing transposon of *Bacteroides fragilis*, *Proc. Natl. Acad. Sci. USA* **86:**5340–5344.

142. Hedén, L-O., and Meynell, E., 1976, Comparative study of R1-specific chromosomal transfer in *Escherichia coli* K-12 and *Salmonella typhimurium* LT2, *J. Bacteriol.* **127:**51–58.

143. Hedges, R. W., and Jacob, A. E., 1977, In vivo translocation of genes of *Pseudomonas aeruginosa* onto a promiscuously transmissible plasmid, *FEMS Microbiol. Lett.* **2:**15–19.

144. Hedges, R. W., Jacob, A. E., and Crawford, I. P., 1977, Wide range plasmid bearing the *Pseudomonas aeruginosa* tryptophan synthase genes, *Nature* **267:**283–284.

145. Herrmann, H., and Günter, E., 1984, High frequency FP2 donor of *Pseudomonas aeruginosa* PAO, *Mol. Gen. Genet.* **197:**286–291.

146. Herrmann, H., Klopotowski, T., and Günter, E., 1986, The Hfr status of *Pseudomonas aeruginosa* is stabilized by integrative suppression, *Mol. Gen. Genet.* **204:**519–523.

147. Herrmann, H., Janke D., Kresja, S., and Roy, M., 1988, In vivo generation of R68.45-pPGH1 hybrid plasmids conferring a Phl+ (meta pathway) phenotype, *Mol. Gen. Genet.* **214:**173–176.

148. Hille, J., van Kan, J., Klasen, I., and Schilperoort, R., 1983, Site-directed mutagenesis in *Escherichia coli* of a stable R772::Ti cointegrate plasmid from *Agrobacterium tumefaciens*, *J. Bacteriol.* **154:**693–701.

149. Hirschel, B. J., Galas, D. J., and Chandler, M., 1982, Cointegrate formation by Tn5, but not transposition, is dependent on *recA*, *Proc. Natl. Acad. Sci. USA* **79:**4530–4534.

150. Holloway, B. W., 1978, Isolation and characterization of an R′ plasmid in *Pseudomonas aeruginosa*, *J. Bacteriol.* **133:**1078–1082.

151. Holloway, B. W., 1979, Plasmids that mobilize bacterial chromosome, *Plasmid* **2:**1–19.

152. Holloway, B. W., 1983, *Pseudomonas* genetics and its application to other bacteria, in: *Genetics of Industrial Microorganisms* (Y. Ikeda and T. Beppu, eds.), Kodansha Ltd., Tokyo, pp. 41–45.

153. Holloway, B. W., and Low, K. B., 1987, F-prime and R-prime factors, in: *Escherichia coli and Salmonella typhimurium. Cellular and Molecular Biology* (F. C. Neidhardt, J. L. Ingraham, K. B. Low, B. Magasanik,

M. Schaehter, and H. E. Umbarger, eds.), American Society for Microbiology, Washington, D.C., pp. 1145–1153.

154. Holloway, B. W., Crowther, C., Dean, H., Hagedorn, J., Holmes, N., and Morgan, A. F., 1982, Integration of plasmids into the *Pseudomonas* chromosome, in: *Drug Resistance in Bacteria* (S. Mitsuhashi, ed.), Japan Scientific Societies Press, Tokyo, pp. 231–242.

155. Holloway, B. W., Kearney, P. P., and Lyon, B. R., 1987, The molecular genetics of C1 utilizing microorganisms—An overview, *Antonie van Leeuwenhoek J. Microbiol.* **53**:47–53.

156. Holloway, B. W., O'Hoy, K., and Matsumoto, H., 1987, *Pseudomonas aeruginosa* PAO, in: *Genetic Maps*, Vol. 4 (S. J. O'Brien, ed.), Cold Spring Harbor Laboratory, Cold Spring Harbor, pp. 213–221.

157. Holsters, M., Silva, B., Genetello, C., Engler, G., van Vliet, F., De Block, M., Villarroel, R., van Montagu, M., and Schell, J., 1978, Spontaneous formation of cointegrates of the oncogenic Ti-plasmid and the wide-host-range P-plasmid RP4, *Plasmid* **1**:456–467.

158. Holsters, M., Silva, B., van Vliet, F., Hernalsteens, J. P., Genetello, C., van Montagu, M., and Schell, J., 1978, In vivo transfer of the Ti-plasmid of *Agrobacterium tumefaciens* to *Escherichia coli*, *Mol. Gen. Genet.* **163**:335–338.

159. Hooykaas, P. J. J., Klapwijk, P. M., Nuti, M. P., Schilperoort, R. A., and Rörsch, A., 1977, Transfer of the *Agrobacterium tumefaciens* TI plasmid to avirulent Agrobacteria and to Rhizobium *ex planta*, *J. Gen. Microbiol.* **98**:477–484.

160. Hooykaas, P. J. J., Den Dulk-Ras, H., and Schilperoort, R. A., 1980, Molecular mechanism of Ti plasmid mobilization by R plasmids: Isolation of Ti plasmids with transposon-insertions in *Agrobacterium tumefaciens*, *Plasmid* **4**:64–75.

161. Hooykaas, P. J. J., Den Dulk-Ras, H., and Schilperoort, R. A., 1982, Method for the transfer of large cryptic, non-self-transmissible plasmids: *Ex planta* transfer of the virulence plasmid of *Agrobacterium rhizogenes*, *Plasmid* **8**:94–96.

162. Hooykaas, P. J. J., Peerbolte, R., Regensburg-Tuink, A. J. G., de Vries, P., and Schilperoort, R. A., 1982, A chromosomal linkage map of *Agrobacterium tumefaciens* and a comparison with the maps of *Rhizobium* spp, *Mol. Gen. Genet.* **188**:12–17.

163. Horowitz, B., and Deonier, R. C., 1985, Formation of Δtra F′ plasmids: Specific recombination at *oriT*, *J. Mol. Biol.* **186**:267–274.

164. Ichige, A., Matsutani, S., Oishi, K., and Mizushima, S., 1989, Establishment of gene transfer systems for and construction of the genetic map of a marine *Vibrio* strain, *J. Bacteriol.* **171**:1825–1834.

165. Ichikawa, H., and Ohtsubo, E., 1990, *In vitro* transposition of transposon Tn*3*, *J. Biol. Chem.* **265**:18829–18832.

166. Iida, S., 1977, Directed integration of an F′ plasmid by integrative suppression, *Mol. Gen. Genet.* **155**:153–162.

167. Ikeda, H., Aoki, K., and Naito, A., 1982, Illegitimate recombination mediated *in vitro* by DNA gyrase of *Escherichia coli*: Structure of recombinant DNA molecules, *Proc. Natl. Acad. Sci. USA* **79**:3724–3728.

168. Ikeda, H., Kawasaki, I., and Gellert, M., 1984, Mechanism of illegitimate recombination: Common sites for recombination and cleavage mediated by *E. coli* DNA gyrase, *Mol. Gen. Genet.* **196**:546–549.

169. Isberg, R. R., and Syvanen, M., 1985, Tn*5* transposes independently of cointegrate resolution: Evidence for an alternative model for transposition, *J. Mol. Biol.* **182**:69–78.

170. Jacob, F., and Wollman, E. L., 1956, Recombinaison génétique et mutants de fertilité chez *Escherichia coli*, *Compt. Rend. Acad. Sci. Paris* **242**:303–306.

171. Jaoua, S., Letouvet-Pawlak, B., Monnier, C., and Guespin-Michel, J. F., 1990, Mechanism of integration of the broad-host-range plasmid RP4 into the chromosome of *Myxococcus xanthus*, *Plasmid* **23**:183–193.

172. Johnson, D. A., 1988, Construction of transposons carrying the transfer function of RP4, *Plasmid* **20**:249–258.

173. Johnson, D. A., and Willetts, N. S., 1980, Tn*2301*, a transposon construct carrying the entire transfer region of the F plasmid, *J. Bacteriol.* **143**:1171–1178.

174. Johnson, M. S., McClure, M. A., Feng, D.-F., Gray, J., and Doolittle, R. F., 1986, Computer analysis of retroviral *pol* genes: Assignment of enzymatic functions to specific sequences and homologies with nonviral enzymes, *Proc. Natl. Acad. Sci. USA* **83**:7648–7652.

175. Johnson, S. R., and Romig, W. R., 1979, Transposon-facilitated recombination in *Vibrio cholerae*, *Mol. Gen. Genet.* **170**:93–101.

176. Johnston, A. W. B., and Beringer, J. E., 1977, Chromosomal recombination between *Rhizobium* species, *Nature* **267**:611–613.

177. Johnston, A. W. B., Setchell, S. M., and Beringer, J. E., 1978, Interspecific crosses between *Rhizobium leguminosarum* and *R. meliloti*: Formation of haploid recombinants and of R-primes, *J. Gen. Microbiol.* **104**:209–218.

178. Jones, K. S., Coleman, J., Merkel, G. W., Laue, T. M., and Skalka, A. M., 1992, Retroviral integrase functions as a multimer and can turn over catalytically, *J. Biol. Chem.* **267**:16037–16040.

179. Julliot, J. S., and Boistard, P., 1979, Use of RP4-prime plasmids constructed in vitro to promote a polarized transfer of the chromosome in *Escherichia coli* and *Rhizobium meliloti*, *Mol. Gen. Genet.* **173**:289–298.

180. Julliot, J. S., Dusha, I., Renalier, M. H., Terzaghi, B., Garnerone, A. M., and Boistard, P., 1984, An RP4-prime containing a 285 kb fragment of *Rhizobium meliloti* pSym megaplasmid: Structural characterization and utilization for genetic studies of symbiotic functions controlled by pSym, *Mol. Gen. Genet.* **193**:17–26.

181. Kahn, P. L., 1968, Isolation of high-frequency recombining strains from *Escherichia coli* containing the V colicinogenic factor, *J. Bacteriol.* **96**:205–214.

182. Kameneva, S. V., Polivtseva, T. P., Belavina, N. V., and Shestakov, S. V., 1986, Hfr donor of phototrophic nitrogen fixing bacterium *Rhodopseudomonas sphaeroides*: Localization of mutations in genes controlling nitrogen fixation, *Genetika* (Russ.) **22**:2664–2672.

183. Kaney, A. R., and Atwood, K. C., 1972, Incompatibility of integrated sex factors in double male strains of *Escherichia coli*, *Genetics* **70**:31–39.

184. Katz, L., Brown, D. P., and Donadio, S., 1991, Site-specific recombination in *Escherichia coli* between the *att* sites of plasmid pSE211 from *Saccharopolyspora erythraea*, *Mol. Gen. Genet.* **227**:155–159.

185. Katz, R. A., Merkel, G., Kulkosky, J., Leis, J., and Skalka, A. M., 1990, The avian retroviral IN protein is both necessary and sufficient for integrative recombination in vitro, *Cell* **63**:87–95.

186. Khan, E., Mack, J. P. G., Katz, R. A., Kulkowsky, J., and Skalka, A. M., 1991, Retroviral integrase domains: DNA binding and the recognition of LTR sequences, *Nucleic Acids Res.* **19**:851–860.

187. Kieser, H. M., Kieser, T., and Hopwood, D. A., 1992, A combined genetic and physical map of the *Streptomyces coelicolor* A3(2) chromosome, *J. Bacteriol.* **174**:5496–5507.

188. Kilbane, J. J., and Malamy, M. H., 1980, F Factor mobilization of non-conjugative chimeric plasmids in *Escherichia coli*: General mechanisms and a role for site-specific *recA*-independent recombination at *oriV1*, *J. Mol. Biol.* **143**:73–93.

189. Kiss, G. B., Dobo, K., Dusha, I., Breznovits, A., Orosz, L., Vincze, E., and Kondorosi, A., 1980, Isolation and characterization of an R-prime plasmid from *Rhizobium meliloti*, *J. Bacteriol.* **141**:121–128.

190. Kleckner, N., 1989, Transposon Tn*10*, in: *Mobile DNA* (D. E. Berg and M. M. Howe, eds.), American Society for Microbiology, Washington, D. C., pp. 227–268.

191. Kleckner, N., Roth, J., and Botstein, D., 1977, Genetic engineering *in vivo* using translocatable drug-resistance elements: New methods in bacterial genetics, *J. Mol. Biol.* **116**:125–159.

192. Klein, S., Lohman, K., Clover, R., Walker, G. C., and Signer, E. R., 1992, A directional, high-frequency chromosomal mobilization system for genetic mapping of *Rhizobium meliloti*, *J. Bacteriol.* **174**:324–326.

193. Koonin, E. V., 1992, DnaC protein contains a modified ATP-binding motif and belongs to a novel family of ATPases including also DnaA, *Nucleic Acids Res.* **20**:1997.

194. Kondorosi, A., Kiss, G. B., Forrai, T., Vincze, E., and Banfalvi, Z., 1977, Circular linkage map of *Rhizobium meliloti* chromosome, *Nature* **268**:525–527.

195. Konforti, B. B., and Davis, R. W., 1990, The preference for a 3′ homologous end is intrinsic to RecA-promoted strand exchange, *J. Biol. Chem.* **265**:6916–6920.

196. Krishnapillai, V., Royle, P., and Lehrer, J., 1981, Insertions of the transposon Tn*1* into the *Pseudomonas aeruginosa* chromosome, *Genetics* **97**:495–511.

197. Laban, A., and Cohen, A, 1981, Interplasmidic and intraplasmidic recombination in *Escherichia coli* K-12, *Mol. Gen. Genet.* **184**:200–207.

198. Leemans, J., Villarroel, R., Silva, B., Van Montagu, M., and Schell, J., 1980, Direct repetition of a 1.2 Md DNA sequence is involved in site-specific recombination by the P1 plasmid R68, *Gene* **10**:319–328.

199. Lejeune, P., and Mergeay, M., 1980, R-plasmid-mediated chromosome mobilization in *Pseudomonas fluorescens* 6.2, *Arch. Int. Physiol. Biochim.* **88**:B289–B290.

200. Lejeune, P., Mergeay, M., Van Gijsegem, F., Faelen, M., Gerits, J., and Toussaint, A., 1983, Chromosome transfer and R-prime plasmid formation mediated by plasmid pULB113 (RP4::mini-Mu) in *Alcaligenes eutrophus* CH34 and *Pseudomonas fluorescens* 6.2, *J. Bacteriol.* **155**:1015–1026.

201. Leonardo, J. M., and Goldberg, R. B., 1980, Regulation of nitrogen metabolism in glutamine auxotrophs of *Klebsiella pneumoniae*, *J. Bacteriol.* **142**:99–110.
202. Lichens-Park, A., and Syvanen, M., 1988, Cointegrate formation by IS*50* requires multiple donor molecules, *Mol. Gen. Genet.* **211**:244–251.
203. Liu, L., and Berg, C. M., 1990, Mutagenesis of dimeric plasmids by the transposon γδ (Tn*1000*), *J. Bacteriol.* **172**:2814–2816.
204. Lockshon, D., and Morris, D. R., 1985, Sites of reaction of *Escherichia coli* DNA gyrase on pBR322 *in vivo* as revealed by oxolinic acid-induced plasmid linearization, *J. Mol. Biol.* **181**:63–74.
205. Low, K. B., 1972, *Escherichia coli* K-12 F-prime factors, old and new, *Bacteriol. Rev.* **36**:587–607.
206. Low, K. B., 1987, Hfr strains of *Escherichia coli* K-12, in: *Escherichia coli and Salmonella typhimurium. Cellular and Molecular Biology* (F. C. Neidhardt, J. L. Ingraham, K. B. Low, B. Magasanik, M. Schaechter, and E. Umbarger, eds.), American Society for Microbiology, Washington, D.C., pp. 1134–1137.
207. Luisi-DeLuca, C., Lovett, S. T., and Kolodner, R. D., 1989, Genetic and physical analysis of plasmid recombination in *recB recC sbcB* and *recB recC sbcA Escherichia coli* K-12 mutants, *Genetics* **122**:269–278.
208. Lycett, G. W., and Pritchard, R. H., 1986, Functioning of the F-plasmid origin of replication in an *Escherichia coli* K12 Hfr strain during exponential growth, *Plasmid* **16**:168–174.
209. Machida, Y., Machida, C., and Ohtsubo, E., 1982, A novel type of transposon generated by insertion element IS*102* present in a pSC101 derivative, *Cell* **30**:29–36.
210. Magnin, J.-P., 1987, Isolement d'une souche Hfr de la bactérie photosynthetique *Rhodobacter capsulatus* et cartographie du chromosome, Ph.D. Thesis, Université de Grenoble.
211. Manis, J., Kopecko, D., and Kline, B., 1980, Cloning of a Lac⁺ BamHI fragment into transposon Tn*3* and transposition of the Tn*3*(*lac*) element, *Plasmid* **4**:170–174.
212. Marrs, B., 1981, Mobilization of the genes for photosynthesis from *Rhodopseudomonas capsulata* by a promiscuous plasmid, *J. Bacteriol.* **146**:1003–1012.
213. Martin, R. R., Thorlton, C. L., and Unger, L., 1981, Formation of *Escherichia coli* Hfr strains by integrative suppression with the P group plasmid RP1, *J. Bacteriol.* **145**:713–721.
214. Martinez, E., and de la Cruz, F., 1990, Genetic elements involved in Tn*21* site-specific integration, a novel mechanism for the dissemination of antibiotic resistance genes, *EMBO J.* **9**:1275–1281.
215. Marvo, S. L., King, S. R., and Jaskunas, R. R., 1983, Role of short regions of homology in intermolecular illegitimate recombination events, *Proc. Natl. Acad. Sci. USA* **80**:2452–2456.
216. Matsutani, S., Ohtsubo, H., Maeda, Y., and Ohtsubo, E., 1987, Isolation and characterization of IS elements repeated in the bacterial chromosome, *J. Mol. Biol.* **196**:445–455.
217. Mazodier, P., Petter, R., and Thompson, C., 1989, Intergeneric conjugation between *Escherichia coli* and *Streptomyces* species, *J. Bacteriol.* **171**:3583–3585.
218. Mazodier, P., and Davies, J., 1991, Gene transfer between distantly related bacteria, *Annu. Rev. Genet.* **25**:147–171.
219. McCormick, M., Wishart, W., Ohtsubo, H., Heffron, F., and Ohtsubo, E., 1981, Plasmid cointegrates and their resolution mediated by transposon Tn*3* mutants, *Gene* **15**:103–118.
220. McLaughlin, W., and Ahmad, M. H., 1986, Transfer of plasmids RP4 and R68.45 and chromosomal mobilization in cowpea rhizobia, *Arch. Microbiol.* **144**:408–411.
221. Meade, H. M., and Signer, E. R., 1977, Genetic mapping of *Rhizobium meliloti*, *Proc. Natl. Acad. Sci. USA* **74**:2076–2078.
222. Megias, M., Caviedes, M. A., Palomares, A. J., and Perez-Silva, J., 1982, Use of plasmid R68.45 for constructing a circular linkage map of the *Rhizobium trifolii* chromosome, *J. Bacteriol.* **149**:59–64.
223. Menou, G., Mahillon, J., Lecadet, M. M., and Lereclus, D., 1990, Structural and genetic organization of IS*232*, a new insertion sequence of *Bacillus thuringiensis*, *J. Bacteriol.* **172**:6689–6696.
224. Mercenier, A., and Chassy, B. M., 1988, Strategies for the development of bacterial transformation systems, *Biochimie* **70**:503–517.
225. Mergeay, M., Lejeune, P., Sadouk, A., Gerits, J., and Fabry, L., 1987, Shuttle transfer (or retrotransfer) of chromosomal markers mediated by plasmid pULB113, *Mol. Gen. Genet.* **209**:61–70.
226. Meyer, R., 1989, Site-specific recombination at *oriT* of plasmid R1162 in the absence of conjugative transfer, *J. Bacteriol.* **171**:799–806.
227. Michel, B., Niaudet, B., and Ehrlich, S. D., 1983, Intermolecular recombination during transformation of *Bacillus subtilis* competent cells by monomeric and dimeric plasmids, *Plasmid* **10**:1–10.

228. Miller, I. S., Fox, D., Saeed, N., Borland, P. A., Miles, C. A., and Sastry, G. R. K., 1986, Enlarged map of *Agrobacterium tumefaciens* C58 and the location of chromosomal regions which affect tumorigenicity, *Mol. Gen. Genet.* **205:**153–159.

229. Miller, J. H., 1992, *A Short Course in Bacterial Genetics*, Cold Spring Harbor Laboratory Press, Plainview, N.Y., pp. 2.1–2.67.

230. Miller, R. V., and Kokjohn, T. A., 1990, General microbiology of *recA*: Environmental and evolutionary significance, *Annu. Rev. Microbiol.* **44:**365–394.

231. Mills, D., 1985, Transposon mutagenesis and its potential for studying virulence genes in plant pathogens, *Annu. Rev. Phytopathol.* **23:**297–320.

232. Miura-Masuda, A., and Ikeda, H., 1990, The DNA gyrase of *Escherichia coli* participates in the formation of a spontaneous deletion by *recA*-independent recombination *in vivo*, *Mol. Gen. Genet.* **220:**345–352.

233. Mizuuchi, K., 1992, Transpositional recombination: Mechanistic insights from studies of Mu and other elements, *Annu. Rev. Biochem.* **61:**1011–1051.

234. Mötsch, S., and Schmitt, R., 1984, Replicon fusion mediated by a single-ended derivative of transposon Tn*1721*, *Mol. Gen. Genet.* **195:**281–287.

235. Mötsch, S., Schmitt, R., Avila, P., de la Cruz, F., Ward, E., and Grinsted, J., 1985, Junction sequences generated by "one-ended transposition," *Nucleic Acids Res.* **13:**3335–3342.

236. Moody, E. E. M., and Runge, R., 1972, The integration of autonomous transmissible plasmids into the chromosome of *Escherichia coli* K12, *Genet. Res. (Camb.)* **19:**181–186.

237. Moore, A. T., Nayudu, M., and Holloway, B. W., 1983, Genetic mapping in *Methylophilus methylotrophus* AS1, *J. Gen. Microbiol.* **129:**785–799.

238. Morgan, A. F., 1982, Isolation and characterization of *Pseudomonas aeruginosa* R′ plasmids constructed by interspecific mating, *J. Bacteriol.* **149:**654–661.

239. Müller, B., Jones, C., Kemper, B., and West, S. C., 1990, Enzymatic formation and resolution of Holliday junctions in vitro, *Cell* **60:**329–336.

240. Nag, D. K., DasGupta, U., Adelt, G., and Berg, D. E., 1985, IS*50*-mediated inverse transposition: Specificity and precision, *Gene* **34:**17–26.

241. Naito, A., Naito, S., and Ikeda, H., 1984, Homology is not required for recombination mediated by DNA gyrase of *Escherichia coli*, *Mol. Gen. Genet.* **193:**238–243.

242. Nakazawa, T., Hayashi, E., Yokota, T., Ebina, Y., and Nakazawa, A., 1978, Isolation of TOL and RP4 recombinants by integrative suppression, *J. Bacteriol.* **134:**270–277.

243. Nakazawa, T., Inouye, S., and Nakazawa, A., 1980, Physical and functional mapping of RP4-TOL plasmid recombinants: Analysis of insertion and deletion mutants, *J. Bacteriol.* **144:**222–231.

244. Nakazawa, T., Kimoto, M., and Abe, M., 1990, Cloning, sequencing, and transcriptional analysis of the *recA* gene of *Pseudomonas cepacia*, *Gene* **94:**83–88.

245. Nassif, X., Fournier, J-M., Arondel, J., and Sansonetti, P. J., 1989, Mucoid phenotype of *Klebsiella pneumoniae* is a plasmid-encoded virulence factor, *Infect. Immun.* **57:**546–552.

246. Nayudu, M., and Holloway, B. W., 1981, Isolation and characterization of R-plasmid variants with enhanced chromosomal mobilization ability in *Escherichia coli* K-12, *Plasmid* **6:**53–66.

247. Nishimura, Y., Caro, L., Berg, C. M., and Hirota, Y., 1971, Chromosome replication in *Escherichia coli*. IV. Control of chromosome replication and cell division by an integrated episome, *J. Mol. Biol.* **55:**441–456.

248. Nishimura, A., Nishimura, Y., and Caro, L., 1973, Isolation of Hfr strains from R⁺ and ColV2⁺ strains of *Escherichia coli* and derivation of an R′*lac* factor by transduction, *J. Bacteriol.* **116:**1107–1112.

249. Nordeen, R. O., and Holloway, B. W., 1990, Chromosome mapping in *Pseudomonas syringae* pv. *syringae* strain PS224, *J. Gen. Microbiol.* **136:**1231–1239.

250. Novick, R. P., Clowes, R. C., Cohen, S. N., Curtiss, III, R., Datta, N., and Falkow, S., 1976, Uniform nomenclature for bacterial plasmids: A proposal, *Bacteriol. Rev.* **40:**168–189.

251. Okahashi, N., Sasakawa, C., Okada, N., Yamada, M., Yoshikawa, M., Tokuda, M., Takahashi, I., and Koga, T., 1990, Construction of a *Not*I restriction map of the *Streptococcus mutans* genome, *J. Gen. Microbiol.* **136:**2217–2223.

252. Olasz, F., and Arber, W., 1989, A model for IS*30*-mediated transposition in *E. coli*, *Experientia* **45:**A39.

253. O'Connor, M. B., and Malamy, H. M., 1984, Role of the F factor *oriV1* region in *recA*-independent illegitimate recombination: Stable replicon fusions of the F derivative pOX38 and pBR322-related plasmids, *J. Mol. Biol.* **175:**263–284.

254. O'Connor, M. B., and Malamy, M. H., 1985, Mapping of DNA gyrase cleavage sites in vivo: Oxolinic acid induced cleavages in plasmid pBR322, *J. Mol. Biol.* **181:**545–550.

255. O'Connor, M. B., Kilbane, J. J., and Malamy, M. H., 1986, Site-specific and illegitimate recombination in the *oriV1* region of the F factor: DNA sequences involved in recombination and resolution, *J. Mol. Biol.* **189:**85–102.

256. O'Hoy, K., 1987, Genetic and physical analysis of the *Pseudomonas aeruginosa* PAO chromosome, Ph.D. Thesis, Monash University, Melbourne, Australia.

257. O'Hoy, K., and Krishnapillai, V., 1985, Transposon mutagenesis of the *Pseudomonas aeruginosa* PAO chromosome and the isolation of high frequency of recombination donors, *FEMS Microbiol. Lett.* **29:** 299–303.

258. O'Hoy, K., and Krishnapillai, V., 1987, Recalibration of the *Pseudomonas aeruginosa* strain PAO chromosome map in time units using high-frequency-of-recombination donors, *Genetics* **115:**611–618.

259. Pansegrau, W., Miele, L., Lurz, R., and Lanka, E., 1987, Nucleotide sequence of the kanamycin resistance determinant of plasmid RP4: Homology to other aminoglycoside 3′-phosphotransferases, *Plasmid* **18:** 193–204.

260. Pearce, L. E., and Meynell, E., 1968, Specific chromosomal affinity of a resistance factor, *J. Gen. Microbiol.* **50:**159–172.

261. Pemberton, M. J., and Bowen, A. R. S. G., 1981, High-frequency chromosome transfer in *Rhodopseudomonas sphaeroides* promoted by broad-host-range plasmid RP1 carrying mercury transposon Tn*501*, *J. Bacteriol.* **147:**110–117.

262. Perkins, J. B., and Youngman, P. J., 1984, A physical and functional analysis of Tn*917*, a *Streptococcus* transposon in the Tn*3* family that functions in *Bacillus*, *Plasmid* **12:**119–138.

263. Peterson, B. C., Hashimoto, H., and Rownd, R. H., 1982, Cointegrate formation between homologous plasmids in *Escherichia coli*, *J. Bacteriol.* **151:**1086–1094.

264. Pischl, D. L., and Farrand, S. K., 1983, Transposon-facilitated chromosome mobilization in *Agrobacterium tumefaciens*, *J. Bacteriol.* **153:**1451–1460.

265. Pittard, J., Loutit, J. S., and Adelberg, E. A., 1963, Gene transfer by F′ strains of *Escherichia coli* K-12, *J. Bacteriol.* **85:**1394–1401.

266. Plasota, M., Piechucka, E., Kauc, B., and Wlodarczyk, M., 1984, R68.45 plasmid mediated conjugation in *Thiobacillus* A2, *Microbios* **41:**81–89.

267. Plasterk, R. H. A., 1991, Frameshift control of IS*1* transposition, *Trends Genet.* **7:**203–204.

268. Prère, M.-F., Chandler, M., and Fayet, O., 1990, Transposition in *Shigella dysenteriae*: Isolation and analysis of IS*911*, a new member of the IS*3* group of insertion sequences, *J. Bacteriol.* **172:**4090–4099.

269. Priebe, S. D., and Lacks, S. A., 1989, Region of the streptococcal plasmid pMV158 required for conjugative mobilization, *J. Bacteriol.* **171:**4778–4784.

270. Priefer, U. B., 1989, Genes involved in lipopolysaccharide production and symbiosis are clustered on the chromosome of *Rhizobium leguminosarum* biovar *viciae* VF39, *J. Bacteriol.* **171:**6161–6168.

271. Ratnaningsih, E., Dharmsthiti, S., Krishnapillai, V., Morgan, A., Sinclair, M., and Holloway, B. W., 1990, A combined physical and genetic map of *Pseudomonas aeruginosa* PAO, *J. Gen. Microbiol.* **136:**2351–2357.

272. Read, H. A., Sarma, S. D., and Jaskunas, S. R., 1980, Fate of donor insertion sequence IS*1* during transposition, *Proc. Natl. Acad. Sci. USA* **77:**2514–2518.

273. Reimmann, C., and Haas, D., 1986, IS*21* insertion in the *trfA* replication control gene of chromosomally integrated plasmid RP1: A property of stable *Pseudomonas aeruginosa* Hfr strains, *Mol. Gen. Genet.* **203:**511–519.

274. Reimmann, C., and Haas, D., 1987, Mode of replicon fusion mediated by the duplicated insertion sequence IS*21* in *Escherichia coli*, *Genetics* **115:**619–625.

275. Reimmann, C., and Haas, D., 1990, The *istA* gene of insertion sequence IS*21* is essential for cleavage at the inner 3′ ends of tandemly repeated IS*21* elements *in vitro*, *EMBO J.* **9:**4055–4063.

276. Reimmann, C., Rella, M., and Haas, D., 1988, Integration of replication-defective R68.45-like plasmids into the *Pseudomonas aeruginosa* chromosome, *J. Gen. Microbiol.* **134:**1515–1523.

277. Reimmann, C., Moore, R., Little, S., Savioz, A., Willetts, N. S., and Haas, D., 1989, Genetic structure, function and regulation of the transposable element IS*21*, *Mol. Gen. Genet.* **215:**416–424.

278. Riess, G., Holloway, B. W., and Pühler, A., 1980, R68.45, a plasmid with chromosome mobilizing ability (Cma) carries a tandem duplication, *Genet. Res. (Camb.)* **36:**99–109.

279. Riess, G., Masepohl, B., and Pühler, A., 1983, Analysis of IS21-mediated mobilization of plasmid pACYC184 by R68.45 in *Escherichia coli*, *Plasmid* **10**:111–118.
280. Romero, D. A., Slos, P., Robert, C., Castellino, I., and Mercenier, A., 1987, Conjugative mobilization as an alternative vector delivery system for lactic streptococci, *Appl. Environ. Microbiol.* **53**:2405–2413.
281. Rosner, J. L., and Guyer, M. S., 1980, Transposition of IS1-λBIO-IS1 from a bacteriophage λ derivative carrying the IS1-cat-IS1 transposon (Tn9), *Mol. Gen. Genet.* **178**:111–120.
282. Sanderson, K. E., and MacLachlan, P. R., 1987, F-mediated conjugation, F+ strains, and Hfr strains of *Salmonella typhimurium* and *Salmonella abony*, in: *Escherichia coli and Salmonella typhimurium. Cellular and Molecular Biology*, Vol. 2 (F. C. Neidhardt, J. L. Ingraham, K. B. Low, B. Magasanik, M. Schaechter, and H. E. Umbarger, eds.), American Society for Microbiology, Washington, D. C., pp. 1138–1144.
283. Sanderson, K. E., Ross, H., Ziegler, L., and Mäkelä, P. H., 1972, F+, Hfr, and F' strains of *Salmonella typhimurium* and *Salmonella abony*, *Bacteriol. Rev.* **36**:608–637.
284. Sano, Y., and Kageyama, M., 1984, Genetic determinant of pyocin AP41 as an insert in the *Pseudomonas aeruginosa* chromosome, *J. Bacteriol.* **158**:562–570.
285. Scaife, J., and Gross, J. D., 1963, The mechanism of chromosome mobilization by an F-prime factor in *Escherichia coli* K12, *Genet. Res. (Camb.)* **4**:328–331.
286. Schäfer, A., Kalinowski, J., Simon, R., Seep-Feldhaus, A-H., and Pühler, A., 1990, High-frequency conjugal plasmid transfer from gram-negative *Escherichia coli* to various gram-positive coryneform bacteria, *J. Bacteriol.* **172**:1663–1666.
287. Schmitt, R., Mötsch, S., Rogowsky, P., de la Cruz, F., and Grinsted, J., 1985, On the transposition and evolution of Tn1721 and its relatives, in: *Plasmids in Bacteria* (D. R. Helinski, S. N. Cohen, D. B. Clewell, D. A. Jackson, and A. Hollaender, eds.), Plenum Press, New York, pp. 79–91.
288. Schoonejans, E., and Toussaint, A., 1983, Utilization of plasmid pULB113 (RP4::mini-Mu) to construct a linkage map of *Erwinia carotovora* subsp. *chrysanthemi*, *J. Bacteriol.* **154**:1489–1492.
289. Schurter, W., and Holloway, B. W., 1986, Genetic analysis of promoters on the insertion sequence IS21 of plasmid R68.45, *Plasmid* **15**:8–18.
290. Seeberg, A. H., and Wiedemann, B., 1984, Transfer of the chromosomal *bla* gene from *Enterobacter cloacae* to *Escherichia coli* by RP4::mini-Mu, *J. Bacteriol.* **157**:89–94.
291. Serebrijski, I. G., Kazakova, S. M., and Tsygankov, Y. D., 1989, Construction of Hfr-like donors of the obligate methanol-oxidizing bacterium *Methylobacillus flagellatum* KT, *FEMS Microbiol. Lett.* **59**:203–206.
292. Shah, K. S., Kuykendall, L. D., and Kim, C.-H., 1989, R-prime plasmids from *Bradyrhizobium japonicum* and *Rhizobium fredii*, *Arch. Microbiol.* **152**:550–555.
293. Shen, P., and Huang, H. V., 1986, Homologous recombination in *Escherichia coli*: Dependence on substrate length and homology, *Genetics* **112**:441–457.
294. Sherman, P. A., and Fyfe, J. A., 1990, Human immunodeficiency virus integration protein expressed in *Escherichia coli* possesses selective DNA cleaving activity, *Proc. Natl. Acad. Sci. USA* **87**:5119–5123.
295. Sherratt, D., 1989, Tn3 and related transposable elements: Site-specific recombination and transposition, in: *Mobile DNA* (D. E. Berg and M. M. Howe, eds.), American Society for Microbiology, Washington, D. C., pp. 163–184.
296. Shinomiya, T., Shiga, S., Kikuchi, A., and Kageyama, M., 1983, Genetic determinant of pyocin R2 in *Pseudomonas aeruginosa* PAO: II. Physical characterization of pyocin R2 genes using R-prime plasmids constructed from R68.45, *Mol. Gen. Genet.* **189**:382–389.
297. Simon, R., 1984, High frequency mobilization of gram-negative bacterial replicons by the *in vitro* constructed Tn5-Mob transposon, *Mol. Gen. Genet.* **196**:413–420.
298. Simon, R., Priefer, U., and Pühler, A., 1983, A broad host range mobilization system for *in vivo* genetic engineering: Transposon mutagenesis in gram-negative bacteria, *Bio/Technology* **1**:784–790.
299. Simon, R., O'Connell, M., Labes, M., and Pühler, A., 1986, Plasmid vectors for the genetic analysis and manipulation of rhizobia and other gram-negative bacteria, *Meth. Enzymol.* **118**:640–659.
300. Simon, R., Quandt, J., and Klipp, W., 1989, New derivatives of transposon Tn5 suitable for mobilization of replicons, generation of operon fusions and induction of genes in gram-negative bacteria, *Gene* **80**:161–169.
301. Sistrom, W. R., 1977, Transfer of chromosomal genes mediated by plasmid R68.45 in *Rhodopseudomonas sphaeroides*, *J. Bacteriol.* **131**:526–532.
302. Skotnicki, M. L., Warr, R. G., Goodman, A. E., and Rogers, P. L., 1983, Development of genetic

techniques and strain improvement in *Zymomonas mobilis*, in: *Genetics of Industrial Microorganisms* (Y. Ikeda and T. Beppu, eds.), Kodansha Ltd., Tokyo, pp. 361–365.

303. Smith, C. J., Coote, J. G., and Parton, R., 1986, R-plasmid-mediated chromosome mobilization in *Bordetella pertussis*, *J. Gen. Microbiol.* **132**:2685–2692.

304. Smith, C. L., and Condemine, G., 1990, New approaches for physical mapping of small genomes, *J. Bacteriol.* **172**:1167–1172.

305. Smith, G. R., 1988, Homologous recombination in procaryotes, *Microbiol. Rev.* **52**:1–28.

306. Smith, G. R., 1991, Conjugational recombination in *E. coli*: Myths and mechanisms, *Cell* **64**:19–27.

307. Smith, M. D., and Clewell, D. B., 1984, Return of *Streptococcus faecalis* DNA cloned in *Escherichia coli* to its original host via transformation of *Streptococcus sanguis* followed by conjugative mobilization, *J. Bacteriol.* **160**:1109–1114.

308. Soldati, L., Crockett, R., Carrigan, J. M., Leisinger, T., Holloway, B. W., and Haas, D., 1984, Revised locations of the *hisI* and *pru* (proline utilization) genes on the *Pseudomonas aeruginosa* chromosome map, *Mol. Gen. Genet.* **193**:431–436.

309. Sosio, M., Madon, J., and Hütter, R., 1989, Excision of pIJ408 from the chromosome of *Streptomyces glaucescens* and its transfer into *Streptomyces lividans*, *Mol. Gen. Genet.* **218**:169–176.

310. Sotomura, M., and Yoshikawa, M., 1975, Reinitiation of chromosome replication in the presence of chloramphenicol under an integratively suppressed state by R6K, *J. Bacteriol.* **122**:623–628.

311. Spielmann-Ryser, J., Moser, M., Kast, P., and Weber, H., 1991, Factors determining the frequency of plasmid cointegrate formation mediated by insertion sequence IS*3* from *Escherichia coli*, *Mol. Gen. Genet.* **226**:441–448.

312. Srivastava, R., Sinha, V. B., and Srivastava, B. S., 1989, Chromosomal transfer and in vivo cloning of genes in *Vibrio cholerae* using RP4::mini-Mu, *Gene* **75**:253–259.

313. Steele, J. L., and McKay, L. L., 1989, Conjugal transfer of genetic material in lactococci: A review, *J. Dairy Sci.* **72**:3388–3397.

314. Stewart, G. J., and Carlson, C. A., 1986, The biology of natural transformation, *Annu. Rev. Microbiol.* **40**:211–235.

315. Stout, V. G., and Iandolo, J. J., 1990, Chromosomal gene transfer during conjugation by *Staphylococcus aureus* is mediated by transposon-facilitated mobilization, *J. Bacteriol.* **172**:6148–6150.

316. Strom, A. D., Hirst, R., Petering, J., and Morgan, A., 1990, Isolation of high frequency of recombination donors from Tn*5* chromosomal mutants of *Pseudomonas putida* PPN and recalibration of the genetic map, *Genetics* **126**:497–503.

317. Summers, D. K., and Withers, H. L., 1990, Electrotransfer: Direct transfer of bacterial plasmid DNA by electroporation, *Nucleic Acids Res.* **18**:2192.

318. Suwanto, A., and Kaplan, S., 1989, Physical and genetic mapping of the *Rhodobacter sphaeroides* 2.4.1 genome: Presence of two unique circular chromosomes, *J. Bacteriol.* **171**:5850–5859.

319. Suwanto, A., and Kaplan, S., 1992, Chromosome transfer in *Rhodobacter sphaeroides*: Hfr formation and genetic evidence for two unique circular chromosomes, *J. Bacteriol.* **174**:1135–1145.

320. Tatra, P. K., and Goodwin, P. M., 1983, R-plasmid-mediated chromosome mobilization in the facultative methylotroph *Pseudomonas* AM1, *J. Gen. Microbiol.* **129**:2629–2632.

321. Taylor, A. F., 1992, Movement and resolution of Holliday junctions by enzymes from *E. coli*, *Cell* **69**:1063–1065.

322. Terawaki, Y., Kishi, H., and Nakaya, R., 1975, Integration of R plasmid R*ts*1 to the *gal* region of the *Escherichia coli* chromosome, *J. Bacteriol.* **121**:857–862.

323. Thompson, T. L., Centola, M. B., and Deonier, R. C., 1989, Location of the nick at *oriT* of the F plasmid, *J. Mol. Biol.* **207**:505–512.

324. Tomcsanyi, T., Berg, C. M., Phadnis, S. H., and Berg, D. E., 1990, Intramolecular transposition by a synthetic IS*50* (Tn*5*) derivative, *J. Bacteriol.* **172**:6348–6354.

325. Torres, O. R., Korman, R. Z., Zahler, S. A., and Dunny, G. M., 1991, The conjugative transposon Tn*925*: Enhancement of conjugal transfer by tetracycline in *Enterococcus faecalis* and mobilization of chromosomal genes in *Bacillus subtilis* and *E. faecalis*, *Mol. Gen. Genet.* **225**:395–400.

326. Tortolero, M., Santero, E., and Casadesus, J., 1983, Plasmid transfer and mobilization of *nif* markers in *Azotobacter vinelandii*, *Microbios Lett.* **22**:31–35.

327. Towner, K. J., 1978, Chromosome mapping in *Acinetobacter calcoaceticus*, *J. Gen. Microbiol.* **104**:175–180.

328. Towner, K. J., and Vivian, A., 1977, Plasmids capable of transfer and chromosome mobilization in *Acinetobacter calcoaceticus*, *J. Gen. Microbiol.* **101**:167–171.

329. Trieu-Cuot, P., Carlier, C., Martin, P., and Courvalin, P., 1987, Plasmid transfer by conjugation from *Escherichia coli* to gram-positive bacteria, *FEMS Microbiol. Lett.* **48**:289–294.

330. Trieu-Cuot, P., Carlier, C., and Courvalin, P., 1988, Conjugative plasmid transfer from *Enterococcus faecalis* to *Escherichia coli*, *J. Bacteriol.* **170**:4388–4391.

331. Tsygankov, Y. D., Kazakova, S. M., and Serebrijski, I. G., 1990, Genetic mapping of the obligate methylotroph *Methylobacillus flagellatum*: Characteristics of prime plasmids and mapping of the chromosome in time-of-entry units, *J. Bacteriol.* **172**:2747–2754.

332. Tucker, W. T., and Pemberton, J. M., 1979, Conjugation and chromosome transfer in *Rhodopseudomonas sphaeroides* mediated by W and P group plasmids, *FEMS Microbiol. Lett.* **5**:173–176.

333. Umeda, M., and Ohtsubo, E., 1989, Mapping of insertion elements IS*1*, IS2 and IS*3* on the *Escherichia coli* K-12 chromosome, *J. Mol. Biol.* **208**:601–614.

334. Van Gijsegem, F., and Toussaint, A., 1982, Chromosome transfer and R-prime formation by an RP4::mini-Mu derivative in *Escherichia coli*, *Salmonella typhimurium*, *Klebsiella pneumoniae*, and *Proteus mirabilis*, *Plasmid* **7**:30–44.

335. Van Larebeke, N., Genetello, Ch., Hernalsteens, J. P., De Picker, A., Zaenen, I., Messens, E., Van Montagu, M., and Schell, J., 1977, Transfer of Ti plasmids between *Agrobacterium* strains by mobilisation with the conjugative plasmid RP4, *Mol. Gen. Genet.* **152**:119–124.

336. Van de Putte, P., and Gruijthuijsen, M., 1972, Chromosome mobilization and integration of F factors in the chromosome of *recA* strains of *E. coli* under the influence of bacteriophage Mu-1, *Mol. Gen. Genet.* **118**:173–183.

337. Van der Lelie, D., Bron, S., Venema, G., and Oskam, L., 1989, Similarity of minus origins of replication and flanking open reading frames of plasmids pUB110, pTB913 and pMV158, *Nucleic Acids Res.* **17**:7283–7294.

338. Van der Lelie, D., Wösten, H. A. B., Bron, S., Oskam, L., and Venema, G., 1990, Conjugal mobilization of streptococcal plasmid pMV158 between strains of *Lactococcus lactis* subsp. *lactis*, *J. Bacteriol.* **172**:42–52.

339. Varmus, H., and Brown, P., 1989, Retroviruses, in: *Mobile DNA* (D. E. Berg and M. M. Howe, eds.), American Society for Microbiology, Washington, D. C., pp. 53–108.

340. Vivian, A., 1987, *Acinetobacter calcoaceticus*, in: *Genetic Maps*, Vol. 4 (S. J. O'Brien, ed.), Cold Spring Harbor Laboratory, Cold Spring Harbor, N. Y., pp. 240–241.

341. Warren, G., and Sherratt, D., 1977, Complementation of transfer deficient ColE1 plasmids, *Mol. Gen. Genet.* **151**:197–201.

342. Watson, J. M., and Holloway, B. W., 1978, Chromosome mapping in *Pseudomonas aeruginosa* PAT, *J. Bacteriol.* **133**:1113–1125.

343. Watson, M. D., and Scaife, J. G., 1978, Chromosomal transfer promoted by the promiscuous plasmid RP4, *Plasmid* **1**:226–237.

344. Watson, M. D., and Scaife, J. G., 1980, Integrative compatibility: Stable coexistence of chromosomally integrated and autonomous derivatives of plasmid RP4, *J. Bacteriol.* **142**:462–466.

345. West, S. C., 1992, Enzymes and molecular mechanism of genetic recombination, *Annu. Rev. Biochem.* **61**:603–640.

346. West, S. C., and Connolly, B., 1992, Biological roles of *Escherichia coli* RuvA and RuvC proteins revealed, *Mol. Microbiol.* **6**:2755–2759.

347. Willetts, N. S., 1972, Location of the origin of transfer of the sex factor F, *J. Bacteriol.* **112**:773–778.

348. Willetts, N. S., and Johnson, D., 1981, pED100, a conjugative F plasmid derivative without insertion sequences, *Mol. Gen. Genet.* **182**:520–522.

349. Willetts, N., and Skurray, R., 1987, Structure and function of the F factor and mechanism of conjugation, in: *Escherichia coli and Salmonella typhimurium. Cellular and Molecular Biology*, Vol. 2 (F. C. Neidhardt, J. L. Ingraham, K. B. Low, B. Magasanik, M., Schaechter, and H. E. Umbarger, eds.), American Society for Microbiology, Washington, D. C., pp. 1110–1133.

350. Willetts, N., and Wilkins, B., 1984, Processing of plasmid DNA during bacterial conjugation, *Microbiol. Rev.* **48**:24–41.

351. Willetts, N. S., Crowther, C., and Holloway, B. W., 1981, The insertion sequence IS*21* of R68.45 and the molecular basis for mobilization of the bacterial chromosome, *Plasmid* **6**:30–52.

352. Willison, J. C., Haddock, B. A., and Paraskeva, C., 1980, Transfer of chromosomal genes mediated by plasmid R68-45 in *Paracoccus denitrificans*, *Soc. Gen. Microbiol. Q.* **8**:14.

353. Willison, J. C., Ahombo, G., Chabert, J., Magnin, J-P., and Vignais, P. M., 1985, Genetic mapping of the *Rhodopseudomonas capsulata* chromosome shows nonclustering of genes involved in nitrogen fixation, *J. Gen. Microbiol.* **131:**3001–3015.

354. Yakobson, E. A., and Guiney, Jr., D. G., 1984, Conjugal transfer of bacterial chromosomes mediated by the RK2 plasmid transfer origin cloned into transposon Tn5, *J. Bacteriol.* **160:**451–453.

355. Yoshikawa, M., 1974, Identification and mapping of the replication genes of an R factor, R100-1, integrated into the chromosome of *Escherichia coli* K-12, *J. Bacteriol.* **118:**1123–1131.

356. Youvan, D. C., Elder, J. T., Sandlin, D. E., Zsebo, K., Alder, D. P., Panopoulos, N. J., Marrs, B. L., and Hearst, J. E., 1982, R-prime site-directed transposon Tn7 mutagenesis of the photosynthetic apparatus in *Rhodopseudomonas capsulata*, *J. Mol. Biol.* **162:**17–41.

357. Yu, P-L., Cullum, J., and Drews, G., 1981, Conjugational transfer systems of *Rhodopseudomonas capsulata* mediated by R plasmids, *Arch. Microbiol.* **128:**390–393.

358. Zennaro, E., Marconi, A. M., Fochesato, N., Castelli, F., and Ruzzi, M., 1991, Identification and sequencing of a new IS element in *Pseudomonas fluorescens* strain ST, *Pseudomonas 1991*, Trieste (Italy), p. 235.

359. Zhang, C., and Holloway, B. W., 1992, Physical and genetic mapping of the *catA* region of *Pseudomonas aeruginosa*, *J. Gen. Microbiol.* **138:**1097–1107.

360. Zieg, J., Maples, V. F., and Kushner, S. R., 1978, Recombination levels of *Escherichia coli* K-12 mutants deficient in various replication, recombination, or repair genes, *J. Bacteriol.* **134:**958–966.

Chapter 7

Conjugative Pili and Pilus-Specific Phages

LAURA S. FROST

1. Introduction

In the early 1960s, a number of bacteriophages specific for "male" strains of *Escherichia coli* were discovered both at the Rockefeller University (126, 127, 217) and in Germany (104, 135). In 1964, Crawford and Gesteland (61) observed the R17 phage (163) attached to the sex hair of the F+ cell, a filamentous appendage that was named, more decorously, the F pilus by Brinton and shown to be genetically determined by the fertility factor F of *E. coli* (48).

These first F-specific bacteriophages fell into two morphological groups: the spherical RNA phages and the filamentous DNA phages. These two groups still define the majority of phages that use conjugative pili as receptors. The spherical phages became model systems for the study of RNA replication and translation (195), and the symmetry functions in their capsids were used to develop the principles of icosahedral symmetry and quasi-equivalence (105). The filamentous phage were found to contain single-stranded DNA and became model systems for the study of DNA replication (144). In addition, these phages do not lyse the cell but live in a symbiotic relationship with their host and extrude progeny phage particles through the cell envelope. Thus they have provided an excellent system for the study of membrane-protein interactions (203). Their DNA is packaged as a collapsed circle encased in a helical arrangement of coat proteins, and the structure of the DNA inside these particles, which does not require base pairing, has provided a simple system for studying the packing of DNA in higher order structures (125).

LAURA S. FROST • Department of Microbiology, M330 Biological Sciences Building, University of Alberta, Edmonton, Alberta T6G 2E9.
Bacterial Conjugation, edited by Don B. Clewell. Plenum Press, New York, 1993.

The one area of the life cycles of these pages that has not been investigated in any detail is that of the infectious process. This is primarily because of the complexity of the genetics, structure, and function of the conjugative pilus, the receptor for these phages. Initially, the F plasmid, the related R factors, and the I complex of plasmids were studied for their ability to express antigenically distinct but morphologically related pili that were receptors for phage. The focus then turned to the genetic characterization of the F transfer region, which was instrumental in developing modern genetic techniques and which is reviewed in Chapter 1 of this book. In addition, it became clear that a separate function of plasmids was the control of their own replication and partitioning to daughter cells. This has stimulated a great deal of research and has developed the idea of incompatibility, where two plasmids in the same cell with a similar replication control system inhibit each other's growth, leading to a loss of one of the plasmids.

During the last 30 years, scores of incompatibility groups among plasmids and dozens of transfer regions specifying pili involved in either limited host range or broad host range (promiscuous) transfer have been described. One interesting correlation exists, where the plasmids in a given incompatibility group usually specify related transfer systems. This has simplified the classification of pili and their transfer regions somewhat, since they are currently grouped according to their incompatibility group (27, 56).

Useful information about pili can be obtained from studying their serological relationships. However, only pili in the F and I complexes (120, 136) have been studied in detail. A second, simple approach has been the isolation of pilus-specific bacteriophages. These have been used to determine the degree of relatedness between pili in the same or different incompatibility groups, since the parameter most affected in these experiments is the receptor for the phage, reflecting sequence differences in the pilin subunit. This chapter will deal principally with the use of pilus-specific phage to determine relatedness among various transfer systems as well as the attachment and penetration stages of the phages' life cycles, involving the pilus and other *tra*-related functions.

2. Classification of the Pilus-Specific Bacteriophages

Because of the studies of Bradley in Canada and Coetzee and colleagues in South Africa, a large number of bacteriophages specific for cells carrying conjugative plasmids have been identified. These studies are also responsible for the description of the pili expressed by these plasmids and have established the relationship between incompatibility (Inc) groups and transfer regions. The following discussion of the characteristics of pili and their phages is a summary of these investigations.

2.1. Factors Affecting Plaque Formation

The isolation of plasmid-specific phages is comparatively easy considering their prevalence in the natural world (87). Complications do arise, since many plasmids are repressed for pilus expression (discussed below) or the phages fail to propagate in certain genera of bacteria, a problem encountered in the case of plasmids with a broad host range. Sensitivity to a particular phage can be determined by the familiar spot test or plaque formation, by electron microscopic visualization of phage attached to the pili (see references in Table 7.1), or by titering the propagation of progeny phage in the event of poor plaque formation. Examples of these methodologies are given in Coetzee et al (52, 60).

Table 7.1 enumerates the phages that have been isolated and studied in any detail and for which phage stocks are thought to exist. Unlike the F-specific phages, there is very little information about the other plasmid-specific phages with regard to the size of their RNA or their structural features, much less the number and arrangement of genes in their genome. Unfortunately, there is no generalization that neatly sums up the probability of a cell harboring a plasmid from a particular Inc group being sensitive to a given phage. An example is the IncT group of plasmids and the phage φt (40). The prototype IncT plasmid is Rts1, which has a thermosensitive mode of replication (189). Its pili are serologically related to the pili of the other IncT plasmids, R394, R401, and R402, where the latter two plasmids are also temperature sensitive for replication (36). Rts1 is unusual, since sensitivity to φt can be demonstrated for Rts1 only in *E. coli* CSH2 (40), while other IncT plasmids confer sensitivity to phage φt in both *Salmonella* and most strains of *E. coli* K12. Many other examples exist; for instance, *Proteus morganii* harboring N plasmids are resistant to IKe but can still transfer these plasmids to other strains, which become IKe sensitive (70). *Caulobacter* carrying the IncP-1 plasmid RP1 appears to express pili with an altered configuration of pilin subunits, rendering the cells resistant to PRR1 phage but still transfer-proficient (108). As mentioned previously, *Salmonella* is resistant to filamentous DNA phages, and *Pseudomonas aeruginosa* strain O is resistant to the filamentous phage Pf1 whose host is *P. aeruginosa* K, which expresses similar pili (34, 170). Thus, it becomes imperative to empirically determine the relationship between phage and plasmid in a variety of hosts.

In addition, phage propagation can be affected by temperature, either at the level of expression of the transfer region (pilHα and IncHI plasmids [188]; Hgal phage and IncH plasmids [153]; φM and IncM plasmids [112]; φC-1 and IncC plasmid RA1 [30]) or in plaque formation (φD and IncD plasmids [56]). These temperature-sensitive systems usually reflect the life-style of the host bacterium; for example, RA1 was isolated from *Aeromonas liquefaciens* in snails, and Hgal phages were isolated from river water in Galway, Ireland. Another factor is the state of repression of the transfer region; for instance, IncC pili are expressed constitutively in *Salmonella typhimurium* LT2 but are repressed in *E. coli* strains (30). The control of expression of transfer regions at the molecular level has been described for the F plasmid alone and is discussed in Chapter 2 in this volume. The interested reader is referred to the papers listed in Table 7.1 for information on the complex subject of host range for phage specific for plasmids from other Inc groups.

2.2. The Spherical (Isometric) RNA Phages

The spherical RNA phages include the phages that use the pilus encoded by transmissible plasmids and the phages that infect *Pseudomonas* (161) and *Caulobacter* (130, 173) via chromosomally encoded pili. The phages from both groups are morphologically and structurally similar, suggesting an evolutionary relationship. An exception is the double-stranded, lipid-containing RNA phage φ6 that infects *P. syringae* via pili (141). This similarity could also be extended to the pili to which they attach, since the conjugative and

TABLE 7.1 Biological and Physical Characteristics of Plasmid-Specific Bacteriophages

Phage	Inc group pili[a]	Plasmid for propagation[b]	Dimensions (nm)[c]	HCCl3 DEE[d]	Attachment site[e]	References
		Isometric single-stranded RNA phages				
F-specific[f]	FI-FIV	F	28	−/−	Shaft, 2.	(195)
Folac	FV(S)[g]	pED208[h]	28	−/−	Point, 2.	(41)
Folach	FV, S	R71, pPLS::Tn5	28	(−)/−	Shaft, 2.	(58)
SR	FV, S	pPPLS::Tn5	28	(−)/−	Shaft, 2.	(58)
UA6	FV(S)[g]	pED208	20		Shaft, 2.	(4)
φC-1	C	RA1	27	−/−	Shaft, Base, 2.	(30, 179)
φD	D	R711b(26°C)	27	+/−	Point, 2.	(56)
pilHα	HI[i]	R478(26°C)	25	+/−	Shaft, 2.	(57)
	HII	pHH1508a			Shaft, 2.	
Hgal	HI, HII	R478(22°C)	20	−/	Shaft, 1.	(153)
Iα	I1 (thin), B	R621a	24	−/−	Shaft, 1.	(54)
φM	M	RIP69	27	+/−	Shaft, 3.	(55)
PRR1	P-1	RP4	26		Shaft, Base, 3.	(18, 157)
φt	T	R402(30°C)	25	−/−	Point, 2.	(40)
Ff (e.g., f1)[f]	FI-FV	F	890 × 6.0	+/−	Tip, 2.	(50, 144)
SF	FI-FV, S, D	pPLS::Tn5	925 ×	+/−	Tip, 2.	(58)
f711	D	R711b	870 × 6.0	+/−	Tip, 2.	(20)
φC-2	C	P-lac[j]	1235 × 8.0	+/−	Shaft, 2.	(42)
If1 (If2)	I1, I2 (thin)	R64drd11	1300 × 5.5	+/−	Tip, 1.	(91, 92, 101, 138)
PR64FS	I complex[k]	R64drd11	917 × 6.9	+/−	Tip, 1.	(28, 53, 54)
I2-2	I2 (thick)	R721	910[m] × 7.0	+/−	Tip, 3.	(54)
IKe	I2 (thick), N	N3	1100[m] × 6.6	+/−	Tip, 3.	(22, 45, 113)
IKeh	I2 (thick), N, P	N3		+/−	Tip, 3.	(45, 89)
X	X, I2 (thick), M, N, P-1, U, W, (P-10)	R6K	1177 × 9.0	+/+	Tip, 2, 3.	(39, 45, 59)
X-2	X, R775[n]	R6K	950 × 10	+/−	Tip, 2.	(60)
Pf3	P-1	RP1	760 × 6.0	+/+	Shaft, 3.	(37, 182)
φtf-1	T	pIN25 (30°C)	800 × 10	+/+	Point, 2.	(59)

Lipid-Containing, Double-Stranded, and/or Tailed Phages

φJ	J, C, D	R997	40, 18-nm tail	−/−	Shaft, 2.	(42)
PR4	P-1, N, W, I₂ (thick)	RP1	65, 47-nm tail	+/(−)	Tip, 3.	(18, 19, 22, 31) (35, 51, 141, 182)
PRD1	P-1, N, W	RP1	65, no tail	+/	Tip, 3.	(141, 156, 180)
PR772	P-1, N, W	R772	53, no tail	+/+	Tip, 3.	(51, 52)

[a]All members of an incompatibility group are not necessarily sensitive to a given phage. In addition, the host affects sensitivity. For details, the reader should consult the references.

[b]The temperature in brackets indicates the optimum for phage production, because of either pilus expression or phage replication.

[c]The dimensions represent the diameter for RNA phage, the length and width of filamentous phage, and the diameter of double-stranded phage head.

[d]The sensitivity to chloroform is reported above the slash, while sensitivity to diethyl ether is below the slash, where known. Brackets indicate partial sensitivity.

[e]The attachment site can be on the sides (shaft); the tip; the point, which is the tapered region at the tip of some pili; or the base of the pilus. 1. is thin, flexible pili; 2. is thick, flexible pili; and 3. is rigid pili (24).

[f]The F-specific RNA phages include f2, fr, R17, MS2, M12, and Qβ, and the Ff filamentous phages include f1, fd, M13, ZJ/2, which have been classified as described in the text.

[g](S) refers to F₀lac and pED208, which are considered to be IncFV (79) or S (58).

[h]pED208 is the derepressed, multipiliated form of F₀lac (76).

[i]HI plasmids include HI1, HI2, and HI3. pilHα propagates but does not plaque on the HI3 plasmid MIP233.

[j]φC-2 is propagated in Salmonella typhimurium LT2trpA8/P-lac.

[k]The I complex includes I₁, I+B, B, K, I₅, Z, and I₂.

[m]IfI and I₂-2 appear to vary in length; the median length is reported.

[n]R775 has not been classified with respect to incompatibility.

chromosomally encoded pili are morphologically similar and apparently share the property of pilus retraction.

2.2.1. The F-Specific Coliphages

The RNA phages have a diameter of 27 nm and contain a single-stranded RNA of 3500 to 4700 bases enclosed in an icosahedral shell of 180 copies of a coat protein with a single copy of the attachment protein exposed, in part, on the surface of the virion. Usually, four genes encoded by the RNA are involved in virion assembly, RNA replication, and lysis. Excellent reviews of these phages, especially the phages specific for the F pilus, are available (75, 195).

The most extensively studied of these bacteriophages are the coliphages that infect *Escherichia coli* carrying plasmids of the F complex. These include MS2 (69), R17 (163), f2 (27), fr (135), M12 (104), Qβ (199), GA (198), and SP (143), which are the most studied of the coliphages. The term "coliphage" is a misnomer, since it has been shown that these phages can infect F^+ bacteria from a number of genera; for instance, the F plasmid can be introduced into *Shigella*, *Proteus*, or *Salmonella* and confer sensitivity to these phages. Havelaar et al (99) have used these phages to screen sewage for the presence of conjugative plasmids, taking advantage of the fact that *Salmonella* is resistant to DNA phages.

These phages have been classified into four groups; Group I, MS2, R17, fr, f2; Group II, GA; Group III, Qβ; Group IV, SP on the basis of serological cross-reactivity, physical parameters of the phage particle, and sedimentation velocity of the RNA, where MS2, GA, Qβ, and SP are the representative phages for each group. Recently, the sequence of the RNA of a number of these phages has been completed: Group I: MS2, fr, R17, and M12 (partially completed); Group II: GA; Group III: Qβ; and Group IV; SP. The sequence has been extremely helpful in determining relatedness (summarized by Van Duin [195]). The presence of a readthrough protein, so named because it is translated by readthrough of the coat protein gene, has been used to simplify these groups into A Group phages (I and II) and B Group phages (III and IV). The presence of this gene usually results in a slightly longer RNA, which may be helpful in classifying new RNA phages.

Antiserum raised against one phase is capable of inactivating phages of the same group but not phages of other groups. A phylogenetic tree for RNA phages can be established using serological typing (195). This could be further refined using sequence data as they become available. Treatment with antiphage antisera usually prevents the phage from attaching to the pilus, which is the function of the A protein. This suggests that a domain on the A protein is involved in attachment and is immunogenic and exposed on the surface of the phage particle and that its sequence varies and defines the specificity of the receptor. The sequence of the maturation protein of MS2, Qβ, and GA has been obtained and compared: Only the last 21 residues are homologous in all three phages, while MS2 and GA share 38% homology and Qβ is not homologous to either MS2 or GA for the rest of its sequence. Unfortunately, the nature of the antigenic determinants in the A protein, and the identity of the determinant that is exposed on the surface of the particle has not been determined. This could suggest which domain in the A protein is involved in attachment to the pilus. Similarly, the MS2 and GA coat proteins have sequence identity at 60% of their residues in their coat proteins, whereas Qβ shares only limited homology with either one.

2.2.2. Other Plasmid-Specific RNA Phages

The RNA phages are usually specific for the pili of a single Inc group, an exception being SR, which infects cell carrying plasmids from IncFV, IncS, and IncD, which apparently specify related pilus systems (see below). A host range mutant of $F_O lac$, which is specific for IncFV, can also infect cells carrying IncS and IncD plasmids (52). Host range mutants of RNA phages are rare, and a mutant of SP has been reported that has a wider host range among the IncF complex of plasmids (71). The F-specific phage MS2 can be seen attaching to the pili of R711b (IncD) if the sample is treated with formalin (Figure 7.2), extending the IncFV, S, D complex to include IncFI-IV plasmids as well (33). Similarly, the phage Iα can propagate on cells carrying IncI_1 and B plasmids, again suggesting closely related transfer systems for these plasmids (54). With the exception of the F-specific RNA phages (see above) and $F_O lac$ and UA6, there is no serological cross-reactivity between any of these phages.

The RNA phages attach along the sides of the pilus, sometimes in great numbers, with certain interesting exceptions. Wild-type $F_O lac$, φD and φt attach to the distal end of the pilus along the tapered point (Figure 7.2), while φC-1 (30) and PRR1 (18) are found intermittently along the sides, with a noticeable accumulation at the base of the pilus. $F_O lac$,a host range mutant of $F_O lac$, is able to attach along the shaft of the pilus like other RNA phages instead of at the pointed tip of the pilus (58). Bradley has suggested that this is due to pilus retraction, bringing the phage particle into contact with the cell. Phages D, M, and pilHα are unusually sensitive to chloroform, while $F_O lac$h and SR, but not $F_O lac$, are partially sensitive. The significance of this is not understood.

2.3. The Filamentous Phages

The filamentous phages, as their name implies, are long, thin rods of 6 to 7 nm diameter and vary in length from 1 to 2 μm. They include a number of F-specific phages. The best studied are fl (217), fd (135), M13 (104), and ZJ/2 (14), as well as the phages specific for one or more Inc groups, as listed in Table 7.1. The phages usually attach to the tips of pili specified by either conjugative plasmids or the polar pili of *Pseudomonas aeruginosa* K. However, the phage Pf3 has been reported to attach to the sides of RP1 pili (37); phage C-2 also attaches to the sides of C pili (42); and φtf-1 attaches to the tapered points of T pili (59).

The filamentous phages have been divided into two classes based on symmetry elements in the helical arrangement of coat proteins in the virion as determined by X-ray fiber diffraction. Class I includes the F-specific phages (fl, M13, fd, ZJ/2), Ifl, and IKe, which have five-start helices, rotation angles of 36°, and axial translations of 1.6 nm. Class II includes the Pfl, Pf3, and Xf phages, which have one-start helices, rotation angles of 67°, and translations of 0.3 nm. Unfortunately, many of the filamentous phages have not been classified using this system. The biology of the F-specific filamentous phages (fl, M13, fd) is well understood, however. The Pfl phage, which attaches to the polar pili of *Pseudomonas aeruginosa* K (PAK), is amenable to X-ray crystallographic techniques and is the best studied phage at the structural level. Several recent reviews on the biology of the filamentous phage include Webster and Lopez (201), Rasched and Oberer (165), and Model and Russel (144), while reviews and articles that deal with structural aspects of these phages include Makowski (131) and Marvin (133).

The complete sequences of the Ff phages fd (11, 171), fl (10, 103), and M13 (196) have been reported as well as the sequences of Pf3 (129) and IKe (164). Consistent with its being a Class II phage, Pf3 has little homology with the other phages. However, IKe has 55% homology with the Ff phages and has a similar genetic organization.

The filamentous phages that attach to conjugative pili are not as specific for pili expressed by plasmids of a single incompatibility group compared with the RNA phages, which are useful in identifying the Inc group of a plasmid (Table 7.1). This suggests that there may be a common receptor on these pili that is the site of attachment for these phages; however, it is not known whether there is a conserved sequence in the pilin subunit that is recognized by these phages or if another moiety, such as a minor pilus protein or a posttranslational modification of the pilin subunit, is responsible for their wide host range. Figure 7.1 attempts to summarize the interrelationships between the various incompatibility groups using phages to measure relatedness. Along with the tailed phage listed in Table 7.1 the filamentous phages are the most informative for establishing a pattern of relationships between the incompatibility groups using the pili as a marker.

The X phage, which has a curly appearance (not to be confused with phage Xf, which was isolated from the plant pathogen *Xanthomonas oryzae* [117]), was isolated from *E. coli* harboring the IncX plasmid R6K and was found to give plaques or multiply on cells carrying plasmids from IncM, N, P-1, U, W, and X and the unclassified plasmid R775. It was also shown to attach to but not infect cells carrying the IncP-10 plasmid R91.5 (25). The phage IKe (113), which was originally described as an N-specific phage, has been shown to attach to the thick, rigid pili of $IncI_2$, and a host range mutant, IKeh, extended the host range to include IncP-1 (89). IKe phages are serologically related to I_2-2 phage, which is specific for the rigid pili of $IncI_2$ plasmids (45). The Ifl phage (68, 138) was used to identify plasmids in the I complex (74, 101). Ifl attached to the thin pili specified by both $IncI_1$ and $IncI_2$ plasmids (28), which are serologically distinct from the pili of the F complex. However, antisera to either F or I pili affected the transfer ability of the IncFII plasmid R538 (97). The Ifl phage genome is homologous to that of the Ff phages (91, 92), which are in turn homologous to IKe as determined by a comparison of their sequences (164). The phage PR64FS attaches to the thin pili expressed by the whole IncI complex, including plasmids in $IncI_5$, I_1, and B, B, K, and Z as well as $IncI_1$ and $IncI_2$ (53). Rees et al (166) have suggested that the $IncI_1$ plasmid, ColIb-P9, represents a fusion of two transfer regions: One is derived from a broad host range plasmid related to IncM, N, P, and W that expresses rigid pili most resembling M pili, and the other is derived from a narrow host range plasmid related distantly to F that expresses thin, flexible pili.

The F-specific filamentous phages, known collectively as Ff phages, attach to the pili of the F complex (IncFI-IV); to the pili expressed by the IncD plasmids R687, R778b, and R711b (20); and to the pili of the IncS plasmids R71, TP224::Tn10, pPLS::Tn5 (58). The Ff phages infect cells harboring F_Olac, or its derepressed form pED208, that have been included in the F_Olac incompatibility group (33), the IncS group (58), and the IncFV group, first by morphology (67) and then by sequence analysis (79). The filamentous phage SF can attach and propagate on cells containing IncF, D, or S plasmids. The results of these

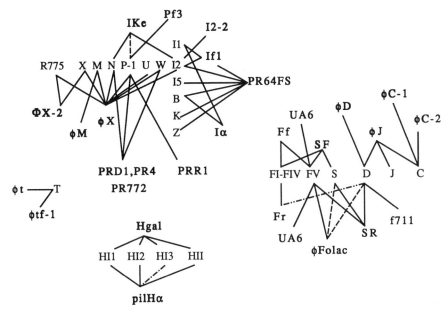

FIGURE 7.1. Interrelationships between various incompatibility groups as deduced by phage sensitivity. The names of the phages are given in bold type, and the incompatibility groups are given in capital letters in horizontal (R775, X, M, N, P-1, U, W; FI-FIV, FV, S, D, J, C; HI1, HI2, HI3, HII; T) and vertical (I1, I2, I5, B, K, Z) columns. Dashed lines indicate the sensitivity pattern of a host range mutant. A dash-dot line represents phage propagation with no visible plaque formation, and a dash-double dot line indicates phage attachment in the presence of formalin.

experiments suggest that the transfer systems of plasmids in many incompatibility groups are related and that the number of clearly different systems is probably rather small.

2.4. The Lipid-Containing and/or Tailed Phages

Phages that contain lipid but are not tailed include PR722 (51, 52), PR3 (182), and PRD1 (156), while PR4 resembles them but has a short tail that is difficult to visualize (31, 51, 141), perhaps because it is extended during the ejection process (D.E. Bradley, personal communication). All four phages infect cells harboring plasmids from the IncP-1, N, and W groups. The phage PR722 is distantly related serologically to PR4, while no serological relationship could be found among the other phages (52). The hexagonal head is embedded in a heavy layer of lipid, and the heads of these phages resemble PM2, a DNA phage that is not pilus specific (73). The J phage, which is specific for pili specified by the IncC, J, and D plasmids, has a hexagonal head, contains no lipid, and has a short noncontractile tail (42). Except for φJ, which attaches laterally on the pilus, these phages attach to the pilus tip. The best studied of the lipid-containing phages is φ6, a double-stranded RNA phage that attaches to the pili of *Pseudomonas syringae*. However, this phage has its layer of lipid inside its hexagonal shell, and its tripartite double-stranded RNA genome makes it rather unique. The lipid-containing phages are reviewed by Mindich and Bamford (141) and are summarized in Table 7.1.

The broad host range of these phages once against suggests that the pili, if not the transfer systems, of IncP, N, and W plasmids are related. By virtue of its ability to infect IncD plasmid-containing cells, φJ connects the transfer systems of IncD, C, and J with those of IncF and S.

3. Classification of Conjugative Pili

The classification of pili has been reviewed in detail by Paranchych and Frost (161). Pili have been classified using morphology and serology as criteria as well as their sensitivity to pilus-specific phages. Currently, the most accepted method of classification is by the incompatibility group of the plasmid expressing the pili, since there is a fair degree of consistency among the members of an incompatibility group with respect to their transfer regions and consequently their pilus types.

The detection of conjugative pili can be quite difficult, since many plasmids are repressed for pilus synthesis, and plasmids that express pili constitutively often express them at low levels. Bradley (24) has reported that all the plasmids studied from Inc I1, D, T, N, P, U, and W expressed pili constitutively in *E. coli*, while members of other Inc groups varied in their level of expression. Derepressed mutants can be isolated by mutagenesis and examined in bald, flagellaless strains (169). Transient derepression, which occurs for a few generations after mating, can be used to increase the number of pili/cell (139) and can be coupled with a spot test using phage to identify the incompatibility group (32). Another method of identifying related pili is using immune labeling (119). The few pili produced by some strains can be concentrated on a microscope grid from a loopful of bacteria (23). Some plasmids that encode a primase can suppress a temperature-sensitive *dnaG3* mutation because of derepressed expression of the primase (169, 204), which is a useful test for derepression.

The most simple classification of pili is on the basis of morphology, where they can be divided into three groups: (a) the thin, flexible pili, which are usually expressed in addition to another pilus type; (b) the thick, flexible pili; and (c) the rigid pili (24).

3.1. Thin, Flexible Pili

The thin, flexible pili are expressed by IncI complex (including IncB, K, and Z) plasmids, which also express rigid pili (28); by IncP-13 plasmids (27); and by IncX plasmids, which also express thick, flexible pili (24). The thin, flexible pili of IncX plasmid R485 are very thin and wavy in appearance, and no phage has been identified that is specific to them (21, 26). Similarly, the pili of the P-13 plasmid pMG26 remain uncharacterized.

The thin, flexible pili of the IncI plasmids were first described by Meynell (136) and Datta (68) on the basis of sensitivity to the phage If1 and serological relatedness. They were grouped into five Inc groups (Iα, β, γ, ω, and ζ), and eventually IncB and IncK plasmids were added to the complex (24). It was reduced to the IncI₁ (Iα, Iζ), IncI₂ (Iγ, Iε), IncB, and

IncK groups (28). Subsequently, $IncI_5$ and IncZ groups were added to the I complex (191, 192). The thin, flexible pili have a diameter of about 6 nm, which is similar to the polar pili of *Pseudomonas*. These pili are sensitive to the filamentous phage PR64FS and vary in sensitivity to the filamentous phage If1 ($IncI_1$, $IncI_2$) or the isometric phage Iα ($IncI_1$, B) (Table 7.1).

The thin, flexible pili of the IncI complex ($IncI_1$, I_2, I_5, B, K, Z) can be serologically divided into two groups, I_1 and I_2, where the pili of $IncI_1$, I_5, B, K, and Z are related, while the pili of $IncI_2$ form a separate group (28, 91, 92). Members of the IncI complex, when derepressed, are capable of mating both in liquid and on solid media at comparable levels. Mating type shifts to surface obligatory if the thin, flexible pili are lost by mutation, suggesting that they have a role in mating stabilization. This was emphasized by their ability to covert the surface obligatory mating types of the IncP plasmid RP4 and IncW plasmid Sa to the universal mating type (mating efficiency is high in both liquid and solid media) if the recipient cell expressed thin, flexible pili (28).

3.2. Thick, Flexible Pili

The thick, flexible pili are specified by members of the IncF, HI1 and 2, HII, C, D, J, S, T, V, X, P-3, P-5, P-8, and P-13 (24, 25, 27). They are characterized by basal knobs, which, in the case of F pili, represent a pool of unassembled pilin subunits derived from the membrane (86). An axial line, representing an inner channel running the length of the pilus, can usually be discerned in these pili (21, 23), which, with the exception of IncF pili, have pointed tips, suggesting that there is a conical formation of either pilin subunits or another minor subunit at the pilus tip (summarized in Table 7.1; Paranchych and Frost, 1988 [161]). This is reminiscent of spinae, which are conical proteinaceous appendages on the surface of some marine bacteria (72). In the cases of pili that apparently are blunt at the tip, the conical arrangement of subunits could be so shallow as to be invisible under the electron microscope.

Serologically, the pili of the IncF complex (IncFI-IV) are related, and Meynell (136) was able to discern six serotypes among the members of these incompatibility groups. The basis of these relationships was determined by DNA sequence analysis of the pilin genes from members of this group (discussed below). Although the pili of $F_O lac$ and its derepressed form, pED208, are morphologically similar to those of F and are sensitive to the Ff phages, they have been grouped with the IncS plasmids by Coetzee et al (58). Finlay et al (79) found the sequence of the pilin gene of pED208, as well as its long leader sequence and N-terminal modification by acetylation, to be highly homologous at the protein level, even though there was no homology at the DNA level. For this reason, $F_O lac$ and pED208 have been reclassified as IncFV by these authors. Homology at the DNA level was investigated by Southern hybridization at high stringency among F, pED208, R711b (IncD), and R71 (IncS). A fragment of the F plasmid containing *traALE* was used as a probe, which hybridized strongly to a fragment in R711b and weakly to a fragment in R71, while it was completely unreactive with pED208. A probe derived from the pilin gene of pED208 was homologous to a fragment in R71 but was not homologous to either F or R711b (unpublished observations). Considering the surprising homology at the protein level between F and pED208, it is not unreasonable to suggest that the pili and perhaps the transfer systems of IncF, D, and S plasmids are related and may one day be classified as subgroups of the IncF complex.

Similarly, the pili of the IncC and IncJ are serologically related and share sensitivity to the phage ϕJ, suggesting that they may share very similar transfer systems but differ in their replication mechanism, leading to a compatible phenotype (42). It should be pointed out that IncP-3 plasmids in *Pseudomonas* are IncC plasmids in *E. coli* and that they specify a surface preferred mating type in *E. coli* and a surface obligatory mating type in *Pseudomonas* (27). The mating type of transfer systems specifying flexible, thick pili is universal or surface preferred, indicating that a mechanism for mating-pari stabilization is functioning. The only exceptions noted to date include the IncC/IncP-3 plasmids and the IncX plasmid R485 mentioned previously. The thick, flexible pili of the IncX plasmid R6K are recognized by the broad host range phage, ϕX, which also recognizes the rigid pili of IncP, W, U, M, N, and I$_2$, suggesting a relationship between thick and rigid pili.

The pili of the IncHI complex are specified by temperature-sensitive transfer systems where transfer is optimal at 30°C (reviewed in Maher and Colleran [130] and Taylor [187]). An exception is the lone member of the IncHI3 group, MIP233, which specifies rigid pili (29). The three subgroups in IncHI (HI1, HI2, HI3) are based on the degree of incompatibility between members of different subgroups and their degree of DNA homology (202). The IncHII group of plasmids specify H pili, which are serologically related to IncHI1 and HI2 pili and share sensitivity to the H-specific phage pilHα (57); however, they are not thermosensitive for transfer and have no DNA homology with the IncHI plasmids (43). IncHII plasmids confer a surface preferred mating type, while the other members of the IncH complex specify universal mating systems (38). Thermosensitive transfer regions are also expressed by some of the IncT plasmids (36), which specify thick, flexible pili at 30°C and have a universal mating type but express short pili at 37°C and switch to a surface obligatory mating type. The IncM plasmids are also temperature sensitive for rigid pili formation (112).

3.3. Rigid Pili

The rigid pili of the IncI complex, IncHI3, M, N, P, U, and W, are characterized by a nail-like appearance and usually have a basal knob at one end and a distinct axial line (Figure 7.2). They have a diameter of 10 to 11 nm and, with the exception of IncU pili, they have very tapered, obvious points (44). The IncI$_1$ pili have exaggerated, unique, terminal knob structures of a membranous appearance (28). Rigid pili are usually associated with a surface obligatory mating type but may acquire a universal mating type in the presence of thin, flexible pili as in the IncI complex, discussed previously. In the case of IncP-7 and IncP-9-11, they specify a surface preferred mating type in *Pseudomonas* (27). The broad host range plasmids of the IncP, N, and W groups are up to 50% homologous with one another, although their pili do not cross-react serologically (109).

Studies with phage have connected the rigid pili from many Inc groups with each other, although the pili are otherwise serologically unrelated. The rigid pili of IncI$_2$, P-1, and N are related by attachment of the phage IKe and its host range mutant Ikeh (discussed previously). IncI$_2$, N, P-1, and W pili are the sites of attachment for PR4 and X phages, and

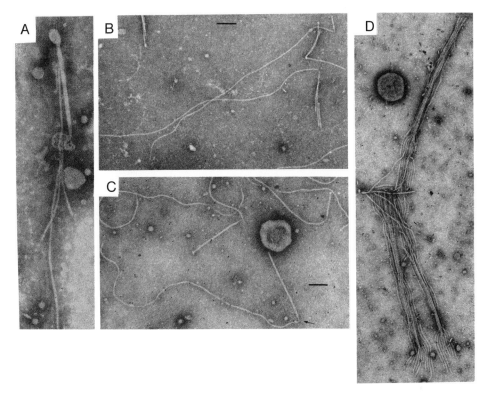

FIGURE 7.2. Electron micrographs of pili and pili-phage complexes. A. The thin and thick pili of an $IncI_1$ plasmid (28). B. φX attached to short X pili. C. φX attached to rigid N pili (35). D. φC-2 phage attached to the shaft of C pili (42). (Figure continued on next page.)

the latter also attaches M and U rigid pili and the thick, flexible pili of the IncX plasmid R6K (Table 7.1).

Figure 7.1 illustrates the pattern of relationships that can be found among known plasmid-phage systems. Currently, the IncF (S, D, C, J, I thin), IncN (P, W, U, M, X, I_2 rigid), IncT, and IncH groups define pilus systems that have not been found to be related. The IncF and IncN groups can be considered to be related if the homology between IKe and the Ff phages is taken into account. φt has many of the characteristics of F_olac, and the H-specific phages pilHα and Hgal are morphologically similar to the other RNA phages. Thus there appears to be a degree of relatedness running through all the plasmid-specified pilus systems, which probably reflects evolution from a distant ancestor. As more information is acquired, the phage-sensitive pilus systems of other bacteria will also be linked into this slowly growing phylogenetic tree.

The classification of pili by morphology is useful in that it points out the three different sorts of transfer systems that plasmids seem to use. The elaboration of thick, flexible pili gives either a universal mating type, where conjugation is equally successful in liquid or solid media, or a surface preferred phenotype (ration of mating efficiencies on solid to liquid media of 45 to 450). The rigid pili are specified by transfer systems that work well

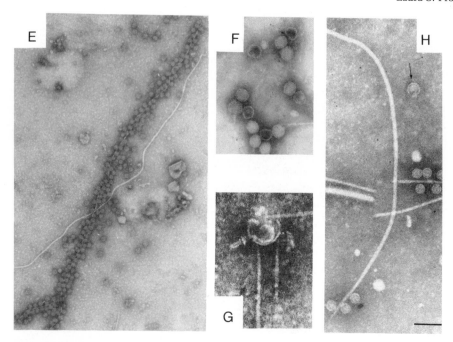

FIGURE 7.2. (*continued*) E. MS2 phage attached to an F pilus with an $F_o lac$ pilus, with no phage attached, in the background (39). F. φJ showing a double disk on the tail at its point of attachment to the head (42). G. An empty PR4 virion, tail exposed, attached by a vertex on its head to a rigid I_2 pilus (32). H. φt virions attached to the pointed tip of T pili (34). Reprinted with the permission of D.E. Bradley, Memorial University, Newfoundland, and the *Journal of General Microbiology*.

only on solid surfaces (ratio is >2000 times better on solid surfaces). The thin, flexible pili appear to be a mating-pair stabilization device that increases the rate of transfer to universal mating type levels (38). Conjugation does not occur in the absence of the thicker pili and the presence of the thin pili alone (28). These differences may reflect the presence or absence of a transfer-encoded mating-pair stabilization system such as the one that has been shown to exist in the F transfer operon (*traNG*) (2, 3).

4. F Pilus Structure

The obvious similarity between the structure of pili and filamentous phages prompted Brinton in 1971 to call pili "epiviruses" (47). The information on filamentous phages in the details of both their genetics and structure is far ahead of that for conjugative pili. The transfer regions of only a handful of plasmids have been mapped, and the gene for the pilin subunit has been identified for the IncF plasmids alone. Maps of IncF plasmids that indicate the location of the transfer region include F (FI) (206); ColB2 (FII) (77); NR1 (FII) (209); R1 (FII); R6 (FII), R100 (FII) (175); R538 (FII) (197); and $F_o lac$/pED208 (formerly EDP208) (76). Maps of plasmids of other Inc groups include R27 (HI1); pHH1508a (HI1) (187); ColIb-P9 (166); pKM101 (N) (208); R46 (N) (49); pCU1 (N) (114); R388 (W) (7); pSa

(W) (193); RK2/RP1/RP4 (P-1α) and R751 (P-1β) (96); R91-5 (IncP-10) (146, 147); and the Ti plasmid (106).

4.1. Identification of the Pilin Gene

The location of the pilin gene, as well as its sequence, is known for the IncF plasmids, and the general location of the genes for the thick and thin pili of the IncI1 plasmid ColIb-P9 has been mapped to two different transfer regions (166). The complete sequence of the transfer region of the F plasmid has been achieved, and the complete sequence of Tra1 and parts of Tra2 of RP4 have been finished (reviewed in Chapter 5 in this volume). The pilin subunit of RP1 appears to have a molecular weight of 8000 and has a blocked N-terminus that is reminiscent of F pilin, which has a molecular weight of 7200 and an acetylated N-terminus (discussed below). An amino acid composition has been obtained, and it has not identified any of the transfer genes in RP1, RP4, or RK2 as the pilin gene to date (unpublished observations).

4.2. The Pilin Gene of the F Transfer Region

The gene for the pilin subunit is the *traA* gene, which encodes a polypeptide of 121 amino acids. The mature pilin subunit is processed to 70 amino acids (7.2 kDa) by cleavage of the first 51 amino acids from the propilin, which is a leader sequence and presumably aids in insertion of the subunit into the pool of unassembled subunits in the inner membrane of the cell (84). This processing is facilitated by the *traQ* gene product (145); however, the actual proteolytic cleavage reaction appears to be the function of host enzymes, since cleavage of a cloned pilin gene can occur in the absence of the F plasmid and the cleavage site in the propilin is highly homologous to the consensus sequence for signal peptidase sites in other membrane proteins. The pilin subunit is further modified by the addition of an acetyl group to the N-terminal residue, an alanine for F and ColB2 (77, 84), and threonine for pED208 (83). The *traG* product has been implicated in this acetylation reaction (85, 118).

4.3. The Modification of the F-Pilin Subunit

There are many reports that purified F-pilin subunits are further modified by the addition of phosphate and glucose residues (9, 47, 65, 161, 190). While almost all the glucose and phosphate could be removed from pED208 pilin by methanol:chloroform extraction, approximately 1 mole of each was associated with 1 mole of pilin from ColB2, an F-like plasmid, when the molecular weight of pilin was corrected from 11,200 to 7,200 (5). 31P-NMR was unable to discern the nature of the phosphodiester linkage, since it did not resemble any known phosphoamino acid signatures. It was concluded that the phosphate was probably due to a tightly associated, noncovalent moiety. However, the sequence of the F-pilin gene indicated that the phosphoglucopeptide reported by Brinton (47) was derived from the first nine amino acids of F pilin purified from cells expressing the *traD8* mutation. Unfortunately, details of this experiment were never reported in full.

Recently, attempts to visualize the presence or absence of phosphate in pili prepara-tions using ESI (electron spectroscopic imaging) have suggested that there is no phosphate arranged in a regular manner within the pilus or the terminal knob of F, FtraD8, ColB2, or pED208 pili (David Bazett-Jones, personal communication). This is especially compelling when compared with results obtained with the filamentous phage f1, where the phosphates in the DNA packaged within the phage particle are clearly visible. If there were one phosphate per pilin subunit, the density of phosphate within the pilus would be approx-imately that for f1 phage, making the phosphate in the pilus detectable. Again, this suggests that the reports of phosphate and glucose in pili are illusory. However, the connection between pili and energy is intriguing, and the presence of phosphate continues to elicit debate. Perhaps the phosphate is involved in a transitory high-energy bond that rapidly dissipates during pilus purification.

4.4. The Serology of F-like Pili

DNA sequence analysis of the different serotypes of F-like pili, as outlined by Lawn and Meynell (120), revealed that the differences in sequence were at the N- and C-terminal of the mature pilin subunits (85). The sequence of the pED208 pilin gene revealed remarkable homology to other IncF pilin genes (79). Studies on the location of the major antigenic determinants in F-like pili (including pED208) indicated that the major epitope was located at the N-terminus and that the acetyl group was important for antigenicity (78, 86, 212, 218). The C-terminus was deduced to be involved in the minor epitope, as determined by assaying the cross-reactivity of the F-like pili using polyclonal and monoclo-nal antisera. Electron microscopic studies on the location of the major and minor epitopes on the pilus surface showed that the major epitope was probably at the tip of pED208 pili and F pili and that the minor epitope was on the sides of the pilus. This confirmed the observations of Meynell et al (140) that the ends and shafts of F pili were antigenically different, and it confirmed the complex interrelationships deduced by Meynell (136) on the basis of serological cross-reactivity. Shearing the pili increased the number of sites for a monoclonal antibody (JEL92) that reacts with an epitope near met-9, but there was no increase in the number of sites for a second monoclonal antibody (JEL 93 or 188) that is specific for the N-terminus. This suggests that the N-terminus is either not exposed on the pilus or is exposed at the tip in a unique configuration. Both monoclonal antibodies reacted with the knobs at the base of the pilus, indicating that the knob is made up of pilin sub-units (86).

4.5. Domains in the F-Pilin Subunit Involved in Phage Attachment

The different sensitivity patterns of F-specific phages (f1, R17, Qβ) of cells carrying F-like plasmids, or F plasmids bearing point mutations in the traA pilin gene, have been attributed to differences in the sequence of the pilin subunit, affecting the site of phage attachment (85, 205, 207). DNA sequence analysis of the pilin genes of these plasmids (82, 85) revealed differences in sequence in three domains of the pilin subunit that could be associated with phage attachment. The N-terminus itself, which is thought to be exposed at the pilus tip and defines the major epitope (see above), appeared not to be involved in f1 filamentous phage attachment, since both F and pED208, which have very different N-termini, are equally sensitive to this phage. Instead, a region just beyond the N-terminus (near residue met-9) was implicated as the receptor for f1 phage. A region spanning residues 12–22 and the last few residues at the C-terminus appeared to be involved in attachment of the isometric RNA phages, where R17 was more affected by changes in sequence in the domain defined by residues 12–22 while Qβ was more sensitive to the addition of charged residues at the C-terminus (161).

Thus, if the α-helical character of F pilin is high (65 to 70%) (4, 65), the secondary structure predictions are taken into consideration (84, Ken Usher, personal communication), and the length of an α-helix is 0.54 nm per turn, it seems plausible that F-like pilin can be represented as a three- or four-helix bundle, with the bends between the helices facing the interior and the exterior of the pilus. This would mean that the helices would lie perpendicular to the axis of the pilus, in a manner reminiscent of tobacco mosaic virus (TMV) virus (148). No evidence for a minor protein at the tip of the F pilus exists; however, there is circumstantial evidence that the configuration of the pilin subunits at the tip is different than at the ends of broken pili (86). Because most conjugative pili have pointed tips, it may be that the tips represent a conical rather than helical arrangement of subunits and that this exposes new sites on the pilin subunit for receptor recognition in phage attachment, mating-pair formation, and surface exclusion.

Recently, Grossman and Silverman (93) cloned the genes for F pilus formation behind an isopropyl thiogalactoside (IPTG)-inducible promoter. The pili produced by induction of this construct, pTG801, attached R17 phage weakly, suggesting that the phage attachment site on the pilus (residues 12–22) was obscured. Unlike wild-type F pili, these pili were coated on their sides with monoclonal antibodies (86) that were thought to be specific for the tip of the F pilus. The conformation of the pilin subunits within the pilus returned to normal when the rest of the F transfer operon was introduced into the cells, suggesting that a gene (known not to be *traD*), usually associated with DNA transfer, is important in the correct assembly of the pilus (94).

4.6. Physical Parameters of F Pili

Fiber diffraction has provided the most information on the structure of the pilus (80, 134). The reassessment of the initial data, using the corrected molecular weight of F pilin (7200 as determined by DNA sequence analysis, as opposed to 11,200 as previously determined by SDS-PAGE, or sodium dodecyl sulphate polyacrylamide gel electrophoresis), gave a picture of the F pilus that was extremely reminiscent of the Ff phage and Class I phages in general. F pili have a mass per unit length of 30 kDa per nm and a subunit rise of 1.28 nm. F pili can be described as a five-start helix with 25 units in two turns of the helix with a rise of 32 nm (Figure 7.3). Each unit consists of a disk of five subunits, whose dimensions are given in Figure 7.3. Class I phages also have a five-start helix, and it is feasible that both the pilus and the phage assemble and disassemble in the membrane using a similar mechanism. The polar pili of *Pseudomonas* and Pf1 phage (Class II) have a similar symmetry match with a single-start helix. It is not known whether all conjugative pili have a five-start helix motif or whether filamentous phages that attach to the tip of the pilus always

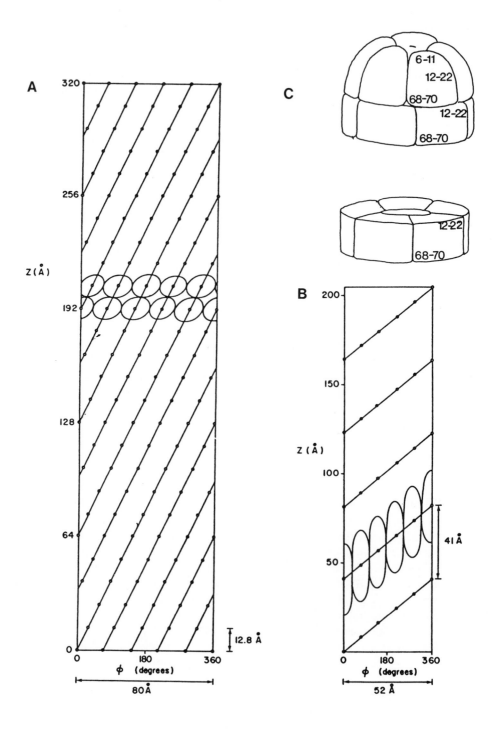

have matching symmetry elements with the pilus. However, the Class I phages Ff (f1, fd, M13), IKe, and If1 phages attach at the pilus tip, while Pf3, which is a Class II phage, attaches to the sides of the RP1 pilus (36).

Treatment of pili with chaotropic agents or antibodies has revealed some information on the ultrastructure of pili. Pili are resistant to treatment with urea, guanidine hydrochloride, Brij 58, deoxycholic acid, heat, or alkali and form vesicles at pH 1. They disaggregate in the presence of SDS or Sarkosyl (66). Treatment with HCl (190), pyrophosphate (47), or Triton X-100 (136) causes the unwinding of the F pilus along its longitudinal axes into thinner filaments. Exposure to sodium pyrophosphate, pH 2 at 55°C (47) or prolonged exposure to cesium chloride (unpublished observations) causes the pili to form vesicles. These vesicles can also be found along the F-pilus filament in an F mutant that has lowered ability to attach RNA phage (137). Antisera can be used to demonstrate a regular transverse periodicity in F-like pili and a diagonal periodicity in F-like and I-like pili (120, 123, 136). Sonication of F pili causes disruption of the pilus into disks, presumably equivalent to one layer of five subunits (unpublished observations).

4.7. The Assembly of Filamentous Phages and Pili

The subject of Ff filamentous phage assembly has been studied in depth, and several reviews are available (144, 165, 201). Salient features of the model for Ff phage assembly include the inner membrane being the assembly site and the irreversible transition of the gene VIII protein or coat protein from 50% α-helix in lipid bilayers to nearly 100% α-helix in the mature phage particle. The process of assembly during phage growth and disassembly during the infectious process is thought to occur at the inner membrane, which may be involved in "assembly sites" or adhesion zones where the inner membrane is fused with the outer membrane (8, 128), and may be triggered by the presence of the gene I protein (107). The process by which the phage particle emerges from the cell has been illustrated by Marvin (132), where the transition from a planar array of elongated, α-helical subunits to a filamentous arrangement of subunits is shown for Pf1 virus. Marvin has also demonstrated that his model could be applied to both classes of phage, since the packing of the asymmetric gene VIII subunits in either a five start (I) or single-start (II) helix results in a similar appearance for the mature phage particles (133).

FIGURE 7.3. A presentation of the physical parameters of pili. A. The F pilus, outer diameter 8 nm, inner diameter 2 nm. The F pilus can be represented as a five-start helix with a rise per subunit of 1.28 nm and a helix symmetry of 25 subunits in two turns of the helix and crystallographic repeat of 32 nm (134). B. The polar pili of *Pseudomonas* represent a filamentous structure with a single-start helix, five subunits per turn, with a rise per subunit of 4.1 nm. The diameter of these pili is 5.2 nm (81, 200). The sketch of the subunits illustrates the packing of the subunits in the pilus; however, there is no information on the shape of the subunits in either system. C. A sketch of a single layer of five subunits in the F pilus as well as a representation of a slightly different packing of the subunits at its tip. The numbers 6–11 represent the amino acids in the pilin subunit (70 residues) that form a domain thought to be involved in filamentous phage attachment; numbers 12–22 represent the site of Group I RNA phage attachment, and 68–70 represent the site of Group III phage attachment. Residues 6–11 may be exposed on the upper surface of broken pili, and the acetylated N-terminus may be exposed at the tip.

Folkhard et al (80) suggested that pilus assembly/disassembly involved a transition from a planar arrangement of hexamers of pilin to a helical, pentameric arrangement in the mature pilus. The F-pilin subunit has a high potential to assume both β structure and α-helices for large portions of its sequence; however, it is known that the α-helical conformation predominates in the assembled pilus. It has been suggested that a conformational flip-flop between β sheet and α helical structure may drive pilus assembly and disassembly or retraction (84).

5. Pilus Involvement in Phage Infection

5.1. Pilus Retraction

A controversial aspect of pilus function is the proposed property of retraction. A popular model involves pilus outgrowth and retraction, which are thought to be in equilibrium where pilus assembly requires energy and any perturbation to the system results in either pilus loss or a spurt of pilus synthesis. The principal evidence that retraction exists is based on the electron microscopic observation of the accumulation of phage particles at the cell surface after their initial attachment to the tips or sides of the pilus. This has been demonstrated for fl phage attachment to F pili (110), PR4 to W pili (19), Pf3 and PRR1 to P pili (15, 18), and C-1 change to RA1 pili (IncC) (30) and for several phages, for example, PO4, that attach to the polar pili of PAK (16, 17).

Retraction is thought to be important in establishing cell-to-cell contact during conjugation, since mating mixtures involving F+ cells from mating aggregates that become stabilized and resistant to shear or detergents such as low levels of SDS (2, 3). Retraction has also been indirectly demonstrated by disturbing the energy metabolism of the cell by the addition of cyanide (149) or arsenate (155) or by heating to 50°C (150) or cooling to 20°C (151). Pilus outgrowth has been encouraged by the addition of antipilus antisera (121) or RNA phage (15) and rapid washing of the cells (122). Increased pilus production also occurs in cells carrying the *cya* mutation (136). Pilus outgrowth as measured by the formation of fd-F pili complexes was affected by changes in temperature and the addition of metabolic poisons and was tied to the membrane potential, but not the pH gradient, of the cell (214, 215). O'Callaghan et al (154) reported an increase in pili fragments in the media following infection with M13.

There is evidence that pilus retraction is not essential and is used to increase the efficiency of mating or phage infection. Ou and Anderson (158) separated mating cells, connected by an F pilus, with a micromanipulator and demonstrated F transfer into the recipient cell. Harrington and Rogerson (98) demonstrated that mating could take place at low frequencies when the donor and recipient cells were separated by a nucleopore filter. Thus, conjugation may occur by passage of the nucleic acid through the lumen of the pilus, and retraction may shorten the pilus and stabilize the process. Stabilization of mating aggregates by the *traNG* genes of F may be an additional step, and the presence of retraction and/or mating-pair stabilization may be responsible for the differences among the mating types (universal, surface preferred, surface obligatory) seen with various plasmids.

5.2. The Filamentous Phages

With the exception of Pf3, C-2, and tf-1, all of the filamentous phages isolated attach to the tips of conjugative pili, while Pf3 and C-2 attach to the sides of pili and tf-1 attaches to the tapered portion of the pilus at the pilus tip (Table 7.1). The attachment process is energy independent, although subsequent stages of infection do require an energy source. Attachment requires an intact gene III protein, a minor component of the phage particle that is required for phage stability and infectivity, which forms an adsorption structure that can be visualized in the electron microscope (90). Other filamentous phages have no obvious structure at the tip except ϕC2, which has a disklike appendage that attaches to the pilus (42). Up to three phage particles can attach per pilus (50), suggesting that only one to three of the gene III proteins are required to secure the phage to the pilus. The upper limit of three phage per pilus is probably due to steric hindrance. The gene III proteins appear as five lollipops protruding from the end of the phage, and these "balls" can be removed by treatment of the phage with subtilisin, rendering the phage noninfectious. The "balls" correspond to the globular N-terminal domain of the gene III protein, which can block phage infection (6). Stengele et al (183) have shown that the N-terminal half of the gene III protein of fd phage consists of two domains—one involved in attachment and the other in penetration of the DNA into the host cell—and that gene III undergoes proteolytic cleavage on entry into the cell. Glaser-Wuttke et al (88) have demonstrated that gene III forms aqueous pores in lipid bilayers and, when expressed by a phage or plasmid, has a number of pleiotropic effects on the cell, including superinfection immunity to Ff phage, inability to form mating pairs, and sensitivity to detergents (12, 13).

Sun and Webster (185) showed that a chromosomal gene *fii* was required for fl phage infection, and this was subsequently identified as *tolQ* of the *tolQRAB* locus, which is responsible for the import of macromolecules such as colicins. The *tol* mutation does not prevent receptor recognition but does block uptake, leading to a tolerant phenotype (168, 186). Ff phage infection requires the *tolQRA* loci, but not *tolB*. Mutations in *tolQRA* result in tolerance to colicins E1, E2, E3, and K. The mechanism by which the DNA is unsheathed and transported to the cytoplasm via the *tol* system is unknown. Jakes et al (111) reported an interesting experiment where a fusion protein of gene III and the RNase activity of colicin E3 was able to kill F$^+$ cells using the pilus for attachment and *tol* gene products for penetration into the cell, suggesting that little phage or transfer region genetic information is required.

Bradley and Whelan (37) demonstrated that filamentous phages that adsorb to the pilus tips also use the *tol* system during infection, and the *tolQ* mutants were resistant to these phages (fd, IKe, PR64, X) but were still competent for transfer. Pf3, which adsorbs to the shaft of P pili, and tf-1, which adsorbs to the pointed portion of the tip of T pili, infected *tolQ* cells. *tolQ* cells were resistant to the X-2 phage and, although attachment of this phage has not been demonstrated (57), these results suggest that ϕX-2 attaches to the pilus tip.

Thus, it appears that the filamentous phages use the pilus as a primary attachment site and that, perhaps, retraction brings the phage particle into close proximity with the cell surface where penetration occurs using auxiliary, chromosomally encoded systems. Infection of F$^-$ cells has been achieved at low levels using transducing particles where antibiotic resistance could be selected for instead of plaque formation and the unwieldy titer increase assay was avoided (168). Thus, the transfer system of conjugative plasmids appears to be unimportant in the infectious process of filamentous phage, beyond providing an initial receptor for phage attachment.

Experiments showing that cells, denuded of their pili by blending (152) or centrifugation (214), were not able to attach Ff phage, suggesting that a receptor, either the pilus tip or the *tol* system, was unavailable to the phage. Schandel et al (172) showed that the F mutant, *traC1044*, did not express pili but that cells carrying the mutant F plasmid attached and were infected by f1 phage. This suggests that the pilus tip is exposed on the pilus surface and is available to the phage and that, therefore, blended cells with stubby pili do not provide a suitable attachment site for the phage. Thus, it seems plausible that the configuration of pilin subunits in a point at the pilus tip is the true receptor for filamentous phage, and broken pili are not adequate substitutes.

5.3. The Spherical RNA Phages

Because most of the research reported on the life cycle of RNA phages concerns only the F-specific phages, the following discussion will be limited to them. The spherical RNA phages attach to the sides of their respective pili, and the receptor on the F pilus for Group I phage differs from that for Group III phage. The attachment protein or A protein is involved in recognition of the pilus, and infectious RNA-A protein complexes of Group I phages have been isolated (124, 177), while Group III phages require the presence of the readthrough protein in addition to the RNA and A protein (176).

The most complete studies on the infectious process have been carried out on Group I phages. Attachment requires an ionic strength of 0.1 or greater (62) and can occur at low temperatures, suggesting that attachment is energy independent. RNA phages attach equally well to free or cell-bound pili. Studies on the kinetics of MS2 attachment to F pili found that monovalent cations promoted weak attachment, while divalent cations gave a stronger interaction (64).

Attachment is followed by "eclipse" of the phage particle; the RNA becomes ribonuclease sensitive (159, 216), and the viral capsid is quickly released (163, 178). The RNA retains its secondary structure (210) and remains bound to the A protein. The RNA-A protein complex is injected into the host cell, and the A protein (39 kDa) undergoes a cleavage into two polypeptides of 24 and 15 kDa (115). The eclipse reaction occurs optimally at 37°C in the presence of divalent cations (159) and is energy-requiring (63). It does not occur in the cold and does not occur with free pili (194), suggesting that an energy source must be available either at the base of the pilus or along its length. Penetration of the RNA-A protein complex is accompanied by a sharp drop in intracellular nucleoside triphosphates and a loss of F pili from the cell surface (163). Compounds that disturb the energy metabolism of the cell (and affect the level of piliation) also prevent infection by the RNA phages (63).

In studies employing R17, Wong and Paranchych (211) demonstrated that the A protein was attached to the 3' end of the RNA and that it was this end that entered the cell first. In MS2 phage, it has been determined that the A protein is bound to two sequences in the RNA—one at the 5' end (near position 393) and one at the 3' end at position 3515 (176)—and that the cleavage of the A protein liberates the two ends (195). Phage particles that have no A protein are missing the 5' end of the RNA, suggesting that the A protein protects this region from ribonuclease in the intact particle.

Bradley has observed the accumulation of RNA phage particles (phage C-1 and the IncC plasmid RA1 [30]; PRR1 phage and P-1 pili [18]; PP7 phage and *Pseudomonas* polar pili [15]) at the base of the pilus, suggesting that the pilus may retract until the particle comes into contact with the cell surface, where penetration occurs. Krahn and Paranchych (116) have shown that the eclipse reaction can occur at sites removed from the cell surface, since a single cell is capable of eclipsing a large number of phage particles.

Unlike the filamentous phages, which use a host transport system (*tolQRAB*) to penetrate the cell, the Class I F-specific RNA phages require a functional F *traD* gene product for penetration of the RNA-A protein complex (1, 174). The *traD* mutation does not affect phage attachment or eclipse and does not affect the penetration of the Class III phage, Qβ. The TraD protein is a large (81.7-kDa) protein found in the inner membrane that is thought to pump the nucleic acid from the donor to the recipient cell during conjugation (see Chapter 2, this volume). Its participation in the penetration of the RNA of Class I phages suggests that it might be capable of transporting nucleic acids in either direction and that it may interact with the A protein or a protein expressed by the transfer region during phage infection and conjugation, respectively (160). Unfortunately, almost nothing is known about the attachment, eclipse, and penetration of other RNA phages, and the information gleaned for Class I phages (f2, fr, R17, MS2) is taken as a paradigm for all spherical RNA phages.

5.4. The Lipid-Containing Phages

Almost nothing is known about the interaction of the lipid-containing and/or tailed bacteriophages with pili. The pilus receptors for these phages are usually very short and scarce, and convincing evidence using the electron microscope is almost impossible to achieve. The interaction of phage φ6 with the pili of *P. phaseolicola* is the best studied system and has been reviewed in Mindich and Bamford (141). There appears to be a requirement for pilus retraction (167), and the adsorption protein has been identified as P3 (142).

6. Conclusions

Recent evidence that the F plasmid can be mated into *Pseudomonas* (95) or yeast (102), if a replication system compatible with the recipient organism is provided, suggests that the host range for transfer functions is much broader for all plasmids than originally thought. It is possible that plasmid transfer occurs throughout the bacterial world and between kingdoms, promoting gene exchange, without easily being detected, because of the failure to maintain the plasmid after conjugation. The related processes of natural transformation and competence, which involve the uptake of DNA, have also been implicated in prokaryotic evolution. The mosaic nature of genes echoes this genetic diversification and argues for mechanisms that promote interspecies recombination among prokaryotes, as suggested by Smith et al (181). Nowhere is this mosaic pattern more evident than in the plasmids and phages that reside in bacteria, where homology studies at the gene level are revealing a complex patchwork of genetic information.

An examination of the similarities in structure and function of the plasmid-encoded pili and the pili of *Pseudomonas* and *Caulobacter*, as well as a comparison of the phages specific for these pili, suggests that these systems are evolutionarily related and that the pilus systems of *Pseudomonas* and *Caulobacter* are derived from transfer regions integrated into the chromosome that have lost the ability to transfer. Interestingly, one of the competence genes involved in the uptake of DNA by *Bacillus subtilis* has homology to the pilin of *Pseudomonas* (46) and a number of pathogenic bacteria including the spp. of *Neisseria* and *Moraxella* (161). *Neisseria gonorrhoeae* and *Haemophilus influenza* have the property of transformation (184), which in the case of the former organism is related to the presence of pili. Thus, the properties of genetic competence and natural transformation are probably related to conjugation, where all three processes are involved in transfer of DNA through the bacterial envelope and where the genetic organization, at least for competence and conjugation, is proving to be very complex. Similarly, correlations have been made between the transfer of T-DNA from *Agrobacterium tumefaciens*, containing the Ti plasmid, to plants and conjugation in bacteria. The existence of conjugative transposons (reviewed in Chapter 15, this volume), suggests that conjugative functions have melded together with transpositional elements to achieve a wide host range.

Perhaps the most striking similarity exists between conjugative transfer and phage biology. Brinton (47) pointed out that the F pilus could be considered an epivirus where the pilus resembles a filamentous phage and the process of conjugation is an infectious process. While pilus assembly and retraction is superficially similar to the penetration and outgrowth of filamentous phages, the process of plasmid nicking and transfer through a conjugation pore resembles the packaging reaction of the single-stranded phage φX174 (100). In this reaction, a single strand of DNA is "packaged" into the recipient cell, which substitutes for the phage head, from a specific initiation point, OriT, using the rolling-circle form of DNA replication. When transfer of one genome has been completed, the single strand is cut, released, and recircularized.

This chapter has attempted to draw together the scattered information on pili and their phages and to demonstrate that the pili of many systems are related and that the classification of pili by incompatibility may be artificial and may be useful only in imposing order on a chaotic array of plasmids. As information accumulates on the organization and function of transfer regions, including the role of the pilus and retraction in conjugation, it is hoped that the classification of conjugative plasmids will rely more heavily on transfer functions than on the unrelated, but better understood, property of incompatibility.

ACKNOWLEDGMENTS. Helpful discussions with Bill Paranchych, Loren Day, and David Bradley are gratefully acknowledged. I also wish to thank David Bazett-Jones for the unpublished observations on the ESI of pili and phage.

References

1. Achtman, M., Willetts, N.S., and Clark, A.J., 1971, Beginning a genetic analysis of conjugational transfer determined by the F factor in *E. coli* by isolation and characterization of transfer-deficient mutants, *J. Bacteriol.* 106:529–538.

2. Achtman, M., Morelli, G., and Schwuchow, S., 1978, Cell-cell interactions in conjugating *Escherichia coli*: role of F pili and fate of mating aggregates, *J. Bacteriol.* 135:1053–1061.

3. Achtman, M., Schwuchow, S., Helmuth, R., Morelli, G., and Manning, P.A., 1978, Cell-cell interactions in conjugating *Escherichia coli*: con⁻ mutants and stabilization of mating aggregates, *Mol. Gen. Genet.* 164:171–183.

4. Armstrong, G.D., Frost, L.S., Sastry, P.A., and Paranchych, W., 1980, Comparative biochemical studies on F and EDP208 conjugative pili, *J. Bacteriol.* 141:333–341.

5. Armstrong, G.D., Frost, L.S., Vogel, H.J., and Paranchych, W., 1981, Nature of the carbohydrate and phosphate associated with ColB2 and EDP208 pilin, *J. Bacteriol.* 145:1167–1176.

6. Armstrong, J., Perham, R., and Walker, J., 1981, Domain structure of bacteriophage fd adsorption protein, *FEBS Lett.* 135:167–172.

7. Avila, P., and De La Cruz, F., 1988, Physical and genetic map of the IncW plasmid R388, *Plasmid* 20:155–157.

8. Bayer, M.E., and Bayer, M.H., 1986, Effects of bacteriophage fd infection on *Escherichia coli* HB11 envelope: a morphological and biochemical study, *J. Virol.* 57:258–266.

9. Beard, J.P., Howe, T.G.B., and Richmond, M.H., 1972, Purification of sex pili from *Escherichia coli* carrying a derepressed F-like R factor, *J. Bacteriol.* 111:814–820.

10. Beck, E., and Zink, B., 1981, Nucleotide sequence and genome organization of filamentous bacteriophages f1 and fd, *Gene* 16:35–58.

11. Beck, E., Sommer, R., Auerswald, E.A., Kurz, C., Zink, B., Osterburg, G., Schaller, H., Sugimoto, K., Sugisaki, H., Okamoto, T., and Takanami, M., 1978, Nucleotide sequence of bacteriophage fd DNA, *Nucl. Acids Res.* 5:4495–4503.

12. Boeke, J., and Model, P., 1982, A prokaryotic membrane anchor sequence: carboxyl terminus of bacteriophage f1 gene 3 protein retains it in the membrane, *Proc. Natl. Acad. Sci. USA* 79:5200–5204.

13. Boeke, J., Model, P., and Zinder, N., 1982, Effects of bacteriophage f1 gene 3 protein on the host cell membrane, *Mol. Gen. Genet.* 186:185–192.

14. Bradley, D.E., 1964, The structure of some bacteriophages associated with male strains of *Escherichia coli*, *J. Gen. Microbiol.* 35:471–478.

15. Bradley, D.E., 1972, Shortening of *Pseudomonas aeruginosa* pili after RNA-phage adsorption, *J. Gen. Microbiol.* 72:303–319.

16. Bradley, D.E., 1972, Evidence for the retraction of *Pseudomonas aeruginosa* RNA phage pili, *Biochim. Biophys. Res. Commun.* 47:142–149.

17. Bradley, D.E., 1973, Basic characterization of a *Pseudomonas aeruginosa* pilus-dependent bacteriophage with a long noncontractile tail, *J. Virol.* 12:1139–1148.

18. Bradley, D.E., 1974, Adsorption of bacteriophages specific for *Pseudomonas aeruginosa* R factors RP1 and R1822, *Biochem. Biophys. Res. Commun.* 57:893–900.

19. Bradley, D.E., 1976, Adsorption of the R-specific bacteriophage PR4 to pili determined by a drug resistance plasmid of the W compatibility group, *J. Gen. Microbiol.* 95:181–185.

20. Bradley, D.E., 1977, Characterization of pili determined by drug resistance plasmids R711b and R778b, *J. Gen. Microbiol.* 102:349–363.

21. Bradley, D.E., 1978, Determination of very thin pili by the bacterial drug resistance plasmid R485, *Plasmid* 1:376–387.

22. Bradley, D.E., 1979, Morphology of pili determined by the N incompatibility group plasmid N3 and interaction with bacteriophages PR4 and IKe, *Plasmid* 2:632–636.

23. Bradley, D.E., 1980, Determination of pili by conjugative bacterial drug resistance plasmids of incompatibility groups B, C, H, J, K, M, V, and X, *J. Bacteriol.* 141:828–837.

24. Bradley, D.E., 1980, Morphological and serological relationships of conjugative pili, *Plasmid* 4:155–169.

25. Bradley, D.E., 1981, Conjugative pili of plasmids in *Escherichia coli* K-12 and *Pseudomonas*, in: *Molecular Biology, Pathogenicity, and Ecology of Bacterial Plasmids* (S.B. Levy, R.C. Clowes, and E.L. Koenig, eds.), Plenum Publishing, New York, pp. 217–226.

26. Bradley, D.E., 1982, Further characterization of R485, and IncX plasmid that determines two kinds of pilus, *Plasmid* 7:95–100.

27. Bradley, D.E., 1983, Specification of the conjugative pili and surface mating systems of *Pseudomonas* plasmids, *J. Gen. Microbiol.* 129:2545–2556.

28. Bradley, D.E., 1984, Characteristics and function of thick and thin conjugative pili determined by transfer-derepressed plasmids of incompatibility groups I₁, I₂, I₅, B, K and Z, *J. Gen. Microbiol.* 130:1489–1502.

29. Bradley, D.E., 1986, The unique conjugation system of IncHI3 plasmid MIP233, *Plasmid* 16:63–71.

30. Bradley, D.E., 1989, Conjugation system of IncC plasmid RA1, and the interaction of RA1 pili with specific RNA phage C-1, *Res. Microbiol.* 140:439–446.

31. Bradley, D.E., and Coetzee, J.N., 1982, The determination of two morphologically distinct types of pilus by plasmids of incompatibility group I2, *J. Gen. Microbiol.* 128:1923–1926.

32. Bradley, D.E., and Fleming, J., 1983, Incompatibility group identification for repressed plasmids using host cell lysis by specific bacteriophages, *J. Microbiol. Methods* 1:171–176.

33. Bradley, D.E., and Meynell, E., 1978, Serological characteristics of pili determined by the plasmids R711b and F$_o$*lac*, *J. Gen. Microbiol.* 108:141–149.

34. Bradley, D.E., and Pitt, T.L., 1974, Pilus-dependence of four *Pseudomonas aeruginosa* bacteriophages with non-contractile tails, *J. Gen. Virol.* 24:1–15.

35. Bradley, D.E., and Rutherford, E.L., 1975, Basic characterization of a lipid-containing bacteriophage specific for plasmids of the P, N, and W compatibility groups, *Can. J. Microbiol.* 21:152–163.

36. Bradley, D.E., and Whelan, J., 1985, Conjugation systems of IncT plasmids, *J. Gen. Microbiol.* 131: 2665–2671.

37. Bradley, D.E., and Whelan, J., 1989, *Escherichia coli tolQ* mutants are resistant to filamentous bacterio-phages that adsorb to the tips, not the shafts, of conjugative pili, *J. Gen. Microbiol.* 135:1857–1863.

38. Bradley, D.E., Taylor, D.E., and Cohen, D.R., 1980, Specification of surface mating systems among conjugative drug resistance plasmids in *Escherichia coli* K-12, *J. Bacteriol.* 143:1466–1470.

39. Bradley, D.E., Coetzee, J.N., Bothma, T., and Hedges, R.W., 1981, Phage X: a plasmid-dependent, broad host range, filamentous bacterial virus, *J. Gen. Microbiol.* 126:389–396.

40. Bradley, D.E., Coetzee, J.N., Bothma, T., and Hedges, R.W., 1981, Phage t: a group T plasmid-dependent bacteriophage, *J. Gen. Microbiol.* 126;397–403.

41. Bradley, D.E., Coetzee, J.N., Bothma, T., and Hedges, R.W., 1981, Phage F$_o$*lac*: an F$_o$*lac* plasmid-dependent bacteriophage, *J. Gen. Microbiol.* 126:405–411.

42. Bradley, D.E., Hughes, V.M., Richards, H., and Datta, N., 1982, R plasmids of a new incompatibility group determine constitutive production of H pili, *Plasmid* 7:230–238.

43. Bradley, D.E., Aoki, T., Kitao, T. Arai, T., and Tschape, H., 1982, Specification and characteristics for the classification of plasmids in incompatibility group U, *Plasmid* 8:89–93.

44. Bradley, D.E., Sirgel, F.A., Coetzee, J.N., Hedges, R.W., and Coetzee, W.F., 1982, Phages C-2 and J: IncC and IncJ plasmid-dependent phages, respectively, *J. Gen. Microbiol.* 128:2485–2498.

45. Bradley, D.E., Coetzee, J.N., and Hedges, R.W., 1983, IncI$_2$ plasmids specify sensitivity to filamentous bacteriophage IKe, *J. Bacteriol.* 154:505–507.

46. Breitling, R., and Dubnau, D., 1990, A membrane protein with similarity to N-methylphenylalanine pilins is essential for DNA binding by competent *Bacillus subtilis*, *J. Bacteriol.* 172:1499–1508.

47. Brinton, C. C., Jr., 1971, The properties of sex pili, the viral nature of "conjugal" genetic transfer systems, and some possible approaches to the control of drug resistance, *Crit. Rev. Microbiol.* 1:105–160.

48. Brinton, C.C., Jr., Gemski, P., Jr., and Carnahan, J., 1964, A new type of bacterial pilus genetically controlled by the fertility factor of *E. coli* K12 and its role in chromosome transfer, *Proc. Natl. Acad. Sci. USA* 52:776–783.

49. Brown, A.C., and Willetts, N.S., 1981, A physical and genetic map of the IncN plasmid R46, *Plasmid* 5:188–201.

50. Caro, L.G., and Schnos, M., 1966, The attachment of the male-specific bacteriophage f1 to sensitive strains of *Escherichia coli*, *Proc. Natl. Acad. Sci. USA* 56:126–132.

51. Coetzee, W.F., and Becker, P.J., 1979, Pilus-specific, lipid-containing bacteriophages PR4 and PR772: comparison of physical characteristics of genomes, *J. Gen. Microbiol.* 45:195–200.

52. Coetzee, J.N., Lecatsas, G., Coetzee, W.F., and Hedges, R.W., 1979, Properties of R plasmid R772 and the corresponding pilus-specific phage PR772, *J. Gen. Microbiol.* 110:263–273.

53. Coetzee, J.N., Sirgel, F.A., and Lecatsas, G., 1980, Properties of a filamentous phage which adsorbs to pili coded by plasmids of the IncI complex, *J. Gen. Microbiol.* 117:547–551.

54. Coetzee, J.N., Bradley, D.E., and Hedges, R.W., 1982, Phages I alpha and I2-2: IncI plasmid-dependent bacteriophages, *J. Gen. Microbiol.* 128:2797–2804.

55. Coetzee, J.N., Bradley, D.E., Hedges, R.W., Fleming, J., and Lecatsas, G., 1983, Bacteriophage M: an incompatibility group M plasmid-specific phage, *J. Gen. Microbiol.* 129:2271–2276.

56. Coetzee, J.N., Bradley, D.E., Lecatsas, G., du Toit, L., and Hedges, R.W., 1985, Bacteriophage D: an IncD group plasmid-specific phage, *J. Gen. Microbiol.* 131:3375–3383.

57. Coetzee, J.N., Bradley, D.E., Fleming, J., du Toit, L., Hughes, V.M., and Hedges, R.W., 1985, Phage pilH alpha: a phage which adsorbs to IncHI and IncHII plasmid-coded pili, *J. Gen. Microbiol.* 131:1115–1121.

58. Coetzee, J.N., Bradley, D.E., Hedges, R.W., Hughes, V.M., McConnell, M.M., du Toit, L., and Tweehuysen, M., 1986, Bacteriophages F$_o$lach, SR, SF: phages which adsorb to pili encoded by plasmids of the S-complex, *J. Gen. Microbiol.* 132:2907–2917.

59. Coetzee, J.N., Bradley, D.E., Hedges, R.W., Tweehuizen, M., and du Toit, L., 1987, Phage tf-1: a filamentous bacteriophage specific for bacteria harboring the IncT plasmid pIN25, *J. Gen. Microbiol.* 133:953–960.

60. Coetzee, J.N., Bradley, D.E., du Toit, L., and Hedges, R.W., 1988, Bacteriophage X-2: a filamentous phage lysing IncX-plasmid-harboring bacterial strains, *J. Gen. Microbiol.* 134:2535–2541.

61. Crawford, E.M., and Gesteland, R.F., 1964, The adsorption of bacteriophage R-17, *Virology* 22:165–167.

62. Danziger, R.E., and Paranchych, W., 1970, Stages in phage R17 infection. 3. Energy requirements for the F-pili mediated eclipse of viral infectivity, *Virology* 40:554–564.

63. Danziger, R.E., and Paranchych, W., 1970, Stages in phage R17 infection. II. Ionic requirements for phage R17 attachment to F-pili, *Virology* 40:547–553.

64. Date, T., 1979, Kinetic studies of the interaction between MS2 phage and F pilus of *Escherichia coli*, *Eur. J. Biochem.* 96:167–176.

65. Date, T., Inuzuka, M., and Tomoeda, M., 1977, Purification and characterization of F pili from *Escherichia coli*, *Biochemistry* 16:5579–5585.

66. Date, T., Inuzuka, M., and Tomoeda, M., 1978, Structure of F pili in *Escherichia coli*, in: *Pili* (D.E. Bradley, E. Raizen, P. Fives-Taylor, and J. Ou, eds.), International Conferences on Pili, Washington, DC, pp. 301–318.

67. Datta, N., 1975, Epidemiology and classification of plasmids, in: *Microbiology* (D. Schlessinger, ed.), American Society for Microbiology, Washington, DC, pp. 9–15.

68. Datta, N., 1979, Plasmid classification: incompatibility grouping, in: *Plasmids of Medical, Environmental and Commercial Importance* (K.N. Timmis and A. Puhler, eds.), Elsevier/North-Holland, Amsterdam, pp. 3–12.

69. Davis, J.E., Strauss, H.J., and Sinsheimer, R.L., 1961, Bacteriophage MS2: another RNA phage, *Science* 134:1427.

70. Dennison, S., and Baumberg, S., 1975, Conjugational behavior of N plasmids in *Escherichia coli*, *Mol. Gen. Genet.* 138:323–331.

71. Dennison, S., and Hedges, R.W., 1972, Host specificities of RNA phages, *J. Hygiene* 70;55–61.

72. Easterbrook, K.B., and Rao, D.V.S., 1984, Conical spinae associated with a picoplanktonic procaryote, *Can. J. Microbiol.* 30:716–718.

73. Espejo, R.T., and Canelo, E.S., 1968, Properties of bacteriophage PM2: a lipid-containing bacterial virus, *Virology* 34:738–747.

74. Falkow, S., Guerry, P., Hedges, R.W., and Datta, N., 1974, Polynucleotide sequence relationships among plasmids of the I compatibility complex, *J. Gen. Microbiol.* 85:65–76.

75. Fiers, W., 1979, RNA bacteriophages, in: *Comprehensive Virology*, Vol. 13 (H. Fraenkel-Conrat and R.R. Wagner, eds.), Plenum Press, New York, pp. 69–204.

76. Finlay, B.B., Paranchych, W., and Falkow, S., 1983, Characterization of conjugative plasmid EDP208, *J. Bacteriol.* 156:230–235.

77. Finlay, B.B., Frost, L.S., and Paranchych, W., 1984, Localization, cloning, and sequence determination of the conjugative plasmid ColB2 pilin gene, *J. Bacteriol.* 160:402–407.

78. Finlay, B.B., Frost, L.S., Paranchych, W., Parker, J.M., and Hodges, R.S., 1985, Major antigenic determinants of F and ColB2 pili, *J. Bacteriol.* 163:331–335.

79. Finlay, B.B., Frost, L.S., and Paranchych, W., 1986, Nucleotide sequence of the *traYALE* region from IncFV plasmid pED208, *J. Bacteriol.* 168:990–998.

80. Folkhard, W., Leonard, K.R., Malsey, S., Marvin, D.A., Dubochet, J., Engel, A., Achtman, M., and Helmuth, R., 1979, X-ray diffraction and electron microscope studies on the structure of bacterial F pili, *J. Mol. Biol.* 130:145–160.

81. Folkhard, W., Marvin, D.A., Watts, T.H., and Paranchych, W., 1981, Structure of polar pili from *Pseudomonas aeruginosa* strains K and O, *J. Mol. Biol.* 149:79–93.

82. Frost, L.S., and Paranchych, W., 1988, DNA sequence analysis of point mutations in *traA*, the F pilin gene, reveal two domains involved in F-specific bacteriophage attachment, *Mol. Gen. Genet.* 213:134–139.

83. Frost, L.S., Armstrong, G.D., Finlay, B.B., Edwards, B.F., and Paranchych, W., 1983, N-terminal amino acid sequencing of EDP208 conjugative pili, *J. Bacteriol.* 153:950–954.

84. Frost, L.S., Paranchych, W., and Willetts, N.S., 1984, DNA sequence of the F *traALE* region that includes the gene for F pilin, *J. Bacteriol.* 160:395–401.

85. Frost, L.S., Finlay, B.B., Opgenorth, A., Paranchych, W., and Lee, J.S., 1985, Characterization and sequence analysis of pilin from F-like plasmids, *J. Bacteriol.* 164:1238–1247.

86. Frost, L.S., Lee, J.S., Scraba, D.G., and Paranchych, W., 1986, Two monoclonal antibodies specific for different epitopes within the amino-terminal region of F-pilin, *J. Bacteriol.* 168:192–198.

87. Furuse, K., 1987, Distribution of coliphages in the environment: general considerations, in: *Phage Ecology* (S.M. Goyal, ed.), John Wiley and Sons, New York, pp. 87–105.

88. Glaser-Wuttke, G., Keppner, J., and Rasched, I., 1989, Pore-forming properties of filamentous phage fd, *Biochim. Biophys. Acta* 985:239–247.

89. Grant, R.B., Whiteley, M.H., and Shapley, A.J., 1978, Plasmids of incompatibility group P code for the capacity to propagate bacteriophage IKe, *J. Bacteriol.* 136:808–811.

90. Gray, C., Brown, R., and Marvin, D.A., 1981, Adsorption complex of filamentous fd virus, *J. Mol. Biol.* 146:621–627.

91. Grindley, N.D.F., Grindley, J.N., and Anderson, E.S., 1972, Molecular studies of R factor compatibility groups, *Mol. Gen. Genet.* 119:287–297.

92. Grindley, N.D.F, Humphreys, G.O., and Anderson, E.S., 1973, Molecular studies of R factor compatibility groups, *J. Bacteriol.* 115:387–389.

93. Grossman, T., and Silverman, P.M., 1989, Structure and function of conjugative pili: inducible synthesis of functional F pili by *Escherichia coli* K-12 containing a *lac-tra* operon fusion, *J. Bacteriol.* 171:650–656.

94. Grossman, T.H., Frost, L.S., and Silverman, P.M., 1990, Structure and function of conjugative pili: monoclonal antibodies as probes for structural variants of F pili, *J. Bacteriol.* 172:1174–1179.

95. Guiney, D.G., 1982, Host range of conjugation and replication functions of the *Escherichia coli* sex plasmid F′*lac*: comparison with the broad-host-range plasmid RK2, *J. Mol. Biol.* 162:699–703.

96. Guiney, D.G., and Lanka, E., 1989, Conjugative transfer of IncP plasmids, in: *Promiscuous Plasmids of Gram-Negative Bacteria* (C.M. Thomas, ed.), Academic Press, London, p. 2756.

97. Harden, V., and Meynell, E., 1972, Inhibition of gene transfer by antiserum and identification of serotypes of sex pili, *J. Bacteriol.* 109:1067–1074.

98. Harrington, L.C., and Rogerson, A.C., 1990, The F pilus of *Escherichia coli* appears to support stable DNA transfer in the absence of wall-to-wall contact between cells, *J. Bacteriol.* 172:7263–7264.

99. Havelaar, A.H., Hogeboom, W.M., and Pot, R., 1984, F specific RNA bacteriophages in sewage: methodology and occurrence, *Wat. Sci. Tech.* 17:645.

100. Hayashi, M., Aoyama, A., Richardson, D.L., Jr., and Hayashi, M.N., 1988, Biology of the bacteriophage φX174, in: *The Bacteriophages*, Vol. 2 (R. Calendar, ed.), Plenum Press, New York, pp. 1–71.

101. Hedges, R.W., and Datta, N., 1973, Plasmids determining I pili constitute a compatibility complex, *J. Gen. Microbiol.* 77:19–25.

102. Heinemann, J.A., and Sprague, G.F., Jr., 1989, Bacterial conjugative plasmids mobilize DNA transfer between bacteria and yeast, *Nature* 340:205–209.

103. Hill, D.F., and Petersen, G.B., 1982, Nucleotide sequence of bacteriophage f1 DNA, *J. Virol.*, 44:32–46.

104. Hofschneider, P.H., 1963, Unterschungen uber kleine *E. coli* K12 Bakteriophagen 1. und 2. Mitteilung, *Z. Naturforsch.* 18b:203–210.

105. Hohn, T., and Hohn, B., 1970, Structure and assembly of simple RNA bacteriophages, *Adv. Virol. Res.* 16:43–98.

106. Holsters, M., Silva, B., Van Vliet, F., Genetello, C., De Block, M., Dhaese, P., Depicker, A., Inze, D., Engler, G., Villroel, R., Van Montagu, M., and Schell, J., 1980, The functional organization of the nopaline *A. tumefaciens* plasmid pTiC58, *Plasmid* 3:212–230.

107. Horabin, J.I., and Webster, R.E., 1986, Morphogenesis of f1 filamentous bacteriophage. Increased expression of gene I inhibits bacterial growth, *J. Mol. Biol.* 188:403–413.

108. Hua, T.-C., Scholl, D.R., and Jollick, J.D., 1981, Functional modification of the plasmid RP1-specified pilus by *Caulobacter vibrioides*, *J. Gen. Microbiol.* 124:119–128.

109. Ingram, L.C., 1973, Deoxyribonucleic acid-deoxyribonucleic acid hybridization of R factors, *J. Bacteriol.* 115:1130–1134.

110. Jacobson, A., 1972, Role of F pili in the penetration of bacteriophage f1, *J. Virol.* 10:835–843.

111. Jakes, K.S., Davis, N.G., and Zinder, N., 1988, A hybrid toxin from bacteriophage f1 attachment protein and colicin E3 has altered cell receptor specificity, *J. Bacteriol.* 170:4231–4238.

112. Karste, G., Adler, K., and Tschape, H., 1987, Temperature sensitivity of M pilus formation as demonstrated by electron microscopy, *J. Basic Microbiol.* 27:225–228.

113. Khatoon, H., Iyer, R.V., and Iyer, V.N., 1972, A new filamentous bacteriophage with sex-factor specificity, *Virology* 48:145–155.

114. Konarska-Kozlowska, M., and Iyer, V.N., 1981, Physical and genetic organization of the IncN-group plasmid pCU1, *Gene* 14:195–204.

115. Krahn, P.M., O'Callaghan, R. J., and Paranchych, W., 1972, Stages in phage R17 infection. VI. Injection of A protein and RNA into the host cell, *Virology* 47:628–637.

116. Krahn, P.M., and Paranchych, W., 1971, Heterogeneous distribution of A protein in R17 phage preparations, *Virology* 43:533–535.

117. Kuo, T.-T., Huang, T.-C., and Chow, T.-Y., 1969, A filamentous bacteriophage in *Xanthomonas oryzae*, *Virology* 39:548–550.

118. Laine, S., Moore, D., Kathir, P., and Ippen-Ihler, K., 1985, Genes and gene products involved in the synthesis of F-pili, in: *Plasmids in Bacteria* (D.R. Helinski, S.N. Cohen, D.B. Clewell, D.A. Jackson, and A. Hollaender, eds.), Plenum Publishing, New York, pp. 535–553.

119. Lawn, A.M., 1967, Simple immunological labelling method for electron microscopy and its application to the study of filamentous appendages of bacteria, *Nature* 214:1151–1152.

120. Lawn, A.M., and Meynell, E., 1970, Serotypes of sex pili, *J. Hygiene* 68:683–694.

121. Lawn, A.M., and Meynell, E., 1972, Antibody-stimulated increase in sex pili in R+ enterobacteria, *Nature* 235:441–442.

122. Lawn, A.M., and Meynell, E., 1975, Extrusion of sex pili by rapidly washed R+ *Escherichia coli*, *J. Gen. Microbiol.* 86:188–190.

123. Lawn, A.M., Meynell, E., and Cooke, M., 1971, Mixed infections with bacterial sex factors: sex pili of pure and mixed phenotype, *Ann. Institut Pasteur* 120:3–8.

124. Leipold, B., and Hofschneider, P.H., 1975, Isolation of an infectious RNA-A protein complex from the bacteriophage M12, *FEBS Lett.* 55:50.

125. Lin, T.C., Webster, R.E., and Konigsberg, W., 1980, Isolation and characterization of the C and D proteins coded by gene IX and gene VI in the filamentous bacteriophage f1 and fd, *J. Biol. Chem.* 255:10331–10337.

126. Loeb, T., 1960, Isolation of a bacteriophage specific for the F+ and Hfr mating types of *Escherichia coli* K-12, *Science* 131:932–933.

127. Loeb, T., and Zinder, N., 1961, A bacteriophage containing RNA, *Proc. Natl. Acad. Sci. USA* 47:282–285.

128. Lopez, J, and Webster, R.E., 1983, Morphogenesis of filamentous bacteriophage f1: orientation of extrusion and production of polyphage, *Virology* 127:177–193.

129. Luiten, R.G.M., Putterman, D.G., Schoenmakers, J.G.G., Konings, R.N.H., and Day, L.A., 1985, Nucleotide sequence of the genome of Pf3, and IncP-1 plasmid-specific filamentous bacteriophage of *Pseudomonas aeruginosa*, *J. Virol.* 56:268–276.

130. Maher, D., and Colleran, E., 1987, The environmental significance of thermosensitive plasmids of the H incompatibility group, in: *Bioenvironmental Systems* (D.L. Wise, ed.), CRC Press, Boca Raton, Florida, pp. 1–35.

131. Makowski, L., 1984, Structural diversity in filamentous bacteriophages, in: *The Structures of Biological Macromolecules and Assemblies, Vol. 1: The Viruses* (A. McPherson and F. Jurnak, eds.), Wiley, New York, pp. 203–253.

132. Marvin, D.A., 1989, Dynamics of telescoping *Inovirus*: a mechanism for assembly at membrane adhesions, *Int. J. Biol. Macromol.* 11:159–164.

133. Marvin, D.A., 1990, Model-building studies of *Inovirus*: genetic variations on a genetic theme, *Int. J. Biol. Macromol.* 12:125–138.

134. Marvin, D.A., and Folkhard, W., 1986, Structure of F-pili: reassessment of the symmetry, *J. Mol. Biol.* 191:299–300.

135. Marvin, D.A., and Hoffmann-Berling, H., 1963, Physical and chemical properties of two new small bacteriophages, *Nature* 219:485–486.

136. Meynell, E., 1978, Experiments with sex pili: an investigation of the characters and function of F-like and I-like sex pili based on their reactions with antibody and phage, in: *Pili* (D.E. Bradley, E. Raizen, P. Fives-Taylor, and J. Ou, eds.), International Conferences on Pili, Washington, DC, pp. 207–233.

137. Meynell, G.G., and Aufreiter, E., 1969, Selection of mutant bacterial sex factors determining altered sex pili, *J. Gen. Microbiol.* 59:429–431.

138. Meynell, G.G, and Lawn, A.M., 1968, Filamentous phages specific for the I sex factor, *Nature* 217:1184–1186.

139. Meynell, E., Meynell, G.G., and Datta, N., 1968, Phylogenetic relationships of drug resistance factors and other transmissible bacterial plasmids, *Bacteriol. Rev.* 32:35–83.

140. Meynell, E., Matthews, R.A., and Lawn, A.M., 1974, Antigenic differences between the ends and shafts of sex pili, *J. Gen. Microbiol.* 82:203–205.

141. Mindich, L., and Bamford, D.H., 1988, Lipid-containing bacteriophages, in: *The Bacteriophages*, Vol. 2 (R. Calendar, ed.), Plenum Press, New York, pp. 475–520.

142. Mindich, L., Sinclair, J.F., and Cohen, J., 1976, The morphogenesis of bacteriophage φ6: particles formed by nonsense mutants, *Virology* 75:224–231.

143. Miyake, T., Shiba, T., Sakurai, T, and Watanabe, I., 1969, Isolation and properties of two new RNA phages SP and FI, *Jpn. J. Microbiol.* 13:375–382.

144. Model, P., and Russel, M., 1988, Filamentous bacteriophage, in: *The Bacteriophages*, Vol. 2 (R. Calendar ed.), Plenum Press, New York, pp. 375–456.

145. Moore, D., Sowa, B.A., and Ippen-Ihler, K., 1982, A new activity in the Ftra operon which is required for F-pilin synthesis, *Mol. Gen. Genet.* 188:459–464.

146. Moore, R.J., and Krishnapillai, V., 1982, Tn7 and Tn*501* insertions into *Pseudomonas aeruginosa* plasmid R91-5: mapping of two transfer regions, *J. Bacteriol.* 149:276–283.

147. Moore, R.J., and Krishnapillai, V., 1982, Physical and genetic analysis of deletion mutants of plasmid R91-5 and the cloning of transfer genes in *Pseudomonas aeruginosa*, *J. Bacteriol.* 149:284–293.

148. Namba, K., and Stubbs, G., 1986, Structure of tobacco mosaic virus at 3.6 A resolution: implications for assembly, *Science* 231:1401–1406.

149. Novotny, C.P., and Fives-Taylor, P., 1974, Retraction of F pili, *J. Bacteriol.* 117:1306–1311.

150. Novotny, C.P. and Fives-Taylor, P., 1978, Effects of high temperature on *Escherichia coli* F pili, *J. Bacteriol.* 133:459–464.

151. Novotny, C.P., and Lavin, K., 1971, Some effects of temperature on the growth of F pili, *J. Bacteriol.* 107:671–682.

152. Novotny, C.P., Raizen, E., Knight, W.S., and Brinton, C.C., Jr., 1969, Functions of F pili in mating-pair formation and male bacteriophage infection studied by blending spectra and reappearance kinetics, *J. Bacteriol.* 98:1307–1319.

153. Nuttall, D., Maher, D., and Colleran, E., 1987, A method for the direct isolation of IncH plasmid-dependent bacteriophages, *Letts. Applied Microbiol.* 5:37–40.

154. O'Callaghan, R., Bradley, R., and Paranchych, W., 1973, The effect of M13 phage infection upon the F pili of *E. coli*, *Virology* 54:220–229.

155. O'Callaghan, R.J., Bundy, L., Bradley, R., and Paranchych, W., 1973, Unusual arsenate poisoning of the F pili of *Escherichia coli*, *J. Bacteriol.* 115:76–81.

156. Olsen, R.H., Siak, J., and Gray, R.H., 1974, Characteristics of PRD1, a plasmid-dependent broad host range DNA bacteriophage, *J. Virol.* 14:689–699.

157. Olsen, R.H., and Thomas, D.D., 1973, Characteristics and purification of PRR1, an RNA phage specific for the broad host range *Pseudomonas* R1822 drug resistance plasmid, *J. Virol.* 12:1560–1567.

158. Ou, J.T., and Anderson, T.F., 1970, Role of pili in bacterial conjugation, *J. Bacteriol.* 102:648–654.

159. Paranchych, W., 1966, Stages in phage R17 infection: the role of divalent cations, *Virology* 28:90–99.

160. Paranchych, W., 1975, Attachment, ejection and penetration stages of the RNA phage infectious process, in: *RNA Phages* (N. Zinder, ed.), Cold Spring Harbor, Cold Spring Harbor, New York, pp. 85–111.

161. Paranchych, W., and Frost, L.S., 1988, The physiology and biochemistry of pili, *Adv. Microbiol. Physiol.* 29:53–114.

162. Paranchych, W., and Graham, A.F., 1962, Isolation and properties of an RNA-containing bacteriophage, *J. Cell. Comp. Physiol.* 60:199–208.

163. Paranchych, W., Ainsworth, S.K., Dick, A.J., and Krahn, P.M., 1971, Stages in phage R17 infection. V. Phage eclipse and the role of F pili, *Virology* 45:615–628.

164. Peeters, B.P.H., Peters, R.M., Schoenmakers, J.G.G., and Konings, R.N.H., 1985, Nucleotide sequence and genetic organization of the genome of the N-specific filamentous bacteriophage IKe: comparison with the genome of the F-specific filamentous phage M13, fd and f1, *J. Mol. Biol.* 181:27–39.

165. Rasched, I., and Oberer, E., 1985, Ff coliphages: structural and functional relationships, *Microbiol. Rev.* 50:401–427.

166. Rees, C.E.D., Bradley, D.E., and Wilkins, B.M., 1987, Organization and regulation of the conjugation genes of IncI₁ plasmid CoIIb-P9, *Plasmid* 18:223–236.

167. Romantschuk, M., and Bamford, D.H., 1985, Function of pili in bacteriophage φ6 penetration, *J. Gen. Virol.* 66:2461–2469.

168. Russel, M., Whirlow, H., Sun, T.P., and Webster, R.E., 1988, Low-frequency infection of F-bacteria by transducing particles of filamentous bacteriophages, *J. Bacteriol.* 170:5312–5316.

169. Sasakawa, C., and Yoshikawa, M., 1978, Requirements for suppression of a *dnaG* mutation by an I-type plasmid, *J. Bacteriol.* 133:485–491.

170. Sastry, P.A., Finlay, B.B., Pasloske, B.L., Paranchych, W., Pearlstone, J.R., and Smillie, L.B., 1985, Comparative studies of the amino acid and nucleotide sequences of pilin derived from *Pseudomonas aeruginosa* PAK and PAO, *J. Bacteriol.* 164:571–577.

171. Schaller, H., Beck, E., and Takanami, M., 1978, Sequence and regulatory signals of the filamentous phage genome, in: *The Single-Stranded DNA Phages* (D.T. Denhardt, D. Dressler, and D.S. Ray, eds.), Cold Spring Harbor Laboratory, Cold Spring Harbor, New York, pp. 139–163.

172. Schandel, K.A., Maneewannakul, S., Ippen-Ihler, K., and Webster, R.E., 1987, A traC mutant that retains sensitivity to f1 bacteriophage but lacks F pili, *J. Bacteriol.* 169:3151–3159.

173. Schmidt, J.M., 1966, Observations on the adsorption of *Caulobacter* bacteriophages containing RNA, *J. Gen. Microbiol.* 45:347–352.

174. Schoulaker-Schwarz, R. and Engelberg-Kulka, H., 1983, Effect of an *Escherichia coli traD* (ts) mutation on MS2 RNA, *J. Gen. Virol.* 64:207–210.

175. Sharp, P.A., Cohen, S.N., and Davidson, N., 1973, Electron microscope heteroduplex studies of sequence relations among plasmids of *Escherichia coli*, *J. Mol. Biol.* 75:235–255.

176. Shiba, R., and Suzuki, Y., 1981, Localization of A protein in the RNA-A protein complex of RNA phage MS2, *Biochim. Biophys. Acta* 654:249.

177. Shiba, T., and Miyake, T,. 1975, New type of infectious complex of *E. coli* RNA phage, *Nature* 254: 157–159.

178. Silverman, P.M., and Valentine, R.C., 1969, The RNA injection step of bacteriophage f2 infection, *J. Gen. Virol.* 4:111–124.

179. Sirgel, F.A., Coetzee, J.N., Hedges, R.W., and Lecatsas, G., 1981, Page C-1: an IncC group; plasmid-specific phage, *J. Gen. Microbiol.* 122:155–160.

180. Smart, W., Sastry, P.A., Paranchych, W., and Singh, B., 1988, Mapping of the T-cell recognition sites of *Pseudomonas aeruginosa* PAK polar pili, *Infect. Immun.* 56:18–23.

181. Smith, J.M., Dowson, C.G., and Spratt, B.G., 1991, Localized sex in bacteria, *Nature* 349:29–31.

182. Stanisich, V., 1974, The properties and host range of male-specific bacteriophages of *Pseudomonas aeruginosa*, *J. Gen. Microbiol.* 84:332–342.

183. Stengele, I., Bross, P., Garces, X., Giray, J., and Rasched, I., 1990, Dissection of functional domains in phage fd adsorption protein: discrimination between attachment and penetration sites, *J. Mol. Biol.* 212: 143–149.

184. Stewart, G.J., and Carlson, C.A., 1986, The biology of natural transformation, *Ann. Rev. Microbiol.* 40:211–235.

185. Sun, T.P., and Webster, R.E., 1986, fii, a bacterial locus required for filamentous phage infection and its relation to colicin-tolerant tolA and tolB, *J. Bacteriol.* 165:107–115.

186. Sun, T.P., and Webster, R.E., 1987, Nucleotide sequence of a gene cluster involved in entry of E colicins and single-stranded DNA of infecting filamentous bacteriophages into *Escherichia coli*, *J. Bacteriol.* 169:2667–2674.

187. Taylor, D.E., 1989, General properties of resistance plasmids, in: *Handbook of Experimental Pharmacology*, Vol. 91 (L.E. Bryan, ed.), Springer-Verlag, Berlin, pp. 325–357.

188. Taylor, D.E., and Levine, J.G., 1980, Studies of the temperature-sensitive transfer and maintenance of H incompatibility group plasmids, *J. Gen. Microbiol.* 116:475–484.

189. Terawaki, Y., and Rownd, R., 1972, Replication of the R factor Rts1 in *Proteus mirabilis, J. Bacteriol.* 109:492–498.

190. Tomoeda, M., Inuzuka, M., and Date, T., 1975, Bacterial sex pili, *Prog. Biophys. Mol. Biol.* 30: 23–56.

191. Tschape, H., and Tietze, E., 1981, Genetische und molekulare grundlagen der plasmid-Spezies-Hypothese, *Biologisches Zentralblatt* 100:353–384.

192. Tschape, H., and Tietze, E., 1983, Characterization of conjugative plasmids belonging to a new incompatibility group (IncZ), *Zeitschrift fur Allgemeine Mikrobiologie* 23:393–401.

193. Valentine, C.R.I., and Kado, C.I., 1989, Molecular genetics of IncW plasmids, in: *Promiscuous Plasmids of Gram-Negative Bacteria* (C.M. Thomas, ed.), Academic Press, London, pp. 125–163.

194. Valentine, R.C., and Strand, M., 1965, Complexes of F-pili and RNA bacteriophage, *Science* 148:511.

195. Van Duin, J., 1988, Single-stranded RNA bacteriophages, in: *The Bacteriophages*, Vol. 1 (R. Calendar, ed.), Plenum Press, New York, pp. 117–167.

196. Van Wezenbeek, P.M.G.F., Hulsebos, T.J.M., and Schoenmakers, J.G.G., 1980, Nucleotide sequences of the filamentous bacteriophage M13 DNA genome; comparison with phage fd, *Gene* 11:129–148.

197. Vapnek, D., 1977, *Eco*RI, *Hind*III, and *Bam*HI cleavage map of R538-1, in: *DNA Insertion Elements, Plasmids, and Episomes* (A.I. Bukhari, J.A. Shapiro, and S.L. Adhya, eds.), Cold Spring Harbor Laboratory, Cold Spring Harbor, New York, p. 674.

198. Watanabe, I., Sakurai, T., Shiba, T., and Ohno, T., 1967, Isolation and grouping of RNA phages, *Proc. Jpn. Acad.* 43:204–209.

199. Watanbe, I., Miyake, T, Sakurai, T., Shiba, T,. and Ohno, T., 1967, Group characteristics of RNA phages, *Proc. Jpn. Acad.* 43:210–213.

200. Watts, T.H., Kay, C.M., and Paranchych, W., 1983, Spectral properties of three quaternary arrangements of *Pseudomonas* pilin, *Biochemistry* 22:3640–3646.

201. Webster, R.E., and Lopez, J., 1985, Structure and assembly of the Class I filamentous bacteriophage, in: *Virus Structure and Assembly* (S. Casjens, ed.), Jones and Bartlett, Boston, pp. 236–267.

202. Whiteley, M., and Taylor, D.E., 1983, Identification of DNA homologies among H incompatibility group plasmids by restriction enzyme digestion and southern transfer hybridization, *Antimicrob. Agents Chemother.* 24:194–200.

203. Wickner, W.T., and Lodish, H.F., 1985, Multiple mechanisms of protein insertion into and across membranes, *Science* 230;400–407.

204. Wilkins, B.M., 1975, Partial suppression of the phenotype of *Escherichia coli* K-12 *dnaG* mutants by some I-like conjugative plasmids, *J. Bacteriol.* 122:899–904.

205. Willetts, N.S., and Maule, J., 1986, Specificities of IncF plasmid conjugation genes, *Genet. Res.* 47:1–11.

206. Willetts, N.S., and Skurray, R., 1987, Structure and function of the F factor and mechanism of conjugation, in: *Escherichia coli and Salmonella typhimurium: Cellular and Molecular Biology* (F.C. Neidhardt, J.B. Ingraham, K.B. Low, B. Magasanik, M. Schaecter, and H.E. Umbarger, eds.), American Society for Microbiology, Washington, DC, pp. 1110–1133.

207. Willetts, N.S., Moore, P.M., and Paranchych, W., 1980, Variant pili produced by mutants of the Flac plasmid, *J. Gen. Microbiol.* 117:455–464.

208. Winans, S.C., and Walker, G.C., 1985, Conjugal transfer system of the IncN plasmid pKM101, *J. Bacteriol.* 161:402–410.

209. Womble, D.D., and Rownd, R.H., 1988, Genetic and physical map of plasmid NR1: comparison with other IncFII antibiotic resistance plasmids, *Microbiol. Rev.* 52:433–451.

210. Wong, K., and Paranchych, W., 1976, The preservation of the secondary structure of R17 RNA during penetration into host bacteria, *Virology* 73:476–488.

211. Wong, K., and Paranchych, W., 1976, The polarity of penetration of phage R17 RNA, *Virology* 73:489–497.

212. Worobec, E.A., Taneja, A.K., Hodges, R.S., and Paranchych, W., 1983, Localization of the major antigenic determinant of EDP208 pili at the N-terminus of the pilus protein, *J. Bacteriol.* 153:955–961.

213. Worobec, E.A., Paranchych, W., Parker, J.M., Taneja, A.K., and Hodges, R.S., 1985, Antigen-antibody interaction. The immunodominant region of EDP208 pili, *J. Biol. Chem.* 260:938–943.

214. Yamamoto, M., Kanegasaki, S., and Yoshikawa, M., 1980, Effects of temperature and energy inhibitors on complex formation between *Escherichia coli* male cells and filamentous phage fd, *J. Gen. Microbiol.* 119:87–93.
215. Yamamoto, M., Kanegasaki, S., and Yoshikawa, M., 1981, Role of membrane potential and ATP in complex formation between *Escherichia coli* male cells and filamentous fd, *J. Gen. Microbiol.* 123:343–349.
216. Zinder, N., 1963, Properties of a bacteriophage containing RNA, *Perspec. Virol.* 3:58–63.
217. Zinder, N.D., Valentine, R.C., Roger, M., and Stoekenius, W., 1963, f1, a rod-shaped male-specific bacteriophage that contains DNA, *Virology* 20:638–640.

Chapter 8

Plasmid Incompatibility and Replication Control

BARBARA LEWIS KITTELL and DONALD R. HELINSKI

1. Introduction

Not all plasmids can stably coexist in a bacterial host cell. Two genetically distinguishable plasmids that cannot be stably maintained within a particular host are designated as members of the same incompatibility (Inc) group (19). Generally, if two plasmids are members of the same incompatibility group, the introduction of one of the two plasmids by conjugation, transformation, or transduction into a cell carrying the other plasmid destabilizes the inheritance of the resident plasmid. The placement of plasmids in various Inc groups has proved useful in classifying plasmid elements and identifying plasmids that are genetically related. The phenomenon of plasmid incompatibility is a consequence of two plasmids sharing common elements responsible for plasmid maintenance, namely, replication control and/or partitioning systems (63). It is also dependent on the fact that at least for those plasmids examined, representing plasmids from both gram-positive and gram-negative bacteria, the selection of individual plasmid molecules for replication and partitioning is carried out randomly from a common pool of molecules (5, 63, 82, 84).

Plasmids vary considerably in size and copy number. However, in a particular bacterial host, a given plasmid is maintained at a defined copy number. The key to the existence of a plasmid in an extrachromosomal state at this defined copy number is its ability to regulate the frequency of initiation of replication from a specific origin sequence. As with bacterial and animal viruses, plasmids employ different strategies to assure their maintenance in the host cell at this characteristic copy number. The various mechanisms of regulation of plasmid copy number have in common the ability to correct stochastic fluctuations in copy

BARBARA LEWIS KITTELL and DONALD R. HELINSKI • Department of Biology and Center for Molecular Genetics, University of California, San Diego, La Jolla, California 92093-0634.
Bacterial Conjugation, edited by Don B. Clewell. Plenum Press, New York, 1993.

number that occur during cell division. In addition, many plasmids carry genetic information that specifies a *cis*-acting partitioning mechanism, further ensuring that each daughter cell receives at least one copy of the plasmid molecule after cell division. Mechanisms of replication and partitioning can function independent of each other. Interference with either the copy-number control mechanism or the partitioning mechanism of one plasmid element by a second genetically distinguishable element can result in incompatibility and, consequently, the loss of one or the other plasmid from individual cells during growth of the bacterial population. This chapter will concentrate on the role of common elements of replication control in plasmid incompatibility. Reviews (4, 62) have recently been published dealing with the structure of plasmid partitioning systems and incompatibility as a result of interaction between common components of these systems.

2. General Features of Replication Control

Basic replicons, consisting of an origin (or origins) of replication and one or more adjoining controlling elements, often are less than 3 kilobases (kb) in size. This compactness of replicon elements has greatly facilitated their isolation and the construction of miniplasmids that express all or many of the replication features of the intact plasmid, including incompatibility properties. The construction of miniplasmids has, in turn, greatly facilitated the analysis of the structure of replicons and partitioning systems. Origins of replication and elements involved in copy-number control have been identified and, in many instances, have been found to express incompatibility. In addition, the majority of replicons also carry a structural gene (*rep*) that codes for a protein that is required specifically for the initiation of plasmid replication. The *rep* gene product often acts *in trans* on an origin of replication, but in some cases it functions only in an *in cis* orientation. An increasing number of plasmids have been shown to contain more than one replicon within their overall structure (17). This is particularly the case for plasmids in the IncF group (6). Composite, multireplicon plasmids present a complication to the use of incompatibility grouping as a means of classifying plasmids with similar replicons, since plasmids carrying more than one replicon generally will not be destabilized by the presence of a second plasmid carrying only a single replicon that is homologous to any one of the replicons of the multireplicon plasmid (17).

A variety of plasmid replication control systems have been examined in considerable detail. In every case, one can identify a negative feedback control system that is capable of responding to changes in plasmid copy number during cell growth (44, 59, 63, 86). These systems must respond to either upward or downward fluctuations in plasmid copy number and establish a steady-state level of plasmid copy number that is a hereditary characteristic of the particular plasmid. This feature of the replication control systems not only is critical to maintenance of the plasmid during exponential growth of the bacterial population but also functions to increase plasmid copy number over a relatively short period of time under circumstances where a plasmid enters a cell as a single copy via the process of conjugation, transformation, or transduction. Studies on a variety of plasmid elements have elucidated two basic schemes that plasmids use to negatively control their own replication. In one, plasmids encode a small, diffusible RNA molecule that acts as an antisense transcript to negatively regulate some aspect of plasmid replication (Figure 8.1). These small regulatory

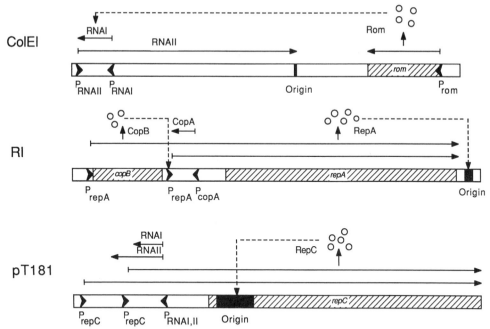

FIGURE 8.1. Replication control of RNA-regulated plasmids. The bars represent the replication regions of ColE1, R1, and pT181, respectively. The solid lines above the bars are RNA transcripts; the direction of transcription is designated by the arrowheads. The hatched regions within the bars are protein-coding regions, and the proteins produced are depicted as circles. The dotted lines delineate the site of action for each protein. The details of replication control for each plasmid are described in the text.

RNA molecules are transcribed in a region overlapping a transcript that is either directly involved or whose product is involved in replication initiation. Because transcription of the regulatory RNA occurs in the direction opposite to transcription of the RNA required for replication initiation, the two RNAs are homologous and can hybridize. The formation of these RNA:RNA hybrids can then effectively regulate plasmid copy number. The replication control mechanisms of plasmids regulated by small RNAs differ, depending on how the actual target RNA functions to bring about replication initiation (i.e., is it a replication primer itself, or does it code for a replication initiation protein?) and the means by which the formation of the RNA:RNA hybrid prevents replication. However, in every case examined, the regulatory RNA is constitutively transcribed and is present at a level that is proportional to the plasmid copy number.

Many plasmids that do not regulate their replication using small RNAs have replication origins with several features in common that appear to be involved in replication control (Figure 8.2). These plasmids contain a series of direct repeats, called iterons, that are located at the origin and are necessary both for replication initiation and for copy-number control. Members of this group of plasmids also code for a *trans*-acting initiation protein. Selected plasmids using each type of replication control will be discussed below in an attempt to understand how the different elements function mechanistically to bring about a stable copy number.

A.

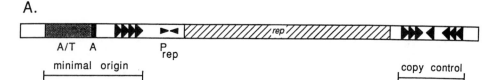

A/T   A

P rep

minimal origin

copy control

B.

Iteron-containing plasmid	# Iterons in minimal origin	#Iterons in copy control	#bp/ iteron	Rep protein size (kdal) [a]	refs
E. coli					
Narrow host range					
pSC101	3	0	18	37	3,16
R6K	7	0	22	35	90,91
F	4	5	19	29	58,99
P1	5	9	19	32	2
Rts1	3	9	18-24	33	37,38,67
pColV-K30	5	5	18	39	74
pCU1	13	ND	37	ND	43
λdv	4	0	19	34	29,110
Broad host range					
R1162/RSF1010	3.5	0	20	31	31,47
RK2	5	3	17	44/33	87,89
pSa	2	0	13	35	93
Other bacteria					
pCTTI (C. trachomatis)	4	ND	22	30	88
pCl305 (L.lactis)	3.5	ND	22	46	33
pFA3 (N. gonorrhoeae)	4	ND	22	39	28
pIP404 (C. perfringens)	14(6)	ND	8(16)	49	26

ND- not determined

[a]Some of the Rep proteins are identified only on the basis of the position of their genes in the replicon, and their sizes are determined from their amino acid sequences derived from nucleotide sequence analysis.

FIGURE 8.2. Replication regions of iteron-containing plasmids. A. Composite structure of iteron-containing replicons. This figure represents many of the structural features found in the replication regions of iteron-containing plasmids; it does not depict any specific origin, and the replication regions of individual plasmids will vary. Within the minimal origin, there is an A/T rich region (the shaded box), one or more DnaA boxes (the solid box), and a cluster of iterons (arrowheads). The *rep* gene sequence is often located near the origin (lined box) and is often preceded by a pair of inverted repeats that serve as sites for autoregulation. The copy control region is not essential for replication but contains a cluster of iterons, often irregularly spaced, that are similar in sequence to those located at the origin. B. Characteristics of specific iteron-containing plasmids. The number of iterons in the minimal origin and in the copy control region is specified for each plasmid as well as the number of bp in each iteron. An additional number in () for pIP404 indicates there are 14 iterons 8 bp in length and 6 iterons 16 bp in length.

3. Replication Control of RNA-Regulated Plasmids

3.1. Co1EI-Type Plasmids

The *Escherichia coli* ColE1-type plasmids (related plasmids include p15A, pMB1, RSF1030, NTP1, NTP16, ColD13, pBR322) are perhaps the best understood group of RNA-regulated plasmids (for reviews see 44, 63, 80). In this group, the regulatory RNA interacts with an RNA primer that is necessary for replication initiation (Figure 8.1). For example, in the case of ColE1, replication requires the synthesis of a 700-base preprimer RNA (designated RNAII) whose transcription is initiated 555 bases upstream of the origin of replication. The 3' end of RNAII forms a hybrid with the DNA template near the replication origin in a process referred to as coupling. Coupling is dependent on the formation of an appropriate secondary structure at the 5' end of RNAII that allows a stem loop to form just 3' to the origin. After coupling has occurred, the RNA-DNA hybrid is cleaved by RNaseH at the origin, leaving a free 3' hydroxyl group to serve as a primer for the initiation of DNA synthesis by DNA polymerase I (PolI). After replication has proceeded, RNaseH then removes the RNA primer remaining at the origin. ColE1 has two alternative modes of replication that allow replication to proceed in the absence of either RNaseH alone or both RNaseH and PolI (18); however, both of these modes also require RNAII in the correct structural form to act as a replication primer.

The small RNA molecule that controls ColE1 replication is a 108-base antisense transcript, designated RNAI, that initiates 445 bases upstream from the replication origin and proceeds in a direction opposite to RNAII (105, 107). It is, therefore, homologous to the 5' end of RNAII. When RNAI molecules hybridize to RNAII, RNAII is unable to assume the secondary structure that is required for it to interact at the origin and form an RNA:DNA hybrid (50, 51). RNAI concentration, therefore, determines replication initiation rate and plasmid copy number by controlling the number of RNAII molecules that will be able to achieve a replication-proficient conformation. Levels of both RNAs in vivo have recently been measured and are reported to be 400 molecules per cell of RNAI and three molecules per cell of RNAII (10). Both RNAs have a relatively short half-life of approximately 2 minutes, thus allowing rapid adjustments to copy-number deviation (10).

The mechanism of RNAI:RNAII hybrid formation has been studied in detail (101, 102, 104). Both RNAs exhibit significant secondary structure, consisting of a series of imperfect stems with single-stranded loops at the ends. Properly folded RNAI includes three such stems with single-stranded loops at the ends and a 5' single-stranded tail. The initial interaction between RNAI and RNAII occurs at the loops on the ends of the stems and is called "kissing." This is followed by pairing at the 5' end of RNAI (nucleation) and, finally, progressive pairing along the entire length of the RNAI molecule. The initial kissing reactions are unstable and reversible; however, as hybrid formation proceeds, each step produces a more stable complex until finally a completely paired duplex is irreversibly formed.

ColE1 also encodes a protein, called Rom (or Rop) for RNAI modulator, that plays a role in plasmid copy number. This protein is not essential for replication; however, in its absence, plasmid copy number is increased two- to fivefold (92, 112). Rom is a 63 amino acid, dimeric protein that affects copy number by enhancing RNAI:RNAII hybrid formation (12, 34, 103, 106). Detailed analyses of the kinetics of complex formation have demonstrated that Rom interacts with the early, relatively unstable, kissing intermediate (102, 106, 108). It has been shown to bind to both RNA molecules simultaneously, thereby increasing the stability of that complex about 100-fold. This increased stability results in more of the initial RNAI:RNAII interactions proceeding to form complete hybrids and, therefore, causes a decreased copy number.

Because RNAI is a diffusible, negative replication regulator, it is the primary incompatibility determinant of the ColE1 plasmid. Consequently, a compatible plasmid with a DNA sequence insert that produces RNAI cannot coexist with a ColE1 replicon dependent on the same RNAI for copy-number control (94). Many single-base-pair change mutations in RNAI have been isolated, and their characterization has confirmed that RNAI has the central role in determining copy number and incompatibility properties (45, 105, reviewed in 63). For example, single-base changes in RNAI with the concomitant change in the primer, RNAII, can alter the specificity of the regulatory RNAI molecule and result in the creation of a new incompatibility group (45). The Rom protein, although involved in copy-number control, is not an incompatibility determinant because it does not function independently in the control of copy number. It does, however, enhance the incompatibility exerted by RNAI (68, 106). The ColE1 origin of replication, because it does not titrate a diffusible inhibitor or activator molecule, also is not an incompatibility determinant (32).

3.2. IncFII Plasmids

In another group of RNA-regulated plasmids maintained in *E. coli*, the regulatory RNA controls the synthesis of a *cis*-acting replication initiation protein. This group includes the IncFII plasmids whose prototypes are R1, NR1, and R6-5 (for reviews see 60, 63, 85, 116). These plasmids are generally large (i.e., approximately 100 kb) and exist at one to two copies per cell. Their replication origin region has been narrowed to a 2.5-kb region that includes the coding sequence for a replication initiation protein called RepA (Figure 8.1). RepA is a *cis*-acting protein that is present at levels rate limiting for plasmid replication. Therefore, control of plasmid replication is at the level of control of RepA expression. Transcription of the *repA* gene is controlled by two promoters. The upstream promoter is constitutive, and the downstream promoter is repressed by a small protein known as the CopB protein. The *copB* gene is located within the RepA transcript between the upstream promoter and the CopB-regulated promoter. At a normal plasmid copy number, the regulated promoter is typically repressed. Deletion of the *copB* gene leads to a plasmid copy-number increase of 8- to 10-fold. The presence of the CopB-repressible promoter as well as the constitutive promoter for the *repA* gene allows for a rapid increase in *repA* transcription and, subsequently, in plasmid copy number when the copy number falls too low, an adjustment that may be particularly important to assure that low-copy-number plasmids are not lost from the cell.

RepA expression is controlled posttranscriptionally by a 90-base antisense RNA called CopA (60). CopA is transcribed within the leader region of the RepA mRNA in the opposite direction of RepA mRNA transcription. CopA is, therefore, homologous to the RepA mRNA leader region, which is referred to as CopT or CopA target, and interacts with it to form an RNA:RNA duplex (60). CopA is synthesized at levels higher than RepA mRNA, and the concentration of CopA is inversely proportional to the expression of RepA as well as being proportional to plasmid copy number.

Both CopA and CopT exhibit significant secondary structure and, as is the case with RNAI:RNAII duplex formation in the ColE1 plasmid, formation of a CopA:CopT hybrid requires several steps (77, 78). These steps include the formation of a kissing intermediate that involves loop II of CopA as well as the upper parts of the corresponding stems followed by the formation of a fully hybridized duplex. Full duplex formation requires the formation of the kissing intermediate, which is the rate-limiting step, and a single-stranded stretch of nucleotides at least 10 bases in length 5' to the stem-loop II in CopA.

It is still not clear mechanistically how CopA:CopT duplex formation regulates RepA expression. The initial hypothesis based on computer modeling suggested that CopA:CopT duplex formation resulted in the formation of an alternate structure near the start of RepA translation that sequestered the Shine-Dalgarno sequence within a stem, thus blocking translation (85). Attempts to find evidence supporting structural changes in the RNA near the start of translation after duplex formation has occurred have failed, however (69). More recently, Blomberg et al (7) have demonstrated that RNaseIII cleavage of the CopA:CopT duplex is involved in the regulation of RepA expression. They were able to isolate processed RNAs both in vivo and in vitro that were dependent on hybrid formation and the presence of RNaseIII and found that in RNaseIII mutants, *repA-lacZ* fusions showed a 10-fold increase in expression. Although RNaseIII processing of the CopA:CopT duplex may indeed be involved in regulation of RepA expression, the mechanism of this is still unclear. It is postulated that cleavage of the duplex may substantially increase the rate at which the remaining RepA mRNA decays.

CopA, like RNAI, is the major incompatibility determinant of the IncFII plasmids (54). Mutations in CopA that result in the inability of CopA to interact with CopT cause a loss of incompatibility (8, reviewed in 63). The CopA genes of the IncFII plasmids R1, NR1, and R6 are identical (116) and, therefore, these plasmids cannot stably coexist in an *E. coli* cell. As shown for the ColE1 plasmids, changes in base positions in CopA that are responsible for CopA:CopT interactions presumably would shift the plasmid to a new incompatibility group. A divergence in the sequence of CopA most likely accounts for the compatibility observed between the IncFII plasmids and several naturally occurring plasmids that apparently have a similar structure and mechanism of regulation of plasmid replication (113). CopB is not involved in incompatibility by itself because the CopB repressible promoter is usually fully repressed and, therefore, extra CopB has little effect (55). If the RepA constitutive promoter is deleted, however, then CopB can become involved in incompatibility (61). Because the RepA protein is *cis*-acting, the IncFII origin is not an incompatibility determinant (97).

3.3. pT181-Type Plasmids

Major progress has been made, particularly recently, in our understanding of the regulation of plasmid replication in gram-positive bacteria (30, 63). The best studied example is plasmid pT181 of *Staphylococcus aureus*, representing a third group of RNA-regulated plasmids in which the RNA regulates the synthesis of a *trans*-acting replication inhibition protein (27, for reviews, see 30, 63). Plasmid pT181 is a 4.4-kb plasmid maintained at a copy number of 22 plasmids per cell. This plasmid replicates via a rolling-circle mode of replication, and single-stranded intermediates can be isolated. Plasmid replication requires a replication initiation protein, RepC, that is rate limiting for replication (Figure 8.1). Initiation involves RepC binding to a cruciform structure at the replication origin, followed by the nicking of the DNA template, the covalent linkage of RepC to the DNA, and finally the formation of an initiation complex with host proteins.

RepC expression is controlled by two antisense RNAs (RNAI and RNAII) that are homologous to the leader sequence of the RepC mRNA (64, 66). These two antisense RNAs have a common 5′ end and different 3′ ends, and both are capable of forming an RNA:RNA hybrid with the RepC mRNA. The details of hybrid formation have not yet been determined. However, the formation of this RNA:RNA hybrid has recently been shown to induce the formation of a transcriptional terminator hairpin just 5′ to the RepC translational start (65). Terminated or attenuated RNAs can be detected in vivo at levels proportional to antisense RNA levels. Furthermore, fusions have demonstrated that transcriptional termination can account for all of the inhibitory activity of the antisense RNAs and that all but 3% of the primary transcripts are terminated. Therefore, in the case of pT181, it appears that RNA:RNA hybrid formation functions mechanistically like a transcriptional attenuator to regulate the expression of RepC.

The antisense RNAs are, as expected, incompatibility determinants and, when cloned, exert strong incompatibility against pT181 replicons (64). Mutations in the antisense RNA sequence that alter incompatibility have also been identified (11, 35). Unlike ColE1 and IncFII plasmids, however, the pT181 origin is also an incompatibility determinant (36, 64). When cloned alone, the origin expresses strong incompatibility against other pT181 replicons because it can effectively bind and titrate the rate-limiting, *trans*-acting RepC protein. In an interesting comparative analysis of four naturally occurring staphylococcal plasmids that are related to plasmid pT181, it was observed that each of the plasmids showed considerable similarity in the nucleotide sequence of the basic elements of their replicons, in their copy control mechanism, and in their overall structural organization (81). Despite these similarities in sequence of the antisense RNAs and origin regions of the pT181 family of plasmids, there is sufficient sequence divergence to place each of these plasmids in a separate incompatibility group.

4. Replication Control of Iteron-Containing Plasmids

4.1. Structural Features of the Replication Origin Region

The iteron-containing plasmids are a large, diverse group of plasmids that, as previously mentioned, have a number of features in common at their replication origins (Figure 8.2A). These include one or more clusters of direct repeats, or iterons, the coding sequence for a *trans*-acting replication initiation protein, one or more consensus DnaA binding sites, and an A + T rich region adjacent to the iterons (9, 24). One iteron cluster is always essential *in cis* for replication and contains at least three iterons, in the case of pSC101 (16), and up to seven iterons, as seen with R6K (90). These iterons range in size from 17 to 24 base pairs (bp) and are regularly spaced. Many iteron-containing plasmids, such as P1 (2), F (58), RK2 (89), and Rts1 (37, 67), have a second cluster of iterons that is nonessential for replication but plays a role in copy-number control. Deletion of this nonessential cluster of iterons leads to a substantial increase in plasmid copy number and in some cases destabilization of the plasmid. Nonessential iteron clusters also vary in size among the different plasmids and have been shown to contain from three to nine iterons. Both the orientation of these iterons within the cluster and the spacing between the iterons in these nonessential iteron clusters are more variable than is seen with the iterons that are required for replication. Proper spacing of iterons at the origin of replication may be essential for orienting the iterons on one face of the helix for binding to the plasmid-specified replication initiation protein.

The gene for the *trans*-acting replication initiation protein is usually located near the replication origin (24). These proteins are generally 30 to 40 kilodaltons (kD) in size and, in the cases thus far examined, exist as dimers in solution (1, 22, 49). The expression of the Rep protein is often autoregulated, as is the case with the Rep proteins of P1 (14), R6K (23), F (83), pSC101 (114, 117), and Rts1 (96), or under plasmid-encoded transcriptional control, as seen in RK2 and RSF1010 (R1162) (reviewed in 44). It has been demonstrated that these proteins initiate replication by binding to the iterons within the origin of replication (1, 22, 39, 48, 75, 79, 98). Autoregulation is achieved by binding to similar full or partial iteron sequences located near the start of transcription of the *rep* genes (39, 40, 98, 114). There is evidence that, as well as being positive activators of replication, some of the Rep proteins may also play a negative role in controlling plasmid replication. Overexpression *in trans* of the π replication protein (25) and the RepA protein of P1 (70) in *E. coli* results in an inhibition of R6K and P1 replication, respectively. The physiological significance of this negative effect of excess levels of these replication proteins is unclear. Other support for a negative activity for plasmid-specified replication proteins comes from the isolation of mutations in the R6K π protein (25) and the RK2 TrfA proteins (20) that result in a substantially elevated and specific increase in copy number. This copy-up mutant phenotype in the case of several of the mutants can be fully complemented by low levels of wild-type protein. As described below, models have been proposed that require both positive and negative activities of the plasmid initiation protein for the control of plasmid replication.

It has been known for some time that the iteron sequences both at the origin and in the copy control regions are incompatibility determinants that are involved in the negative control of replication of the iteron-containing plasmids. It was initially demonstrated for the plasmid F (100) that a cluster of five 22-bp iterons, when cloned on a compatible plasmid, exerted incompatibility against an F replicon. This observation has now been demonstrated for a number of iteron-containing plasmids (2, 37, 46, 52, 76, 111, 118). Furthermore, with the possible exception of plasmid P1, it has been demonstrated that the severity of the incompatibility phenotype correlates directly with the number of iterons present on the normally compatible plasmid.

4.2. Titration Model of Regulation

Although it is well established that the iterons are incompatibility determinants, the mechanism(s) by which the iterons exert replication control is not fully understood. It was initially postulated (111) that the function of the iterons was to titrate a rate-limiting initiation protein (Figure 8.3). Early work supported this theory when it was demonstrated that the replication initiation proteins did, indeed, bind to the iterons and that deleting various numbers of nonessential iterons led to a copy-number increase that was proportional to the number of iterons deleted (71). In the case of R1162, it was further shown in vivo that increasing the level of the *rep* gene products can increase plasmid copy number and can also overcome the incompatibility exerted by the iterons (41), lending further support to the titration model. In the case of plasmids R6K and RK2, however, the provision of excess plasmid-specified replication protein does not affect the incompatibility expressed by iterons (42, 53).

As more information has been obtained about the replication control of other iteron-containing plasmids, it appears that the titration model in its simplest form is no longer sufficient to explain a number of observations made with several plasmid systems. For instance, it is now known, as discussed earlier, that many of the replication initiation proteins are autoregulated, while certain other plasmids (e.g., RK2) carry regulatory elements outside of the basic replicon region that regulate the level of expression of the plasmid-specified replication initiation proteins. An autoregulated protein is derepressed for its own synthesis when its levels become too low, making simple titration by iterons virtually impossible. It has also been demonstrated for plasmids R6K (25), P1 (70), RK2 (21), and Rts1 (95) that overexpression of the replication protein does not lead to a parallel increase in plasmid copy number. Although a small increase can be observed with overexpression of the initiation protein, the increases in copy number that would have been predicted by the titration model simply were not achieved. For example, increasing the level of TrfA proteins 170-fold above the normal *E. coli* level increased the copy number of an intact RK2 plasmid or a minimal RK2-origin plasmid only 30%. In the case of the R6K

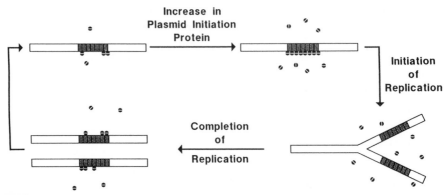

FIGURE 8.3. The titration model for regulation of plasmid copy number. Details of this model are discussed in the text. The bar represents a plasmid replicon, and the shaded regions within the bar represent iterons. Plasmid-encoded initiation protein molecules, depicted by filled circles, bind to the iterons. When sufficient replication protein, generally in the form of dimers, is present to saturate the iteron-binding sites, replication initiates. Duplication of the iteron sequences results in insufficient replication initiation protein to associate with a full set of iterons at the origin. Replication cannot occur again until protein levels are increased to the level required for fully loading a complete set of origin iterons.

plasmid, dropping the level of the π protein to 1% of the normal *E. coli* level had no effect on plasmid copy number. Furthermore, in the case of R6K, it has been demonstrated, using antibodies against the π protein, that this protein is present at a level 10,000 times what would be necessary to saturate the repeats. Finally, as stated previously, in some cases, including R6K, P1, and Rts1, high levels of replication protein were actually inhibitory to replication rather than stimulatory.

To accommodate these observations, several modifications to the titration model have been proposed. One such modification is that the iteron binding sites in the replication origin have a different affinity for the Rep protein than do the iterons at the start of *rep* gene transcription (13). Necessarily, the autoregulatory site would have the highest affinity for Rep protein; the copy-number control region, if present, would have a somewhat lower affinity; and the iterons at the replication origin would have the lowest affinity. Masson and Ray (48) presented in vitro binding results that suggest the RepE protein of the F plasmid binds more strongly to the *repE* promoter region than to the iterons within *ori2*, lending support to the proposal that different iteron binding sites may have differing affinities for the Rep protein. However, further evidence to support this model has not been forthcoming.

A second modification of the titration model was initially proposed by Trawick and Kline (109), who suggested that the Rep protein can exist in two forms. In one form, the Rep protein functions as an autorepressor to limit the amount of Rep protein synthesized. A small amount of Rep protein, however, would be irreversibly converted from the auto-repressor form to a form that would be active for replication initiation. This form would have a high affinity for iterons at the origin and, therefore, would be subject to titration by excess iterons. In this fashion, the Rep protein functions as both a positive and negative regulator of replication. This model allows for the autoregulatory nature of most Rep proteins. The observation that overexpression of the Rep protein does not lead to increases in copy number could be explained by proposing that only a fixed amount of protein is converted to an active form, regardless of how much is overproduced. The finding of Rep protein mutations that led to increased plasmid copy numbers was used to support this hypothesis by demonstrating that the Rep protein has a negative as well as a positive role in replication. Although the two-form model has certain attractive features, there is no direct evidence for two physically distinct or functionally distinguishable forms of Rep protein. Recently, Wickner et al (115) have demonstrated that the association of RepA from plasmid P1 with the host proteins DnaJ and DnaK, in a temperature- and ATP-dependent manner, can enhance the affinity of RepA for the origin iteron binding sites approximately 100-fold. They suggest that this interaction facilitates some change in the protein that enhances its binding capabilities, although an actual alteration in the RepA protein itself was not demonstrated in this study.

4.3. Coupling Model of Regulation

An alternative model to explain how iterons regulate plasmid copy number in plasmid P1 was proposed by Pal and Chattoraj (70) and Chattoraj et al (15). They found that both the origin and the copy control region iterons could compete with the RepA promoter autoregulation site for RepA binding, suggesting that there were not two different forms of RepA. Furthermore, when the copy control region (*incA*) was present *in cis* to the origin, overexpression of RepA decreased plasmid copy number. When *incA* was present *in trans* to the origin, P1 plasmid replication was inhibited, and this inhibition was insensitive to RepA concentration. When examining RepA protein:DNA complexes by electron microscopy, they observed protein-mediated DNA looping between the *incA* region and the origin. They proposed that the RepA-mediated, *incA*:origin looping can result in steric hindrance and effectively block replication initiation. This physical interaction between *incA* and the origin was observed intramolecularly and intermolecularly. The steric hindrance model suggests that origins that were looped to *incA* or bound to one another by RepA would be inactive for replication and that an equilibrium would exist between inactive and active origins that would result in a constant copy number.

A similar model, termed the handcuffing, or coupling, model (Figure 8.4), was independently put forth by McEachern et al (53) to explain how the R6K iterons and their initiation protein, π, function to control plasmid replication. This model proposes that π binds to the origin iterons and that the protein-bound origin undergoes one of two possible fates, depending on the concentration of π-bound iterons present in the cell. When the concentration is low, replication is initiated at this origin after association with host replication proteins. The process of initiation continues until the normal copy number is achieved. When the plasmid copy number is at or above the normal level, then the

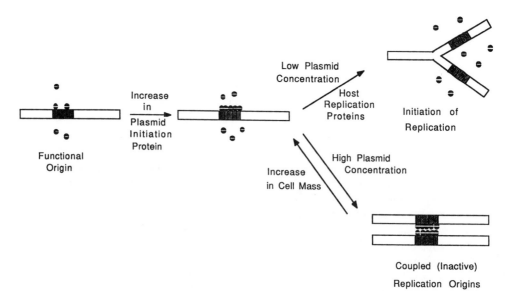

FIGURE 8.4. The coupling model for regulation of plasmid copy number. Details of this model are discussed in the text. The bar represents a plasmid replicon, and the shaded regions within the bar represent iterons. As plasmid initiation protein (depicted by the filled circles) concentration increases, the iterons become saturated with bound protein (shown as a dimer). If the plasmid (origin iteron) concentration is low, host replication proteins can bind to the iteron-replication protein complex at the origin and replication is initiated. If plasmid concentration is at or above the plasmid's copy number, two protein-bound iteron molecules interact to form a coupled origin complex that is unable to replicate. Increasing the cell mass lowers the concentration of the coupled origin complexes, resulting in dissociation of the complexes. This dissociation releases the initiation protein-bound iterons for association with host replication proteins and the initiation of replication.

concentration of π-bound iterons is sufficiently high to result in binding of virtually all of the protein-bound iterons to each other. The coupled, or handcuffed, origins are inactive; this is, they are not accessible to host replication proteins required to initiate replication. This plasmid-sensing mechanism involving iterons and the plasmid Rep protein relies on direct communication between plasmid molecules. This communication between plasmid molecules determines whether a plasmid will associate with host replication proteins and, therefore, be duplicated, or whether the plasmid is blocked for replication until the plasmid (iteron) concentration is sufficiently low to release the origins from their coupled state.

There are both genetic and biochemical data supporting the handcuffing model for R6K. First, as discussed previously, increasing π levels does not increase plasmid copy number, nor does it suppress iteron-mediated incompatibility. This is because it is not the absolute π level that is important in plasmid regulation in this model but rather the level of protein-bound iterons. Origin mutations that weaken incompatibility also weaken π binding (53), consistent with the central role of protein-bound iterons in the regulation of replication. Furthermore, it was shown that addition of a second iteron cluster to a functional R6K replicon *in cis* can affect the relative copy number specifically of that origin in a manner that is dependent on both the position and the orientation of the extra iterons, implying that a physical interaction between groups of iterons is occurring. According to the coupling model, copy-up π proteins are less able than the wild-type protein to handcuff origins together. Evidence supporting this interpretation includes the fact that these copy-up proteins do not inhibit replication of an R6K origin when overexpressed, as observed for the wild-type protein, and that the copy-up mutant proteins can suppress the altered replication phenotype seen with the extra iteron plasmids. Biochemical evidence in support of the handcuffing model for R6K includes electron microscopy analyses demonstrating π-mediated coupling between two origin fragments (53); an in vitro ligation assay that demonstrates that two origin fragments have an increased likelihood to be ligated together when the π protein is bound to the origin (53); and an in vitro catenation assay demonstration the ability of the π protein to couple supercoiled DNA molecules containing R6K origin iterons (56).

Recently, a coupling model for the replication control of the broad host range plasmid RK2 that is similar to the handcuffing model put forth for R6K has been proposed (42). In the case of RK2, the replication protein TrfA binds to the origin iterons. After binding to TrfA, the origin may associate with host proteins, leading to subsequent replication initiation as proposed for the R6K system, or may associate with a second, TrfA-bound origin in a coupling reaction that results in a block in replication initiation. Unlike the R6K replication protein, overexpression of TrfA 170-fold in vivo does not result in a reduction of RK2 plasmid copy number (21).

Genetic evidence supporting the coupling model of replication control for RK2 is similar to that obtained for R6K and includes the observation that increasing the level of TrfA above the normal concentration of TrfA in *E. coli* does not result in a concomitant increase in plasmid copy number (21). The properties of copy-up mutants of the TrfA protein, like those of the π protein of R6K, also support a negative role for TrfA in controlling copy number in addition to its positive role in the initiation of plasmid replication (20).

An in vitro replication system for RK2 has been used to obtain further evidence for the coupling mechanism of plasmid copy-number control (42). Supercoiled pUC19 DNA containing RK2 origin iterons was shown to inhibit the replication in vitro of an RK2 template. This inhibition was not relieved on the addition of excess amounts of purified TrfA protein, suggesting that the iterons were not merely titrating active Rep protein. If the two-form Rep protein titration model was correct and two forms of TrfA existed in the purified protein preparation, one might expect that the addition of excess active TrfA protein would eventually allow iteron inhibition to be overcome, but this was not the observed result. Furthermore, when a copy-up mutant protein was substituted for wild-type protein in the in vitro replication system, no increase in replication kinetics was observed; however, the RK2 replicon was no longer sensitive to iteron inhibition *in trans*. These data are consistent with the predictions of the coupling model that TrfA-bound iterons should be able to couple with active RK2 origins, thus preventing their replication, and that copy-up TrfA proteins are defective in this iteron coupling reaction and, therefore, are insensitive to iteron inhibition.

As summarized in Figure 8.2B, a number of plasmid elements from a diverse set of bacteria contain, as the major structural feature of their replicons, iterons at the origin of replication and a structural gene that encodes a replication protein. For the plasmids R6K, P1, and RK2, evidence has been obtained in support of a replication initiation control mechanism that involves sensing of plasmid copy number by the formation of coupled origin sequences mediated specifically by the plasmid initiation protein. While both genetic and biochemical observations have been made that are consistent with this mechanism, more direct proof must be obtained before this mechanism can be established for the control of plasmid copy number for either one or all three of these plasmids. Furthermore, it is likely that not all of the iteron-containing plasmids will satisfy the basic predictions of this model. In fact, it has been shown that the control of initiation of the iteron-containing plasmid RSF1010 is probably a function of the concentration of the plasmid encoded Rep protein (41). In this case, some form of a titration model (Figure 8.3) of control may best satisfy the observations made on the regulation of initiation of plasmid replication. Even in the cases of R6K (72, 73) and P1 (57), evidence has been presented for replication control elements in addition to iterons. It is not clear at this time, however, that any of these elements other than the repeated sequences and the replication initiation protein are significant under normal physiological conditions in the control of copy number for these two plasmids. Nevertheless, much remains to be done before we will fully understand the nature of all of the plasmid factors controlling plasmid copy number and the mechanisms responsible for their interactions.

5. Concluding Remarks

Analyses of the replication control systems for a variety of plasmid elements in both gram-positive and gram-negative bacteria have revealed highly ordered structural interactions between RNA (repressor) and RNA (target) molecules (in some cases facilitated by plasmid-specified proteins), or nucleoprotein complexes involving a plasmid encoded replication initiation protein and direct nucleotide sequence repeats. In each case, it is the concentration of at least one of these components involved in these complex structures that determines the frequency of initiation of plasmid replication. An imbalance as a result of a mutational event in the gene encoding for the regulatory element or due to the introduction into a cell of additional copies of the regulatory gene on an incoming plasmid can result in a loss of the resident plasmid, or an incompatibility phenotype. Interestingly, many naturally occurring plasmids have similar replication control mechanisms and/or are similar in the composition of their regulatory components; they are, therefore, members of the same incompatibility group. It is likely that the phenomenon of incompatibility is operative in natural environments, thus placing limitations on plasmid combinations stably maintained in gram-positive and gram-negative bacteria. Plasmids that contain more than one functional replicon presumably are less restricted in this regard. It is clear that the mechanisms of incompatibility depend on the formation of highly ordered structures involving specific nucleic acid-nucleic acid and/or nucleoprotein complexes. A more detailed understanding of these structures and the factors influencing their formation is necessary for a better understanding of the control of plasmid copy number and stable maintenance.

References

1. Abeles, A., 1986, P1 plasmid replication: purification and DNA-binding activity of the replication protein RepA, *J. Biol. Chem.* 261:3548–3555.
2. Abeles, A.L., Snyder, K.M., and Chattoraj, D.K., 1984, P1 plasmid replication: replicon structure, *J. Mol. Biol.* 173:307–324.
3. Armstrong, K.A., Acosta, R., Ledner, E., Machida, Y., Pancotto, M., McCormick, M., Ohtsubo, H., and Ohtsubo, E., 1984, A 37 kilodalton plasmid-encoded protein is required for replication and copy number control in pSC101 and its Ts derivative pHS1, *J. Mol. Biol.* 175:331–347.
4. Austin, S., and Nordstrom, K., 1990, Partition-mediated incompatibility of bacterial plasmids, *Cell* 60: 351–354.
5. Bazaral, M, and Helinski, D.R., 1970, Replication of a bacterial plasmid and an episome in *Escherichia coli*, *Biochemistry* 9:399–406.
6. Berquist, P.L., Saadi, S., and Maas, W.K., 1986, Distribution of basic replicons having homology with RepFIA, RepFIB and RepFIC among incF group plasmids, *Plasmid* 15:19–34.
7. Blomberg, P., Wagner, E.G.H., and Nordstrom, K., 1990, Control of replication of plasmid R1: the duplex between the antisense RNA, CopA, and its target, CopT, is processed specifically in vivo and in vitro by RNase III, *EMBO J.* 9:2331–2340.
8. Brady, G., Frey, J., Danbara, H., and Timmis, K.N., 1983, Replication control mutations of plasmid R6-5 and their effects on interactions of the RNA-I control element with its target, *J. Bacteriol.* 154:429–436.
9. Bramhill, D., and Kornberg, A., 1988, A model for initiation at origins of DNA replication, *Cell* 54: 915–918.
10. Brenner, M., and Tomizawa, J., 1991, Quantitation of ColEI-encoded replication elements, *Proc. Natl. Acad. Sci. USA* 88:405–409.
11. Carleton, S., Projan, S.J., Highlander, S.K., Moghazeh, S., and Novick, R.P., 1984, Control of pT181 replication. II. Mutational analysis, *EMBO J.* 3:2407–2414.
12. Castagnoli, L., Scarpa, M., Kokkinidis, M., Banner, D.W., Tsernoglou, D., and Cesareni, G., 1989, Genetic and structural analysis of the ColEI Rop (Rom) protein, *EMBO J.* 8:621–629.
13. Chattoraj, D.K., Abeles, A.L., and Yarmolinsky, M.B., 1985, P1 plasmid maintenance: a paradigm of precise control, in: *Plasmids in Bacteria* (D.R. Helinski, S.N. Cohen, D.B. Clewell, D.A. Jackson, and A. Hollaender, eds.), Plenum Publishing, New York, pp. 355–381.
14. Chattoraj, D.K., Synder, K.M., and Abeles, A.L., 1985, P1 plasmid replication: multiple functions of RepA protein at the origin, *Proc. Natl. Acad. Sci. USA* 82:2588–2592.
15. Chattoraj, D.K., Mason, R.J., and Wickner, S.H., 1988, Mini-P1 plasmid replication: the autoregulation-sequestration paradox, *Cell* 52:551–557.
16. Churchward, G., Linder, P., and Caro, L., 1983, The nucleotide sequence of replication and maintenance functions encoded by plasmid pSC101, *Nucl. Acids Res.* 11:5645–5659.
17. Couturier, M., Bex, F., Berquist, P., and Maas, W., 1988, Identification and classification of bacterial plasmids, *Microbiol. Rev.* 52:375–395.

18. Dasgupta, S., Masukata, H., and Tomizawa, J., 1987, Multiple mechanisms for initiation of ColEI DNA replication: DNA synthesis in the presence and absence of ribonuclease H, *Cell* 51:1113–1122.

19. Datta, N., 1979, Plasmid classification: incompatibility grouping, in: *Plasmids of Medical, Environmental and Commercial Importance* (K.N. Timmis and A. Puhler, eds.), Elsevier/North Holland Publishing Co., Amsterdam, pp. 3–12.

20. Durland, R.H., Toukdarian, A., Fang, F. and Helinski, D.R., 1990, Mutations in the *trfA* replication gene of the broad-host-range plasmid RK2 result in elevated plasmid copy numbers, *J. Bacteriol.* 172:3859–3867.

21. Durland, R.H., and Helinski, D.R., 1990, Replication of the broad-host-range plasmid RK2: direct measurement of intracellular concentrations of the essential TrfA replication proteins and their effect on plasmid copy number, *J. Bacteriol.* 172:3849–3858.

22. Filutowicz, M., Uhlenhopp, E., and Helinski, D.R., 1986, Binding of purified wild-type and mutant π initiation proteins to a replication origin region of plasmid R6K, *J. Mol. Biol.* 187:225–239.

23. Filutowicz, M., Davis, G., Greener, A., and Helinski, D.R., 1985, Autorepressor properties of the π initiation protein encoded by plasmid R6K, *Nucl. Acids Res.* 13:103–114.

24. Filutowicz, M., McEachern, M., Greener, A., Mukhopadyay, P., Ulenhopp, E., Durland, R., and Helinski, D., 1985, Role of the π initiation protein and direct nucleotide sequence repeats in the regulation of plasmid R6K replication, in: *Plasmids in Bacteria* (D.R. Helinski, S.N. Cohen, D.B. Clewell, D.A. Jackson, and A. Hollaender, eds.), Plenum Publishing, New York, pp. 125–140.

25. Filutowicz, M., McEachern, M.J., and Helinski, D.R., 1986, Positive and negative roles of an initiator protein at an origin of replication, *Proc. Natl. Acad. Sci. USA* 83:9645–9649.

26. Garnier, T., and Cole, S.T., 1988, Identification and molecular genetic analysis of replication functions of the bacteriocinogenic plasmid pIP404 from *Clostridium perfringens*, *Plasmid* 19:151–160.

27. Gennaro, M.L., Iordanescu, S., Novick, R.P., Murray, R.W., Steck, T.R., and Khan, S.A., 1989, Functional organization of the plasmid pT181 replication origin, *J. Mol. Biol.* 205:355–362.

28. Gilbride, K.A., and Brunton, J.L., 1990, Identification and characterization of a new replication region in the *Neisseria gonorrhoeae* β-lactamase plasmid pFA3, *J. Bacteriol.* 172:2439–2446.

29. Grosschedl, R., and Hobom, G., 1979, DNA sequences and structural homologies of the replication origins of lambdoid bacteriophages, *Nature* 277:621–627.

30. Gruss, A., and Ehrlich, S.D., 1989, The family of highly interrelated single-stranded deoxyribonucleic acid plasmids, *Microbiol. Rev.* 53:231–241.

31. Haring, V., and Scherzinger, E., 1989, Replication proteins of the IncQ plasmid RSF1010, in: *Promiscuous Plasmids of Gram-Negative Bacteria* (C. Thomas, ed.), Academic Press, San Diego, California, pp. 95–124.

32. Hashimoto-Gotoh, T., and Inselburg, J., 1979, ColEI plasmid incompatibility: localization and analysis of mutations affecting incompatibility, *J. Bacteriol.* 139:608–619.

33. Hayes, F., Vos, P., Fitzgerald, G.F., de Vos, W.M., and Daly, C., 1991, Molecular organization of the minimal replicon of the novel, narrow-host-range, lactococcal plasmid pCI305, *Plasmid* 25:16–26.

34. Helmer-Citterich, M., Anceschi, M.M., Banner, D.W., and Cesarini, G., 1988, Control of ColEI replication: low affinity specific binding of Rop (Rom) to RNA I and RNA II, *EMBO J.* 7:557–566.

35. Highlander, S. K., and Novick, R.P., 1990, Mutational and physiological analyses of plasmid pT181 functions expressing incompatibility, *Plasmid* 23:1–15.

36. Iordanescu, S., 1987, The Inc3B determinant of plasmid pT181. A mutational analysis, *Mol. Gen. Genet.* 207:60–67.

37. Kamio, Y., and Terawaki, Y., 1983, Nucleotide sequence of an incompatibility region of mini-Rts1 that contains five direct repeats, *J. Bacteriol.* 155:1185–1191.

38. Kamio, Y., Tabuchi, A., Itoh, Y., Katagiri, H., and Terawaki, Y., 1984, Complete nucleotide sequence of mini-Rts1 and its copy mutant, *J. Bacteriol.* 158:307–312.

39. Kamio, Y., Itoh, Y., and Terawaki, Y., 1988, Purification of Rts1 RepA protein and binding of the protein to mini-Rts1 DNA, *J. Bacteriol.* 170:4411–4414.

40. Kelly, W., and Bastia, D., 1985, Replication initiator protein of plasmid R6K autoregulates its own synthesis at the transcriptional step, *Proc. Natl. Acad. Sci. USA* 82:2574–2578.

41. Kim, K., and Meyer, R.J., 1985, Copy number of the broad host-range plasmid R1162 is determined by the amounts of essential plasmid-encoded proteins, *J. Mol. Biol.* 185:755–767.

42. Kittell, B.L., and Helinski, D.R., 1991, Iteron inhibition of RK2 replication in vitro: evidence for

intermolecular coupling of replication origins as a mechanism for RK2 replication control, *Proc. Natl. Acad. Sci. USA* 88:1389–1391.

43. Kozlowski, M., Thatte, V., Lau, P.C.K., Visentin, L.P., and Iyer, V.N., 1987, Isolation and structure of the replicon of the promiscuous plasmid pCU1, *Gene* 58:217–228.

44. Kues, U., and Stahl, U., 1989, Replication of plasmids in gram-negative bacteria. *Microbiol. Rev.* 53: 491–516.

45. Lacatena, R.M., and Cesareni, G., 1981, Base pairing of RNAI with its complementary sequence in the primer precursor inhibits ColEI replication, *Nature* 294:623–626.

46. Lin, L.-S., and Meyer, R.J., 1986, Directly repeated, 20-bp sequence of plasmid R1162 DNA is required for replication, expression of incompatibility, and copy number control, *Plasmid* 15:35–47.

47. Lin, L.-S., Kim, Y.-J., and Meyer, R.J., 1987, The 20 bp, directly repeated DNA sequence of broad host range plasmid R1162 exerts incompatibility in vivo and inhibits R1162 replication in vitro, *Mol. Gen. Genet.* 208:390–397.

48. Masson, L., and Ray, D.S., 1986, Mechanism of autonomous control of the *Escherichia coli* F plasmid: different complexes of the initiator/repressor protein are bound to its operator and to an F plasmid replication origin, *Nucl. Acids Res.* 14:5693–5711.

49. Masson, L., and Ray, D.S., 1988, Mechanism of autonomous control of the *Escherichia coli* F plasmid: purification and characterization of the *repE* gene product, *Nucl. Acids Res.* 16:413–424.

50. Masukata, H., and Tomizawa, J., 1984, Effects of point mutation on formation and structure of the RNA primer for ColEI replication, *Cell* 36:513–522.

51. Masukata, H., and Tomizawa, J., 1986, Control of primer formation for ColEI plasmid replication: conformational change of the primer transcript, *Cell* 44:125–136.

52. McEachern, M.J., Filutowicz, M., Yang, S., Greener, A., Mukhopadhyay, P., and Helinski, D.R., 1986, Elements involved in the copy number regulation of the antibiotic resistance plasmid R6K, in: *Banbury Report 24: Antibiotic Resistance Genes: Ecology, Transfer and Expression*, Cold Spring Harbor Laboratory, Cold Spring Harbor, New York, pp. 195–204.

53. McEachern, M.J., Bott, M.A., Tooker, P.A., and Helinski, D.R., 1989, Negative control of plasmid R6K replication: possible role of intermolecular coupling of replication origins, *Proc. Natl. Acad. Sci. USA* 86:7942–7946.

54. Molin, S., and Nordstrom, K., 1980, Control of replication of plasmid R1. Functions involved in replication, copy number control, incompatibility, and switch-off of replication, *J. Bacteriol.* 141:111–120.

55. Molin, S., Stougaard, P., Light, J., Nordstrom, M., and Nordstrom, K., 1981, Isolation and characterization of new copy mutants of plasmid R1 and identification of a polypeptide involved in copy number control, *Mol. Gen. Genet.* 181:123–130.

56. Mukherjee, S., Erickson, H., and Bastia, D., 1988, Detection of DNA looping due to simultaneous interaction of a DNA-binding protein with two spatially separated binding sites on DNA, *Proc. Natl. Acad. Sci. USA* 85:6287–6291.

57. Muraiso, K., Mukhopadhyay, G., and Chattoraj, D.K., 1990, Location of a P1 plasmid replication inhibitor determinant within the initiator gene, *J. Bacteriol.* 172:4441–4447.

58. Murotsu, T., Matsubara, K., Sugisaki, H., and Takanami, M., 1981, Nine unique repeating sequences on a region essential for replication and incompatibility of the mini-F plasmid, *Gene* 15:257–271.

59. Nordstrom, K., 1990, Control of plasmid replication—how do DNA iterons set the replication frequency? *Cell* 63:1121–1124.

60. Nordstrom, K., Molin, S., and Light, J., 1984, Control of replication of bacterial plasmids: genetics, molecular biology and physiology of the plasmid R1 system, *Plasmid* 12:71–90.

61. Nordstrom, M., and Nordstrom, K., 1985, Control of replication of FII plasmids: comparison of the basic replicons and of the small copB systems of plasmids R100 and R1, *Plasmid* 31:81–87.

62. Nordstrom, K., and Austin, S., 1989, Mechanisms that contribute to the stable segregation of plasmids, *Ann. Rev. Genet.* 23:37–69.

63. Novick, R.P., 1987, Plasmid incompatibility, *Microbiol. Rev.* 51:381–395.

64. Novick, R.P., Adler, G.K., Projan, S.J., Carleton, S., Highlander, S., Gruss, A., Khan, S.A., and Iordanescu, S., 1984, Control of pT181 replication. I. The pT181 copy control function acts by inhibiting the synthesis of a replication protein, *EMBO J.* 3:2399–2405.

65. Novick, R.P., Iordanescu, S., Projan, S.J., Kornblum, J., and Edelman, I., 1989, pT181 plasmid replication is regulated by a countertranscript-driven transcriptional attenuator, *Cell* 59:395–404.
66. Novick, R.P., Projan, S. J., Kumar, C.C., Carleton, S., Gruss, A., Highlander, S.K., and Kornblum, J., 1985, Replication control for pT181, an indirectly regulated plasmid, in: *Plasmids in Bacteria* (D.R. Helinski, S.N. Cohen, D.B. Clewell, D.A. Jackson, and A. Hollaender, eds.), Plenum Publishing, New York, pp. 299–320.
67. Nozue, H., Tsuchiya, K., and Kamio, Y., 1988, Nucleotide sequence and copy control function of the extension of the *incI* region (*incI-b*) of Rts1, *Plasmid* 17:46–56.
68. Nugent, M.E., Smith, T.J., and Tacon, W.C.A., 1986, Characterization and incompatibility properties of Rom⁻ derivatives of pBR322-based plasmids, *J. Gen. Micro.* 132:1021–1026.
69. Ohman, M., and Wagner, E.G.H., 1989, Secondary structure of the RepA mRNA leader transcript involved in control of replication of plasmid R1, *Nucl. Acid Res.* 17:2557–2579.
70. Pal, S.K., and Chattoraj, D.K., 1988, P1 plasmid replication: initiator sequestration is inadequate to explain control by initiator-binding sites, *J. Bacteriol.* 170:3554–3560.
71. Pal, S.K., Mason, R. J., and Chattoraj, D.K., 1986, P1 plasmid replication. Role of initiator titration in copy number control, *J. Mol. Biol.* 192:275–285.
72. Patel, I., and Bastia, D., 1986, A replication origin is turned off by an origin-"silencer" sequence, *Cell* 47:785–792.
73. Patel, I., and Bastia, D., 1987, A replication initiator protein enhances the rate of hybrid formation between a silencer RNA and an activator RNA, *Cell* 51:455–462.
74. Perez-Casal, J.F., Gammie, A.E., and Crosa, J.H., 1989, Nucleotide sequence analysis and expression of the minimum REPI replication region and incompatibility determinants of pColV-K30, *J. Bacteriol.* 171:2195–2201.
75. Perri, S., Helinski, D.R., and Toukdarian, A., 1991, Interactions of plasmid encoded replication initiation proteins with the origin of DNA replication in the broad host range plasmid RK2, *J. Biol. Chem.* 226:12536–12543.
76. Persson, C. and Nordstrom, K., 1986, Control of replication of the broad host range plasmid RSF1010: the incompatibility determinant consists of directly repeated DNA sequences, *Mol. Gen. Genet.* 203:189–192.
77. Persson, C., Wagner, E.G.H., and Nordstrom, K., 1988, Control of replication of plasmid R1: kinetics of in vitro interaction between the antisense RNA, CopA, and its target, CopT, *EMBO J.* 7:3279–3288.
78. Persson, C., Wagner, E.G.H., and Nordstrom, K., 1990, Control of replication of plasmid R1: structures and sequences of the antisense RNA, CopA, required for its binding to the target RNA, CopT, *EMBO J.* 11:3767–3775.
79. Pickney, M., Diaz, R., Lanka, E., and Thomas, C.M., 1988, Replication of mini RK2 plasmid in extracts of *Escherichia coli* requires plasmid-encoded protein TrfA and host-encoded proteins DnaA, B, G, DNA gyrase and DNA polymerase III, *J. Mol. Biol.* 203:927–938.
80. Polisky, B., 1988, ColEI replication control circuitry: sense from antisense, *Cell* 55:929–932.
81. Projan, S.J., and Novick, R., 1988, Comparative analysis of five related Staphylococcal plasmids, *Plasmid* 19:203–221.
82. Projan, S., and Novick, R.P., 1984, Reciprocal intrapool variation in plasmid copy numbers: a characteristic of segregational incompatibility, *Plasmid* 12:52–60.
83. Rokeach, L.A., Sogaard-Anderson, L., and Molin, S., 1985, Two functions of the E protein are key elements in the plasmid F replication control system, *J. Bacteriol.* 164:1262–1270.
84. Rownd, R., 1969, Replication of a bacterial episome under relaxed control, *J. Mol. Biol.* 44:387–402.
85. Rownd, R.H., Womble, D.D., Dong, X., Luckow, V.A., and Wu, R.P., 1985, Incompatibility and IncFII plasmid replication control, in: *Plasmids in Bacteria* (D.R. Helinski, S.N. Cohen, D.B. Clewell, D.A. Jackson, and A. Hollaender, eds.), Plenum Publishing, New York, pp. 335–354.
86. Scott, J.R., 1984, Regulation of plasmid replication, *Microbiol. Rev.* 48:1–23.
87. Smith, C.A., and Thomas, C.M., 1984, Nucleotide sequence of the *trfA* gene of the broad host-range plasmid RK2, *J. Mol. Biol.* 175:251–262.
88. Sriprakash, K.S., and Macavoy, E.S., 1987, Characterization and sequence of a plasmid from the trachoma biovar of *Chlamydia trachomatis*, *Plasmid* 18:205–214.
89. Stalker, D.M., Thomas, C.M., and Helinski, D.R., 1981, Nucleotide sequence of the origin of replication of the broad host range plasmid RK2, *Mol. Gen. Genet.* 181:8–12.

90. Stalker, D.M., Kolter, R., and Helinski, D.R., 1979, Nucleotide sequence of the region of an origin of replication of the antibiotic resistance plasmid R6K, *Proc. Natl. Acad. Sci. USA* 76:1150–1154.

91. Stalker, D.M., Kolter, R., and Helinski, D.R., 1982, Plasmid R6K DNA replication. I. Complete nucleotide sequence analysis of an autonomously replicating segment, *J. Mol. Biol.* 161:33–43.

92. Summers, D.K., and Sherratt, D. J., 1984, Multimerization of high copy number plasmids causes instability: ColEI encodes a determinant essential for plasmid monomerization and stability, *Cell* 36:1097–1103.

93. Tait, R.C., Kado, C.I., and Rodriguez, R.L., 1983, A comparison of the origin of replication of pSa with R6K, *Mol. Gen. Genet.* 192:32–38.

94. Tamm, J., and Polisky, B., 1983, Structural analysis of RNA molecules involved in plasmid copy number control, *Nucl. Acids Res.* 11:6381–6397.

95. Terawaki, Y., Nozue, H., Zeng, H., Hayashi, T., Kamio, Y., and Itoh, Y., 1990, Effects of mutations in the *repA* gene of plasmid Rts1 on plasmid replication and autorepressor function, *J. Bacteriol.* 172:786–792.

96. Terawaki, Y., Hong, Z., Itoh, Y., and Kamio, Y., 1988, Importance of the C terminus of plasmid Rts1 RepA protein for replication and incompatibility of the plasmid, *J. Bacteriol.* 170:1261–1267.

97. Timmis, K., Andres, L., and Slocombe, P., 1978, Plasmid incompatibility: cloning of an IncFII determinant of R6-5, *Nature* 71:4556–4560.

98. Tokino, T., Murotsu, T., and Matsubara, K., 1986, Purification and properties of the mini-F plasmid-encoded E protein needed for autonomous replication control of the plasmid, *Proc. Natl. Acad. Sci. USA* 83:4109–4113.

99. Tolun, A., and Helinski, D.R., 1982, Separation of the minimal replication region of the F plasmid into a replication origin segment and a *trans*-acting segment, *Mol. Gen. Genet.* 186:372–377.

100. Tolun, A., and Helinski, D.R., 1981, Direct repeats of the plasmid *incC* region express F incompatibility, *Cell* 24:687–694.

101. Tomizawa, J., 1984, Control of ColEI plasmid replication: the process of binding of RNA I to the primer transcript, *Cell* 38:861–870.

102. Tomizawa, J., 1985, Control of ColEI plasmid replication: initial interaction of RNA I and the primer transcript is reversible, *Cell* 40:527–535.

103. Tomizawa, J., 1986, Control of ColEI plasmid replication: binding of RNA I to RNA II and inhibition of primer formation, *Cell* 47:89–97.

104. Tomizawa, J., 1990, Control of ColEI plasmid replication. Intermediates in the binding of RNA I and RNA II, *J. Mol. Biol.* 212:683–694.

105. Tomizawa, J., and Itoh, T., 1981, Plasmid ColEI incompatibility determined by interaction of RNA I with primer transcript, *Proc. Natl. Acad. Sci. USA* 78:6096–6100.

106. Tomizawa, J,. and Som, T., 1984, Control of ColEI plasmid replication: enhancement of binding of RNA I to the primer transcript by the Rom protein, *Cell* 38:871–878.

107. Tomizawa, J., Itoh, T., Selzer, G., and Som, T., 1981, Inhibition of ColEI RNA primer formation by a plasmid-specified small RNA, *Proc. Natl. Acad. Sci. USA* 78:1421–1425.

108. Tomizawa, J., 1990, Control of ColEI plasmid replication. Interaction of Rom protein with an unstable complex formed by RNA I and RNA II, *J. Mol. Biol.* 212:695–708.

109. Trawick, J.D., and Kline, B.C., 1985, A two-stage molecular model for control of mini-F replication, *Plasmid* 13:59–69.

110. Tsurimoto, T., and Matsubara, K., 1981, Purification of bacteriophage λO protein that specifically binds to the origin of replication, *Mol. Gen. Genet.* 181:325–331.

111. Tsutsui, H., Fujiyama, A., Murotsu, T., and Matsubara, K., 1983, Role of nine repeating sequences of the mini-F genome for expression of F-specific incompatibility phenotype and copy number control, *J. Bacteriol.* 155:337–344.

112. Twigg, A.J., and Sherratt, D., 1980, Trans-complementable copy-number mutants of plasmid ColEI, *Nature* 283:216–218.

113. Vanooteghem, J.-C., and Cornelis, G., 1990, Structural and functional similarities between the replication region of the *Yersinia* virulence plasmid and the RepFIIA replicons, *J. Bacteriol.* 172:3600–3608.

114. Vocke, C., and Bastia, D., 1985, The replication initiator protein of plasmid pSC101 is a transcriptional repressor of its own cistron, *Proc. Natl. Acad. Sci. USA* 82:2252–2256.

115. Wickner, S., Hoskins, J., and McKenny, K., 1991, Function of DnaJ and DnaK as chaperones in origin-specific DNA binding by RepA, *Nature* 350:165–167.
116. Womble, D., and Rownd, R., 1988, Genetic and physical map of plasmid NR1: comparison with other IncFII antibiotic resistance plasmids, *Microbiol. Rev.* 52:433–451.
117. Yamaguchi, K., and Masamune, Y., 1985, Autogenous regulation of synthesis of the replication protein in plasmid pSC101, *Mol. Gen. Genet.* 200:362–367.
118. Yamaguchi, K., and Yamaguchi, M., 1984, The replication origin of pSC101: the nucleotide sequence and replication functions of the ori region, *Gene* 29:211–219.

Chapter 9

Agrobacterium-Mediated Transfer and Stable Incorporation of Foreign Genes in Plants

CLARENCE I. KADO

1. Introduction

The transfer of genetic material between a prokaryotic organism to eukaryotic organisms represent a unique phenomenon in biology. The *Agrobacterium*-mediated gene transfer to plants is the classic example of such a phenomenon. The genes required for this transfer are located on a 200-kbp Ti plasmid. As illustrated in Figure 9.1, a specific sector (the T-DNA) of the Ti plasmid is the element transferred. The 25-kbp T-DNA is bordered by 23-bp directly repeated sequences termed left and right borders. These borders are the targets where specific cleavages occur to generate T-intermediates in the *Agrobacterium* cell. The T-intermediates are transferred by still unknown mechanisms to plant cells during infection of the host at wounded sites. A prerequisite for the transmission of the T-intermediates is the close interaction of the bacterial cells with the plant cells. Plant signals in the form of phenolic compounds generated by phenylalanine ammonia lyase- and tyrosine ammonia lyase-initiated lignin biosynthetic pathways are recognized by *A. tumefaciens* through a sophisticated two-component gene regulatory system (recently reviewed in 38). This regulatory system operates through the products of two virulence genes: *virA* and *virG*. These genes are part of a six-operon *vir* regulon on the Ti plasmid (52). On recognition of the plant signal, transcription of the normally silent operons is initiated through the signal transduction pathway. Specific gene products of the *vir* regulon catalyze events that lead to the generation of T-intermediates and to their efficient transfer into the plant host cell. Once

CLARENCE I. KADO • Davis Crown Gall Group, Department of Plant Pathology, University of California, Davis, Davis, California 95616.

Bacterial Conjugation, edited by Don B. Clewell. Plenum Press, New York, 1993.

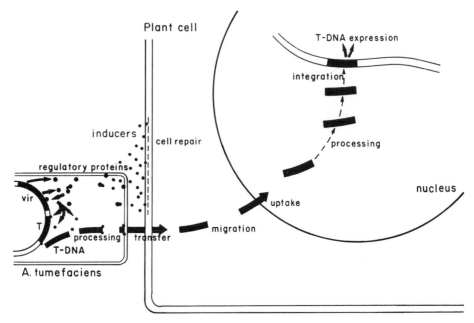

FIGURE 9.1. *Agrobacterium*-mediated gene transfer to a plant cell.

entered, the T-DNA is processed for eventual integration into one or more of the chromosomes of the host cell. Enzymes encoded by genes on the T-DNA catalyze the formation of indole-3-acetic acid, cytokinin, and opines. The first two compounds are growth hormones that promote non-self-limiting tumor formation. The opines are organic acid-basic amino acid conjugates such as octopine [N^2-(D-1-carboxyethyl)-L-arginine], nopaline [N^2-(1,3-dicarboxypropyl)-L-arginine], and phosphorylated disaccharides. The type of opine synthesized in the tumor depends on the type of opine synthesizing gene present in the T-DNA (31, 47). Thus, reference to octopine Ti plasmids indicates that the T-DNA contains the octopine synthase gene (*ocs*), and likewise, nopaline Ti plasmids bear a T-DNA containing the nopaline synthase gene (*nos*). Because the T-DNA growth hormone and opine genes possess eukaryotic promoters (6, 27), transcription of these genes occurs once they are introduced into the plant on the T-DNA. Transcription occurs within 4 days after infection by *A. tumefaciens* (34) and likely before integration of the T-DNA into the plant chromosome (36). Integration of the T-DNA assures stability and constitutive expression of these genes. Hence, non-self-limiting growth ensues, culminating in the formation of tumors at the infection site.

The clever evolutionary host-parasite interaction was promoted by the promiscuity of the conjugation process developed in *A. tumefaciens*. Although the mechanism of T-DNA transfer from the bacterium to the plant cell remains to be elucidated, an interkingdom conjugative DNA transfer mechanism is one possibility. Transducing and viral-like transforming DNA transfer mechanisms are equally possible. This chapter weighs the pros and cons of these mechanisms as analogies to the T-DNA transfer system.

2. The Transferred DNA (T-DNA)

The length of the T-DNA and the genes that it contains varies between different Ti plasmids. In the octopine-type Ti plasmids, the T-DNA is in three segments: a left T-DNA (TL) of 13 kbp, a central T-DNA (TC) of 1.5 kbp, and a right T-DNA (TR) of 7.8 kbp. TL contains the growth hormone producing genes, TC contains genes of unknown function, and TR contains the opine synthase genes. With the nopaline-type Ti plasmids, a single T-DNA element of about 23 kbp is present, simplifying the analysis because of its unit transfer mechanism. All T-DNAs are flanked by direct terminal repeats of 25 bp at their ends. These repeats, called left and right borders, are required *in cis* for T-DNA processing. The processing reaction is catalyzed by enzymes encoded by *virD1* and *virD2* genes of the *virD* operon. VirD1 and VirD2 proteins catalyze the cleavage or nicking within the borders of the T-DNA (3, 20, 56, 59, 67). Single-stranded T-intermediates, initially called the T-strand (3, 56, 57), and double-stranded T-intermediates are formed (20, 59, 67). In addition to linear single- and double-stranded T-intermediates, a small amount of double-stranded circular intermediates have been observed in induced *Agrobacterium* cells (4, 5, 40, 41, 63, 72). Circular forms may not be the transferred molecules because the transferred T-DNA must be in linear form (7). Of the three forms, the double-stranded T-intermediate is the predominant form. Kinetic studies have shown that the double-stranded T-intermediate molecules are formed within 30 minutes, a time that is one-half the generation time of *A. tumefaciens*, peaking 6 hours after induction (59), while the single-stranded intermediate is maximally formed between 16 and 24 hours (16). Although the three types of T-intermediates are generated, the preponderance of the double-stranded form of the T-intermediate is temptingly suggestive that it is the biologically active form.

Formation of the intermediates requires a precise nicking site and protein-DNA complex formation. This is reminiscent of certain plasmid transfer systems. For example, with the broad host range plasmid RK2 transfer origin, there is specific nicking within a 19-bp inverted repeat containing a 10-bp sequence that is very similar to a 10-bp sequence stretch within the T-DNA borders, which is composed of 23-bp direct repeats (30). The RK2 repeats contain the sites for primase binding. These initiation sites are on both strands in the transfer origin (*oriT*) region and entail TraH, TraI, and TraJ proteins that are required for relaxosome formation.

In interbacterial genetic exchange, three mechanisms are well known: conjugation, transformation, and transduction. Because transformation and transduction are usually limited in their host range, conjugatively mediated transfer has predominated in the case of interspecies genetic exchange. Thus, the T-DNA transfer system may likely be using a highly promiscuous transfer system involving a conjugation pilus for plant transformation.

3. Genes Involved in Ti Plasmid Conjugal Transfer

The efficient transfer of the Ti plasmid between donor and recipient bacteria is initiated by certain conjugal opines that are synthesized in crown gall tumors (21, 39, 48). (For details, see Chapter 10.) Specific opines like nopaline, nopalinic acid, and agrocinopines A and B are synthesized in crown gall tumors initiated by *A. tumefaciens* C58 harboring plasmid pTiC58. The phosphorylated disaccharides agrocinopines A and B induce the expression of pTiC58 conjugal transfer genes (21, 22). Although the mechanism of conjugal transfer of the Ti plasmids is not understood, the genes involved in conjugation on the Ti plasmid have been investigated. These genes are located in three distinct regions of the TiC58 plasmid, designated TraI, TraII, and TraIII (68). TraIII is located about 4 kb downstream of the nopaline catabolic operon (*noc*) (53) and about 3 kb from the *oriV/par/inc* locus mapped by Gallie et al (24). TraI and TraII are located opposite TraIII. These loci are near the locus that confers catabolism of agrocinopines (68). On a circular map of pTiC58 (18), TraI and TraII are located at the 7:30–7:45 position, while TraIII is located at the 1:30 position. Mutations in these Tra regions do not affect virulence on *Kalanchoë* (68). Moreover, deletion of all of the Tra regions down to a "mini-Ti" plasmid has no effect on tumorigenicity (51). Thus, if a conjugal transfer mechanism is operating in the transfer of the T-DNA to plants, other loci must be involved. The proposal that the Ti plasmid-mediated transfer of the T-DNA to plant cells might be an adaptation of conjugal plasmid transfer (9, 57, 62) would require genes other than Ti plasmid Tra genes. That the T-DNA transfer is independent of any genes within its borders would leave only the *vir* genes. If some of the *vir* genes indeed function to promote T-DNA transfer conjugatively, then perhaps Ti plasmid conjugal transfer between bacteria may be affected by altering the *vir* genes by mutations. Indeed, the efficiency of Ti plasmid conjugative transfer between *Agrobacterium* strains is affected appreciably by certain mutations in the *vir* genes, particularly *virB* genes (25, 58). Whether the *vir* genes promote T-DNA transfer by a conjugative means into plants remains to be demonstrated. The conjugative analogy used by some workers is that of F-plasmid, since a single-stranded mode of transmission has been recognized only for this plasmid (72). However, conjugative transfer as well as transformation and transfection between bacteria has been recognized to involve double-stranded molecules, whereby one strand is hydrolyzed to provide the driving energy for the transfer of the opposite strand (for review see 38).

4. Genes Involved in the Transfer of T-DNA to Plants

The interaction of *A. tumefaciens* with plant cells initiates a stepwise process involving *vir* gene activation, T-intermediate formation, T-DNA transfer to the bacterial membrane, and its translocation across both bacterial and plant membranes. The genes involved in the nicking and unwinding of the T-DNA to form T-intermediates have been identified to be genes of the *virD* operon (67, 74), specifically *virD1* and *virD2* (26, 37, 59). VirD1 and VirD2 show topoisomerase/endonuclease activity responsible for the nicking and unwinding/ligation reaction (26). Analysis of VirD2 showed that only the amino-terminal half of VirD2 protein was involved in the endonuclease reaction yielding double-stranded T-intermediate formation (60). The carboxy-terminal half has no endonuclease activity but has DNA unwinding activity very much like a helicase. DNA helicase has been reported to have an endonuclease activity associated with its amino-terminal half and a helicase activity in the carboxy-terminal half of the protein (71). For VirD2, both amino- and carboxy-terminal portions are required for virulence. Thus, VirD2 must possess a second function.

That function may rest in its recognition and covalent attachment to the 5' termini of the T-intermediate (20, 32, 69, 75) and possibly a helicase-like activity. Attached VirD2 protein may function to protect the double-stranded T-intermediate from 5' → 3' exonucleolytic attack and may also facilitate transfer across the cell and nuclear membranes like the "pilot" protein of adenovirus, whose genome is a double-stranded DNA of about 36,500 bp bearing the proprimer protein pTP of 50,000 Da (61). VirD2 is 56,000 Da (49) and has two prominent domains with Lys-Arg-Pro/Glu-Arg peptide sequences in the carboxy-terminal half of the protein that suggest involvement in nuclear localization (60, 76). Indeed transgenic tobacco plants bearing the *virD2* gene show that the VirD2 protein is abundantly present in the nuclei (35, 38). This would suggest that VirD2 could be associated with the incoming T-DNA and facilitate its transfer into the nucleus.

In addition to VirD proteins, the genes of the *virC* operon *virC1* and *virC2* play a role in the T-intermediate transfer process. Although the functions of VirC1 and VirC2 proteins are not understood, evidence indicates that the VirC1 protein can bind to a unique sequence called overdrive (64), which is located near the right border of the T-DNA (46). The VirC1 complexed with the overdrive was found to enhance T-intermediate formation (19, 64), perhaps by facilitating VirD1/D2 protein complex formation (50). Because no similar overdrive sequences have been observed for nopaline Ti plasmids and because T-DNAs built with synthetically synthesized borders still function as effectively as their wild-type counterparts, VirC1 protein may have function(s) other than promoting T-intermediate formation. In fact, VirC1 and VirC2 proteins pTiC58 are necessary for the transfer of T-DNA to maize (29). These proteins may prove more efficient than those counterparts encoded by octopine Ti plasmids because nopaline plasmid pTiC58 is more efficient in transforming many different plants (8, 10, 33). The function of VirC1 and VirC2 may be necessary for promoting efficiency of T-DNA transfer as well as T-intermediate formation. Double-stranded T-intermediate concentrations are highest in induced *Agrobacterium* cells containing a mutation in the *ros* chromosomal gene, that these concentrations do not raise the level of virulence lends further support to this view (14, 15). The *ros* gene encodes a repressor that blocks the expression of *virC* and *virD*. Strains bearing a mutated *ros* gene have the ability to infect plants normally refractory to transformation by *A. tumefaciens* (15). The *ros* mutants are derepressed in *virC* and *virD* expression irrespective of function of the two-component regulation mediated by VirA and VirG. Strains bearing both regulatory systems (*virA/G* and a mutation in *ros*) have increased double-stranded T-intermediate formation. These strains have broader host range specificities (M. Cooley and C. I. Kado, unpublished results).

The 69-kDa protein encoded by *virE2* preferentially binds single-stranded DNA nonspecifically (11–13, 17, 28, 54). Because a certain proportion of the population of T-intermediate molecules are single-stranded species, VirE2 protein is thought to bind these DNAs and serve to protect them against nucleases like a viral capsid protein. Such single-stranded DNA binding proteins serve in DNA replication as in the conjugal transfer of F plasmid (72) and in the replication of adenovirus DNA, which encodes a 72-kDa DNA binding protein that cooperatively binds the lagging single strand during DNA replication. Bound adenovirus DNA is partially resistant to hydrolysis by nucleases (44, 66). The adenovirus single-stranded binding protein is directly involved in DNA replication, much like the bacteriophage T4 gene 32 protein (65).

5. Mechanism of T-DNA Transfer to Plants

The formation of T-intermediates requires the presence of *vir* gene products, which are necessary for transfer of these molecules. Some Vir proteins form complexes with the T-intermediate, whereas others form complexes with the membrane to facilitate DNA transfer. Some hints on the identity of the *vir* genes that are involved in this process have surfaced. Besides *virD1* and *virD2* gene products, the genes of the *virB* operon, composed of 11 genes, encode products that are associated with the bacterial membrane at sites where T-intermediate transfer likely occurs. These products contain signal peptides and membrane-spanning domains in, for example, VirB1, VirB2, and VirB3, which contain highly hydrophobic regions for both octopine and nopaline *virB* gene products (55). VirB4, which contains a potential ATP binding site, does not contain a signal peptide or any highly hydrophobic region. It nevertheless was found to be associated with the *Agrobacterium* inner membrane and contains ATPase activity (55a). VirB5 also contains a signal peptide near the N-terminus but no prominent hydrophobic region. VirB6 does not contain a signal peptide but contains a very hydrophobic domain. VirB7 sequence analysis suggests that this protein is similar to lysis proteins encoded by CloDf13, ColA-Ca31, ColE3, and ColE1 (55). Because colicins do not possess signal peptides, their excretion from the bacterial cell is facilitated by the lysis proteins. VirB7 might function like these lysis proteins in the translocation of the hydrophobic proteins bearing no signal peptide, such as VirB2, VirB3, or VirB6. No signal peptide is present in VirB8, whose function remains to be identified. A signal peptide occurs at the N-terminus of VirB9. VirB10 also lacks a signal peptide but contains a hydrophobic periplasmic domain. The protein appears to be associated with the inner membrane (70). VirB11 has homologies to the *comG* ORF1 of *Bacillus subtilis* (2) and the KilB protein of RK2 (43). An ATPase activity is associated with VirB11 (11a). *comG* ORFs possess sequence similarities to a group of pilin genes of Gram-negative bacteria.

Another striking similarity was revealed from sequence comparisons with the conjugative transfer functions of plasmids of the IncP1 and IncW groups, namely RK2 and R388, respectively. Of compelling interest are the similarity between *virB* genes and those genes (*ptl*) involved in toxin secretion in *Bordetella pertussis* (70b), specifically ORFB, C, and D (55b). Table 9.1 shows the comparison between *virG* genes and those at the Tra2 region of RK2, the Trw region of R388 and *ptl* of *B. pertussis*. Furthermore, there are sequence similarities between the Tra1 genes of RK2 such as between the *traG* locus and *virD4*, but the latter gene product is functionally dissimilar to TraG protein since it fails to complement a *traG* mutant (70a). VirB and VirD4 proteins also promote the conjugative transfer of an IncQ plasmid between *A. tumefaciens* (7a). Noteworthy is the strong similarity between the genetic arrangement between the Trw genes of R388 and the *virB* genes of the *vir* regulon (Fig. 9.2).

Based on these sequence analyses, there is the strong possibility that VirB proteins are associated with the membrane complex of *Agrobacterium* and may form structures very much like bacterial conjugation pili to facilitate T-intermediate transfer to plant cells. The precise mechanism for this activity remains to be determined.

Studies dealing with the T-intermediate transfer process have suggested that this intermediate must be in a transcriptionally active double-stranded form before it can integrate into the plant chromosomes. Histochemical analyses of freshly transformed tissues revealed cells expressing the β-glucuronidase gene used as a reporter of T-DNA

TABLE 9.1 Similarities between VirB and VirD Protein
Sequences to Those of the Tra Regions of RK2, R388 and
ptl Toxin Secretion Operon of *Bordetella pertussis*

Gene	R388		RK2	ptl	
VirB1					
VirB2	TrwL	(15%)	Tra2C		
VirB3	TrwM	(26%)	Tra2D	ptlA	[26%]
VirB4	TrwK	(32%)	Tra2E	ptlB	[30%]
VirB5	TrwJ	(25%)	Tra2F		
VirB6	TrwI	(22%)		ptlC	[17%]
VirB7	TrwH	(12%)			
VirB8	TrwG	(22%)		ptlD	[32%]
VirB9	TrwF	(31%)		ptlE	[27%]
VirB10	TrwE	(36%)	Tra2I	ptlF	[32%]
VirB11	TrwD	(36%)	Tra2B	ptlG	[35%]
VirD1					
VirD2	TrwC	(37%)	TraI		
VirD3					
VirD4	TrwB	(21%)	Tra2G		

Numbers in parenthesis represent percent identity, whereas numbers in brackets are
percent selective identity.

transfer before integration conceivably could have occurred (36). Cotransformation experiments using two T-DNAs, each bearing part of an aminoglycoside phosphotransferase gene, were shown to result in kanamycin resistant transformants (45). Such transformants could occur only if homologous recombination were taking place in the transformed cell, a process that requires the presence of double-stranded T-DNA molecules. Recent agroinfection experiments introducing allelic DNA in *Brassica* showed homologous recombination to occur (23). It has been proposed that T-DNA integration uses an illegitimate recombination system (26a, 42). Although evidence is accumulating favoring the double-stranded form, critical evidence is presently lacking that shows unequivocally that double-stranded T-intermediates are indeed the transferred form of the T-DNA. Experimental designs for demonstrating such transfers are difficult, since in contrast to interbacterial studies, there are no plant chromosomeless minicells that can be used.

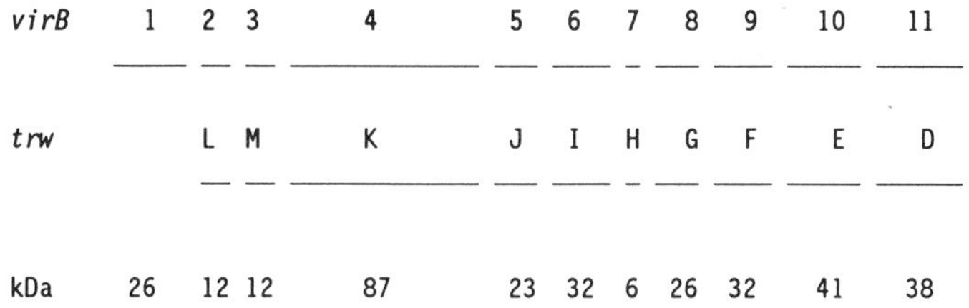

FIGURE 9.2. Similarity of the genetic organization between the *virB* operon of the Ti plasmid virulence regulon and the *trw* operon of IncW plasmid R388. Numbers represent rounded-off molecular weights of each open reading frame.

6. Conclusions

The transfer of T-DNA from *Agrobacterium* to plants requires the genes of the virulence regulon. Of the *vir* genes required for T-intermediate formation, *virD1* and *virD2* encode proteins that catalyze site-specific cleavage at the borders and unwinding of the T-DNA on the Ti plasmid to generate single- and double-stranded linear intermediates. VirD2 may protect *in vivo* the ends of the T-intermediates by covalent linkage and may serve as the protein that facilitates transfer across cell and nuclear membranes. This facilitation of T-intermediate transfer across the membranes likely uses the products of the *virB* genes. Because its genetic arrangement is similar to that of the Trw operon of the broad-host-range plasmid R388, *virB* genes likely encode structural proteins involved in piluslike and porinlike structures that help mediate the transfer of DNA to recipient cells. Enhanced promiscuity of the T-intermediate transfer to plant cells requires *virC1*, *virC2*, and *virE2* genes. Loss of this feature does not render *Agrobacterium* completely avirulent in dicots, but it does in monocots such as maize. The molecular form of the T-intermediate that is transferred to plants remains to be determined. Certainly, both single- and double-stranded T-intermediates are formed in induced *Agrobacterium* cells. The energetics of DNA transfer in other bacterial systems favors the double-stranded molecular form, since it can derive the energy required for transfer from the nontransferred strand (through hydrolysis of the nontransferred strand). The mechanism of the transfer/integration process may follow those used by integrative viral replicons such as adenoviruses, phage P27, and Epstein-Barr virus (1). Like the T-DNA, the genomes of these viruses contain terminally repeated sequences (Epstein-Barr virus DNA contains direct terminally repeated sequences [1]). There may, therefore, be some commonality in transfer/integration mechanisms between these diverse systems.

ACKNOWLEDGMENTS. The author is indebted to all members of the Davis Crown Gall Group, Dr. Laura Frost, and Dr. Fernando de la Cruz, for much of their unpublished information. While on sabbatical leave Dr. de la Cruz and Ken Shirasu discovered the homologies between *virB* genes and R388 *trw* genes. The research by the group was supported by NIH grant GM-45550.

References

1. Adams, A., 1980, Molecular biology of the Epstein-Barr virus, in: *Viral Oncology* (G. Klein, ed.), Raven Press, New York, pp. 683–711.
2. Albano, M., Breitling, R., and Dubnau, D. A., 1989, Nucleotide sequence and genetic organization of the *Bacillus subtilis comG* operon, *J. Bacteriol.* **171:**5386–5404.
3. Albright, L., Yanofsky, M. F., Leroux, B., Ma, D., and Nester, E. W., 1987, Processing of the T-DNA of *Agrobacterium tumefaciens* generates border nicks and linear, single-stranded T-DNA, *J. Bacteriol.* **169:**1046–1055.
4. Alt-Morbe, J., Rak, B., and Schroder, J., 1986, A 3.6 kbp segment from the *vir*-region of Ti-plasmids contains genes responsible for border sequence directed production of T region circles in *E. coli*, *EMBO J* **5:**1129–1135.
5. Alt-Morbe, J., Heeinemeyer, W., and Schroder, J., 1990, The virD genes from the *vir* region of the Ti plasmid: T-region border dependent processing steps in different rec mutants of *Escherichia coli*, *Gene* **96:**43–49.
6. An, G., Ebert P. R., Yi, B.-Y., and Choi, C.-H., 1986, Both TATA box and upstream regions are required for the nopaline synthase promoter activity in transformed tobacco cells, *Mol. Gen. Genet.* **203:**245–250.

7. Bakkeren, G., Koukolikova-Nicola, Z., Grimsley, N., and Hohn, B., 1989, Recovery of *Agrobacterium tumefaciens* T-DNA molecules from whole plants early after transfer, *Cell* **57**:847–857.

7a. Beijersbergen, A., Dulk-Ras, A. D., Schilperoort, R. A., and Hooykaas, P. J. J., 1992, Conjugative transfer by the virulence system of *Agrobacterium tumefaciens*, *Science* **256**:1324–1327.

8. Boulton, M. I., Buchholz, W. G., Marks, M. S., Markham, P. G., and Davies, J. W., 1989, Specificity of *Agrobacterium*-mediated delivery of maize streak virus DNA to members of the Gramineae, *Plant Mol. Biol.* **12**:31–40.

9. Bucahanan-Wollaston, J., Passiatore, E., and Cannon, F., 1987, The *mob* and *oriT* mobilization functions of a bacterial plasmid promote its transfer to plants, *Nature* **328**:172–175.

10. Charest, P. J., Iyer, V. N., and Miki, B. L., 1989, Virulence of *Agrobacterium tumefaciens* strains with *Brassica napus* and *Brassica juncea*, *Plant Cell Repts.* **8**:303–306.

11. Christie, P. J., Ward, J. E., Winans, S. C., and Nester, E. W., 1988, The *Agrobacterium tumefaciens virE* product is a single-stranded DNA binding protein that associates with T-strands, *J. Bacteriol.* **170**:2659–2667.

11a. Christie, P. J., Ward, J. E. Jr., Gordon, M. P., and Nester, E. W., 1989, A gene required for transfer of T-DNA to plants encodes an ATPase with autophosphorylating activity, *Proc. Natl. Acad. Sci. USA* **86**:9677–9681.

12. Citovsky, V., DeVos, G., and Zambryski, P., 1988, Single-stranded DNA binding protein encoded by the *virE* locus of *Agrobacterium tumefaciens*, *Science* **240**:501–504.

13. Citovsky, V., Wong, M. L., and Zambryski, P., 1989, Cooperative interaction of *Agrobacterium* VirE2 protein with single-stranded DNA: implications for the T-DNA transfer process, *Proc. Natl. Acad. Sci. USA* **86**:1193–1197.

14. Close, T. J., Rogowsky, P. M., Kado, C. I., Winans, S. C., Yanofsky, M. F., and Nester, E. W., 1987, Dual control of *Agrobacterium tumefaciens* Ti plasmid virulence genes, *J. Bacteriol.* **169**:5113–5118.

15. Cooley, M. B., D'Souza, M. R., and Kado, C. I., 1991, The *virC* and *virD* operons of the *Agrobacterium* Ti plasmid are regulated by the *ros* chromosomal gene: analysis of the cloned *ros* gene, *J. Bacteriol.* **173**:2608–2616.

16. Culianez-Marcia, F. A., and Hepburn, A. G., 1988, The kinetics of T-strand production in a nopaline-type helper strain of *Agrobacterium tumefaciens*, *Mol. Plant-Microbe Interact.* **1**:207–214.

17. Das, A., 1988, *Agrobacterium tumefaciens virE* operon encodes a single-stranded DNA binding protein, *Proc. Natl. Acad. Sci. USA* **85**:2909–2913.

18. Depicker, A., De Wilde, M., deVos, G., de Vos, R., van Montagu, M., and Schell, J., 1980, Molecular cloning of overlapping segments of the nopaline Ti-plasmid pTiC58 as means to restriction endonuclease mapping, *Plasmid* **3**:193–211.

19. DeVos, G, and Zambryski, P., 1989, Expression of *Agrobacterium* nopaline-specific VirD1, VirD2, and VirC1 proteins and their requirement for T-strand production in *E. coli*, *Mol. Plant-Microbe Interact.* **2**:43–52.

20. Durrenberger, F., Crameri, A., Hohn, B., and Koukolikova-Nicola, Z., 1989, Covalently bound VirD2 protein of *Agrobacterium tumefaciens* protects the T-DNA from exonucleolytic degradation, *Proc. Natl. Acad. Sci. USA* **86**:9154–9158.

21. Ellis, J. G., Kerr, A., Petit, A., and Tempe, J., 1982, Conjugal transfer of nopaline and agropine Ti-plasmids—the role of agrocinopines, *Mol. Gen. Genet.* **186**:269–273.

22. Ellis, J. G., Murphy, P. J., and Kerr, A., 1982. Isolation and properties of transfer regulatory mutants of the nopaline Ti-plasmid pTiC58, *Mol. Gen. Genet.* **186**:275–281.

23. Gall, S., Pisan, B., Hohn, T., Grimsley, N., and Hohn, B., 1991, Genomic homologous recombination *in planta*, *EMBO J* **10**:1571–1578.

24. Gallie, D. R., Hagiya, M., and Kado, C. I., 1985, Analysis of *Agrobacterium tumefaciens* plasmid pTiC58 replication region with a novel high-copy-number derivative, *J. Bacteriol.* **161**:1034–1041.

25. Gelvin, S. B., and Habeck, L. L., 1990, *vir* genes influence conjugal transfer of the Ti plasmid of *Agrobacterium tumefaciens*, *J. Bacteriol.* **172**:1600–1608.

26. Ghai, J., and Das, A., 1989, The *virD* operon of *Agrobacterium tumefaciens* Ti plasmid encodes a DNA-relaxing enzyme, *Proc. Natl. Acad. Sci. USA* **86**:3109–3113.

26a. Gheysen, G., Villaroel, R., and Van Montagu, M., 1991, Illegitimate recombination in plants: a model for T-DNA integration, *Genes Dev.* **5**:287–297.

27. Gielen, J., De Beuckeleer, M., Seurinck, J., Deboeck, F., DeGreve, H., Lemmers, M., Van Montagu, M., and Schell, J., 1984, The complete nucleotide sequence of the TL-DNA of the *Agrobacterium tumefaciens* plasmid pTiAch5, *EMBO J.* **3**:835–846.

28. Gietl, C., Koukolikova-Nicola, Z., and Hohn, B., 1987, Mobilization of T-DNA from *Agrobacterium* to plant cells involves a protein that binds single-stranded DNA, *Proc. Natl. Acad. Sci. USA* **84**:9006–9010.

29. Grimsley, N., Hohn, B., Ramos, C., Kado, C., and Rogowsky, P., 1989, DNA transfer from *Agrobacterium* to *Zea mays* or *Brassica* by agroinfection is dependent on bacterial virulence functions, *Mol. Gen. Genet.* **217**:309–316.

30. Guiney, D. G., 1991, The transfer origin of the IncP plasmid RK2: identification of functional domains and essential sequences, *Plasmid* **25**:227.

31. Guyon, P., Chilton, M. D., Petit, A., and Tempe, J., 1980, Agropine in "null-type" crown gall tumors: evidence for generality of the opine concept, *Proc. Nat. Acad. Sci. USA* **65**:2693–2697.

32. Herrera-Estrella, A., Chen, Z.-M., Van Montagu, M., and Wang, K., 1988, VirD proteins of *Agrobacterium tumefaciens* are required for the formation of a covalent DNA-protein complex at the 5′ terminus of T-strand molecules, *EMBO J.* **7**:4055–4062.

33. Hobbs, S. L. A., Jackson, J. A., and Mahon, J. D., 1989, Specificity of strain and genotype in the susceptibility of pea to *Agrobacterium tumefaciens*, *Plant Cell Repts.* **8**:274–277.

34. Horsch, R. B., and Klee, H. J., 1986, Rapid assay of foreign gene expression in leaf discs transformed by *Agrobacterium tumefaciens*: role of T-DNA borders in the transfer process, *Proc. Natl. Acad. Sci. USA* **83**:4428–4432.

35. Lin, T.-S., 1992, Ph.D. thesis, University of California, Davis, *Agrobacterium* Ti plasmid T-DNA pilot-protein establishes in the nuclei of transformed tobacco cells.

36. Janssen, B.-J., and Gardner, R. C., 1989, Localized transient expression of GUS in leaf discs following cocultivation with *Agrobacterium*, *Plant Mol. Biol.* **14**:61–72.

37. Jayaswal, R. K., Veluthambi, K., Gelvin, S. B., and Slightom, J. L., 1987, Double-stranded cleavage of T-DNA and generation of single-stranded T-DNA molecules in *Escherichia coli* by a virD-encoded border-specific endonuclease from *Agrobacterium tumefaciens*, *J. Bacteriol.* **169**:5035–5045.

38. Kado, C. I., 1991, Molecular mechanisms of crown gall tumorigenesis, *Crit. Revs. Plant Sci.* **10**:1–32.

39. Klapwijk, P. M., Schleuderman, R., and Schilperoort, R. A., 1978, Coordinated regulation of octopine degradation and conjugative transfer of Ti plasmids in *Agrobacterium tumefaciens*: evidence for a common regulatory gene and separate operons, *J. Bacteriol.* **136**:775–785.

40. Koukolikova-Nicola, Z., Shillito, R. D., Hohn, B., Wang, K., Van Montagu, M., and Zambryski, P., 1985, Involvement of circular intermediates in the transfer of T-DNA from *Agrobacterium tumefaciens* to plant cells, *Nature* **313**:191–196.

40a. Lanka, E., Pansegrau, W., Lessl, M., Balzer, D., Ziegelin, G., and Durrenberger, M., 1992, The molecular basis of conjugative IncP plasmid transfer, EMBO Workshop, Las Navas del Marques, Spain (abstract).

41. Machida, Y., Usami, S., Yamamoto, A., Niwa, Y., and Takebe, I., 1986, Plant inducible recombination between the 25 bp border sequences of T-DNA in *Agrobacterium tumefaciens*, *Mol. Gen. Genet.* **204**:374–382.

42. Mayerhofer, R., Koncz-Kalman, Z., Nawrath, C., Bakkeren, G., Crameri, A., Angelis, K., Redi, G. P., Schell, J., Hohn, B., and Koncz, C., 1991, T-DNA integration: a mode of illegitimate recombination in plants, *EMBO J.* **10**:697–704.

43. Motallebi-Veshareh, M., Jagura-Burdzy, G., Williams, D. R., and Thomas, C. M., Identification of the *kilB* gene of broad-host-range plasmid RK2 as a putative transfer function, *Plasmid* **25**.

44. Nass, K., and Frenkel, G., 1980, The adenovirus-specific DNA binding protein inhibits the hydrolysis of DNA by DNase in vitro, *J. Virol.* **35**:314.

45. Offringa, R., deGroot, M. J. A., Haageman, H. J., Does, D. P., van den Elzen, P. J. M., and Hooykaas, P. J. J., 1990, Extrachromosomal homologous recombination and gene targeting in plant cells after *Agrobacterium*-mediated transformation, *EMBO J.* **9**:3077–3084.

46. Peralta, E. G., Hellmiss, R., and Ream, L. W., 1986, Overdrive, a T-DNA transmission enhancer on the *Agrobacterium tumefaciens* tumor-inducing plasmid, *EMBO J.* **5**:1137–1142.

47. Petit, A., Delhaye, S., Tempe, J., and Morel, G., 1970, Recherches sur les guanidines des tissus de crown gall. Mise en evidence d'une relation biochemique specifique entre les souches d'*Agrobacterium tumefaciens* et les tumeurs qu'elles induisen, *Physiol. Veg.* **8**:205–213.

48. Petit, A., Tempe, J., Kerr, A., Holsters, M., van Montagu, M., and Schell, J., 1978, Substrate induction of conjugative activity of *Agrobacterium tumefaciens* Ti plasmids, *Nature* **271**:570–572.

49. Porter, S. G., Yanofsky, M. F., and Nester, E. W., 1987, Molecular characterization of the *virD* operon from *Agrobacterium tumefaciens*, *Nucl. Acids Res.* **15**:7503–7516.

50. Ream, W., 1989, *Agrobacterium tumefaciens* and interkingdom genetic exchange, *Annu. Rev. Phytopath.* **27**:583–618.
51. Rempel, H., 1988, Genetic analysis of the left boundary of the virulence region by deletion mutagenesis of *Agrobacterium tumefaciens* plasmid pTiC58. Ph.D. thesis, University of California, Davis, 112 pp.
52. Rogowsky, P. M., Powell, B. S., Shirasu, K., Lin, T.-S., Morel, P., Zyprian, E. M., Steck, T. R., and Kado, C. I., 1990, Molecular characterization of the *vir* regulon of *Agrobacterium tumefaciens*: complete nucleotide sequence and gene organization of the 28.63-kbp regulon cloned as a single unit, *Plasmid* **23**:85–106.
53. Schardl, C. L., and Kado, C. I., 1983, A functional map of the nopaline catabolism genes on the Ti plasmid of *Agrobacterium tumefaciens* C58, *Mol. Gen. Genet.* **191**:10–16.
54. Sen, P., Pazour, G. J., Anderson, D., and Das, A., 1989, Cooperative binding of *Agrobacterium tumefaciens* VirE2 protein to single-stranded DNA, *J. Bacteriol.* **171**:2573–2380.
55. Shirasu, K., Morel, P., and Kado, C. I., 1990, Characterization of the *virB* operon of an *Agrobacterium tumefaciens* Ti plasmid: nucleotide sequence and protein analysis, *Mol. Microbiol.* **4**:1153–1163.
55a. Shirasu, K., Koukolikova-Nicola, Z., Hohn, B., and Kado, C. I., 1992, An inner-membrane associated virulence protein for T-DNA transfer from *Agrobacterium tumefaciens* to plants exhibits ATPase activity. *Proc. Natl. Acad. Sci, USA* (in press)
55b. Shirasu, K., and Kado, C. I. 1993, The *virB* operon of the *Agrobacterium tumefaciens* virulence regulon has sequence similarities to B, C, and D open reading frames downstream of the pertussis toxin-operon and to the DNA transfer-operons of broad-host-range conjugative plasmids. *Nucleic Acids Res.* **21** (in press).
56. Stachel, S. E., Timmerman, B., and Zambryski, P. C., 1986, Generation of single-stranded T-DNA molecules during the initial stages of T-DNA transfer to plant cells, *Nature* **322**:706–712.
57. Stachel, S. E., and Zambryski, P. C., 1986, *Agrobacterium tumefaciens* and the susceptible plant cell: a novel adaptation of extracellular recognition and DNA conjugation, *Cell* **47**:155–157.
58. Steck, T. R., and Kado, C. I., 1990, Virulence genes promote conjugative transfer of the Ti plasmid between *Agrobacterium* strains, *J. Bacteriol.* **172**:2191–2193.
59. Steck, T. R., Close, T. J., and Kado, C. I., 1989, High levels of double-stranded transferred DNA (T-DNA) processing from an intact nopaline Ti plasmid, *Proc. Natl. Acad. Sci. USA* **86**:2133–2137.
60. Steck, T. R., Lin, T.-S., Powell, B. S., and Kado, C. I., 1990, VirD2 gene product from the nopaline plasmid pTiC58 has at least two activities required for virulence, *Nucl. Acids Res.* **18**:6953–6958.
61. Sussenbach, J. S., 1987, The structure of the genome, in: *The Adenoviruses* (H. S. Ginsberg, ed.), Plenum Press, New York, London, pp. 35–124.
62. Tempe, J., Petit, A., Holsters, M., Van Montagu, M., and Schell, J., 1977, Thermosensitive step associated with transfer of the Ti-plasmid during conjugation: possible relation to transformation in crown gall, *Proc. Natl. Acad. Sci. USA* **74**:2848–2849.
63. Timmerman, B., Van Montagu, M., and Zambryski, P., 1988, *vir*-induced recombination in *Agrobacterium*. Physical characterization of precise and imprecise T-circle formation, *J. Mol. Biol.* **203**:373–384.
64. Toro, N., Datta, A., Carmi, O. A., Young, C., Prusti, R. K., and Nester, E. W., 1989, The *Agrobacterium tumefaciens virC1* gene product binds to overdrive, a T-DNA transfer enhancer, *J. Bacteriol.* **171**:6845–6849.
65. van der Vliet, P., Landberg, J., and Jansz, H. S., 1977, Evidence for a function of the adenovirus DNA-binding protein in initiation of DNA synthesis as well as elongation on nascent DNA chains, *Virology* **80**:98–110.
66. van der Vliet, P., Keegstra, N., and Jansz, H., 1978, Complex formation between the adenovirus type 5 DNA binding protein and single-stranded DNA, *Eur. J. Biochem.* **86**:389–396.
67. Veluthambi, K., Jayaswal, R. K., and Gelvin, S. B., 1987, Virulence genes A, G, and D mediate the double-stranded border cleavage of T-DNA from *Agrobacterium* Ti plasmid, *Proc. Natl. Acad. Sci. USA* **84**:1881–1885.
68. von Bodman, S. B., McCutchan, J. E., and Farrand, S. K., 1989, Characterization of conjugal transfer functions of *Agrobacterium tumefaciens* Ti plasmid pTiC58, *J. Bacteriol.* **171**:5281–5289.
69. Ward, E. R., and Barnes, W. M., 1988, VirD2 protein of *Agrobacterium tumefaciens* very tightly linked to the 5′ end of T-strand DNA, *Science* **242**:927–930.
70. Ward, J. E., Jr., Dale, E. M., Nester, E. W., and Binns, A. N., 1990, Identification of a VirB10 protein aggregate in the inner membrane of *Agrobacterium tumefaciens*, *J. Bacteriol.* **172**:5200–5210.
70a. Waters, V., Pasengrau, W., Lanka, E., and Guiney, D., 1992, Mutational analysis of essential IncPa tra genes traF and traG, Fallen Leaf Lake Conference, 1992 Abstracts, p. 36.
70b. Weiss, A., 1993, Toxin secretion in *Bordetella pertussis*: breaking the Gram-negative barrier, in: C. I. Kado and J. H. Crosa (eds.), *Molecular Biology of Bacterial Virulence*, Kluwer Academic Publishers, Dordrecht, Netherlands.

71. Wessel, R., Muller, H., and Hoffmann-Berling, H., 1990, Electron microscopy of DNA helicase-I complexes in the act of strand separation, *Eur. J. Biochem.* **189:**277–285.

72. Willetts, N., and Wilkins, B., 1984, Processing of plasmid DNA during bacterial conjugation, *Microbiol. Revs.* **48:**24–41.

72a. Winans, S. C., 1992, Two-way chemical signaling in *Agrobacterium*-plant interactions, *Microbiol. Revs.* **56:**12–31.

73. Yamamoto, A., Iwahashi, M., Yanofsky, M. F., Nester, E. W., Takebe, I., and Machida, Y., 1987, The promoter proximal region of the *virD* locus of *Agrobacterium tumefaciens* is necessary for the plant inducible circularization of T-DNA, *Mol. Gen. Genet.* **206:**174–177.

74. Yanofsky, M. F., Porter, S. G., Young, C., Albright, L. M., Gordon, M. P., and Nester, E. W., 1986, The virD operon of *Agrobacterium tumefaciens* encodes a site-specific endonuclease, *Cell* **47:**471–477.

75. Young, C., and Nester, E. W., 1988, Association of the VirD2 protein with the 5′ end of T strands in *Agrobacterium tumefaciens*, *J. Bacteriol.* **170:**3367–3374.

76. Zambryski, P. C., 1992, Chronicles from the *Agrobacterium*-plant cell DNA transfer story, *Annu. Rev. Plant Physiol. Plant Mol. Biol.* **43:**465–490.

Chapter 10

Conjugal Transfer of *Agrobacterium* Plasmids

STEPHEN K. FARRAND

1. Introduction: A Historical Perspective

Agrobacterium and its virulence plasmids have been the center of attention for better than a decade. The capacity of this organism to transfer DNA to plant cells has resulted in a paradigm for transkingdom genetic exchange and also has provided an extraordinarily useful tool for the study of plant processes at the molecular level. The molecular details of the mechanism by which *Agrobacterium* transfers DNA to plants are becoming understood and are reviewed in Chapter 9 of this book. However, only a little more than 15 years ago, the theory that *Agrobacterium* transferred DNA to plants, first seriously proposed by Petit et al (76), was subject to some doubt. To set the scene, early reports on the presence of bacterial DNA in tumor cells based on filter hybridization experiments (82, 92, 93, 97) were being challenged (25, 36) and results from renaturation kinetic analyses (9) provided no evidence to substantiate the model. However, all of these hybridization experiments suffered from one deficiency or another. The renaturation kinetic analyses, for example, although extremely sensitive, by their design could not detect specific sequences representing less than 5% of the probe even if present at a high copy number. Because the probes used in these studies were total bacterial genomic DNA, a set of specific sequences representing less than 250 kilobases (kb) would have gone undetected. These limitations were recognized, but in the days before recombinant DNA, technologies for developing more precise probes were extremely restricted. The problem of how to develop a set of specific candidate probes from a bacterium that lacked genetic systems such as F or bacteriophage λ was daunting; the one

STEPHEN K. FARRAND • Departments of Plant Pathology and Microbiology, University of Illinois at Urbana-Champaign, Urbana, Illinois 61801.
Bacterial Conjugation, edited by Don B. Clewell. Plenum Press, New York, 1993.

hope at the time was that, if they did indeed exist, the specific sequences themselves might be associated with a plasmid in the virulent agrobacteria.

In this light, two sets of experiments performed in the late 1960s and early 1970s were critical to the continued development of the *Agrobacterium* system. Hamilton and Fall (46) reported in 1971 that for one isolate of *A. tumefaciens*, strain C58, the capacity to induce tumors was lost at high frequency when the organism was grown at 37°C, a temperature well above its optimum of 28°C. Conversion to avirulence was permanent and was reminiscent of phenotypic changes associated with the loss of epigenetic elements, such as plasmids or bacteriophages, that had been observed in other bacterial species. At about the same time, Kerr (57, 58) reported that the capacity to induce tumors could be transferred to an avirulent *Agrobacterium* strain following superinfection into tumors induced by and containing viable virulent agrobacteria. This indicated that some factor involved in virulence was transmissible at high frequency. Taken together, these two sets of results suggested that a plasmid or a bacteriophage might encode virulence functions. These two experiments, easily repeatable in other laboratories (45, 66, 110, 120), provided the impetus for the eventual discovery of the Ti plasmid (109, 120, 129). Once in hand, the Ti plasmid provided a logical starting point for the systematic search for specific DNA sequences with which to probe crown gall tumors. Only 2 years after its discovery, Chilton et al (10) were the first to report that a small segment of Ti plasmid DNA, the T-DNA, is present in DNA isolated from crown gall tumor cells. In a sense, then, the history of the discovery of the Ti plasmid, and subsequently of the T-DNA, is the history of the discovery of the conjugal nature of this virulence element.

The in planta work of Kerr pointed to the transmissibility of the Ti plasmid. However, early efforts to demonstrate the self-conjugal nature of these elements in controlled mating experiments were not successful. The Ti plasmids could be mobilized to recipient agrobacteria using broad host range R plasmids such as RP4 (8, 53), but ex planta matings using *Agrobacterium* donors containing only their indigenous plasmids failed to yield transconjugants with consistency or at convincing frequencies. The key to the problem came from elegant experiments performed by Kerr et al (59) and by Genetello et al (41). Both groups reasoned that because transfer occurred in planta but not in culture, tumors must be producing some diffusible substance that induced conjugation. Neoplasias induced by virulent agrobacteria were known to produce new compounds called opines. Because they were thought to be specific to the tumors, the opines were logical candidates to test as conjugal regulators. The results were striking and unequivocal. In the absence of opines, conjugal transfer of the Ti plasmids could not be detected. If the Ti plasmid transferred at all under such growth conditions, it occurred at frequencies of less than one in 10^8 per donor cell. However, pregrowth of the donor with the appropriate opine resulted in transfer of the Ti plasmid at frequencies of around 10^{-2} (41, 59).

As these experiments were being conducted, several laboratories were mapping the Ti plasmids and determining the locations for genes involved in tumorigenesis, opine catabolism, and other demonstrable phenotypes encoded by these elements (14, 51). From these studies it seemed that loci involved in conjugal transfer were neither genetically nor physically associated with Ti plasmid genes responsible for tumorigenesis. Deletion and transposon insertion mutations abolishing tumorigenicity had no demonstrable effects on conjugal transfer, while mutants defective in the latter remained fully tumorigenic. With emphasis being placed on understanding the virulence functions of the Ti plasmid, further characterization of the conjugal transfer system was not actively pursued.

However, several recent developments have renewed interest in functions of the Ti plasmid involved in bacterial self-transfer. First, analysis of the *cis-* and *trans-*acting Ti plasmid-encoded functions required for tumorigenicity, as well as the nature of the transferred DNA intermediate, strongly suggests an evolutionary relationship between systems involved in T-DNA transfer and plasmid conjugation (reviewed in 83, 100, 130). Second, at least one nonagrobacterial plasmid can be mobilized to plants (5a). This heterologous transfer is dependent on functional *mob* determinants and an intact *oriT* region of this plasmid, further emphasizing the relationship between the two transfer mechanisms. Third, two recent reports suggest that there is a direct interaction between Ti plasmid conjugal transfer (Tra) functions and those encoded by the *vir* region that are required for T-DNA transfer to plants (40, 102). Finally, Ti plasmid conjugal transfer and T-DNA mobilization to plants are phenotypes that are both regulated by soluble, plant-produced inducers.

There is an additional reason for investigating the Ti plasmid conjugal transfer system. It is becoming increasingly clear that plasmid transfer in situ represents a major avenue by which bacteria ensure genetic diversity (reviewed in 13). The fact that opines induce Ti plasmid conjugation gives reason to suspect that plasmid transfer in the environment of *Agrobacterium*-induced plant neoplasias is important to the natural biology of this DNA element and of its bacterial host.

Most studies have focused on the *Agrobacterium* virulence elements, but other plasmids present in isolates of this genus also are self-conjugal. Most of these elements are cryptic, encoding no known functions. One exception, the agrocinogenic plasmid, pAgK84, is important with respect to a commercially successful agent marketed to biocontrol crown gall disease in the field. The conjugal nature of this plasmid is of interest from an economic perspective; dissemination of pAgK84 to virulent agrobacteria in the field could lead to the eventual breakdown of this biocontrol system. Furthermore, the conjugal transfer system of pAgK84 presents some intriguing features with respect to its genetic organization.

In this chapter we will concentrate on the features of the Ti plasmids associated with conjugation and on how these traits are related, structurally, genetically, and functionally, to those involved in tumorigenesis. We will also consider the significance of conjugal transfer to the biology of *Agrobacterium*. Finally, we will present the available information concerning the conjugal nature of other *Agrobacterium* plasmids.

2. Characteristics of *Agrobacterium* Virulence Plasmids

2.1. Classification

The virulence elements of *Agrobacterium* can be categorized based on several criteria. In the most general, the plasmids are divided into two classes based on the pathogenic phenotype they impart on their host bacteria. Those that confer the inception of unorganized or teratomatous plant tumors are called Ti for tumor inducing, and those that confer the development of transformed or hairy roots are called Ri for root inducing. Although these two general classes of plasmids can be quite different, among those that have been studied, some share reasonably large regions of homology. Much of this homology is located in the respective *vir* regions. Genes in this region, which are expressed in the bacteria, encode functions required for tumor or hairy root induction. Representative plasmids from both classes are self-conjugal, although little is known about the relationships between the Tra genes among the two groups.

The *Agrobacterium* virulence plasmids also are classified with respect to incompatibility functions. The most thoroughly studied elements, the octopine/mannityl opine-type and the nopaline/agrocinopine-type Ti plasmids (see below), belong to the IncRh1 incompatibility group. This group also includes certain limited host range Ti plasmids as well as some nononcogenic *Agrobacterium* plasmids that encode opine catabolic functions. The agropine/agrocinopine-type Ti plasmid, pTiBo542, belongs to a separate group, IncRh2, while the agropine-type Ri plasmid, pRi1855, is a member of the IncRh3 class. Interestingly, an IncRh3 Ri plasmid shows incompatibility with pSymG1008, the symbiosis plasmid in *Rhizobium trifolii* strain G1008 (72).

Finally, these plasmids are classified on the basis of metabolic properties they confer on both the transformed plant and the host bacterium. As briefly mentioned above, *Agrobacterium*-induced plant neoplasias are characterized, in part, by the production of novel low-molecular-weight carbon compounds called opines. Synthesis of these compounds is dependent on enzymes encoded by the T-DNA inherited from the inducing bacterium (reviewed in 19). There are some 9 or 10 opine families known, but any given T-DNA usually encodes the production of between one and three classes. The Ti plasmids also encode functions that enable their host agrobacteria to use some, but not all, opine classes as sole sources of carbon and energy. Significantly, with but a few exceptions, the Ti or Ri plasmids confer catabolism of precisely those opines found in the tumors induced by that bacterial type. Put another way, the opine biosynthetic genes located on the T-region are matched by specific sets of determinants, encoding catabolism of those same opines, located elsewhere on the virulence plasmid. With this strong correlation, the *Agrobacterium* virulence elements are often classified based on the opine types found in the plant tumors induced by their host bacteria. It should be noted, however, that plasmids of the same opine type may or may not be related. For example, classical wide host range octopine-type Ti plasmids are all very closely related and are virtually indistinguishable with respect to restriction fragment patterns. However, these plasmids are less closely related to the octopine-type Ti plasmids found in some limited host range biovar 3 agrobacteria. Furthermore, while the classical octopine-type Ti plasmids also encode production and catabolism of the mannityl opines, the biovar 3-associated octopine-type Ti plasmids do not. The agropine-type Ti and Ri plasmids also encode mannityl opine determinants (agropine is one of the four mannityl opines) but do not confer catabolism of octopine. Nor are these plasmids related to each other or to the classical wide host range octopine-type elements as judged by incompatibility grouping. Therefore, although opine-type is a useful criterion for describing Ti plasmids, it does not necessarily reflect the degree of relatedness between elements of the same class.

2.2. Physical Structure and Genetic Organization

All Ti and Ri plasmids examined to date are relatively large, covalently closed, circular, double-stranded DNA elements. Most are in the 140- to 250-kb size range. However, the Ti plasmids present in the so-called "lippia" strains are between 400 and 500 kb in size (107). Although these plasmids represent a diverse group of elements, those that have been mapped share a remarkable conservation of overall physicogenetic organization. By convention, the T-region is oriented at the 12 o'clock position. In virtually all Ti and Ri plasmids examined to date, opine catabolic determinants map in the 1 o'clock to 3 o'clock region, and the *vir* operons are located in the 8 o'clock to 11 o'clock area.

Two Ti plasmid classes, the nopaline/agrocinopine-type, as exemplified by pTiC58, and the octopine/mannityl opine-type, as exemplified by pTiAch5 (also pTiA6, pTiI5955, and pTiB6), have been subjected to the most detailed analyses. Maps for these two plasmids are presented in Figure 10.1. The two Ti plasmids strongly cross-hybridize over about 30% of their lengths (31). Two strongly homologous areas, A and B, encode the T-regions and the replication functions of the two plasmids. A third, region D, contains virtually all of *vir*.

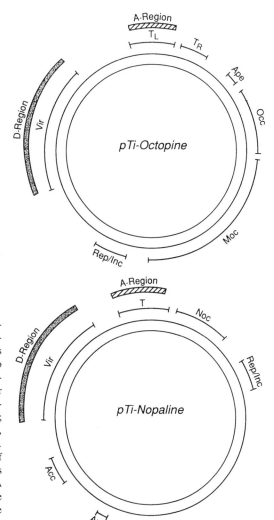

FIGURE 10.1. Organization of typical octopine- and nopaline-type Ti plasmids. These elements are about 200 kb in size. Defined regions include T, the T-region, which is divided into T-left (T_L) and T-right (T_R) in classical octopine-type Ti plasmids; opine catabolic regions for octopine (Occ), mannityl opines (Moc), nopaline (Noc), and agrocinopines A and B (Acc); Rep/Inc, replication and incompatibility; Vir, virulence; Ape, bacteriophage API exclusion. Arcs above the regions represent domains of strong homology between the two plasmids as described by Engler et al (31). Region A (hatched arc) encompasses the portions of the two T-regions that encode the *onc* genes, while region D (stippled arc) corresponds to Vir.

Furthermore, the *vir* genes of the two plasmids are cross-functional; recombinant clones containing *vir* segments of one plasmid will complement *vir* mutations in the other (85).

Substantial areas of each plasmid have been sequenced. Segments in which most, if not all, of the DNA sequence is known include the *vir* regions (69, 80, 86, 118, 125, 127) and the T-DNAs of both plasmids (2, 16, 42, 63, 67, 128). The octopine and nopaline catabolic loci have been almost entirely sequenced (44, 89, 94, 95, 108, 115), along with portions of the region required for agropine catabolism from pTi15955 (S. B. Hong et al, in preparation) and the replicator region of pTiB6S3 (103). Considering the extensive genetic analyses and the DNA sequence determinations, these two Ti plasmids are among the best characterized of the large bacterial epigenetic elements.

3. Conjugal Transfer of Ti and Ri Plasmids

As described previously, the conjugal nature of Ti plasmids preceded their discovery as autonomously replicating DNA elements. Although most of the work with these elements has focused on functions involved in oncogenesis, several early studies characterized their conjugal transfer properties. More recently, a proposal that T-DNA transfer to plant cells evolved from a conjugal transfer system (100) has renewed interest in the conjugal transfer properties of the Ti plasmids.

3.1. Characteristics of the Ti Plasmid Conjugal Transfer System

3.1.1. Inducibility

The most notable characteristic of Ti plasmid conjugation is its inducibility. Early on it was noted that most Ti plasmids failed to transfer at appreciable frequencies when matings were conducted on normal laboratory media. However, these elements appeared to transfer at high frequencies when plant tumors induced by the virulent donor strain were super-inoculated with the recipient *Agrobacterium* (57, 58). These in planta matings, or "Kerr crosses," led to the discovery that certain classes of opines produced by the crown gall tumors function as signals to induce conjugal transfer of the virulence plasmids.

When donors harboring wild-type octopine- or nopaline-type Ti plasmids are mated with recipients in the absence of the inducer opines, rare transconjugants appear. Most of these transconjugants harbor spontaneous transfer-constitutive mutant derivatives of the Ti plasmid (27, 75). These elements, formally analogous to the *drd* mutants of some conjugal R plasmids, now transfer at high frequencies in the absence of relevant conjugal opine. Such mutants have proved valuable in studies on the regulation of conjugation, and their characteristics will be discussed in detail in Section 4.

All of the Ti plasmids examined to date appear to be regulated in their conjugal transfer properties. Such may not be the case for the Ri plasmids. These virulence elements can transfer from *A. rhizogenes* donor strains without addition of opines or any other plant-produced factors (79). However, this observation is complicated by the fact that most *A. rhizogenes* strains contain other large plasmids that readily form cointegrates with the Ri plasmids (79, 122). It has not been determined whether the constitutive transfer of the Ri plasmids represents functions associated with these virulence elements, or results from mobilization by or cointegrative transfer with the additional plasmids present in the donors.

3.1.2. Temperature and pH Effects

An early report showed that conjugal transfer of the octopine-type Ti plasmid pTiB6 was temperature sensitive (TS) with transfer frequencies of 10^{-2} at 27°C, dropping to $<$ 10^{-6} at 33°C and above (106). Similar results, again using pTiB6, were reported by Hooykaas et al (52). It has been known for many years that tumor induction by various *Agrobacterium* isolates contains a temperature-sensitive step, being inhibited at temperatures above 30°C even though both the plants and the pathogen are capable of growth at the elevated temperatures (84). The observations that conjugal transfer also was temperature sensitive suggested that the two Ti plasmid transfer systems shared some common function(s). However, more recent results show that most Ti plasmids are capable of conjugal transfer at temperatures above 35°C (L. Zhang and A. Kerr, personal communication). This is true even for the nopaline-type plasmid pTiC58, which is temperature sensitive for replication and can be cured from strains by growth at 35°C or above (120). Plasmid pTiB6 appears to be the exception and presumably carries a TS mutation in one or more Tra genes.

Conjugal transfer frequencies of pTiA6 remain constant over a pH range of 5.9 to 7.7 (L. Zhang and A. Kerr, personal communication). However, lowering the pH from 5.9 to 5.6 decreases the transfer frequency by almost a factor of 10. Interestingly, induction of *vir* genes by acetosyringone and monosaccharides is dependent on acid conditions. Maximal induction occurs between pH 5.5 and 5.75 (1, 85, 126) and is greatly reduced at pH above 6.0 (1). Thus, induction of T-DNA transfer requires pH conditions that are inhibitory to Ti plasmid conjugal transfer.

3.1.3. Mating Substrate

All published studies on Ti plasmid conjugation have employed matings on membrane filters or on agar surfaces. Under such conditions, when the plasmids are maximally induced, transfer frequencies can approach 10^{-1} per input donor. It is not known if such frequencies can be obtained during matings in liquid media. However, the natural habitat of *Agrobacterium* is soil, and thus particulate or plant surfaces. It would not be surprising if Ti plasmids transferred best during matings in which donor and recipient are present on solid surfaces.

3.1.4. Kinetics of Transfer

There are no published reports concerning kinetics of Ti plasmid transfer *in vitro*. However, Kerr (58) observed that in planta transfer is not detectable until 3 weeks after superinoculation of tumors with the recipients. For 3 weeks to 5 weeks postinoculation, the frequency at which transconjugants appeared increased steadily from 10^{-4} to 10^{-2} per input recipient. However, because the tumors are producing opines, and because transconjugants, but not recipients, can use these compounds, it is not clear if this increase represents continued mating events or a competitive advantage imparted to the transconjugants by inheritance of the Ti plasmid.

We have examined the kinetics of Ti plasmid transfer *in vitro* using a mobilization assay with a transfer-constitutive derivative of pTiC58 (D. M. Cook and S. K. Farrand, unpublished results). Low numbers of transconjugants are detectable within 30 minutes from the time donors and recipients are collected together on membrane filters. The conjugation frequency rises rapidly and peaks at about 10^{-2} per input donor at 2 hours.

3.1.5. Inhibitors and Stimulators

Octopine-type Ti plasmids conjugally transfer when donors are incubated on minimal medium plates containing octopine. However, no transfer is observed if matings are carried out on rich medium, even in the presence of octopine at levels sufficient for induction on minimal medium (52). This suggests that some component of the rich medium inhibits conjugal transfer. Opine-mediated induction on minimal medium is not influenced by the presence of sugars such as glucose, galactose, lactose, rhamnose, or xylose; acids such as citrate, malate, pyruvate, or succinate; or most amino acids. However, the sulfur-containing amino acids methionine, cysteine, and cystine completely block transfer of an octopine-type Ti plasmid (52). These inhibitory amino acids seem to exert their effects at the regulatory level; transfer-constitutive Ti plasmid mutants, which no longer require opines for induction, conjugate at normal levels on rich medium and on minimal medium containing the inhibitory agents.

In addition to the conjugal opines, other factors appear to stimulate Ti plasmid conjugal transfer. It has been known for some time that different octopine-type Ti plasmids, although virtually indistinguishable with respect to genetic and physical structure, vary with respect to inducibility of conjugation by octopine. One group, called the efficient or Tra^e plasmids, requires only small amounts of the opine for maximal induction, giving transfer frequencies of around 10^{-2} (131). A second set, the inefficient or Tra^{ie} plasmids, are not induced for transfer with low levels of octopine, and high levels induce transfer at frequencies about an order of magnitude lower than those obtained with the Tra^e group. Zhang and Kerr (131) observed that strains harboring Tra^e-type Ti plasmids secrete a substance that stimulates the conjugal transfer of Tra^{ie}-type Ti plasmids. This phenomenon will be discussed in detail in Section 4.3 and in the Addendum.

3.1.6. Conjugal Host Range

Ti and Ri plasmids are freely conjugal within the genus *Agrobacterium*. In fact, proof that these plasmids are virulence elements, and that the type of plasmid determines the nature of the transformed tissues, derived from experiments in which various Ti and Ri plasmids were conjugated into different *Agrobacterium* backgrounds (43, 79, 120, 122). Most plasmid transfer experiments have used one of three standard *Agrobacterium* recipients representing two chromosomal backgrounds: Strains A136 (120) and C58C1RS (30) are two antibiotic-resistant, Ti plasmid-cured derivatives of the biovar 1 *A. tumefaciens* strain C58, and strain K57 (58) is a biovar 1 *A. radiobacter* isolate. Transfer from a given donor to various recipients seems to occur at about the same frequencies, suggesting that restriction-modification systems may not exert a strong influence on Ti plasmid transfer within at least some members of the genus.

The Ti plasmids are freely conjugal to several species and biovars of the genus *Rhizobium* (53). Once introduced, the Ti plasmids persist in these bacteria. This is not surprising considering the high degree of relatedness between agrobacteria and the fast-growing rhizobia (121, 123). Early studies showed that the Ti plasmids were not inherited by *Escherichia coli*. However, stable cointegrates between the Ti plasmids and the broad host range R plasmid, RP4, could be conjugally transferred to and maintained in *E. coli* hosts (50, 111). This indicated that the *Agrobacterium* plasmids cannot replicate in the enteric host, and the extent of the conjugal host range remained an unanswered question. Recently, using a bacteriophage fd-marked Ti plasmid, Sprinzl and Geider (99) reported conjugal transfer of this element to *E. coli*. Thus, although the Ti plasmid has a narrow replicative host range, its conjugal transfer range may be considerably broader than originally thought. Thorough conjugal host range studies remain to be done.

3.1.7. Mobilization of Chromosomal Genes

During a study of opine catabolic functions, Dessaux et al (20) discovered that the octopine-type Ti plasmid pTiR10 apparently can mobilize transfer of chromosomal genes. The octopine catabolic operon, *occ*, of this Ti plasmid encodes genes for arginine catabolism that can complement a natural defect in catabolism of this amino acid exhibited by a Ti plasmid-cured derivative of *A. tumefaciens* strain C58 (29). However, in the C58 chromosomal background growth on arginine requires the addition of octopine, since arginine is not a natural inducer of the *occ* operon. In experiments during which pTiR10 was mated to C58C1, a control was included in which the mix was plated on minimal medium containing arginine as the sole carbon source. Surprisingly, colonies arose that could not be accounted for by mutations in either donor or recipient. Further experiments confirmed that these Arg$^+$ transconjugants represented recipient cells that had acquired new traits from the donor. Furthermore, although acquisition of these markers required the induction of the Ti plasmid transfer system, they were unlinked to actual inheritance of the plasmid itself. Remarkably, a wide range of determinants could be transferred, including certain deficiencies associated with strain R10. For example, strain C58, but not strain R10, can use sorbose as a carbon source and nitrate as an anaerobic electron acceptor. Many Arg$^+$ recombinants of the recipient C58 strain inherited the sorbose and nitrate deficiencies of strain R10. However, screening additional markers verified that these transconjugants were indeed derivatives of strain C58. Strains C58 and R10, although both biovar 1 agrobacteria, are only distantly related. Stable inheritance of R10 chromosomal genes by strain C58 in the face of significant diversity points to a novel genetic exchange system.

3.2. Location and Organization of Tra Functions

In the early 1980s, transposon-generated insertion and deletion mutations were used to localize various functions encoded on the octopine-type Ti plasmid pTiB6S3 (14) and the nopaline-type Ti plasmid pTiC58 (51). For the first, two Tn7 insertion mutations were isolated that affected conjugal transfer. The parent plasmid was a transfer-constitutive mutant giving conjugation frequencies of about 10^{-2}, and the two insertions lowered the rate by about five orders of magnitude. The two insertions mapped to the 1 o'clock region of the plasmid just anticlockwise from *occ*, the locus encoding catabolism of octopine, the conjugal opine. Within the precision of the transposon mapping, the two insertions defined an approximately 10-kb segment of the plasmid involved in conjugal transfer (Figure 10.2). The two Tra⁻ mutants were still able to use octopine, and mutations in *occ* had no effect on conjugal transfer, indicating that these two systems, although coordinately regulated, are independently expressed. No other insertions or deletions affecting conjugal transfer were isolated. Koekman et al (64) reported on a series of deletion derivatives of the closely related octopine-type Ti plasmid pTiAch5. Deletions removing the 1 o'clock region of this plasmid completely abolished conjugation.

Similar analyses have been conducted with a transfer-constitutive mutant of the nopaline-type Ti plasmid pTiC58 (51). A single Tn7 insertion mapping between *noc* and Rep at about 2 o'clock abolished conjugal transfer (Figure 10.2), while a deletion extending from *noc* into this region resulted in an attenuated Tra phenotype. In addition, two deletions that removed the 8 o'clock region of the Ti plasmid abolished transfer. Both deletions also removed *acc*, the locus encoding catabolism of the conjugal opines, agrocinopines A and B (see below).

More recently, Beck von Bodman et al (4) isolated Tra-defective mutants following

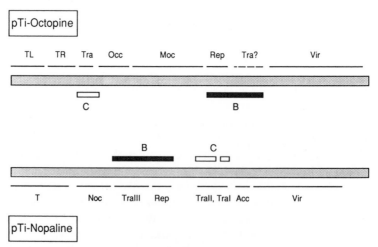

FIGURE 10.2. Linear representations of typical octopine- and nopaline-type Ti plasmids showing relationships and homologies between known Tra regions. Abbreviations are the same as in Figure 10.1, with the addition of Tra for conjugal transfer. TraI, II, and III represent the three known Tra regions of pTiC58 as defined by Beck von Bodman et al (4). The lettered bars between the two maps represent regions of strong heteroduplex homology between the two plasmids as described by Engler et al (31). Region B (solid bar) overlaps the replication regions of the two plasmids, as well as TraIII of pTiC58. For this reason, it is very possible that a corresponding Tra region is located between Rep and Vir on the octopine-type Ti plasmids (Tra?; see text). Homology region C (open bar) covers Tra regions I and II of pTiC58, and a known, but poorly defined, Tra domain of the octopine-type Ti plasmid.

random Tn5 mutagenesis of a strain harboring a transfer-constitutive derivative of pTiC58. Thirteen independent mutants were isolated, and in each case the transposon was located in the Ti plasmid. The insertions clustered to three regions: TraI, TraII, and TraIII. TraI and TraII are closely linked to each other and map just anticlockwise to *acc* (Figure 10.2). Complementation analyses showed that these two Tra regions are genetically distinct. The third set, TraIII, maps to the region lying between *noc* and Rep. All insertions in Tra region III completely abolished transfer. However, one of four insertions in TraI and one of three insertions in TraII resulted in greatly decreased, but still detectable, frequencies of transfer. All insertion mutations except one in TraIII could be complemented *in trans*, indicating that these loci encode *trans*-acting gene products.

The three Tra regions of pTiC58 lie within two of the four homology regions shared between the nopaline- and octopine-type Ti plasmids. TraIII maps within homology region B (Figure 10.2). Interestingly, in both Ti plasmids this region is closely associated with the Rep locus of the two Ti plasmids. Heteroduplexes within this region show no loops or other areas of single strandedness within a 10-kb continuous segment extending from Rep through TraIII of pTiC58. This suggests that the homologous segment of the octopine-type Ti plasmids encodes Tra genes and that the corresponding functions are homologous to those encoded within TraIII of pTiC58.

Tra regions I and II lie within homology region C with the largest segment of strong homology, as determined by heteroduplex analysis, corresponding to TraII of pTiC58. One Tn7 insertion abolishing transfer in the octopine-type Ti plasmid is located within this region. A 1-kb segment of the octopine-type Ti plasmid C-region also shows strong homology to the TraI region of pTiC58. However, heteroduplex analysis within this section of homology region C indicates that the area corresponding to TraI of pTiC58 contains information not present in the corresponding region of the octopine-type Ti plasmid. This is consistent with an analysis of the functions encoded by TraI from pTiC58. At least one activity, regulatory in nature, is encoded within TraI (see below).

Taken together, these observations suggest that the Tra systems of the classical octopine- and nopaline-type Ti plasmids are closely related. However, as will be discussed below, the mechanisms by which the conjugal opines regulate the expression of the Tra systems of these two plasmid types appear to be quite different.

Virtually nothing is known about the locations of Tra genes on other Ti plasmid types. Nor have there been any studies reported on the Tra systems of Ri plasmids. Although homology studies between octopine- or nopaline-type Ti plasmids and other Ti and Ri plasmids have been conducted, most have been confined to analyses of T-DNA and to regions encoding *vir* functions, plasmid replication, and opine catabolism. However, Drummond and Chilton (23) reported a series of cross-hybridization studies between various *Agrobacterium* Ti plasmids and restriction fragments of the octopine-type Ti plasmid pTiB6. All of the nopaline-type plasmids examined showed strong hybridization with fragments representing homology regions B and C. The agropine -type Ti plasmid pTiBo542 showed only weak cross-homologies within these regions. However, the precision of the technique does not allow accurate predictions concerning the relatedness of Tra genes among these plasmids. Interestingly, in a second study, relatively strong homologies were detected between regions B and C of pTiC58 and Nif plasmids from two *Rhizobium* species, *R. leguminosarum* and *R. trifolii* (81). Both of these *Rhizobium* plasmids are self-conjugal (54) and are compatible with pTiC58.

3.3. Conjugal Transfer Functions

The Tra functions of self-transmissible plasmids can be divided into three categories. The first two are those that are required for plasmid DNA processing and include the *cis*-acting *oriT* and the *trans*-acting proteins required for nicking the DNA and producing the copy to be transported to the recipient. The third class are those functions involved in recipient recognition and mating pair, or mating aggregate formation. Included in this last group are the conjugation-specific pili and gene products involved in forming the channel through which the transferred DNA moves from donor to recipient.

Nothing is known concerning Ti plasmid encoded genes or gene products required for mating-pair formation. Stemmer and Sequeira (101) showed that several *Agrobacterium* isolates produce numerous, very thin pili. Pili were easily detected on cells grown in rich or minimal media. Addition of plant cell extracts or of the appropriate conjugal opine had no effect on quantity or type of pilus produced. In addition, no difference in piliation was detected between *Agrobacterium* strains with and without Ti plasmids or between strains with wild-type or transfer-constitutive Ti plasmids. Mutants of *A. tumefaciens* strains unable to specifically attach to plant cells are conjugally proficient (4), indicating that at least some functions involved in attachment associated with T-DNA transfer to plant cells are not required for conjugal transfer of the Ti plasmid to bacterial recipients.

To date, no bacteriophages have been isolated that plaque specifically on *Agrobacterium* strains harboring wild-type or transfer-constitutive Ti plasmids. One agrobacteriophage, AP1, is specifically excluded by host cells containing octopine- or nopaline-type Ti plasmids (53, 90, 111). Intriguingly, the AP1 exclusion function, called *ape*, maps very close to the anticlockwise end of the pTiC58 TraII region. Similarly, *ape* maps within a 2-kb region at the anticlockwise end of homology region C of the octopine-type Ti plasmid pTiB6S3 (14), which corresponds to the TraII region of pTiC58. The mechanistic relationship between exclusion of bacteriophage AP1 and conjugal transfer, if any, is not known. Transposon insertion mutations giving an Ape⁻ phenotype have no effect on conjugal transfer of pTiB6S3 (14), and close linkage between the two loci could be only coincidental.

Recently, the *oriT* region of pTiC58 was localized to a 65-bp fragment located within TraII (12a) (Figure 10.3). Sequence analysis of a 665-bp fragment containing the 65-bp *oriT*-active segment reveals some interesting features. First, the region contains several perfect and imperfect inverted repeats (IR). These structures seem to be hallmarks of *oriT* regions and usually surround the sites at which conjugal and mobilizable plasmids are nicked prior to strand transfer to the recipient (124). Removing three of these inverted repeats decreased the *oriT* activity 50-fold, a finding consistent with that seen with other *oriT* regions (38, 119). Second, the 65-bp *oriT*-active region contains a 35-bp domain showing strong sequence identity with the *nic* region of RSF1010 (Figure 10.3). This domain includes an 8-bp segment showing an exact match with the corresponding regions of RSF1010 and pTF1, known to contain the site at which a single-stranded nick is introduced prior to DNA transfer (22).

Although there is strong conservation within the *nic* regions of RSF1010 and the *oriT* of pTiC58, there are no extended sequence similarities within the domains containing the inverted repeats from the two plasmids. Furthermore, the Ti plasmid does not mobilize RSF1010 (12a), suggesting that the IR sequences may be involved in establishing specificity to the nicking reaction or to some later event dependent on processing at the *nic* site.

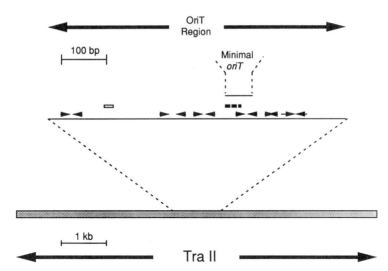

FIGURE 10.3. Organization of the *oriT* region of the nopaline-type Ti plasmid, pTiC58. The ca 600-bp *oriT* region is located within TraII and is composed of a 65-bp minimal mobilizable sequence flanked by a series of indirect perfect and imperfect repeats (arrowheads). The closed box shows the location of an *oriT*-active Ti plasmid domain showing exact base-pair matches with those that define the nick site of RSF1010 *oriT*. The open box represents an 8-bp sequence strongly conserved with the nick site of RK2/RP4 and the left conserved domain of T-region border sequences. See text for details.

The 665-bp *oriT*-active fragment also contains a domain related to the *nic* site of RK2/RP4 and to the left conserved domain of T-region borders. The borders, which flank the T-DNA, serve as the cleavage sites for the *virD*-encoded single-stranded endonuclease (113, 116) and are functionally the T-region equivalents of *oriT nic* sites. However, the pTiC58 domain containing these conserved sequences lacks *oriT* activity. The significance of this sequence conservation remains to be determined.

Clones containing the pTiC58 *oriT* region are mobilized efficiently by other Ti plasmids, including the nopaline/agrocinopine element, pTiT37, and the octopine-type plasmid pTi15955 (12a). Thus, the mobilization and transfer functions of other IncRh1 Ti plasmids are fully cross-functional with the *oriT* of pTiC58. This is consistent with the high degree of cross-hybridization observed between the Tra regions of these plasmids (see above) and suggests that the conjugal transfer systems of these elements derived from a common ancestor.

3.4. Conjugal Opine Catabolic Determinants

Although not strictly Tra determinants, genes encoding catabolism of the conjugal opine may be considered as part of the conjugal transfer systems of the Ti plasmids. First, expression of these genes is generally coregulated with those encoding Tra functions (see below). Second, expression of these genes is required for efficient transport of the conjugal opine, a function probably necessary for it to act as an inducer.

For the classical octopine-type Ti plasmids, octopine is the conjugal opine (Figure 10.4). The *occ* locus, which encodes catabolism of this arginyl opine, maps to the 3 o'clock region of the Ti plasmid (Figure 10.2), just clockwise from a known Tra region. This locus encodes several functions, including an octopine transport system, a two-component octopine oxidase, and ornithine cyclodeaminase (94, 108). Current evidence suggests that the locus is transcribed in an anticlockwise direction from a single inducible promoter (44, 115). The first four open reading frames present in the operon could encode polypeptides that share amino acid sequence similarities with proteins encoded by the histidine transport operon of *Salmonella typhimurium*. This suggests that octopine is transported by an ATP-driven, periplasmic-binding, protein-dependent uptake system (108). Finally, as discussed below, expression of *occ* is positively regulated by the product of *occR*, a gene located at the far right-hand (clockwise) end of the operon (44, 115).

The classical nopaline-type Ti plasmids are induced to transfer by the complex sugar phosphate opine, agrocinopine A (28) (Figure 10.4). Catabolism of this opine and its degradation product, agrocinopine B, is encoded by *acc*, which maps just clockwise from Tra region I (47) (Figure 10.2). The locus is about 7 kb in size and is organized into six gene-sized complementation groups (G. T. Hayman et al, submitted for publication). Mutations in *acc* abolish opine use and also the opine-dependent inducibility of Tra functions (3). Among the nopaline-type Ti plasmids, the structure of *acc* is highly conserved, even though these plasmids show considerable overall restriction site polymorphisms (48). In contrast to the octopine regulatory system, the *acc* locus is negatively regulated by the product of *accR*. This gene is tightly linked to *acc* and maps in the 2-kb region between *acc* and TraI (see below) (3).

Octopine Family

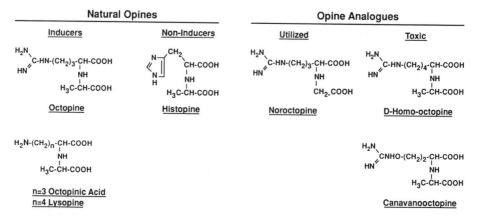

FIGURE 10.4. A. Chemical structures of the octopine family of conjugal inducer opines. The natural opines are those found in crown gall tumors induced by octopine-type *Agrobacterium* strains. The analogues can be synthesized chemically, or, in the case of homo-octopine, can be isolated from tumors grown in the presence of homo-arginine. All of the natural opines, except histopine, induce Ti plasmid-encoded opine catabolism and conjugal transfer when fed to the appropriate *Agrobacterium* strain. None of the octopine analogues are active inducers. (Figure continued on next page.)

Agrocinopine Family

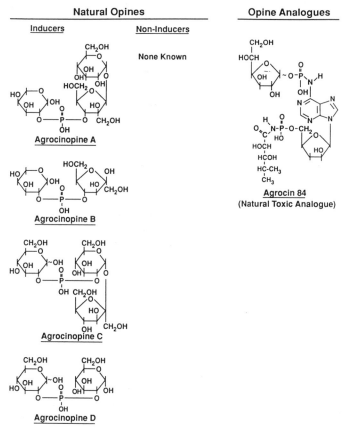

FIGURE 10.4. (*continued*) B. Chemical structures of the agrocinopine family of conjugal inducer opines. Tumors induced by classical nopaline-type *A. tumefaciens* strains contain agrocinopines A and B, while those induced by agropine-type strains produce agrocinopines C and D. These opine families are also found in hairy roots induced by certain *A. rhizogenes* isolates (19). Each of the two members of each subfamily can induce catabolic determinants and conjugal transfer systems encoded by their respective Ti plasmids. No synthetic analogues have been reported for either subfamily. However, the natural antibiotic agrocin 84 (see 33 and 34) is a toxic analogue of both the A/B and the C/D families.

Conjugal transfer of the agropine-type IncRh2 Ti plasmid, pTiBo542, is induced by another sugar phosphate opine, agrocinopine C (Figure 10.4), or by its degradation product, agrocinopine D, but not by agrocinopines A or B (28). The location of the genes encoding catabolism of this opine on the physical map of pTiBo542 is not known. Furthermore, the agrocinopine C/D catabolic locus shows no detectable cross-hybridization with *acc* from pTiC58 (48). If the two systems share a common ancestry, they have undergone considerable divergence.

In summary, it is interesting to note that in the most well-studied systems—that is, the octopine- and nopaline-type Ti plasmids—the genes encoding conjugal opine catabolism, regulation by the conjugal opine, and at least one set of conjugal transfer functions are closely linked. Furthermore, the Tra genes linked to the conjugal opine determinants in the two plasmids appear to be homologues. Tra regions I and II of pTiC58 are linked to *acc*, while homology region C of the octopine-type Ti plasmids, which cross-hybridizes with TraI and II, is linked to *occ*. Interestingly, TraIII of pTiC58 is linked to *noc*, and nopaline is chemically related to octopine. Furthermore, the *occ* and *noc* loci show an overall similar pattern of gene organization (115), and both are regulated by activators (see below) (44, 115). However, nopaline does not induce conjugal transfer of the nopaline-type Ti plasmids (28).

4. Regulation of Ti Plasmid Conjugal Transfer

Although repression of transfer is not novel among conjugal elements, the existence of specific inducer signals in the form of low-molecular-weight carbon compounds is singular to the *Agrobacterium* Ti plasmids. Furthermore, the opines are used as carbon and energy sources via Ti plasmid-encoded functions, the expression of which are themselves regulated by their cognate substrates.

4.1. Regulation of Conjugal Transfer in Octopine-Type Ti Plasmids

Early genetic studies suggested that *occ* and conjugal transfer functions of octopine-type Ti plasmids were coregulated by octopine in a negative fashion. Strains harboring a wild-type octopine Ti plasmid were repressed for transfer, and this repression could be alleviated by addition of octopine (41, 59). Lysopine (Figure 10.4), a homologue of octopine found naturally in octopine-type tumors, also relieves repression of both *occ* and Tra functions (78). On the other hand, D-histopine, another naturally occurring homologue of octopine, does not induce *occ* or conjugal transfer. In a similar fashion, noroctopine (arginine-N-2-acetic acid) (see Figure 10.4), a synthetic analogue of octopine, is not found in tumors and does not induce *occ* or conjugal transfer. However, cells harboring octopine-type Ti plasmids can grow on noroctopine or on histopine if inducing amounts of octopine are included in the medium (75, 78). This indicates that these two compounds are usable but noninducing analogues of octopine.

Petit and Tempé (75) used these noninducing analogues to isolate spontaneous mutants of an octopine-type Ti plasmid that were constitutive for octopine catabolism. Such mutants fell into two classes with respect to conjugal transfer. One group remained repressed for Tra functions, and the other showed opine-independent, constitutive transfer. In the latter class, the efficiency of constitutive transfer corresponded to the degree of derepression at *occ*. Some mutants grew well on octopine analogues, and these transferred their Ti plasmids at high frequencies without opine induction. Others grew poorly on the analogues and showed lower, but measurable, rates of uninduced conjugal transfer.

Similar results were reported by the Leiden group (61,62). Regulatory mutants of the octopine-type Ti plasmid pTiB6S3 were isolated and fell into two classes. One group was constitutive for both octopine use and conjugal transfer, while the second was constitutive for octopine catabolism but remained inducible for conjugal transfer (61, 62). The double-constitutive phenotype could be complemented *in trans* with a compatible plasmid carrying the *occ* region from a wild-type Ti plasmid. However, the complemented Tra phenotype was not completely wild-type; transfer under noninducing conditions was at least two orders of magnitude higher than that of the wild-type control, while transfer under inducing conditions was 1000-fold lower than the uncomplimented Tra^c parent. The opine-constitutive phenotype of the *occ*^{c}/Tra^i mutant class was not complementable. In both studies, analogue-utilizing mutants also were isolated and remained repressed for opine catabolism unless octopine or the opine analogue was included in the medium. Such mutants remain inducible for conjugal transfer, but, like opine catabolism, transfer functions now respond to induction by the analogue.

Mutant plasmids constitutive for conjugal transfer also can be obtained readily. Such spontaneous mutations are isolated by mating strains carrying wild-type Ti plasmids with an appropriate *Agrobacterium* recipient under noninducing conditions (52, 78). Rare transconjugants appear, and most of these contain Ti plasmids that now transfer at high frequency in the absence of inducer opines. As with the opine-constitutive mutants, the Tra-constitutive mutants fall into two classes with respect to opine catabolism. One group remains inducible, while the other now shows constitutive expression of opine catabolism functions. The mutation in this later class is dominant; *occ*^{i}, Tra^c/*occ*^{i}, Tra^i merodiploids remain constitutive for conjugal transfer (52).

From these results, a model proposing a single repressor was developed to account for the coordinate regulation of opine catabolism and conjugal transfer by the cognate conjugal opine (62, 78) (see Figure 10.5). Coregulation by a single element is consistent with the fact that mutants selected for constitutive expression of one trait were often (but not always) also constitutive for the other. That such spontaneous mutants derepressed for both phenotypes can be isolated was taken to indicate that the general control was negative with a single gene product regulating expression of both traits. In the model, each phenotype is expressed from genes whose promoters are regulated by the opine-responsive repressor. Mutants de-repressed for both phenotypes represent repressor nulls, while the variants constitutive for only one trait result from operator mutations in *occ* or in Tra operons. The fact that *occ*^{c}/Tra^c mutants can be *trans*-complemented while, in general, mutants constitutive for only one of the traits cannot is consistent with the model. Furthermore, mutations resulting in analogue-responsive induction were viewed as representing variant repressor proteins that now recognize the opine analogue as a valid inducer.

However, there are some difficulties with this model. First, in several cases, complementation with plasmids encoding the wild-type regulatory system often gave incomplete restoration of regulation or resulted in inexplicable regulatory phenotypes (61, S. K. Farrand, unpublished results). Second, Klapwijk et al (62) reported the isolation of a class of mutants that was *occ*$^-$/Tra$^-$. These mutants were obtained by selecting for resistance to homo-octopine, a toxic analogue of octopine. The mutations in some of these isolates were revertible at reasonable frequencies, indicating that the double-null phenotype did not result from deletion events removing *occ* and Tra structural genes. This suggested that coregulation of the two phenotypes involves a positive function.

The structure and regulation of the *occ* operon recently has been investigated in some detail (see above) (44, 108, 115). Analyses using *lacZ* fusions or promoter probes indicate that *occ* probably contains a single promoter that initiates transcription in an anticlockwise

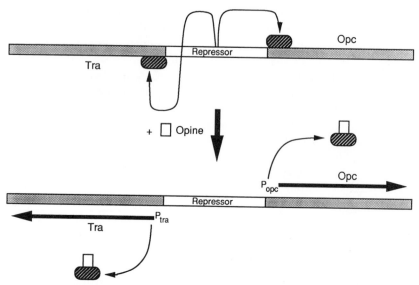

FIGURE 10.5. Classical model for the coordinate regulation of Ti plasmid-encoded opine catabolism and conjugal transfer. In this model, a single repressor gene encodes a protein (striped oval) that negatively regulates transcriptional units for conjugal transfer (Tra) and opine catabolism (Opc). On its addition, the conjugal opine (open box), or a specific degradation product, interacts with the repressor, causing it to disassociate from the Tra and Opc operator regions. Transfer and opine catabolism genes are then transcribed from their respective promoters (P).

direction (Figure 10.6). Furthermore, this promoter is not active when isolated from other *occ* elements. This observation suggests that *occ* is regulated by an activator. A gene whose product regulates *occ* has been cloned and sequenced. Called *occR*, this gene maps just clockwise from the *occ* structural operon but is divergently transcribed (44, 115). The open reading frame corresponding to *occR* can encode a protein, called OccR, with a size of 32.6 kDal. Comparison of the predicted amino acid sequence of OccR with other proteins in various databases showed a similarity with a group of positive gene regulators belonging to the LysR family. These activators, including OccR, all share a conserved domain called the LysR signature. Interestingly, OccR is related to NocR, the positive regulator of the nopaline catabolic operon of nopaline-type Ti plasmids (115).

While it has yet to be shown that OccR positively coregulates Tra, the fact that the protein appears to be an activator accounts for the *occ⁻*/Tra⁻ class of mutants reported by Klapwijk et al (62). Furthermore, the mutant analyses described above are not necessarily inconsistent with regulation by activation. Traᶜ/*occᶜ* Ti plasmids could arise through mutations in *occR* that result in altered regulatory proteins that no longer require the opine coinducer for transcriptional activation (44). Similarly, mutations may alter the activator such that it now recognizes opine analogues as coinducers. This would account for the noroctopine-using mutants that remain *occⁱ*/Traⁱ but now respond to the analogue. The aberrant complementation phenotypes may be due to copy-number effects or, more likely, to the formation of heteromultimeric forms of OccR. At least two members of the LysR

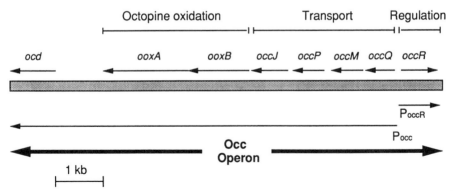

FIGURE 10.6. Organization of the regulon encoding octopine catabolism from the classical octopine-type Ti plasmids. The regulon appears to be organized as two divergent transcriptional units. One encodes the positive regulator, AccR, required for expression of octopine catabolism. Presumably, this activator also regulates the Ti plasmid Tra genes. The other is a multicistronic operon encoding functions required for transport and degradation of members of the octopine family of conjugal opines. The products of *occJ*, *occP*, *occM*, and *occQ* are closely related by deduced amino acid sequence comparisons with the correspondingly lettered gene products of the histidine transport operon from *Salmonella typhimurium* (108). Presumably, this region encodes a periplasmic protein-type ATP-dependent opine transport system. *ooxA* and *ooxB* encode the two components of octopine oxidase, while *ocd* encodes the enzyme ornithine cylcodeaminase, which converts ornithine to proline (35, 94).

activator family function as protein tetramers (7, 91), and it is possible that active OccR is also an α_n multimeric species. Formation of heteromultimers composed of mutant and wild-type subunits in the merodiploid strains might account for incomplete complementation.

4.2. Regulation of Conjugal Transfer in Nopaline-Type Ti Plasmids

Coregulation of conjugal transfer and conjugal opine catabolism appears to be mediated by a different mechanism in the nopaline-type Ti plasmids. In these elements the two traits are induced by the sugar phosphate opines agrocinopines A and B (see above) (28). Ellis et al (27) isolated a number of spontaneous Tra^c derivatives of pTiC58. These mutants fell into two classes with respect to opine catabolism. One class, containing the vast majority of the Tra^c mutants, was also derepressed for agrocinopine catabolism. The second class, represented by only one isolate, had lost the ability to catabolize the conjugal opine.

Agrocinopine A is a complex sugar conjugate, and chemical synthesis of the active opine is difficult (26, 68, 88). To date, no nontoxic analogues of this opine have been reported, so a strategy using such compounds for isolating mutants derepressed for opine catabolism has not been possible. However, mutants unable to catabolize agrocinopines have been isolated (27). Most are unaffected in their conjugal transfer properties. Others, however, are Tra$^-$, and the Ti plasmids in these mutants all contain large deletions. Presumably, the deletions inactivating *acc* extend into a transfer region. This is consistent with the mapping data showing that *acc*, the agrocinopine catabolic locus, is adjacent to TraI (4, 47) (see Figure 10.7). This genetic evidence suggests that conjugal transfer and conjugal

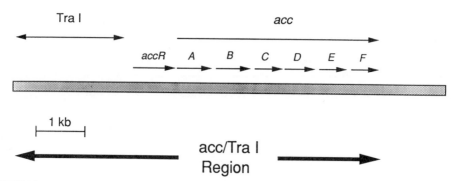

FIGURE 10.7. Organization of the nopaline-type Ti plasmid conjugal transfer-opine catabolism regulatory region. This region, located at about 8 o'clock on the pTiC58 map (see Figures 10.1 and 10.2), encodes the locus for uptake and catabolism of agrocinopines A and B (*acc*), the TraI region, and a gene, called *accR*, the product of which acts as a negative regulator for both opine catabolism and conjugal transfer. The *acc* locus, which is transcribed from left to right (47), is divided into six units based on complementation analysis (G. T. Hayman et al, unpublished). The regulatory gene *accR* is tightly linked to *acc* and to TraI, which is believed to encode an activator that regulates conjugal transfer. See text for details.

opine catabolism in the nopaline-type Ti plasmid are coregulated by a repressor, with agrocinopine acting as the inducer.

Recently, a fragment of pTiC58 has been cloned that *trans*-complements the *acc*c/Trac phenotype of the spontaneous mutant Ti plasmids. Merodiploids containing the mutant Ti plasmid and the clone exhibit repressed opine catabolism and conjugal transfer phenotypes (3). Furthermore, the fragment represses expression of *lacZ* reporter fusions in TraII and in *acc*. Sequence analysis of this fragment identified an open reading frame that could encode a protein of about 28 kDal. The protein, called AccR, has been visualized (S. Beck von Bodman and S.K. Farrand, unpublished results), and it shares extensive amino acid sequence identity with a family of repressor proteins that includes FucR, GlpR, and DeoR from *E. coli*, and LacR from *Lactobacillus casei* (3). These results are consistent with the genetic and biochemical evidence that the two coregulated phenotypes are controlled by a common repressor. The gene encoding this regulatory protein, *accR*, maps just upstream from *acc* (Figure 10.7). Analysis of the single spontaneous Trac/acc− mutant described by Ellis et al (27; also see above) showed that this Ti plasmid had suffered a 1.1-kb deletion extending from *accR* into *acc* (3). This accounts for the unexpected opine-negative phenotype. Thus, although the conjugal transfer systems of the octopine- and nopaline-type Ti plasmids appear to have evolved from a common source, the mechanisms by which they are regulated by the respective conjugal opines are remarkably different. Ironically, the pTiC58 system fits the model proposed to explain the genetic analysis of the octopine-type Ti plasmids (62, 75) (Figure 10.5).

It is clear from these results that primary coregulation of opine catabolism and conjugal transfer is mediated by classical positive or negative regulatory systems. Presumably, in both cases, low-level expression of the opine catabolic operons allows uptake of the conjugal opine. Once in the cell, the opine, or a specific metabolite, in one case activates OccR and in the other case inactivates AccR. In each, alteration of the regulatory protein results in activation or derepression of conjugal transfer and opine catabolic genes. In this respect, regulation of conjugal transfer differs from that of T-DNA transfer. The former senses signals through a classical metabolic uptake system, while the latter senses signals via a two-component regulatory system.

Conjugal transfer of the nopaline-type Ti plasmids appears to be regulated by at least two additional control mechanisms. Beck von Bodman et al (4) described the isolation of a class of Tra^c mutants that differed from the spontaneous mutants reported by Ellis et al (27). These mutants, produced by Tn5 mutagenesis, still show proper regulation of *acc*. Furthermore, the transposon insertions in the two independent mutants of this class both map to a position in TraI, some 4.5 kb from *accR*. Clearly these mutations are not affecting the Tra/opine repressor.

The Tn5 Tra^c mutation is *cis*-dominant; it is not complementable *in trans*, and a Ti plasmid carrying a copy of each allele is Tra^c (4). The mutation is also *trans*-dominant, and merodiploids containing a wild-type pTiC58 and a clone containing the Tn5-mutated TraI region exhibit the mutant Tra^c/acc^i phenotype.

A model accounting for this mutant phenotype arose from analysis of reporter gene fusions. A TraII::*lacZ* fusion on a clone lacking TraI and *accR* produces very little β-galactosidase activity when *in trans* to wild-type pTiC58. Unexpectedly, the clone is also inactive in a background lacking any Ti plasmid. This is inconsistent with Tra genes being regulated solely by AccR. However, high levels of β-galactosidase are produced when the reporter fusion is placed *in trans* to either class of Tra^c Ti plasmid, while intermediate levels of the enzyme are produced in strains containing the TraII::*lacZ* reporter and the cloned Tn5-mutated TraI region (79a).

That the reporter fusion is not active in the absence of *accR* indicates that the repressor does not directly control the expression of genes in TraII. However, the fusion is expressed when in the presence of either Tra^c mutant class. These observations indicate that TraII is positively regulated and that its expression requires an activator function encoded by TraI. Expression of TraI, in turn, could be regulated by *accR*, although there are no data to support this. What then is the nature of the Tn5 Tra^c mutation? The model predicts that the transposon insertion generates constitutive expression of the TraI-encoded activator gene required for expression of the TraII::*lacZ* reporter fusion, perhaps by altering the operator/promoter region, or by transposon-promoted transcription of the gene. Either mechanism is consistent with the *cis*- and *trans*-dominant character of the mutation. Constitutive expression of the activator results in induction of TraII and subsequent conjugal transfer. Because the constitutive mutation is downstream from the Tra/*acc* repressor in the regulatory pathway, *acc* remains repressed.

The major elements of the model are presented in Figure 10.8. While opine catabolism is regulated solely by AccR, conjugal transfer is regulated by negative and positive elements. Primary regulation is mediated through the repressor activity of AccR, thus making conjugation responsive to opine signals. However, expression of genes in TraII is under direct positive control mediated by the activator encoded in TraI.

The nature and mode of action of the activator are unknown. However, a fragment expressing this regulatory element derepresses conjugal transfer of other nopaline-type Ti plasmids such as pTiT37 and also classical octopine-type Ti plasmids, but has no effect on transfer of agropine-type plasmids (K. R. Piper et al, unpublished results).

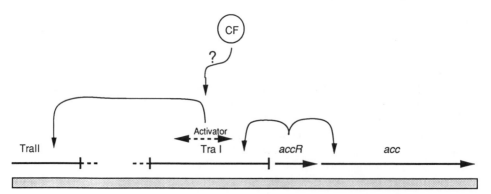

FIGURE 10.8. Model for the regulation of opine catabolism and conjugal transfer by agrocinopines A and B on the nopaline-type Ti plasmid, pTiC58. In this model, the expression of *acc* and TraI are negatively regulated by AccR, a repressor encoded by *accR*. The regulatory gene is tightly linked to *acc* and to TraI. TraI, in turn, encodes an activator required to express genes in TraII that encode functions directly involved in conjugal plasmid transfer. According to the model, in the absence of the conjugal opines, AccR binds to specific operator sequences blocking expression of *acc* and also of TraI. Hence no activator is present to induce TraII. On their addition, the conjugal opines interact with AccR, causing it to disassociate from the operator regions. This results in expression of *acc* and of TraI, which in turn leads to production of the Tra activator. This factor positively regulates expression of TraII. In this model, conjugation factor (CF) enhances the activator function by some unknown mechanism.

4.3. Other Levels of Regulation

There is some evidence that chromosomal functions may be involved in regulating Ti plasmid transfer. Several nopaline-type Ti plasmids, when assayed in their wild-type *A. tumefaciens* strains of origin, are partially derepressed for transfer in the absence of agrocinopines A and B (28). However, when one such plasmid, pTiT37, is present in the *A. tumefaciens* strain C58 background, it is tightly repressed for transfer, and conjugation is dependent on induction with the conjugal opine (28, K. R. Piper and S. K. Farrand, unpublished results). Interestingly, the *acc* locus of pTiT37 is properly regulated in both chromosomal backgrounds, suggesting that the mechanism responsible for the partial derepression in strain T37 operates downstream of the Tra/*acc* repressor. It is conceivable that a chromosomal determinant is required for the TraI-encoded activator-related regulation of transfer. As described above, the pTiC58 activator function derepresses transfer of pTiT37 in the C58 chromosomal background.

As briefly described above, Zhang and Kerr (131) recently reported that certain *A. tumefaciens* donors can influence the efficiency at which other donors transfer their Ti plasmids. They observed that octopine-type Ti plasmids fell into two classes with respect to octopine-inducible transfer frequencies. Efficient or Tra[e] Ti plasmids conjugate at frequencies close to 10^{-2} when induced with low levels of octopine. Inefficient or Tra[ie] Ti plasmids show no response to low levels of inducer but do show transfer at frequencies of about 10^{-5} when the octopine concentration is raised 70-fold above that required for maximal transfer of Tra[e] plasmids. Remarkably, culture supernatants from opine-induced Tra[e] strains will boost transfer frequencies of Tra[ie] strains to wild-type levels under low inducer conditions. Production of this conjugation factor, or CF, by wild-type strains is induced by conjugal opines, but transfer-constitutive mutants produce CF without opine induction. CF is not produced by strains lacking a Ti plasmid. This suggests that CF is synthesized by Ti plasmid-encoded gene products, the expression of which are regulated by the conjugal opine. Both octopine- and nopaline-type strains produce CF, and the factor produced by nopaline-type strains is active on donors harboring Tra[ie] octopine-type Ti plasmids. CF does not substitute for the conjugal opine; transfer still requires primary induction. However, CF appears to potentiate transfer by inefficient donors. Tra[ie] donors produce CF, but only after prolonged incubation with high concentrations of octopine.

CF is dialyzable and sensitive to autoclaving or treatment with strong base (131). It is resistant to treatment with DNase, RNase, and protease. There have been no reports to date concerning its structure, the genetics or biochemistry of its synthesis, or its mode of action. Production of CF by an octopine-type strain is inducible by octopine and also by many other imino acids including nopaline, octopine and nopaline analogues, two of the mannityl opines, and proline (131). Its production is not induced by arginine or pyruvate, the two components of octopine. CF apparently exerts its effect on donor cells. Tra[ie] donors treated with CF and then washed and mated with recipients show stimulated Ti plasmid transfer. Pretreatment of the recipient, on the other hand, did not stimulate transfer from a Tra[ie] donor.

5. Relationship of Ti Plasmid Conjugal Transfer to Other DNA Transfer Processes

Recent comparative studies indicate that DNA transfer systems encoded by diverse plasmids may have had common ancestors. For example, the nick site in the *oriT* region of RK2/RP4 is located in a sequence almost identical to the left conserved domain of T-region borders (119). Furthermore, the positions at which the nicks occur in the two sites are identical. The similarities correlate with the enzymology of the nicking reaction. TraI, the RK2/RP4 relaxase, contains a domain that is conserved in VirD2, the protein thought to be responsible for nicking at T-region borders (74). Finally, the *tra2* region of RP4 contains a group of genes that are closely related in organization and in encoded protein amino acid sequences to genes in the *virB* operon of pTiC58 (65). This virulence operon is believed to encode the functions by which *Agrobacterium* forms a mating bridge with its host plant cells. These intriguing similarities suggest that the RK2/RP4 conjugal transfer system and the Ti plasmid encoded T-DNA transfer system derived from a common ancestor. The validity of this notion was considerably strengthened by a recent report from Hooykaas's group (5). They demonstrated that the *vir* system of the octopine-type Ti plasmid pTiB6 can mobilize the transfer of an RSF1010 derivative from an *Agrobacterium* donor to an *Agrobacterium* recipient. That is, the *vir* system is an extraordinarily broad-host-range conjugation apparatus recognizing as valid recipients, not only plant cells, but also bacteria. Although the frequencies of mobilization were low, the transfer event is dependent upon the *vir* system, occurring only when the system was properly induced. Furthermore, no transfer was observed when the Ti plasmid contained mutations in the *vir* region (5).

5.1. Relationships with Other Plasmid Conjugal Transfer Systems

Is the Ti plasmid conjugal transfer system related to that of any other plasmid? It is still too early to draw firm conclusions. However, as described above, the *oriT* region of pTiC58 contains a domain that is virtually identical to the *nic* region of RSF1010. This is provocative for several reasons. First, RSF1010 is a broad host range element that is itself not self-conjugal. However, it can be mobilized by a number of apparently unrelated conjugal elements. Mobilization requires, in addition to its *oriT*, RSF1010-encoded *mob* functions that are involved in processing the plasmid at the *nic* site (17,18). Several years ago, Buchanan-Wollaston et al (5a) reported that the Ti plasmids can transfer RSF1010 derivatives to plant cells. This transkingdom mobilization requires the RSF1010 *nic* site and intact *mob* functions as well as appropriate Ti plasmid-encoded *vir* functions.

Recently, Ward et al (117) showed that RSF1010 suppresses Ti plasmid-dependent T-DNA transfer to plants. This inhibition occurs in a manner consistent with a competitive interaction between DNA transfer intermediates and products of the *virB* operon. So, although T-region borders and the RSF1010 *nic* site do not share sequences in common, the processed *nic*-Mob complex of RSF1010 is recognized by *vir* functions as a legitimate substrate for transfer to plant cells. On the other hand, the Ti plasmid does not mobilize conjugal transfer of the R plasmid even though the *oriT* regions of the two elements are clearly related. The features of the processed RSF1010 transfer intermediate that allow it to function with the Ti plasmid *vir* system are evidently not recognized by the Ti plasmid conjugal transfer apparatus.

The *oriT* region of pTiC58 also contains domains related to the *nic* site of the IncP1α plasmids RP4 and RK2 (see above) (12a). Although the cross-conserved pTiC58 sequence is not in the domain of the *nic* region, it is interesting to note that this sequence corresponds to that region of the RP4/RK2 *nic* domain that is homologous to T-region borders (119). However, unlike the RSF1010 system, the RP4/RK2 *oriT* will not substitute for a T-region border (15).

There is, then, an interesting perplexity. RSF1010 is mobilized to plants or bacteria by the Ti plasmid *vir* system in a RSF1010 *nic*-Mob-dependent manner, but the R plasmid *oriT* shows no relatedness with T-region borders. The *nic* domain of RP4/RK2, on the other hand, shows homology with T-region borders but does not function with *vir* as a recognizable border for transfer to plants. The *oriT* regions of both RP4/RK2 and RSF1010 contain elements homologous to those present in the *oriT* region of pTiC58, but neither R plasmid *oriT* allows mobilization by the Ti plasmid to recipient bacteria. Clearly, more work is necessary before we understand what factors confer specificity between *nic* complexes and the conjugal transfer machinery.

5.2. Interactions between Determinants of Ti Plasmid Conjugal Transfer and Virulence

There is some question as to whether the Tra functions of Ti plasmids interact with the *vir* system. Early mapping studies yielded Tra⁻/Vir⁺ and Tra⁺/Vir⁻ but no Tra⁻/Vir⁻ transposon insertion mutants of both octopine- and nopaline-type Ti plasmids (14, 51).

Certain deletion mutants were defective in both phenotypes, but the deletions are large and extend through known Tra and Vir regions.

In their mutational analysis of pTiC58 Tra functions, Beck von Bodman et al (4) reported that all Tra⁻ mutants isolated were fully virulent. Furthermore, they showed that insertion mutations in *virA*, *virB*, *virD*, and *virG*, as well as mutations in the chromosomal virulence loci, *chvA* and *chvB*, had no detectable effect on conjugal transfer. They concluded that the two Ti plasmid-encoded gene transfer systems are physically and functionally distinct.

More recently, two reports have appeared demonstrating that mutations in *vir* affect conjugal transfer functions. Gelvin and Habeck (40) reported that insertion mutations in *virA*, *virB*, *virD*, and *virG* of an octopine-type Ti plasmid exerted demonstrable effects on the conjugal transfer frequency of the mutant plasmid following induction with octopine. No *vir* mutation completely abolished conjugal transfer. Rather, several lowered conjugal transfer frequencies by factors of two to three orders of magnitude. For most of the *vir* loci, the effects were allele dependent. Some insertions in a given *vir* operon affected transfer, whereas others in the same operon did not. The suppressive effects of the *vir* mutations could be overcome by supplying the wild-type *vir* allele *in trans*.

In a similar study, Steck and Kado (102) reported that insertion mutations in the *vir* region of pTiC58 lowered conjugal transfer frequencies from 2- to 20-fold as compared to the unmutagenized parent plasmid. Again, the effects of the mutations were allele dependent, with some *vir* insertion mutations in *virC* and in *virD* actually increasing transfer frequencies. Interestingly, these workers noted a 10-fold stimulatory effect of acetosyringone on the conjugal transfer frequency of the parent Ti plasmid.

The relationship between *vir* functions and Tra functions remains to be clarified. There is certainly a discrepancy between the results of Beck von Bodman et al (4) and those of Gelvin and Habeck (40) and Steck and Kado (102). In this regard it is interesting to note that one avirulent mutant of pTiC58 containing a ca. 6-kb deletion that removes much of the 5′ end of the *virB* operon is not affected in conjugal transfer functions (51). Similarly, a deletion in the octopine-type Ti plasmid pTiB6S3, extending from the middle of *virB* through *virG*, *virC*, *virD*, and *virE* into T_L, reportedly had no effect on conjugal transfer (14).

5.3. Involvement of Chromosomal Determinants

Although Beck von Bodman et al (4) did not obtain any Tra⁻ Tn5 insertions that mapped to the bacterial chromosome or to the large cryptic plasmid present in the target strain, there is circumstantial evidence that nonplasmid determinants may influence plasmid conjugal transfer in *Agrobacterium*. First, an RSF1010 derivative was mobilized at low levels from an *Agrobacterium* donor known to lack additional plasmids less than 2000 kb in size (12a). Second, another transmissible *Agrobacterium* plasmid, pAgK84, contains a Tra region only 3.5 kb in size (see below) (37). This is probably too small to encode a complete self-contained conjugal transfer system. However, pAgK84 is transmissible by mating from an *Agrobacterium* donor that contains no other detectable plasmids (37). It seems likely that this plasmid is mobilized by another conjugal element, perhaps associated with the *Agrobacterium* chromosome. Last, chromosomal determinants may influence the mechanism by which conjugal transfer is regulated by opines. As described above, pTiT37 is partially constitutive for conjugal transfer from its host of origin. However, in the C58 chromosomal background, transfer of this Ti plasmid is fully repressed and requires addition of agrocinopine to induce conjugation (28, D. M. Cook and S. K. Farrand, unpublished results).

6. Conjugal Transfer of Other *Agrobacterium* Plasmids

Agrobacterium isolates usually contain more than one plasmid (69, 70, 129). Some of these have known functions, while others remain cryptic. However, several of these plasmids are known to be self-conjugal

6.1. pAtC58

Agrobacterium tumefaciens strain C58 harbors, in addition to its Ti plasmid, a 410-kb plasmid called pAtC58 (6). This plasmid confers no demonstrable phenotype on strain C58; derivatives cured of pAtC58 are indistinguishable from the wild-type strain (55, 87). Using a derivative of pAtC58 marked with a Tn*1* insertion, van Montagu and Schell (112) showed that this element is self-conjugal to an *Agrobacterium* recipient. No information is available concerning the organization or expression of Tra genes on this plasmid. However, pAtC58 can mobilize RSF1010 at frequencies of about 10^{-4} per input donor (12a).

6.2. Opine Catabolic Plasmids

Agrobacterium radiobacter is distinguished from *A. tumefaciens* by its inability to induce plant neoplasias. This is something of an artificial division, since introduction of a Ti plasmid into *A. radiobacter* strains generally converts them to pathogens. However, many *A. radiobacter* isolates harbor large plasmids of their own (70), some of which encode catabolism of opines. Two such plasmids, pAtK84b and pAtK112, both of which encode catabolism of nopaline, are known to be self-conjugal (28). In apparent distinction from nopaline-type Ti plasmids, conjugation of these two elements is induced by both nopaline and agrocinopine (28).

Plasmid pAtK84b has been examined in some detail. This element is indigenous to *A. radiobacter* strain K84, which is used commercially as an inoculum to control crown gall disease in the field (for review, see 33 and 34). The plasmid is 173 kb in size and shares about 80% sequence homology with pTiC58 (11). Regions that hybridize strongly with the nopaline-type Ti plasmid include those encoding nopaline catabolism and replication. However, pAtK84b lacks significant detectable homology with *vir* and T-region sequences from pTiC58 or from an octopine-type Ti plasmid. The nopaline catabolic plasmid also shows strong homology with probes corresponding to the B and C cross-homology regions of the octopine- and nopaline-type Ti plasmids. These regions encode the known Tra loci of the two virulence elements (see above and Figure 10.2). Consistent with this homology at

Tra, pAtK84b mobilizes vectors containing the minimal pTiC58 *oriT* with the same efficiency as does pTiC58 (12a).

Despite the obvious relatedness between pAtK84b and pTiC58, the two plasmids show some interesting differences. Most notably, although both encode use of agrocinopines A and B, the *acc* locus from pTiC58 does not appreciably cross-hybridize with pAtK84b (48). Furthermore, pAtK84b and pTiC58 differ with respect to functions encoded by their *acc* determinants. While both encode uptake of agrocinopines A and B, *acc*(pTiC58), but not *acc*(pAtK84b), confers uptake of and sensitivity to the antiagrobacterial antibiotic, agrocin 84 (48).

The two plasmids also may differ with respect to regulation of conjugal transfer. In both, conjugation is inducible by agrocinopines. However, pAtK84b, but not pTiC58, is inducible also by nopaline (28). This difference may result from factors not encoded by the plasmids; when pAtK84b is in the C58 chromosomal background, conjugal transfer, while remaining responsive to agrocinopine, is no longer inducible by nopaline (D. M. Cook and S. K. Farrand, unpublished). This suggests that chromosomal functions may be involved in regulating opine-inducible conjugal transfer functions of pAtK84b, and perhaps also of pTiC58.

6.3. pAgK84

Strain K84 harbors a second plasmid important to its functioning as a biological control strain. This 48-kb element, called pAgK84, encodes production of a novel antiagrobacterial antibiotic called agrocin 84 (30, 98). This plasmid has been extensively characterized (37, 37a, 88a). In addition to encoding production of and immunity to agrocin 84, pAgK84 is transmissible by mating to *Agrobacterium* and *Rhizobium* recipients, but not to *E. coli* or to *Pseudomonas aeruginosa* (37).

Plasmid pAgK84 has been saturation mutagenized with Tn*5* (37) and a single 3.5-kb Tra region was located. Analysis of deletion derivatives (37a, 56) and of Tn*5*-induced mutants (37) shows this to be the only region of the plasmid required for conjugal transfer. This is an extraordinarily small Tra region and brings into question the self-conjugal nature of this plasmid. As described previously, it may be that this element is mobilized by other transmissible elements present in its *Agrobacterium* host. Because the plasmid is transmissible from an *Agrobacterium* donor lacking any additional detectable plasmids, such ancillary transfer functions could be encoded only by the chromosome or by megaplasmids greater than 2000 kb in size. In this regard it is important to recall that an RSF1010 derivative is mobilized at low frequencies from an *Agrobacterium* host that lacks any other detectable plasmid species (see above).

7. Conjugal Plasmid Transfer and the Natural Biology of *Agrobacterium*

A teleological rationale for the evolution of the Ti plasmid-mediated plant transformation system has been the subject of much speculation (see, for example, 104 and 105). The most popular explanation has, as its fundamental premise, the supposition that opines confer onto *Agrobacterium* strains able to catabolize them a competitive growth advantage in habitats in which these compounds are present (77). There is, to date, no direct proof for the validity of this hypothesis. However, there are a number of positive correlates that support the model (see 19 for details). First, tumors and hairy roots induced by virulent agrobacteria synthesize and secrete opines, sometimes at high levels. Second, there is a strong correlation between the opines produced by a given tumor and the opines that can be catabolized by the *Agrobacterium* strain that induced that tumor. Third, opines are virtually unique to these plant neoplasias, and the capacity to use the compounds is not a widespread trait among microorganisms. A few nonagrobacterial organisms, most notably some soil pseudomonads, can catabolize these compounds. However, even among pseudomonad isolates of the same species, this is an atypical phenotype. Interestingly, several octopine-using pseudomonads have been isolated from marine mollusks (21). As an exception to the rule, octopine commonly is found in the muscle tissues of many mollusks (39). Finally, opines are not merely nutritional sources; they also possess signaling properties. Opines have been shown to potentiate the acetosyringone-dependent induction of Ti plasmid *vir* genes (114). And, as described previously, certain opines induce conjugal transfer.

A key feature of the model (Figure 10.9) is that opines contribute to dissemination of the Ti plasmid by two mechanisms. First, catabolism of opines confers on the Ti plasmid-containing agrobacteria a growth advantage, and thus an edge in their competition with other soil microflora. This results in the vertical expansion of the Ti plasmid population. Second, induction of Ti plasmid conjugal transfer can lead to dissemination of the plasmid to other agrobacteria. This horizontal spread may be advantageous to the long-term survival of the Ti plasmid; such transfer could allow access of these plasmids to chromosomal backgrounds better suited to a particular habitat.

Is there any proof that the Ti plasmids have spread among various agrobacterial backgrounds in nature? That Ti plasmids can conjugally transfer on tumors is indisputable. The conjugal nature of these elements was first demonstrated by planta matings (57–59). That the Ti plasmids *do* transfer in nature is more problematic. Perhaps the best evidence for this comes from examination of the wide host range octopine-type Ti plasmids. Comparison of these elements from a number of independent *Agrobacterium* isolates shows them to be virtually indistinguishable, although their agrobacterial hosts are from different geographical sites and are often very different in their physiological properties (96). Thus, the plasmids are highly conserved, while the bacterial hosts are quite divergent. This implies the horizontal spread of this Ti plasmid and suggests that the dispersal has occurred in the relatively recent past (see also 32).

A similar observation suggests that the agrocinogenic plasmid pAgK84 also has been disseminated among agrobacteria in nature. Although first identified in biovar 2 *A. radiobacter* strain K84, an Australian isolate (71), a closely related plasmid was discovered in biovar 1 *A. tumefaciens* strain Bo542 (98), originally recovered from a dahlia gall in Germany (96). Furthermore, transfer of pAgK84 from strain K84 to indigenous agrobacteria in the field has been reported (73).

The work of Dessaux et al (20) suggests that mobilization of chromosomal genes by the Ti plasmid could play a significant role in the evolution of the *Agrobacterium* taxon. The Ti plasmid transfer system is apparently able to direct substantial changes in the recipient genotype. The genus *Agrobacterium* is a broad and diverse taxon (49, 60), and a

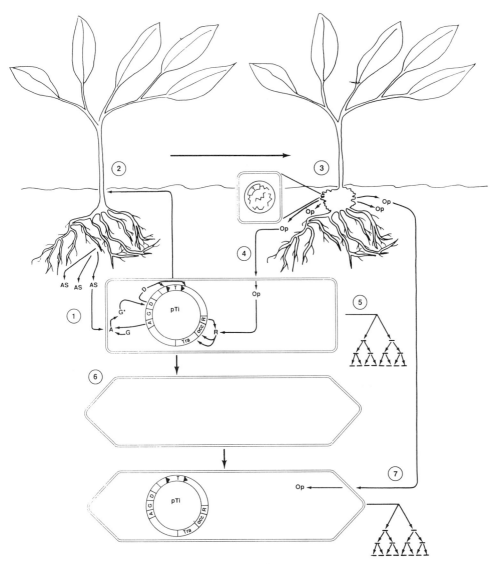

FIGURE 10.9. Proposed role for opines in the vertical and horizontal spread of *Agrobacterium* Ti plasmids. When virulent agrobacteria come in contact with susceptible plants, a series of events is initiated, culminating in the transfer from the bacterium to a plant cell of a portion of the Ti plasmid, called the T-region (step 1). The transferred bacterial DNA becomes integrated into host nuclear DNA (step 2), where it is now called T-DNA. Expression of genes encoded in the integrated T-DNA results in the transformation of a normal plant cell to a crown gall tumor cell (step 3). Characteristically, such tumors produce one or more families of opines, which can be taken up and catabolized by the inciting *Agrobacterium* strain via functions encoded by the resident Ti plasmid (step 4). This results in a selective increase in the *Agrobacterium* population within the environment of the tumor (step 5). In addition, a subclass of the opines produced by a given tumor can induce conjugal transfer of the Ti plasmid from the inducing strain to plasmidless *Agrobacterium* strains present in the soil (step 6). This horizontal spread of the Ti plasmid confers on the recipient bacterium the capacity to use opines produced by the tumor (step 7).

mechanism for mediating wide-ranging gene transfer and stabilization could ensure the emergence of novel genotypes better adapted to a given environment.

8. Perspective and Conclusions

Evidence to date indicates that the conjugal transfer of Ti plasmids probably is not mechanistically different from that of other transmissible plasmids. Why then should we continue investigating the conjugal transfer systems of these *Agrobacterium* plasmids? There is first the question of the ancestral derivations of conjugation systems. It is clear that *vir*-mediated T-DNA transfer to plants occurs by a conjugal mechanism (100), and recent evidence discussed above indicates an evolutionary relationship between this system and the conjugal transfer system of the IncP1α plasmids. There is also reason to suspect that the Ti plasmid conjugal transfer system is related to the Mob system of RSF1010. The three systems tie together through the observation that the RSF1010 Mob system can interact with the Ti plasmid *vir* functions to direct gene transfer to plants. The existence of sequences within the *oriT* region of pTiC58 that are homologous to those present in the *oriT* region of RP4/RK2 also points toward an ancestral relationship between these transfer systems. Yet the three systems are not all cross-functional; some degree of specificity has evolved. It will take a detailed analysis of all three systems to gain a full understanding of any one of them. The key may lie in the specificities.

The Ti plasmid system offers several features not available for study in other conjugal systems. Most notable among these is the regulation of transfer by external signals. We are beginning to understand how opines regulate transfer induction. However, the recent discovery of CF introduces a new component to the system. It will be interesting to determine if CF is novel to the Ti plasmids or if it is a general feature of conjugal plasmid transfer systems. The existence of CF, along with the well-known pheromone phenomenon exhibited by streptococci (12, 24; see also chapter 13), implies a communication system at the community level geared to the dissemination of genetic information.

Finally, the Ti plasmids offer a system for assessing the role of conjugal gene transfer in the natural environment. The retrospective observations described above are consistent with conjugal transfer playing a role in the in situ spread of *Agrobacterium* plasmids. The fact that transfer can take place on crown gall tumors provides a system amenable to laboratory manipulation as well as to experimentation in the field. Galls can be maintained in tissue culture where they can be evaluated as substrates for gene transfer events, and hypotheses developed from such experiments can be tested in the natural ecosystems of galls present on plants in the field.

Addendum

Results from experiments completed just before this review went to the printers have clarified the nature of the Tra activator and provide a role for CF in the regulation of Ti plasmid conjugal transfer. Sequence analysis identified an open reading frame mapping to the TraI region and located 250 bp downstream from the site of the *trac* Tn5 insertion. This ORF, called *traR*, can encode a protein that is related to LuxR, a gene activator required for transcription of the *lux* operon in *Vibrio fischeri* (Piper et al, 79a). LuxR requires a low molecular weight co-inducer, called autoinducer (AI), in order to activate transcription of the *V. fischeri* bioluminescence operon. Autoinducer, which is N-(β-oxo-hexan-1-oyl)-L-homoserine lactone, is synthesized by the *V. fischeri* cells themselves. This system of regulation, called autoinduction, is believed to monitor cell population densities, and to activate gene transcription only under conditions of high cell density (see ref. 23a for review). Our results indicated that CF is required in order for TraR to activate transcription of Ti plasmid *tra* genes. Furthermore, CF has been identified as N-(β-oxo-octan-1-oyl)-L-homoserine lactone, a close homologue of *V. fischeri* AI (Zhang et al, 132). AI will substitute for CF in the TraR-dependent activation of Ti plasmid *tra* gene transcription, although it is not nearly as efficient a co-inducer. Finally, TraR of pTiC58 is 88% identical to TraR of the octopine/mannityl opine-type Ti plasmid, pTi15955 (S. Winans, personal communication) and the two activator genes are cross-functional.

It is clear that Ti plasmid conjugal transfer is regulated by autoinduction and that the *Agrobacterium* regulatory system is closely related to that of an autoinduction system present in a distantly related bacterium. A similar system regulates production of virulence factors by *Pseudomonas aeruginosa*. The *P. aeruginosa* activator, called LasR (39a), is similar to TraR and LuxR, and the *P. aeruginosa* autoinducer, so far uncharacterized, will activate TraR-dependent transcription of Ti plasmid Tra genes (K. R. Piper and S. K. Farrand, unpublished results). These results suggest that autoinduction, mediated by an activator and a cognate substituted homoserine lactone may be a common regulatory mechanism in diverse bacteria.

References

1. Alt-Mörbe, J., Kühlmann, H., and Schröder, J., 1989, Differences in induction of Ti plasmid virulence genes *virG* and *virD*, and continued control of *virD* expression by four external factors, *Mol. Plant Microbe Interact.* **2**:301–308.
2. Barker, R. F., Idler, K. B., Thompson, D. V., and Kemp, J. D., 1983, Nucleotide sequence of the T-DNA region from the *Agrobacterium* I octopine Ti plasmid pTi15955, *Plant Mol. Biol.* **2**:335–350.
3. Beck von Bodman, S., Hayman, G. T., and Farrand, S. K., 1992, Opine catabolism and conjugal transfer of the nopaline Ti plasmid pTiC58 are coordinately regulated by a single repressor, *Proc. Natl. Acad. Sci. USA* **89**:643–647.
4. Beck von Bodman, S., McCutchan, J. E., and Farrand, S. K., 1989, Characterization of conjugal transfer functions of *Agrobacterium tumefaciens* Ti plasmid pTiC58, *J. Bacteriol.* **171**:5281–5289.
5. Beijersbergen, A., Den Dulk-Ras, A., Schilperoort, R. A., and Hooykaas, P. J. J., 1992, Conjugative transfer by the virulence system of *Agrobacterium tumefaciens*, *Science* **256**:1324–1327.
5a. Buchanan-Wollaston, V., Passiatore, J. E., and Cannon, F., 1987, The *mob* and *oriT* mobilization functions of a bacterial plasmid promote its transfer to plants, *Nature* (London) **328**:172–175.
6. Casse, F., Boucher, C., Julliot, J. S., Michel, M., Dénarié, J., 1979, Identification and characterization of large plasmids in *Rhizobium meliloti* using agarose gel electrophoresis, *J. Gen. Microbiol.* **113**:229–242.
7. Chang, M., and Crawford, I. P., 1990, The roles of indoleglycerol phosphate and the TrpI protein in the expression of *trpBA* from *Pseudomonas aeruginosa*, *Nucl. Acid Res.* **18**:979–988.
8. Chilton, M.-D., Farrand, S. K., Levin, R., and Nester, E. W., 1976, RP4 promotion of transfer of a large *Agrobacterium* plasmid which confers virulence, *Genetics* **83**:609–618.
9. Chilton, M.-D., Currier, T. C., Farrand, S. K., Bendich, A. J., Gordon, M. P., and Nester, E. W., 1974, *Agrobacterium tumefaciens* DNA and PS8 bacteriophage DNA not detected in crown gall tumors, *Proc. Natl. Acad. Sci. USA* **71**:3672–3676.

10. Chilton, M.-D., Drummond, M. H., Merlo, D. J., Sciaky, D., Montoya, A. L., Gordon, M. P., and Nester, E. W., 1977, Stable incorporation of plasmid DNA into higher plant cells: the molecular basis of crown gall-tumorigenesis, *Cell* **11:**263–271.

11. Clare, B. G., Kerr, A., and Jones, D. A., 1990, Characterization of the nopaline catabolic plasmid in *Agrobacterium radiobacter* strains K84 and K1026 used for biological control of crown gall disease, *Plasmid* **28:**126–137.

12. Clewell, D. B., and Weaver, K. E., 1989, Sex pheromones and plasmid transfer in *Streptococcus faecalis*, *Plasmid* **21:**175–184.

12a. Cook, D. M., and Farrand, S. K., 1992, The *oriT* region of the *Agrobacterium tumefaciens* Ti plasmid pTiC58 shares DNA sequence identity with the transfer origins of RSF1010 and RK2/RP4 and with T-region borders. *J. Bacteriol.* **174:**6238–6246.

13. DeFlaun, M. F., and Levy, S. B., 1989, Genes and their varied hosts, in: *Gene Transfer in the Environment* (S. B. Levy and R. V. Miller, eds.), McGraw-Hill, New York, pp. 1–32.

14. De Greve, H., Decraemer, H., Seurinck, J., van Montagu, M., and Schell, J., 1981, The functional organization of the octopine *Agrobacterium tumefaciens* plasmid pTiB6S3, *Plasmid* **6:**235–248.

15. DeFramond, A. J., Barton, K. A., and Chilton, M.-D., 1983, Mini-Ti: a new vector strategy for plant genetic engineering, *Bio/Technology* **1:**262–269.

16. DePicker, A., Stachel, S., Dhaese, P., Zambryski, P., and Goodman, H. M., 1982, Nopaline synthase: transcript mapping and DNA sequence, *J. Mol. Appl. Genet.* **1:**561–573.

17. Derbyshire, K. M., and Willetts, N. S., 1987, Mobilization of the non-conjugative plasmid RSF1010: a genetic analysis of its origin of transfer, *Mol. Gen. Genet.* **206:**415–160.

18. Derbyshire, K. M., Hatfull, G., and Willetts, N. S., 1987, Mobilization of the nonconjugative plasmid RSF1010: a genetic and DNA sequence analysis of the mobilization region, *Mol. Gen. Genet.* **206:**161–168.

19. Dessaux, Y., Petit, A., and Tempé, J., 1991, Opines in *Agrobacterium* biology, in: *Molecular Signalling in Plant-Microbe Communication* (D. P. S. Verma, ed.), pp. 137–169, CRC Press, Columbus, Ohio.

20. Dessaux, Y., Petit, A., Ellis, J. G., Legrain, C., Demarez, M., Wiame, J.-M., Popoff, M., and Tempé, J., 1989, Ti plasmid-controlled chromosome transfer in *Agrobacterium tumefaciens*, *J. Bacteriol.* **171:**6363–6366.

21. Dion, P., 1986, Utilization of octopine by marine bacteria isolated from mollusks, *Can. J. Microbiol.* **32:**959–963.

22. Drolet, M., Zanga, P., and Lau, P. C. K., 1990, The mobilization and origin of transfer regions of a *Thiobacillus ferrooxidans* plasmid: relatedness to plasmids RSF1010 and pSC101, *Molec. Microbiol.* **4:**1381–1391.

23. Drummond, M. H., and Chilton, M.-D., 1978, Tumor-inducing (Ti) plasmids of *Agrobacterium* share extensive regions of DNA homology, *J. Bacteriol.* **136:**1178–1183.

23a. Dunlap, P. V., and Greenberg, E. P., 1991, Role of intercellular chemical communication in the *Vibrio fischeri*-Monocentrid fish symbiosis, in: *Microbial Cell–Cell Interactions*, (M. Dworkin, ed.), pp. 219–253, American Society for Microbiology, Washington, D. C.

24. Dunny, G. M., 1990, Genetic functions and cell-cell interactions in the pheromone-inducible plasmid transfer system of *Enterococcus faecalis*, *Molec. Microbiol.* **4:**689–696.

25. Eden, F. C., Farrand, S. K., Powell, J. S., Bendich, A. J., Chilton, M.-D., Nester, E. W., and Gordon, M. P., 1974, Attempts to detect deoxyribonucleic acid from *Agrobacterium tumefaciens* and bacteriophage PS8 in crown gall tumors by complementary ribonucleic acid/deoxyribonucleic acid-filter hybridization, *J. Bacteriol.* **119:**547–553.

26. Ellis, J. G., and Murphy, P. J., 1981, Four new opines from crown gall tumours—their detection and properties, *Mol. Gen. Genet.* **181:**36–43.

27. Ellis, J. G., Murphy, P. J., and Kerr, A., 1982a, Isolation and properties of transfer regulatory mutants of the nopaline Ti plasmid pTiC58, *Mol. Gen. Genet.* **186:**275–281.

28. Ellis, J. G., Kerr, A., Petit, A., and Tempé, J., 1982b, Conjugal transfer of nopaline and agropine Ti plasmids; the role of agrocinopines, *Mol. Gen. Genet.* **186:**269–274.

29. Ellis, J. G., Kerr, A., Tempé, J., and Petit, A., 1979a, Arginine catabolism, a new function of both octopine and nopaline Ti plasmids of *Agrobacterium*, Mol. Gen. Genet. **173:**263–269.

30. Ellis, J. G., Kerr, A., van Montagu, M., and Schell, J., 1979b, *Agrobacterium*: genetic studies on agrocin 84 production and the biological control of crown gall, *Physiol. Plant Pathol.* **15:**311–319.

31. Engler, G., DePicker, A., Maenhaut, R., Villarroel, R., van Montagu, M., and Schell, J., 1981, Physical mapping of DNA base sequence homologies between an octopine and a nopaline Ti plasmid of *Agrobacterium tumefaciens*, *J. Mol. Biol.* **152**:183–208.

32. Farrand, S. K., 1989, Conjugal transfer of bacterial genes on plants, in: *Gene Transfer in the Environment* (S. Levy and R. V. Miller, eds.), McGraw-Hill, New York, pp. 261–286.

33. Farrand, S. K., 1990, *Agrobacterium radiobacter* strain K84: a model biocontrol system, In: *New Directions in Biological Control* (R. R. Baker and P. E. Dunn, eds.), Alan R. Liss, New York, pp. 679–691.

34. Farrand, S. K., 1991, Biological control of microbial plant pathogens through antibiosis, in: *CRC Handbook of Pest Management in Agriculture*, 2nd ed., Vol. II (D. Pimentel, ed.), CRC Press, Boca Raton, pp. 311–329.

35. Farrand, S. K., and Dessaux, Y., 1986, Proline biosynthesis encoded by the *noc* and *occ* loci of *Agrobacterium* Ti plasmids, *J. Bacteriol.* **167**:732–734.

36. Farrand, S. K., Eden, F. C., and Chilton, M.-D., 1975, Attempts to detect *Agrobacterium tumefaciens* and bacteriophage PS8 DNA in crown gall tumors by DNA-DNA filter hybridization, *Biochim. Biophys. Acta* **390**:264–275.

37. Farrand, S. K., Slota, J. E., Shim, J.-S., and Kerr, A., 1985, Tn*5* insertions in the agrocin 84 plasmid: the conjugal nature of pAgK84 and the locations of determinants for transfer and agrocin 84 production, *Plasmid* **13**:106–117.

37a. Farrand, S. K., Wang, C.-L., Hong, S.-B., O'Morchoe, S. B., and Slota, J. E., 1992, Deletion derivatives of pAgK84 and their use in the analysis of *Agrobacterium* plasmid functions, *Plasmid* **28**:201–212.

38. Fürste, J. P., Pansegrau, W., Zieglin, G., and Lanka, E., 1989, Conjugative transfer of promiscuous InP plasmids: interaction of plasmid-encoded products with the transfer origin, *Proc. Natl. Acad. Sci. USA* **86**:1771–1775.

39. Gade, G., 1980, Biological role of octopine formation in marine molluscs, *Mar. Biol. Lett.* **1**:121–135.

39a. Gambello, M. J., and Iglewski, B. H., 1991, Cloning and characterization of the *Pseudomonas aeruginosa lasR* gene, a transcriptional activator of elastase expression, *J. Bacteriol.* **173**:3000–3009.

40. Gelvin, S. B., and Habeck, L. L., 1990, *Vir* genes influence conjugal transfer of the Ti plasmid of *Agrobacterium tumefaciens*, *J. Bacteriol.* **172**:1600–1608.

41. Genetello, C., van Larebeke, N., Holsters, M., De Picker, A., van Montagu, M., and Schell, J., 1977, Ti plasmids of *Agrobacterium* as conjugative plasmids, *Nature* (London) **265**:561–563.

42. Goldberg, S. B., Flick, J. S., and Rogers, S. G., 1984, Nucleotide sequence of the *tmr* locus of *Agrobacterium tumefaciens* pTi T37 T-DNA, *Nucl. Acid Res.* **12**:4665–4677.

43. Guyon, P., Chilton, M.-D., Petit, A., and Tempé, J., 1980, Agropine in "null-type" crown gall tumors; evidence for the generality of the opine concept, *Proc. Natl. Acad. Sci. USA* **77**:2693–2697.

44. Habeeb, L., Wang, L., and Winans, S. C., 1991, Transcription of the octopine catabolism operon of the *Agrobacterium* tumor-inducing plasmid pTiA6 is activated by a LysR-type regulatory protein, *Mol. Plant-Microbe Interact.* **4**:379–385.

45. Hamilton, R. H., and Chopin, M. N., 1975, Transfer of the tumor-inducing factor in *Agrobacterium tumefaciens*, *Biochem. Biophys. Res. Commun.* **63**:349–354.

46. Hamilton, R. H., and Fall, M. Z., 1971, The loss of tumor-initiating ability in *Agrobacterium tumefaciens* by incubation at high temperatures, *Experentia* **27**:229–230.

47. Hayman, G. T., and Farrand, S. K., 1988, Characterization and mapping of the agrocinopine-agrocin 84 locus on the nopaline Ti plasmid pTiC58, *J. Bacteriol.* **170**:1759–1767.

48. Hayman, G. T., and Farrand, S. K., 1990, *Agrobacterium* plasmids encode structurally and functionally different loci for catabolism of agrocinopine-type opines, *Mol. Gen. Genet.* **223**:465–473.

49. Holmes, B., and Roberts, P., 1981, The classification, identification and nomenclature of Agrobacteria, *J. Appl. Bacteriol.* **50**:443–467.

50. Holsters, M., Silva, B., Genetello, C., Engler, G., van Vliet, F., De Block, M., Villaroel, R., van Montagu, M., and Schell, J., 1978, Spontaneous formation of cointegrates of the oncogenic Ti-plasmid and the wide-host range P-plasmid RP4, *Plasmid* **1**:456–467.

51. Holsters, M., Silva, B., van Vliet, F., Genetello, C., De Bock, M., Dhaese, P., De Picker, A., Inze, D., Engler, G., Villarroel, R., van Montagu, M., and Schell, J., 1980, The functional organization of the nopaline *A. tumefaciens* plasmid pTiC58, *Plasmid* **3**:212–230.

52. Hooykaas, P. J. J., Roobol, C., and Schilperoort, R. A., 1979, Regulation of the transfer of Ti plasmids of *Agrobacterium tumefaciens*, *J. Gen. Microbiol.* **110**:99–109.

53. Hooykaas, P. J. J., Klapwijk, P. M., Nuti, M. P., Schilperoort, R. A., and Rörsch, A., 1977, Transfer of the *Agrobacterium tumefaciens* Ti plasmid to avirulent agrobacteria and to *Rhizobium ex planta*, *J. Gen. Microbiol.* **98**:477–484.

54. Hooykaas, P. J. J., van Brussel, A. A. N., den Dulk-Ras, H., van Slogteren, G. M. S., and Schilperoort, R. A., 1981, Sym-plasmid of *Rhizobium trifolii* expressed in different rhizobial species in *Agrobacterium tumefaciens*, *Nature* (London) **291**:351–353.

55. Hynes, M. F., Simon, R., and Pühler, A., 1985, The development of plasmid-free strains of *Agrobacterium tumefaciens* by using incompatibility with a *Rhizobium meliloti* plasmid to eliminate pAtC58, *Plasmid* **13**:99–105.

56. Jones, D. A., Ryder, M. H., Clare, B. G., Farrand, S. K., and Kerr, A., 1988, Construction of a Tra⁻ deletion mutant of pAgK84 to safeguard the biological control of crown gall, *Mol. Gen. Genet.* **212**:207–214.

57. Kerr, A., 1969, Transfer of virulence between isolates of *Agrobacterium*, *Nature* (London) **223**:1175–1176.

58. Kerr, A., 1971, Acquisition of virulence by non-pathogenic isolates of *Agrobacterium radiobacter*, *Physiol. Plant Pathol.* **1**:241–246.

59. Kerr, A., Manigault, P., and Tempé, J., 1977, Transfer of virulence in vivo and in vitro in *Agrobacterium*, *Nature* (London) **265**:560–561.

60. Kersters, K., De Ley, J., Sneath, P. H. A., and Sackin, M., 1973, Numerical taxonomic analysis of *Agrobacterium*, *J. Gen. Microbiol.* **78**:227–239.

61. Klapwijk, P. M., and Schilperoort, R. A., 1979, Negative control of octopine degradation and transfer genes of octopine Ti plasmids in *Agrobacterium tumefaciens*, *J. Bacteriol.* **139**:424–431.

62. Klapwijk, P. M., Scheulderman, T., and Schilperoort, R. A., 1978, Coordinated regulation of octopine degradation and conjugative transfer of Ti plasmids in *Agrobacterium tumefaciens*: evidence for a common regulatory gene and separate operons, *J. Bacteriol.* **136**:775–785.

63. Klee, H., Montoya, A., Horodyski, F., Lichtenstein, C., Garfinkle, D., Fuller, S., Flores, C., Peschon, J., Nester, E. W., and Gordon, M. P., 1984, Nucleotide sequence of the *tms* genes of the pTiA6NC octopine Ti plasmid: two gene products involved in plant tumorigenesis, *Proc. Natl. Acad. Sci. USA* **81**:1728–1732.

64. Koekman, B. P., Ooms, G., Klapwijk, P. M., and Schilperoort, R. A., 1979, Genetic map of an octopine Ti-plasmid, *Plasmid* **2**:347–357.

65. Lessl, M., Schilf, W., and Lanka, E., 1991, Dissection of the transfer region Tra2 of plasmid RP4: the entry exclusion function requires two cistrons. Abstract. *EMBO Workshop: Bacterial Conjugation Systems*, pp. 27–29.

66. Liao, C. H., and Heberlein, G. T., 1978, A method for the transfer of tumorigenicity between strains of *Agrobacterium tumefaciens* in carrot root disks, *Phytopathology* **68**:135–137.

67. Lichtenstein, C., Klee, H., Montoya, A., Garfinkle, D., Fuller, S., Flores, C., Nester, E., and Gordon, M., Nucleotide sequence and transcript mapping of the *tmr* gene of the pTiA6NC octopine Ti-plasmid: a bacterial gene involved in plant tumorigenesis, *J. Mol. Appl. Genet.* **2**:354–362.

68. Lindberg, M., and Norberg, T., 1988, Synthesis of sucrose 4′-(L-arabinose-2-yl phosphate) (agrocinopine A) using an arabinose 2-H-phosphonate intermediate, *J. Carbohydr. Chem.* **7**:749–755.

69. Melchers, L. S., Thompson, D. V., Idler, K. B., Neuteboom, S. T. C., de Maagd, R. A., Schilperoort, R. A., and Hooykaas, P. J. J., 1987a, Molecular characterization of the virulence gene *virA* of the *Agrobacterium tumefaciens* octopine Ti plasmid, *Plant Mol. Biol.* **11**:227–237.

70. Merlo, D. J., and Nester, E. W., 1977, Plasmids in avirulent strains of *Agrobacterium*, *J. Bacteriol.* **129**:76–80.

71. New, P. B., and Kerr, A., 1972, Biological control of crown gall: field measurements and glasshouse experiments, *J. Appl. Bacteriol.* **35**:279–287.

72. O'Connell, M. P., Hynes, M. F., and Puehler, A., 1987, Incompatibility between a *Rhizobium* Sym plasmid and a Ri plasmid of *Agrobacterium*, *Plasmid* **18**:156–163.

73. Panagopoulos, C. G., Psallidas, P. G., and Alvizatos, A. S., 1979, Evidence of a breakdown in the effectiveness of biological control of crown gall, in: *Soil-Borne Plant Pathogens* (B. Schippers and W. Gams, eds.), Academic Press, London, pp. 570–578.

74. Pansegrau, W., and Lanka, E., 1991, Common sequence motifs in DNA relaxases and nick regions from a variety of DNA transfer systems, *Nucl. Acid Res.* **19**:3455.

75. Petit, A., and Tempé, J., 1978, Isolation of *Agrobacterium* Ti plasmid regulatory mutants, *Mol. Gen. Genet.* **167**:147–155.

76. Petit, A., Delhaye, S., Tempé, J., and Morel, G., 1970, Recherches sur les guanidines des tissus de crown

gall. Mise en évedence d'une relation biochemique spécific entre les souches d'*Agrobacterium tumefaciens* et les tumeurs qu'elles induisent, *Physiol. Veg.* **8:**205–213.

77. Petit, A., Dessaux, Y., and Tempé, J., 1978a, The biological significance of opines. I. A study of opine catabolism by *Agrobacterium tumefaciens*, in: *Proc. 4th Internatl. Congr. Plant Pathogen. Bacter.* (M. Ridé, ed.), INRA, Angers, pp. 143–152.

78. Petit, A., Tempé, J., Kerr, A., Holsters, M., van Montagu, M., and Schell, J., 1978b, Substrate induction of conjugative activity of *Agrobacterium tumefaciens* Ti plasmids, *Nature* (London) **271:**570–572.

79. Petit, A., David, C., Dahl, G. A., Ellis, J. G., Guyon, P., Casse-Delbart, F., and Tempé, J., 1983, Further extension of the opine concept: plasmids in *Agrobacterium rhizogenes* cooperate for opine degradation, *Mol. Gen. Genet.* **190:**204–214.

79a. Piper, K. R., Beck von Bodman, S., and Farrand, S. K., 1993. *Nature* (in press).

80. Porter, S. G., Yanofsky, M. F., and Nester, E. W., 1987, Molecular characterization of the *virD* operon from *Agrobacterium tumefaciens*, *Nucl. Acid Res.* **15:**7503–7513.

81. Prakash, R. K., and Schilperoort, R. A., 1982, Relationship between Nif plasmids of fast-growing *Rhizobium* species and Ti plasmids of *Agrobacterium tumefaciens*, *J. Bacteriol.* **149:**1129–1134.

82. Quétier, F., Huguet, T., and Guillé, E., 1969, Induction of crown gall: partial homology between tumor-cell DNA, bacterial DNA, and the GC-rich DNA of stressed normal cells, *Biochem. Biophys. Res. Commun.* **34:**128–133.

83. Ream, W., 1989, *Agrobacterium tumefaciens* and interkingdom genetic exchange, *Annu. Rev. Phytopathol.* **27:**583–618.

84. Riker, A. J., 1926, Studies on the influence of some environmental factors on the development of crown gall, *J. Agricult. Res.* **32:**83–96.

85. Rogowsky, P. M., Close, T. J., Chimera, J., Shaw, J. J., and Kado, C. I., 1987, Regulation of the *vir* genes of *Agrobacterium tumefaciens* plasmid pTiC58, *J. Bacteriol.* **169:**5101–5112.

86. Rogowsky, P. M., Powell, B. S., Shirasu, K., Lin, T.-S., Morel, P., Zyprian, E. M., Steck, T. R., and Kado, C. I., 1990, Molecular characterization of the *vir* regulon of *Agrobacterium tumefaciens*: complete nucleotide sequence and gene organization of the 28.63-kbp regulon cloned as a single unit, *Plasmid* **23:** 85–106.

87. Rosenberg, C., and Huguet, T., 1984, The pAtC58 plasmid of *Agrobacterium tumefaciens* is not essential for tumour induction, *Mol. Gen. Genet.* **196:**533–536.

88. Ryder, M. H., Tate, M. E., and Jones, G. P., 1984, Agrocinopine A, a tumor-inducing plasmid-coded enzyme product, is a phosphodiester of sucrose and L-arabinose, *J. Biol. Chem.* **259:**9704–9710.

88a. Ryder, M. H., Slota, J. E., Scarim, A., and Farrand, S. K., 1987, Genetic analysis of agrocin 84 production and immunity in *Agrobacterium* spp., *J. Bacteriol.* **169:**4184–4189.

89. Sans, N., Schindler, U., and Schröder, J., 1988, Ornithine cyclodeaminase from Ti plasmid pTiC58: DNA sequence, enzyme properties, and regulation of activity by arginine, *Eur. J. Biochem.* **173:**123–130.

90. Schell, J., 1975, The role of plasmids in crown gall formation by *A. tumefaciens*. In: *Genetic Manipulation with Plant Materials* (L. Ledoux, ed.), Plenum Press, New York, pp. 163–181.

91. Schell, M. A., Brown, P. H., and Raju, S., 1990, Use of saturation mutagenesis to localize probable functional domains in the NahR protein, a LysR-type transcriptional activator, *J. Biol. Chem.* **265:**3844–3850.

92. Schilperoort, R. A., Veldstra, H., Warnaar, S. O., Mulder, G., and Cohen, S. A., 1967, Formation of complexes between DNA isolated from tobacco crown gall tumours and RNA complementary to *Agrobacterium tumefaciens*, *Biochim. Biophys. Acta* **145:**523–525.

93. Schilperoort, R. A., van Sittert, N. J., and Schell, J., 1973, The presence of both phage PS8 and *Agrobacterium tumefaciens* A6 DNA base sequences in A6-induced sterile crown gall tissue cultured in vitro, *Eur. J. Biochem.* **33:**1–7.

94. Schindler, U., Sans, N., and Schröder, J., 1989, Ornithine cyclodeaminase from octopine Ti plasmid Ach5: identification, DNA sequence, enzyme properties and comparison with gene and enzyme from nopaline Ti plasmid C58, *J. Bacteriol.* **171:**847–854.

95. Schrell, A., Alt-Mörbe, J., Lanz, T., and Schröder, J., 1989, Arginase of *Agrobacterium* Ti plasmid C58— DNA sequence, properties and comparison with eucaryotic enzymes, *Eur. J. Biochem.* **184:**635–641.

96. Sciaky, D., Montoya, A. L., and Chilton, M.-D., 1978, Fingerprints of *Agrobacterium* Ti plasmids, *Plasmid* **1:**238–253.

97. Shrivastava, B. I. S., 1970, DNA-DNA hybridization studies between bacterial DNA, crown gall tumor cell DNA and normal cell DNA, *Life Sci.* **9:**889–892.

98. Slota, J. E., and Farrand, S. K., 1982, Genetic isolation and physical characterization of pAgK84, the plasmid responsible for agrocin 84 production, *Plasmid* **8:**175–186.

99. Spinzl, M., and Geider, K., 1988, Transfer of the Ti plasmid from *Agrobacterium tumefaciens* into *Escherichia coli* cell, *J. Gen. Microbiol.* **134:**413–424.

100. Stachel, S. E., and Zambryski, P. C., 1986, *Agrobacterium tumefaciens* and the susceptible plant cell: a novel adaptation of extracellular recognition and DNA conjugation, *Cell* **47:**155–157.

101. Stemmer, W. P. C., and Sequeira, L., 1987, Fimbriae of phytopathogenic and symbiotic bacteria, *Phytopathology* **77:**1633–1639.

102. Steck, T. R., and Kado, C. I., 1990, Virulence genes promote conjugative transfer of the Ti plasmid between *Agrobacterium* strains, *J. Bacteriol.* **172:**2191–2193.

103. Tabata, S., Hooykaas, P. J. J., and Oka, A., 1989, Sequence determination and characterization of the replicator region in the tumor-inducing plasmid pTiB6S3, *J. Bacteriol.* **171:**1665–1672.

104. Tempé, J., and Petit, A., 1983, La piste des opines, in: *Molecular Genetics of the Bacteria-Plant Interaction* (A. Pühler, ed.), Springer-Verlag, Berlin, pp. 14–32.

105. Tempé, J., Petit, A., and Farrand, S. K., 1984, Induction of cell proliferation by *Agrobacterium tumefaciens* and *A. rhizogenes*: a parasite's point of view, in: *Plant Gene Research, Genes Involved in Microbe-Plant Interactions* (D. P. S. Verma and T. Hohn, eds.), Springer-Verlag Wein, New York, pp. 271–286.

106. Tempé, J., Petit, A., Holsters, M., van Montagu, M., and Schell, J., 1977, Thermosensitive step associated with transfer of the Ti plasmid during conjugation: possible relation to transformation, *Proc. Natl. Acad. Sci. USA* **74:**2848–2849.

107. Unger, L., Ziegler, S. F., Huffman, G. A., Knauf, V. C., Peet, R., Moore, L. W., Gordon, M. P., and Nester, E. W., 1985, New class of limited host-range *Agrobacterium* mega-tumor-inducing plasmids lacking homology to the transferred DNA of a wide host-range, tumor-inducing plasmid, *J. Bacteriol.* **164:**723–730.

108. Valdivia, R. H., Wang, L., and Winans, S. C., 1991, Characterization of a putative periplasmic transport system for the accumulation of octopine encoded by the *Agrobacterium tumefaciens* Ti plasmid pTiA6, *J. Bacteriol.* **173:**6398–6405.

109. Van Larebeke, N., Engler, G., Holsters, M., van den Elsacker, S., Zaenen, I., Schilperoort, R. A., and Schell, J., 1974, Large plasmid in *Agrobacterium tumefaciens* essential for crown gall-inducing ability, *Nature* (London) **252:**169–170.

110. Van Larebeke, N., Genetello, C., Schell, J., Schilperoort, R. A., Hermans, A. K., Hernalsteens, J. P., and van Montagu, M., 1975, Acquisition of tumor-inducing ability by non-oncogenic bacteria as a result of plasmid transfer, *Nature* (London) **255:**742–743.

111. Van Larebeke, N., Genetello, C., Hernalsteens, J. P., DePicker, A., Zaenen, I., Messens, E., van Montagu, M., and Schell, J., 1977, Transfer of Ti-plasmids between *Agrobacterium* strains by mobilization with the conjugative plasmid RP4, *Mol. Gen. Genet.* **152:**119–124.

112. Van Montagu, M., and Schell, J., 1979, The plasmids of *Agrobacterium tumefaciens*, in: *Plasmids of Medical, Environmental and Commercial Importance* (K. N. Timmis and A. Pühler, eds.), Elsevier/North Holland Biomedical Press, Amsterdam, pp. 71–95.

113. Veluthambi, K., Jayaswal, R. K., and Gelvin, S. B., 1987, Virulence genes *A, G,* and *D* mediate the double-stranded border cleavage of T-DNA from the *Agrobacterium* Ti plasmid, *Proc. Natl. Acad. Sci. USA* **84:**1881–1885.

114. Veluthambi, K., Krishnan, M., Gould, J. H., Smith, R. H., and Gelvin, S. B., 1989, Opines stimulate induction of the *vir* genes of the *Agrobacterium tumefaciens* Ti plasmid, *J. Bacteriol.* **171:**3696–3703.

115. Von Lintig, J., Zanker, H., and Schröder, J., 1991, Positive regulators of opine-inducible promoters in the nopaline and octopine catabolism regions of Ti plasmids, *Mol. Plant-Microbe Interact.* **4:**370–378.

116. Wang, K., Stachel, S. E., Timmerman, B., van Montagu, M., and Zambryski, P. C., 1987, Site-specific nick in the T-DNA border sequence as a result of *Agrobacterium vir* gene expression, *Science* **235:**587–591.

117. Ward, J. E., Dale, E. M., and Binns, A. N., 1991, Activity of the *Agrobacterium* T-DNA transfer machinery is affected by *virB* gene products, *Proc. Natl. Acad. Sci. USA* **88:**9350–9354.

118. Ward, J. E., Akiyoshi, D. E., Regier, D., Datta, A., Gordon, M. P., and Nester, E. W., 1988, Characterization of the *virB* operon from an *Agrobacterium tumefaciens* Ti plasmid, *J. Biol. Chem.* **263:**5804–5814.

119. Waters, V. L., Hirata, K. H., Pansegrau, W., and Guiney, D. G., 1991, Sequence identity in the nick regions of IncP plasmid transfer origins and T-DNA borders of *Agrobacterium* Ti plasmids, *Proc. Natl. Acad. Sci. USA* **88:**1456–1460.

120. Watson, B., Currier, T. C., Gordon, M. P., Chilton, M.-D., and Nester, E. W., 1975, Plasmid required for virulence of *Agrobacterium tumefaciens*, *J. Bacteriol.* **123:**255–264.

121. Weisburg, W. G., Woese, C. R., Dobson, M. E., and Weiss, E., 1985, A common origin of Rickettsiae and certain plant pathogens, *Science* **230:**556–558.

122. White, F. F., and Nester, E. W., 1980, The relationship of the plasmids responsible for hairy root and crown gall tumorigenicity, *J. Bacteriol.* **144:**710–720.

123. White, L. O., 1972, The taxonomy of the crown-gall organism *Agrobacterium tumefaciens* and its relationship to rhizobia and other agrobacteria, *J. Gen. Microbiol.* **72:**565–574.

124. Willetts, N., and Wilkins, B., 1984, Processing of plasmid DNA during bacterial conjugation, *Microbiol. Rev.* **48:**24–41.

125. Winans, S. C., Ebert, P. R., Stachel, S. E., Gordon, M. P., and Nester, E. W., 1986, A gene essential for *Agrobacterium* virulence is homologous to a family of positive regulatory loci, *Proc. Natl. Acad. Sci. USA* **83:**8278–8282.

126. Winans, S. C., Kerstetter, R. A., and Nester, E. W., 1988, Transcriptional regulation of the *virA* and *virG* genes of Agrobacterium tumefaciens, *J. Bacteriol.* **170:**4047–4054.

127. Winans, S., Allenza, P., Stachel, S., McBride, K., and Nester, E., 1987, Characterization of the *virE* operon of the *Agrobacterium* Ti plasmid pTiA6, *Nucl. Acid Res.* **15:**825–837.

128. Yadav, N. S., Vanderleyden, J., Bennett, D. R., Barnes, W. M., and Chilton, M.-D., 1982, Short direct repeats flank the T-DNA on a nopaline Ti plasmid, *Proc. Natl. Acad. Sci. USA* **79:**6322–6326.

129. Zaenen, I., van Larebeke, N., Teuchy, N., van Montagu, M., and Schell, J., 1974, Supercoiled circular DNA in crown gall inducing *Agrobacterium* strains, *J. Mol. Biol.* **86:**109–127.

130. Zambryski, P., 1988, Basic processes underlying *Agrobacterium*-mediated DNA transfer to plant cells, *Annu. Rev. Genet.* **22:**1–30.

131. Zhang, L., and Kerr, A., 1991, A diffusible compound can enhance conjugal transfer of the Ti plasmid in *Agrobacterium tumefaciens*, *J. Bacteriol.* **173:**1867–1872.

132. Zhang, L., Murphy, P. J., Kerr, A., and Tate, M. E., *Agrobacterium* conjugation and gene regulation by *N*-acyl-L-homoserine lactones. *Nature* (in press).

Chapter 11

Conjugative Plasmids of *Streptomyces*

DAVID A. HOPWOOD and TOBIAS KIESER

1. Introduction

Genetic exchange in *Streptomyces* was first revealed when prototrophic recombinants were recovered from mixed cultures of pairs of auxotrophic derivatives of several wild-type strains (1, 8, 25, 63, 65). A conjugative mechanism, rather than transformation or transduction, was invoked to account for gene exchange because no recombinants were detected without prolonged physical contact of the parental strains (indeed, a period of mixed *growth* was needed) and because the pattern of inheritance of groups of markers was consistent only with recombination of large segments of the parental genomes (25). Plasmids were first clearly implicated in this conjugative process in the most studied strain, *Streptomyces coelicolor* A3(2), when certain derivatives of the wild-type isolate were found to differ in their "fertility" properties—that is, in the frequency with which they generated chromosomal recombinants when mated with various other derivatives—and this ability was inherited "infectiously" (2, 28, 72). These experiments led to the genetic definition of two conjugative plasmids—SCP1 and SCP2—that were deduced to be present in an autonomous state in the wild-type A3(2) strain and to be lost, or in the case of SCP1 sometimes chromosomally integrated, in various of its derivatives. Plasmids responsible for "fertility" (or "chromosome mobilizing ability" [Cma]) (24) were also identified genetically in some other strains, including *Streptomyces rimosus* (18), *Streptomyces lividans* (29), *Streptomyces erythreus* (now called *Saccharopolyspora erythrea*) (15), *Streptomyces venezuelae* (17), and *Streptomyces ambofaciens* (66).

Can plasmids account for all the recombination detected in these organisms? In *S. coelicolor* A3(2), pairs of derivatives that lack both SCP1 and SCP2 produce a very low (usually $10^{-7}-10^{-8}$) frequency of nonparental genotypes in mixed culture. These progeny

DAVID A. HOPWOOD and TOBIAS KIESER • John Innes Institute, John Innes Centre, Norwich NR4 7UH, England.

Bacterial Conjugation, edited by Don B. Clewell. Plenum Press, New York, 1993.

are deduced to arise by recombination, rather than by reverse mutation of the markers, because some of the colonies inherit recessive, nonselected alleles from *both* parents. However, all A3(2) derivatives carry a stably integrated plasmid, SLP1 (see below), which is known to lead to Cma after transfer by mating to *S. lividans* (4). SLP1 may, therefore, be responsible for the low level of recombination observed in SCP1⁻ SCP2⁻ matings in *S. coelicolor*. The situation is clearer in *S. lividans* itself, in which a careful search for genuine recombinants in mixtures of pairs of strains lacking the two conjugative plasmids, SLP2 and SLP3, of the wild-type strain (29) failed to reveal any. The rare colonies that appeared on selective media from such matings belonged to those genotypes that could arise from one or the other parent by reverse mutation rather than to those that would represent the most probable recombinant classes (D. A. Hopwood and H. M. Kieser, unpublished results). Our working assumption is, therefore, that all conjugative DNA transfer in *Streptomyces* depends on the activity of plasmids. However, plasmid-mediated conjugation is a clearly different process from that in gram-negative bacteria, as we see below.

2. The Wide Variety of Sizes and Structures of Conjugative Plasmids in *Streptomyces*

Many plasmids, with diverse properties, have been characterized physically in streptomycetes, and almost all those tested are conjugative and have Cma; the only exceptions we know of are naturally occurring deletions of conjugative forms. All the physically characterized plasmids are double-stranded DNA molecules. Most are covalently closed, circular (CCC) DNAs, ranging in size from less than 4 kb to more than 100 kb, and with copy numbers between one and several hundred (reviewed in 31). The smallest that has been proven to be conjugative is pTA4001, which is 6 kb (46). Interestingly, streptomycetes also contain double-stranded linear DNA plasmids of a wide range of sizes. The smallest reported to date is pSCL1, which is 12 kb (34, 65a, 74a), while the largest is the genetically defined SCP1 (44, 45), recently characterized physically by Kinashi and his colleagues as a 350-kb element (42, 43). Another speciality of the streptomycetes is the widespread occurrence of elements such as SLP1 and pSAM2, which occur integrated in the host chromosome but can excise to become autonomous plasmids. If such elements occur outside of the actinomycetes, they are rare: e14 of *E. coli* (9) has some features in common with the *Streptomyces* elements, but there is no evidence that it is conjugative. There is also a phage of *Streptomyces* (φSF1) whose prophage is a conjugative plasmid (12). *Streptomyces* plasmids seem rarely to code for other properties affecting the host phenotype besides those associated with conjugation: Methylenomycin production and resistance by SCP1 (11, 44, 45) may be the only well-established example in *Streptomyces* (but hydrogen autotrophy in the actinomycete *Nocardia opaca* is also determined by large linear plasmids) (33). It follows from this fact, and from the simplicity of the genetic control of conjugation (see below), that many *Streptomyces* plasmids carry much DNA of unknown function.

In the next three sections we review current knowledge of three diverse kinds of *Streptomyces* plasmids—pIJ101 to represent multicopy CCC plasmids, SLP1 and pSAM2 as examples of integrating elements, and SCP1 as a "giant" linear plasmid—to illustrate the range of properties displayed by conjugative plasmids of this group of bacteria.

3. pIJ101: A Small, Wide Host Range, CCC Conjugative Plasmid of Actinomycetes

pIJ101 is an 8829-bp (37), self-transmissible, high-copy-number plasmid. It was discovered (41) in *Streptomyces lividans* strain ISP5434, where it coexisted with several deletion derivatives (three of which were designated pIJ102-104); a seemingly identical plasmid was also found in *S. coelicolor* ATCC10147 (57). pIJ101 has a very wide host range in actinomycetes: Derivatives of it have been successfully introduced (by artificial transformation) into numerous *Streptomyces* spp (41) as well as members of the actinomycete genera *Micromonospora* (35, 50), *Thermomonospora* (58), *Saccharopolyspora* (75), and *Amycolatopsis* (51). Many derivatives of pIJ101 have become widely used cloning vectors, and its properties have been investigated in greater detail than those of any other *Streptomyces* plasmid. Although pIJ101 is considered here as being representative of the many small, multicopy *Streptomyces* plasmids, these are perhaps not closely related. As far as is known, none of the other plasmids fall into the same incompatibility group as pIJ101.

Figure 11.1 shows a physical map of pIJ101 deduced from the DNA sequence of Kendall and Cohen (37), with a selection of restriction sites chosen to facilitate reference to the literature. pIJ101 contains the typical 73% G + C of *Streptomyces* DNA, which imposes a strong bias on codon usage in translated sequences. This allows good predictions to be made about which potential open reading frames (ORFs) are actually translated (5). Nine ORFs have been deduced from the sequence, seven of them running in counterclockwise direction with respect to the standard map. The most significant exception is *korA*, which forms a divergent, coregulated pair with the *tra* operon (see below). Comparison of the deduced protein sequences with the various databases revealed no obvious similarities with known proteins, except for some conserved sequence motifs (see below). This was rather unexpected because many other *Streptomyces* proteins clearly resemble their counterparts in other prokaryotic and eukaryotic organisms. Thus all the proteins of pIJ101 may perhaps represent unfamiliar types.

3.1. Replication Functions and Copy-Number Control

The 456-aa (50-kDa) Rep protein is absolutely required for plasmid replication. Insertions in the *rep* region were invariably lethal to the plasmid. The 2.1-kb *Sst*II fragment containing *rep* linked to a resistance gene can replicate in *S. lividans* 66 (41) and must therefore also contain a replication origin, which was tentatively placed upstream of *rep* in a segment of DNA deduced to be noncoding by analysis using the FRAME program (5). The 2.1-kb minimal replication fragment also contains an ORF potentially coding for a 56-aa protein. This ORF is nonessential for plasmid replication in *S. lividans* (M. Brasch, personal communication). Two promoters, one upstream and one within the ORF that is thought to encode the *rep* protein, have been identified on the minimal replication fragment by *in vivo* promoter-probing experiments (M. Brasch, personal communication).

Plasmids containing only the minimal replication fragment of pIJ101 (including widely used cloning vectors like pIJ702) are very different from the original plasmid. They lack all transfer capability and have approximately a sixfold reduced copy number (approximately 50 instead of the normal 300 copies per chromosome). (They are mobilized by other

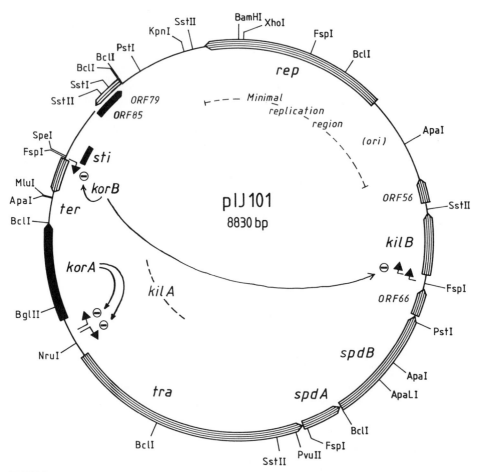

FIGURE 11.1. Functional map of pIJ101 deduced from the sequence (37) and from functional studies (see text). Open reading frames (ORFs) reading in clockwise orientation are drawn as black bars, and those reading in the counterclockwise direction are shaded. The ORFs are designated as follows: *rep*, replication; *tra*, transfer protein; *spdA,B*, spread functions; *kilB, korA,B*, kill and kill-override functions (the product of *kilA* is not a recognized protein); ORFaa, open reading frames possibly specifying proteins of the corresponding number of amino acids. Other symbols: *ori*, origin of replication for plus strand (tentative position); *sti*, strong incompatibility, minus (second) strand origin; ◄┐, promoters; → ⊖, *trans*-acting repression of promoter activity; *ter*, transcriptional terminator.

plasmids [e.g., SCP2] only with the same low frequency as chromosomal markers.) These minimal replicon plasmids accumulate substantial amounts of single-stranded (ss) plasmid DNA (14, 60). (This is the strand synthesized in the counterclockwise direction that would not hybridize with the mRNAs for the majority of the ORFs, perhaps therefore avoiding a potential problem in sequestering them.) Thus, pIJ101 probably replicates via a ss intermediate, by a rolling-circle mechanism, like many plasmids from low G + C unicellular gram-positive bacteria (reviewed in 20). The accumulation of ss plasmid DNA may be caused by lack of an efficient second-strand origin. Such an origin was identified outside the minimal replication fragment c. 1 kb downstream of the *rep* ORF (14). The site was named *sti* (for strong incompatibility) because Sti⁻ plasmids cannot coexist (i.e., are incompatible with) wild-type pIJ101 or derivatives of it carrying *sti* in the correct orientation. This probably explains why all natural deletion derivatives of pIJ101 are Sti⁺. As expected for a second-strand origin, *sti* functions only in one orientation. It was effective in reducing the amount of ssDNA and, therefore, increasing the copy number of dsDNA, of pWOR120, another *Streptomyces* multicopy minimal replicon plasmid compatible with pIJ101, when cloned into it. The *sti* function seems to interact with the product of an ORF described in the next section, *korB* (called *cop* by Deng et al) (14). Artificial constructs containing *sti* but lacking *korB* had a copy number of ~1000, but when *korB* was introduced *in trans* on a compatible plasmid, the copy number of such Sti⁺ plasmids returned to normal (c. 300 copies per chromosome), while Sti⁻ plasmids retained a copy number of ~1000. Plasmid instability and deletion formation, observed frequently with artificial pIJ101 derivatives (41), does not correlate with the presence or absence of *sti*. (Deletion formation in pIJ101 derivatives is dramatically reduced in a mutant derivative of *S. lividans* [71] that was isolated on the basis of being deficient in intraplasmid recombination but found to be normal in homologous chromosomal or chromosome-plasmid recombination [40].)

Any functions of sequences between *sti* and the *rep* gene are unknown. The two overlapping ORFs that might specify proteins of 79 and 85 aa can be interrupted by insertion into the *Sst*I site without obvious effect. The sequence counterclockwise of these ORFs (including the 200-bp *sti* region) has a uniform G + C content in the three positions in all three reading frames, suggesting that it is not translated, while the region next to it in the clockwise direction has the nonrandom G + C content characteristic of translated regions. However, the latter potential ORFs are interrupted by stop codons. This region has been implicated in stable plasmid inheritance (41), but the observed effects may have been nonspecific. Perhaps a piece of DNA has been deleted from this region.

3.2. Partition

It is not known if pIJ101 has specific functions for partition during vegetative growth of the host. Nor is it known how the large pool of plasmid copies present in each multi-chromosomal hyphal compartment is correctly partitioned into the unichromosomal compartments formed during sporulation by production of a series of closely spaced sporulation septa (22). The idea that this process, which would be vital for a successful *Streptomyces* plasmid, is not an automatic process is supported by the finding that the pIJ702 derivative of pIJ101 can replicate in the vegetative mycelium of *Saccharopolyspora erythrea* but cannot be inherited stably through a sporulation cycle (75).

3.3. Conjugation and Fertility Functions and the *kil/kor* Systems

Conjugative functions of pIJ101 (and of many other *Streptomyces* plasmids) require very little genetic information. No sex pili have been implicated, and only one gene, *tra*, is essential for transfer itself. Nevertheless, conjugation is very efficient, producing up to 100% plasmid inheritance and 1% recombinants for chromosomal markers in mixtures of pIJ101+ and pIJ101− strains. The *tra* gene probably specifies a 621-aa (66-kD) protein (there are several possible start codons). Neither DNA binding motifs nor hydrophobic domains characteristic of membrane-associated proteins are recognizable in the aa sequence, but a possible ATP binding domain is present. The *Fsp*I fragment of pIJ101, containing *tra*, *korA*, and *korB*, inserted into the chromosome of a plasmid-free *S. lividans* strain gave Cma (but no pocks). This proved that the *rep* gene of pIJ101 is not required for Cma (G. Pettis, personal communication).

"Pock" formation is a phenomenon unique to conjugative plasmids of actinomycetes. It is seen (on appropriate media) when a plasmid-containing spore develops within a confluent culture of a strain lacking the same plasmid (but not of a plasmid-containing derivative). Under these conditions, transfer of the plasmid leads to a circular zone or "pock" (Figure 11.2) where growth and/or development of the newly infected recipient culture is retarded. Sometimes growth inhibition is accompanied by an obvious change in physiology—perhaps a "stress" response or a reaction to reduced growth rate—most clearly reflected in precocious production of the pigmented antibiotics (normally produced after the main phase of vegetative growth) in the pocks of *S. coelicolor* or *S. lividans*.

Downstream of *tra* follow three ORFs—*spdA*, *spdB*, and "66-aa ORF"—without significant intervening noncoding sequences (there is a 1-bp overlap between *tra* and *spdA*, a 2-bp gap between *spdA* and *spdB* and 4-bp overlap between *spdB* and the "66-aa ORF"). Disruption of *spdA* or *spdB* reduces the pock size (hence the name *spd* for plasmid "spread") but does not affect either the frequency of chromosomal recombination in the standard type of mating in which two parental strains are intimately mixed or the 100% plasmid transfer observed in such mixed cultures. The 66-aa ORF could be disrupted without obvious effect. The products of all three ORFs may be membrane associated, since they are deduced to have strong hydrophobic domains.

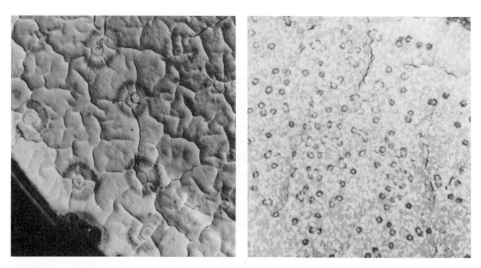

FIGURE 11.2. Pocks caused by plating spores of *S. lividans* carrying pIJ101 derivatives in a lawn of a plasmid-free *S. lividans* culture. Left: wild-type pocks. Right: small pocks caused by a Spd− mutant.

Previously we have speculated that spreading may perhaps involve migration of plasmid copies *within* the hyphae of the recipient (41). We find this an attractive idea because it relates the biology of *Streptomyces* plasmids to the unique mycelial growth habit of their hosts. Presumably, occasional contacts between the sparse hyphal growth in the natural soil environment could lead to plasmid transfer into particular cells of the recipient mycelium. A considerable selective advantage would doubtless accrue to a plasmid capable of intramycelial colonization of the rest of the interconnected series of hyphae. Such a process would involve passage through the cross-walls that occur at intervals in the substrate mycelium. Other roles for the *spd* genes could, however, be imagined. These could include quicker establishment of hyphal contacts and intermycelial pore formation (for the passage of DNA), or extension of the period of competence for plasmid transfer, either of which might account for the influence of the *spd* functions on pock size.

The regulatory circuits of pIJ101 have been analyzed by Stein et al (70) and Stein and Cohen (69). The *tra*, *spdA*, and *spdB* genes and the 66-aa ORF probably form an operon, transcribed from the *tra* promoter (identified as the pIJ101 B promoter of Buttner and Brown) (10). Transcription from the *tra* promoter is repressed by the product of the *korA* gene, a 241-aa (25-kD) protein with a distinctive helix-turn-helix motif for potential DNA binding, and deduced to be diffusible because it acts *in trans*. *korA* also represses its own transcription. The *tra* and *korA* promoters are close together, transcribing in opposite orientations, and transcription from both promoters is about fivefold repressed by normal levels of KorA protein. This regulatory circuit probably provides for a burst of *tra* operon expression when the plasmid enters a new compartment. Perhaps a resulting physiological change could account for the formation of pocks as areas of retarded development, while continuing unregulated expression could explain why plasmids lacking *korA* are not viable (hence the name *kor* for "kill override"). Attempts to localize an expected *kilA* gene, which would be the target of KorA, revealed no protein-coding gene. Instead, "*kilA*" seems to consist of the *korA/tra* promoter region, together with the N-terminus of the *tra* protein (the 1131-bp BglII-BclI fragment), which cannot be cloned by itself in *Streptomyces* (37, 70).

Downstream of the *tra* operon is another ORF involved in plasmid spread. It specifies a 148-aa (15-kD) protein with a hydrophobic N-terminus. It was designated *kilB* because it was the second region (after *kilA*) that could not be cloned on a pIJ101 derivative without the presence of either *korA* or *korB* in the same cells (but cloning of *kilB* onto a SLP1 derivative with a lower copy number was possible) (36). Transcription from the *kilB* promoter (the pIJ101 A promoter of Buttner and Brown) (10) is repressed by a factor of about 10 by KorB. Two forms, 10 and 6 kD, of the KorB protein have been identified in a T7 expression system in *E. coli*. The 10-kD form binds to both the *kilB* and *korB* promoters, while the 6-kD form seems to be specific for the *kilB* promoter (J. Tai, personal communication). A second, unregulated, weak promoter is present upstream of the *kilB* promoter. *korA* has no effect on *kilB* transcription, and it must therefore prevent the lethal *kilB* action in some other way. This latter interaction would be a good candidate for a poison–antidote interaction observed with killer systems of gram-negative plasmids (54).

The DNA sequences of the *korB* and *kilB* promoters, which are both repressed by the *korB* protein, are very similar (24/32 identical bases). The binding of the 10 kD KorB protein has been studied using DNAse I footprinting experiments (76). *korA* and *korB* are transcribed convergently and are separated by a bidirectional transcriptional terminator that was characterized by Deng et al (13). No other pIJ101 terminators have so far been identified.

pSN22, an 11-kb, pock-forming, multicopy plasmid from *Streptomyces nigrifaciens*, has also been studied in great detail (33a, 33b). The DNA of the pSN22 replication region hybridizes to that of pIJ101 and evidence has been obtained that this plasmid also replicates via a single-stranded DNA intermediate (33c). The transfer regions of the two plasmids appear to be completely different (T. Seki, personal communication). On the other hand, the transfer genes of pSAM2 (see below) are similar at the protein level (deduced from the DNA sequence) to the *tra*, *korA*, and *spdB* proteins of pIJ101 (J. Hagège, personal communication).

4. SLP1 and pSAM2: Integrated Elements of *S. coelicolor* and *S. ambofaciens*

SLP1 of *S. coelicolor* (Figure 11.3) and pSAM2 of *S. ambofaciens* (Figure 11.4) are members of a group of at least eight actinomycete elements (reviewed in 32, 41a) that can exist as autonomous plasmids with their own replication, transfer, and pock-forming functions, but they can also integrate by site-specific recombination into the host genome. The plasmid *attP* and host *attB* sites have at least 44 bp of DNA in common. Each element has its specific *att* site, but all the known sites resemble, and probably are, Class II tRNA genes that do not encode the 3' CCA of the acceptor arm (integration into tRNA sequences occurs also in phages P4, P22, I5, and HP1c1 and the archebacterial element SSv1 (48a, 59, 62). The SLP1 and pSAM2 *att* sites would specify tyrosine and proline tRNAs, respectively (53, 59). On integration of the plasmid, the tRNA structures are regenerated because of the identity between *attP* and *attB* and because the *att* sites overlap the 3' ends of the proposed tRNAs but not the 5' ends. Thus, only the terminators of the tRNA genes, not the promoters, are changed (Figure 11.5). The *attB* sequences of various *S. ambofaciens* strains were found to be highly conserved and were also present, with only minor changes, in other actinomycetes (6a, 53). This gives these integrating plasmids a potentially wider host range than that observed with autonomously replicating plasmids (integration of pSAM2 into *Mycobacterium smegmatis* has already been achieved (49a). Integration allows the plasmid to exist stably at unit copy number, imposing a minimal burden on the host. The integration and excision reactions for pSAM2 are catalyzed by the products of plasmid-borne genes, which resemble the *int* and *xis* genes of phage P22, as well as those of λ and φ80, whose *attB* sites do not resemble tRNA genes. This, of course, suggests that these elements might have been derived from temperate phages (7), but no phagelike particles have been found. Perhaps these elements have exchanged dispersal as phage virions for their conjugative mode of transfer.

No derivatives of *S. coelicolor* A3(2) have been found to have lost SLP1 from its integration site near the chromosomal *strA* locus. When *S. coelicolor* is grown in mixed culture with *S. lividans*, c. 1% of the *S. coelicolor* spores donate SLP1 to *S. lividans*. About 90% of the resulting pock-forming *S. lividans* strains contain CCC plasmid DNAs with a copy number of 4 to 10 and of varying sizes (9.5 to 14.5 kb), because of deletion of part of the SLP1 sequence. The remaining 10% contain the complete 17.2-kb element stably integrated (55, 56) at a tRNA tyr locus in *S. lividans* (presumably corresponding to the *attB*

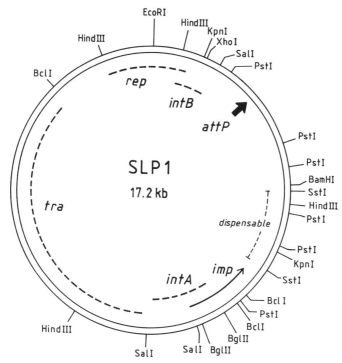

FIGURE 11.3. Map of SLP1 drawn as the circular, autonomously replicating molecule, from information in references 19 and 56. Integration of SLP1 via the *attP* site into the host chromosome, as shown in Figure 11.5, requires *intB*. *imp*, for inhibition of maintenance of the plasmid (19), consists of three open reading frames reading in the direction of the arrow and can be supplied *in trans* to force integration of *imp⁻*, *intB⁺*, *att⁺* plasmids. *rep* (replication function) and *intA* (not needed for integration; see text) are required for maintenance of *imp⁻* plasmids, which occur as CCC forms. *tra* designates the transfer region. Deletion of the dispensable region (*SstI* fragment) does not affect integration, transfer, or maintenance of CCC forms.

site found in the closely related *S. coelicolor*). Both free and integrated forms of SLP1 confer Cma on *S. lividans*, and the integrated form readily conjugates to other SLP1⁻ *S. lividans* strains without further deletions being observed (i.e., deletion formation is peculiar to interspecific conjugation). The integrated element preferentially mobilizes the *str-6* marker near the SLP1 *att* site in *S. lividans* (T. Kieser, unpublished).

Free CCC forms have never been detected in strains containing integrated SLP1. This has been attributed to a specific *imp* (inhibition of maintenance of plasmid) function that is deleted from all autonomously replicating SLP1 derivatives. The *imp* region consists of three ORFs, one of which (*orfC*) has some similarity to the repressor proteins Cro and C1 of λ and LacI of *E. coli* (19). The presence of an integrated SLP1 plasmid prevents transformation with an autonomously replicating SLP1 derivative (occasional transformants—c. 50 per μg plasmid DNA—result from integration of the plasmid into the resident sequence to produce tandem insertions) (4).

Lee et al (48) constructed pIJ101 derivatives carrying a cloned *att* site (*attP* and *attB* were interchangeable in these experiments) into which SLP1 could integrate. When a strain

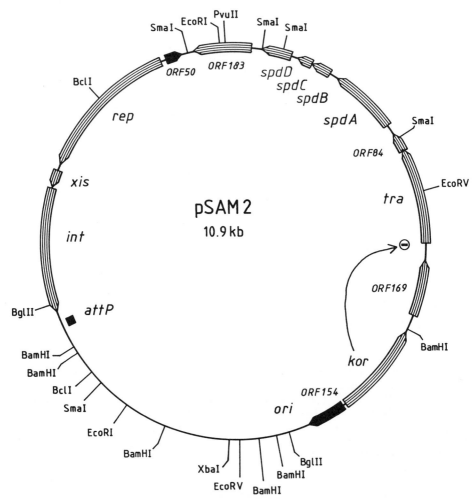

FIGURE 11.4. Functional map of pSAM2 (68) drawn from the DNA sequence (J. Hagège, personal communication). Open reading frames are drawn shaded or black as in Fig. 9.2, depending on their orientation. The region containing *attP*, *int* and *xis* is sufficient for stable integration into the host chromosome (7). *rep* and *ori* (plus-origin), which are located almost opposite from each other, are required for autonomous replication. *tra* is under the negative control of *kor* (indicated by →⊖), and the two genes are similar in their functions and amino acid sequences to *tra* and *korA* of pIJ101. The spread genes *spdA-C* are involved in pock formation. Open reading frames without known function are labelled ORF, followed by the number of amino acids of the hypothetical protein.

containing SLP1ᴵⁿᵀ was mated with a strain containing one of the above *att* site plasmids, SLP1 readily integrated into the cloned *att* site. Similarly, when a hybrid plasmid, consisting of SLP1ᴵⁿᵀ integrated into the *att* site on a pIJ101 derivative, was introduced by transformation into SLP1-free protoplasts, excision of the element from the plasmid and reinsertion into the chromosome were observed. (The excision regenerated the *att*-site plasmid, demonstrating that the "transposition" of SLP1 is conservative.) When, however,

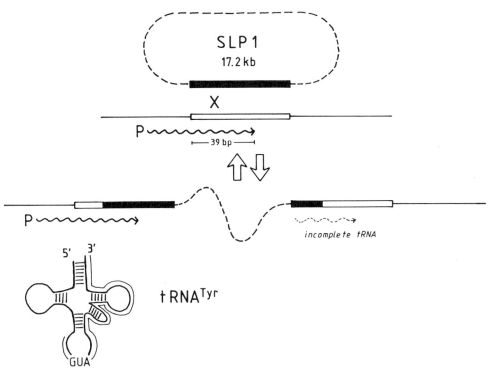

FIGURE 11.5. Integration of SLP1 (dashed lines) into the host chromosome (continuous line) by site-specific recombination between the *attP* site (solid box) and the *attB* site (open box) (53). The *attB* site overlaps the 3′ end of a tRNA^{Tyr} gene, indicated by its promoter P and tRNA product (wavy line). There is a 1-bp difference between the *attB* and *attP* sequences just to the right of the 3′ end of the tRNA sequence, which made it possible to prove that recombination occurs in the indicated 39-bp sequence. Integration of SLP1 regenerates one tRNA^{Tyr} gene (left) and produces a truncated promoterless gene (right). The fine line around the cloverleaf structure shows the extent to which *attP* overlaps the sequence of the mature tRNA^{Tyr}.

the *att*-site plasmid was introduced by transformation into a strain carrying SLP1 already integrated in the chromosome, SLP1 did not jump onto the plasmid-borne *att* site. This indicates that the SLP1 excision function is active only for a short time after entering a new compartment and raises the interesting question about the sequence of events that occurs when a strain carrying SLP1^{INT} conjugates with a SLP1$^-$ strain. Does the SLP1-carrying strain "sense" the presence of a plasmid-free partner, resulting in excision of SLP1, or is SLP1^{INT} transferred to a new host together with flanking chromosomal sequences, followed by excision and reintegration? The latter possibility seems rather unlikely in view of the high frequency of the event.

In pSAM2, the 11-kb integrating plasmid from *S. ambofaciens*, all the functions needed for integration (*attP*, *int*, *xis*), and maintenance of the integrated state are clustered on c. 2.5-kb DNA (facilitating the construction of small integrating vectors) (7, 47, 67). Contrary to initial reports, all the integration functions of SLP1 are also clustered and constitute the *intB-attP* region; the *intA* function (56a) is probably identical to *imp* rather than being a second gene required for integration (M. Brasch, personal communication). The commonly used version of pSAM2 seems to lack an equivalent of the SLP1 *imp* functions, because the integrated form coexists with free CCC forms of the plasmid (about 10 copies per chromosome) and does not prevent the introduction of CCC plasmids by transformation. It is likely, however, that these are the properties of a mutant derivative of the original pSAM2 found in *S. ambofaciens* ATCC15154 strain B2, which does not produce CCC forms either in its original host or when transferred to *S. lividans* (6).

5. SCP1: A "Giant" Linear Plasmid of *S. coelicolor*

SCP1 was shown to be a plasmid by its infectious transfer in mixed culture from an SCP1$^+$ strain of *S. coelicolor* A3(2) to an SCP1$^-$ derivative, which thereupon regained the characteristic SCP1-determined abilities of the original SCP1$^+$ donor: to produce and be resistant to methylenomycin (therefore inhibiting SCP1$^-$ strains) and to generate chromosomal recombinants (without a marked preference for specific markers) at a frequency of c. 10^{-5}–10^{-6} with an SCP1$^-$ strain (72). SCP1$^-$ strains give rise to highly fertile donor forms (giving up to 100% recombinants for specific regions of the chromosome) by integration of SCP1 into the chromosome or pickup of chromosomal segments by SCP1 to form SCP1-prime strains (reviewed in 30). The chromosomal integrants were very stable, with loss of SCP1 from the chromosome or from the cell being rarely, if ever, observed; they included the original NF donor type and certain of the later discovered "new donors" (73). Other "new donors" lost the SCP1-associated phenotypes with high frequency but could give rise to stable donor derivatives. In view of the moderate instability of a proven SCP1-prime (SCP1'-*cysB*) carrying a single identified chromosomal gene (about 5% of SCP1$^-$ derivatives were present in the culture), and the extreme instability of a second (SCP1'-*argA,uraB*), which carried a larger group of linked genes (27), it seems probably that the unstable "new donors" were also SCP1-prime strains that carried relatively large but unmarked segments of chromosome. Their mobilization of chromosomal markers in matings would be analogous to that of artificially constructed SCP2*-prime strains, which were found to cause efficient transfer of chromosomal genes close to the segment of DNA carried by SCP2*, presumably by "donor crossing-over" between homologous sequences on plasmid and chromosome (30). One of the most intriguing properties of NF and other "NF-like" donor strains produced by integration of SCP1 is that they donate chromosomal markers lying on *both* sides of the integration point in matings to SCP1$^-$ cultures (Figure 11.6); and that all recombinants inherit the donor state from the SCP1-containing parent.

SCP1 appeared in pulsed-field gel electrophoresis (PFGE) as a DNA species with a mobility corresponding to a linear molecule of some 350 kb (43), which hybridized with the cloned *mmr* methylenomycin resistance gene (3). NF strains, as expected, lacked the SCP1 band, and the *mmr* gene hybridized instead to the high-molecular-weight chromosomal DNA (43). In the SCP1'-*cysB* strain, the extrachromosomal band appeared to be about 100 kb larger than SCP1 (T. Garbe, personal communication). A recent painstaking restriction analysis of SCP1, using six enzymes, generated an unambiguously linear map of ~350 kb, including two long terminal repeats (TIRs) of some 80 kb each (42) (Figure 11.7). Like the smaller (17-kb) *S. rochei* pSLA2 plasmid (23), and other linear elements from a variety of organisms (64), SCP1 carries a protein covalently bound to each 5' end of the DNA. The involvement of such proteins and the TIRs in replication of this class of elements has been

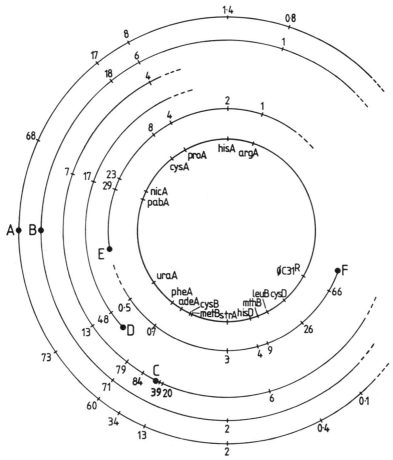

FIGURE 11.6. Percentage frequencies of inheritance of donor markers by recombinants in matings of various types of *S. coelicolor* SCP1 donors with SCP1⁻ strains. A. NF; B. NF-like donor A634; C. bidirectional donor derived from SCP1'-*cysB*; D. unstable *uraA* donor A610; E. combined data for unstable *pabA* donor A607 and stable *pabA* donor A608; F. stable *cysD* donor 2106. ●, presumed position of SCP1, inherited by approximately 100% of the recombinant progeny. (From reference 30, which contains references to the original data.)

discussed by Sakaguchi (64), who has termed such elements "invertrons." The idea is that the two ends of the element would come together in a "racket-frame" configuration, with "juxataposing" proteins holding the DNA together in a four-stranded complex. A new DNA strand would then be initiated from a newly synthesized terminal protein attaching to a free 3' end of the complex.

A current PFGE restriction analysis of the chromosome of *S. coelicolor* A3(2) (40a) indicated the occurrence of a large deletion on or after SCP1 integration into the chromosome to give the NF state. In fact, Kendall and Cullum (39) had already deduced that a deletion had occurred in an NF strain, including the chromosomal gene for an extracellular agarase, *dagA* (38). They identified a ~1-kb element (later named IS*466*) (16) present on

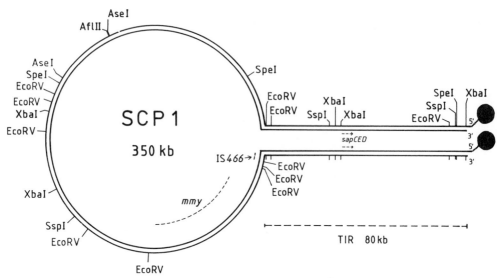

FIGURE 11.7. Map of SCP1, redrawn from reference 42. SCP1 is a 350-kb linear DNA molecule consisting of 190 kb of unique sequence (circle) flanked by 80-kb terminal inverted repeats (TIR) with identical restriction sites. ●, protein molecules bound to the protruding 5′ end. IS466 and the *mmy* (methylenomycin biosynthesis and resistance) genes are in the unique sequence, while the *sapC, D, E* genes, for spore-associated proteins (arrows show direction of transcription), are in the TIRs (J. McCormick, personal communication).

SCP1 and as two closely linked copies on the chromosome, which they suggested to be involved in integration of SCP1. Kinashi et al (43a,b) located the SCP1 copy of IS466 in the position shown in Figure 11.7 and suggested that recombination between this copy of IS466 and a chromosomal copy, *after* integration, could account for deletion of the right-hand end of SCP1 (c. 80 kb) and a chromosomal segment, which would be ca. 40 kb (40a). Perhaps loss of the right-hand TIR could account for the stability of the NF state. Hanafusa and Kinashi (21a) have sequenced the ends of integrated (NF) and autonomous SCP1 and also the unoccupied *S. coelicolor* integration site. These sequences show that one end of SCP1 became inserted, with the loss of only 3 to 5 bp, into the *S. coelicolor* chromosome (Figure 11.8).

A further intriguing fact about SCP1 is that it carries (in the TIRs and therefore duplicated) genes for three out of five so far identified spore-associated proteins (Sap A-E), which occur on the outside of aerial mycelium and spores of *S. coelicolor* (21) (R. Losick, personal communication). Because SCP1⁻ strains sporulate apparently normally, the significance of these three proteins (Sap C, D, and E) is entirely unknown. Better understood is the methylenomycin gene cluster, located on the physical map in a unique part of the SCP1 DNA (Figure 11.7), and shown by Chater and Bruton (11) to contain biosynthetic genes on either side of the *mmr* resistance gene. In common with several other antibiotic gene clusters in *Streptomyces* (reviewed in 49), the cluster also contains regulatory genes. Thus, on conjugal transfer to a new host, SCP1 confers appropriately regulated antibiotic production as well as self-protection to avoid death of the transconjugant strain.

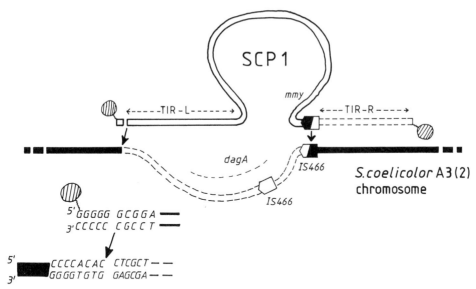

FIGURE 11.8. Integration of SCP1 into the *S. coelicolor* A3(2) chromosome. SCP1 and the host chromosome (not drawn to scale) recombine at the IS*466* sequences present in both, and also at one end of SCP1 (arrows) that joins with a nonhomologous chromosomal sequence (3 to 5 bp are lost from the integrating end of SCP1). One of the 80-kb terminal-inverted repeats (TIR-R) of SCP1 and c. 40 kb of chromosomal sequence containing the *dagA* gene, together with an additional copy of IS*466* (dashed lines), are lost on integration (21a, 40a, 43b).

6. Concluding Remarks

As will be apparent from the above account, many uncertainties still surround the conjugal properties of *Streptomyces* plasmids. Moreover, the cytological and molecular events accompanying conjugal DNA transfer remain very obscure. For example, it is not known whether plasmid-encoded functions are needed for the establishment of the physical connection between mating hyphae, or whether these depend on chromosomal genes with the plasmid merely specifying one or more functions necessary for DNA transfer through already established unions. That the entire process is not independent of plasmid genes is clearly shown by the existence of variants of conjugative plasmids that have lost the capacity for transfer. It is not known whether DNA is transferred in double- or single-stranded form.

Conjugation involving *Streptomyces* plasmids extends outside *Streptomyces* spp. themselves. Interspecific plasmid transfer within the genus *Streptomyces* was established some time ago (26). *S. lividans* has now been shown to act as recipient in matings with an *E. coli* strain containing plasmids proficient for conjugation in *E. coli* itself (5a, 18a, 52). Conversely, *S. lividans* carrying a conjugative *Streptomyces* plasmid (an SCP2* derivative) can mate with and transfer the plasmid to *Mycobacterium smegmatis* (H.M. Kieser, R.E. Melton, T. Kieser, and D.A. Hopwood, unpublished results). These results, together with the discovery of plasmid transfer between streptomycetes in soil (61, 74), indicate a potentially wide role for plasmid-mediated DNA transfer involving *Streptomyces* in an ecological and evolutionary context.

ACKNOWLEDGMENTS. We thank Keith Chater and Mervyn Bibb for helpful comments on the manuscript and Michael Brasch, Juliette Hagège, Haruyasu Kinashi, Greg Pettis, Tatsuji Seki, Julie Tai, and Martin Vögtli for unpublished results.

References

1. Alikhanian, S. I., and Mindlin, S. Z., 1957, Recombinations in *Streptomyces rimosus*, *Nature* **180:**1208–1209.
2. Bibb, M. J., and Hopwood, D. A., 1981, Genetic studies of the fertility plasmid SCP2 and its SCP2* variants in *Streptomyces coelicolor* A3(2), *J. Gen. Microbiol.* **126:**427–442.
3. Bibb, M. J., Schottel, J. L., and Cohen, S. N., 1980, A DNA cloning system for interspecies gene transfer in antibiotic-producing *Streptomyces*, *Nature* **284:**526–531.
4. Bibb, M. J., Ward, J. M., Kieser, T., Cohen, S. N., and Hopwood, D. A., 1981, Excision of chromosomal DNA sequences from *Streptomyces coelicolor* forms a novel family of plasmids detectable in *Streptomyces lividans*, *Mol. Gen. Genet.* **184:**230–240.
5. Bibb, M. J., Findlay, P. R., and Johnson, M. W., 1984, The relationship between base composition and codon usage in bacterial genes and its use for the simple and reliable identification of protein-coding sequences, *Gene* **30:**157–166.
5a. Bierman, M., Logan, R., O'Brien, K., Seno, E. T., and Schoner, B. E., 1992, Plasmid cloning vectors for the conjugal transfer of DNA from *Escherichia coli* to *Streptomyces* spp., *Gene* **116:**43–49.
6. Boccard, F., Pernodet, J.-L., Friedmann, A., and Guérineau, M., 1988, Site-specific integration of plasmid pSAM2 in *Streptomyces lividans* and *S. ambofaciens*, *Mol. Gen. Genet.* **212:**432–439.
6a. Boccard, F., Smokvina, T., Pernodet, J.-L., Friedmann, A., and Guerineau, M., 1989, Structural analysis of loci involved in pSAM2 site-specific integration in *Streptomyces*, *Plasmid* **21:**59–70.
7. Boccard, F., Smokvina, T., Pernodet, J.-L., Friedmann, A., and Guérineau, M., 1989, The integrated conjugative plasmid pSAM2 of *Streptomyces ambofaciens* is related to temperate bacteriophages, *EMBO J.* **8:**973–980.
8. Braendle, D. H., and Szybalski, W., 1959, Heterokaryotic compatibility, metabolic cooperation, and genetic recombination in *Streptomyces*, *Ann. NY Acad. Sci.* **81:**824–851.
9. Brody, H., Greener, A., and Hill, C. W., 1985, Excision and reintegration of the *Escherichia coli* K12 chromosomal element e14, *J. Bacteriol.* **161:**1112–1117.
10. Buttner, M. J., and Brown, N. L., 1987, Two promoters from the *Streptomyces* plasmid pIJ101 and their expression in *Escherichia coli*, *Gene* **51:**179–186.
11. Chater, K. F., and Bruton, C. J., 1985, Resistance, regulatory and production genes for the antibiotic methylenomycin are clustered, *EMBO J.* **4:**1893–1897.
12. Chung, S.-T., 1982, Isolation and characterization of *Streptomyces fradiae* plasmids which are prophage of actinophage φSF1, *Gene* **17:**239–246.
13. Deng, Z., Kieser, T., and Hopwood, D A., 1987, Activity of a *Streptomyces* transcriptional terminator in *Escherichia coli*, *Nucl. Acids Res.* **15:**2665–2675.
14. Deng, Z., Kieser, T., and Hopwood, D. A., 1988, "Strong incompatibility" between derivatives of the *Streptomyces* multi-copy plasmid pIJ101, *Mol. Gen. Genet.* **214:**286–294.
15. Dewitt, J. P., 1985, Evidence for a sex-factor in *Streptomyces erythreus*, *J. Bacteriol.* **164:**969–971.
16. Di Guglielmo, R., Conzelmann, C., Flett, F., and Cullum, J., 1990, *J. Cellular Biochem.* Suppl. 14A, Abstract CC 402, p. 124.
17. Doull, J. L., Vats, S., Chaliciopoulos, M., Stuttard, C., Wong, K., and Vining, L. C., 1986, Conjugational fertility and location of chloramphenicol biosynthesis genes on the chromosomal linkage map of *Streptomyces venezuelae*, *J. Gen. Microbiol.* **132:**1327–1338.
18. Friend, E. J., Warren, M., and Hopwood, D. A., 1978, Genetic evidence for a plasmid controlling fertility in an industrial strain of *Streptomyces rimosus*, *J. Gen. Microbiol.* **106:**201–206.

18a. Gormley, E. P., and Davies, J., 1991, Transfer of plasmid RSF1010 by conjugation from *Escherichia coli* to *Streptomyces lividans* and *Mycobacterium smegmatis*, *J. Bacteriol.* **173**:6705–6708.

19. Grant, S. R., Lee, S. C., Kendall, K., and Cohen, S. N., 1989, Identification and characterization of a locus inhibiting extrachromosomal maintenance of the *Streptomyces* plasmid SLP1, *Mol. Gen. Genet.* **217**:324–331.

20. Gruss, A., and Ehrlich, S. D., 1989, The family of highly interrelated single-stranded deoxyribonucleic acid plasmids, *Microbiol. Rev.* **53**:231–241.

21. Guijarro, J., Santamaria, R., Schauer, A., and Losick, R., 1988, Promoter determining the timing and spatial localization of transcription of a cloned *Streptomyces coelicolor* gene encoding a spore-associated polypeptide, *J. Bacteriol.* **170**:1895–1901.

21a. Hanafusa, T., and Kinashi, H., 1991, The structure of an integrated copy of the giant linear plasmid SCP1 in the chromosome of *Streptomyces coelicolor* 2612, *Mol. Gen. Genet.* **231**:363–368.

22. Hardisson, C., and Manzanal, M. B., 1976, Ultrastructural studies of sporulation in *Streptomyces*, *J. Bacteriol.* **127**:1443–1454.

23. Hirochika, H., Nakamura, K., and Sakaguchi, K., 1984, A linear DNA plasmid from *Streptomyces rochei* with an inverted terminal repetition of 614 base pairs, *EMBO J.* **3**:761–766.

24. Holloway, B. W., 1979, Plasmids that mobilize bacterial chromosome, *Plasmid* **2**:1–19.

25. Hopwood, D. A., 1959, Linkage and the mechanism of recombination in *Streptomyces coelicolor*, *Ann. NY Acad. Sci.* **81**:887–898.

26. Hopwood, D. A., and Wright, H. M., 1973, A plasmid of *Streptomyces coelicolor* carrying a chromosomal locus and its inter-specific transfer, *J. Gen. Microbiol.* **79**:331–342.

27. Hopwood, D. A., and Wright, H. M., 1976, Genetic studies on SCP1-prime strains of *Streptomyces coelicolor* A3(2), *J. Gen. Microbiol.* **95**:107–120.

28. Hopwood, D. A., Chater, K. F., Dowding, J. E., and Vivian, A., 1973, Advances in *Streptomyces coelicolor* genetics, *Bacteriol. Rev.* **37**:371–405.

29. Hopwood, D. A., Kieser, T., Wright, H. M., and Bibb, M. J., 1983, Plasmids, recombination and chromosome mapping in *Streptomyces lividans* 66, *J. Gen. Microbiol.* **129**:2257–2269.

30. Hopwood, D. A., Lydiate, D. J., Malpartida, F., and Wright, H. M., 1985, Conjugative plasmids in *Streptomyces*, in: *Plasmids in Bacteria* (D. Helinski, S. N. Cohen, D. B. Clewell, D. A. Jackson, and A. Hollaender, eds.), Plenum Press, New York, pp. 615–634.

31. Hopwood, D. A., Kieser, T., Lydiate, D. J., and Bibb, M. J., 1986, *Streptomyces* plasmids: their biology and use as cloning vectors, in: *The Bacteria. A Treatise on Structure and Function*, Vol. IX (S. W. Queener and L. E. Day, eds.), Academic Press, Orlando, pp. 159–229.

32. Hütter, R., and Eckhardt, T., 1988, Genetic manipulation, in: *Actinomycetes in Biotechnology* (M. Goodfellow, S. T. Williams, and M. Mordarski, eds.), Academic Press, London, pp. 89–184.

33. Kalkus, J., Reh, M., and Schlegel, H. G., 1990, Hydrogen autotrophy of *Nocardia opaca* strains is encoded by linear megaplasmids, *J. Gen. Microbiol.* **136**:1145–1151.

33a. Kataoka, M., Seki, T., and Yoshida, T., 1991, Five genes involved in self-transmission of pSN22, a *Streptomyces* plasmid, *J. Bacteriol.* **173**:4220–4228.

33b. Kataoka, M., Seki, T., and Yoshida, T., 1991, Regulation and function of the *Streptomyces* plasmid pSN22, genes involved in pock-formation and inviability, *J. Bacteriol.* **173**:7975–7981.

33c. Kataoka, M., Kuno, N., Horiguchi, T., Seki, T., and Yoshida, T., 1993, Replication of a *Streptomyces* plasmid pSN22 through single-stranded intermediates, *J. Bacteriol.*, in press.

34. Keen, C. L., Mendelovitz, S., Cohen, G., Aharonowitz, Y., and Roy, K. L., 1988, Isolation and characterization of a linear DNA plasmid from *Streptomyces clavuligerus*, *Mol. Gen. Genet.* **212**:172–176.

35. Kelemen, G H., Financsek, I., and Járai, M., 1989, Efficient transformation of *Micromonospora purpurea* with pIJ702 plasmid, *J. Antibiot.* **42**:325–328.

36. Kendall, K., and Cohen, S. N., 1987, Plasmid transfer in *Streptomyces lividans*: identification of a *kil-kor* system associated with the transfer region of pIJ101, *J. Bacteriol.* **169**:4177–4183.

37. Kendall, K., and Cohen, S. N., 1988, Complete nucleotide sequence of the *Streptomyces lividans* plasmid pIJ101 and correlation of the sequence with genetic properties, *J. Bacteriol.* **170**:4634–4651.

38. Kendall, K., and Cullum, J., 1984, Cloning and expression of an extracellular-agarase gene from *Streptomyces coelicolor* A3(2) in *Streptomyces lividans* 66, *Gene* **29**:315–321.

39. Kendall, K., and Cullum, J., 1986, Identification of a DNA sequence associated with plasmid integration in *Streptomyces coelicolor* A3(2), *Mol. Gen. Genet.* **202**:240–245.

40. Kieser, H. M., Henderson, D. J., Chen, C. W., and Hopwood, D. A., 1989, A mutation of *Streptomyces*

lividans that prevents intraplasmid recombination has no effect on chromosomal recombination, *Mol. Gen. Genet.* **220**:60–64.

40a. Kieser, H. M., Kieser, T., and Hopwood, D. A., 1992, A combined genetic and physical map of the *Streptomyces coelicolor* A3(2) chromosome, *J. Bacteriol.* **174**:5496–5507.

41. Kieser, T., Hopwood, D. A., Wright, H. M., and Thompson, C. J., 1982, pIJ101, a multi-copy broad host-range *Streptomyces* plasmid: functional analysis and development of DNA cloning vectors, *Mol. Gen. Genet.* **185**:223–238.

41a. Kieser, T., and Hopwood, D. A., 1991, Genetic manipulation of *Streptomyces*: integrating vectors and gene replacement. *Meth. in Enzymol.* **204**:430–458.

42. Kinashi, H., and Shimaji-Murayama, M., 1991, Physical characterization of SCP1, a giant linear plasmid from *Streptomyces coelicolor*, *J. Bacteriol.* **173**:5123–1529.

43. Kinashi, H., Shimaji, M., and Sakai, A., 1987, Giant linear plasmids in *Streptomyces* which code for antibiotic biosynthesis genes, *Nature* **328**:454–456.

43a. Kinashi, H., Shimaji-Murayama, M., and Hanafusa, T., 1991, Nucleotide sequence analysis of the unusually long terminal inverted repeats of a giant linear plasmid, SCP1, *Plasmid* **26**:123–130.

43b. Kinashi, H., Shimaji-Murayama, M., and Hanafusa, T., 1992, Integration of SCP1, a giant linear plasmid, into the *Streptomyces coelicolor* chromosome. *Gene* **115**:35–41.

44. Kirby, R., and Hopwood, D. A., 1977, Genetic determination of methylenomycin synthesis by the SCP1 plasmid of *Streptomyces coelicolor* A3(2), *J. Gen Microbiol* **98**:239–252.

45. Kirby, R., Wright, L. F., and Hopwood, D. A., 1975, Plasmid-determined antibiotic synthesis and resistance in *Streptomyces coelicolor*, *Nature* **254**:265–267.

46. Kobayashi, T., Shimotsu, H., Horinouchi, S., Uozumi, T., and Beppu, T., 1984, Isolation and characterization of a pock-forming plasmid pTA4001 from *Streptomyces lavendulae*, *J. Antibiot.* **37**:368–375.

47. Kuhstoss, S., Richardson, M. A., and Rao, R. N., 1989, Site-specific integration in *Streptomyces ambofaciens*: localization of integration functions in *S. ambofaciens* plasmid pSAM2, *J. Bacteriol.* **171**:16–23.

48. Lee, S. C., Omer, C. A., Brasch, M. A., and Cohen, S. N., 1988, Analysis of recombination occurring at SLP1 *att* sites, *J. Bacteriol.* **170**:5806–5813.

48a. Lee, M. H., Pascopella, L., Jacobs, W. R., and Hatfull, G. R., 1991, Site-specific integration of mycobacteriophage L5 integration-proficient vectors for *Mycobacterium smegmatis*, *Mycobacterium tuberculosis* and Bacille Calmette-Guerin. *Proc. Natl. Acad. Sci. USA* **88**:3111–3115.

49. Martín, J. F., and Liras, P., 1989, Organization and expression of genes involved in the biosynthesis of antibiotics and other secondary metabolites, *Annu. Rev. Microbiol.* **43**:173–206.

49a. Martin, C., Mazodier, P., Mediola, M. V., Gicquel, B., Smokvina, T., Thompson, C. J., and Davies, J., 1991, Site-specific integration of the *Streptomyces* plasmid pSAM2 in *Mycobacterium smegmatis*, *Molec. Microbiol.* **5**:2499–2502.

50. Matsushima, P., and Baltz, R. H., 1988, Genetic transformation of *Micromonospora rosaria* by the *Streptomyces* plasmid pIJ702, *J. Antibiot.* **41**:583–585.

51. Matsushima, P., McHenney, M. A., and Baltz, R. H., 1987, Efficient transformation of *Amycolatopsis orientalis* (*Nocardia orientalis*) protoplasts by *Streptomyces* plasmids, *J. Bacteriol.* **169**:2298–2300.

52. Mazodier, P., Petter, R., and Thompson, C., 1989, Intergeneric conjugation between *Escherichia coli* and *Streptomyces* species, *J. Bacteriol.* **171**:3585–3585.

53. Mazodier, P., Thompson, C., and Boccard, F., 1990, The chromosomal integration site of the *Streptomyces* element pSAM2 overlaps a putative tRNA gene conserved among actinomycetes, *Mol. Gen. Genet.* **222**:431–434.

54. Nordström, K., and Austin, S. J., 1989, Mechanisms that contribute to the stable segregation of plasmids, *Annu. Rev. Genet.* **23**:37–69.

55. Omer, C. A., and Cohen S. N., 1984, Plasmid formation in *Streptomyces*: excision and integration of the SLP1 replicon at a specific chromosomal site, *Mol. Gen. Genet.* **196**:429–438.

56. Omer, C. A., and Cohen, S. N., 1986, Structural analysis of plasmid and chromosomal loci involved in site-specific excision and integration of the SLP1 element of *Streptomyces coelicolor*, *J. Bacteriol.* **166**:999–1006.

56a. Omer, C. A., Stein, D., and Cohen, S. N., 1988, Site-specific insertion of biologically functional adventitious genes into the *Streptomyces lividans* chromosome, *J. Bacteriol.* **170**:2174–2184.

57. Pernodet, J.-L., and Guerineau, M., 1981, Isolation and physical characterization of streptomycete plasmids, *Mol. Gen. Genet.* **182**:53–59.

58. Pidcock, K. A., Montenecourt, B. S., and Sands, J. A., 1985, Genetic recombination and transformation in protoplasts of *Thermomonospora fusca*, *Appl. Environment. Microbiol.* **50:**693–695.

59. Pierson, L. S., and Kahn, M. L., 1987, Integration of satellite bacteriophage P4 in *Escherichia coli*: DNA sequences of the phage and host regions involved in site-specific recombination, *J. Mol. Biol.* **196:**487–496.

60. Pigac, J., Vujaklija, C., Toman, Z., Gamulin, V., and Schrempf, H., 1988, Structural instability of a bifunctional plasmid pZG1 and single-stranded DNA formation in *Streptomyces*, *Plasmid* **19:**222–230.

61. Rafii, F., and Crawford, D. L., 1989, Donor/recipient interactions affecting plasmid transfer among *Streptomyces* species: a conjugative plasmid will mobilize nontransferable plasmids in soil, *Curr. Microbiol.* **19:**115–121.

62. Reiter, W.-D., Palm, P., and Yeats, S., 1989, Transfer RNA genes frequently serve as integration sites for prokaryotic genetic elements, *Nucl. Acids Res.* **17:**1907–1914.

63. Saito, H., and Ikeda, Y., 1959, Cytogenetic studies on *Streptomyces griseoflavus*, *Ann. NY Acad. Sci.* **81:** 862–878.

64. Sakaguchi, K., 1990, Invertrons, a class of structurally and functionally related genetic elements that includes linear DNA plasmids, transposable elements, and genomes of adeno-type viruses, *Microbiol. Rev.* **54:**66–74.

65. Sermonti, G., and Spada-Sermonti, I., 1955, Genetic recombination in *Streptomyces*, *Nature* (London) **176:**121.

65a. Shiffman, D., and Cohen, S. N., 1992, Reconstruction of a *Streptomyces* linear replicon from separately cloned DNA fragments—existence of a cryptic origin of circular replication within the linear plasmid, *Proc. Natl. Acad. Sci. USA* **89:**6129–6133.

66. Smokvina T., Francou, F., and Luzzati, M., 1988, Genetic analysis in *Streptomyces ambofaciens*, *J. Gen. Microbiol.* **134:**395–402.

67. Smokvina, T., Mazodier, P., Boccard, F., Thompson, C. J., and Guérineau, M., 1990, Construction of a series of pSAM2-based integrative vectors for use in actinomycetes, *Gene* **94:**53–59.

68. Smokvina, T., Boccard, F., Pernodet, J.-L., Friedmann, A., and Guérineau, M., 1991, functional analysis of the *Streptomyces ambofaciens* elements pSAM2, *Plasmid* **25:**40–52.

69. Stein, D. S. and Cohen, S. N., 1990, Mutational and functional analysis of the *korA* and *korB* gene products of *Streptomyces* plasmid pIJ101, *Mol. Gen. Genet.* **222:**337–344.

70. Stein, D. S., Kendall, K. J., and Cohen, S. N., 1989, Identification and analysis of transcriptional regulatory signals for the *kil* and *kor* loci of *Streptomyces* plasmid pIJ101, *J. Bacteriol.* **171:**5768–5775.

71. Tsai, J. F.-Y., and Chen, C. W., 1989, Isolation and characterization of *Streptomyces lividans* mutants deficient in intraplasmid recombination, *EMBO J.* **8:**973–980.

72. Vivian, A., 1971, Genetic control of fertility in *Streptomyces coelicolor* A3(2): plasmid involvement in the interconversion of UF and IF strains, *J. Gen. Microbiol.* **69:**353–364.

73. Vivian, A., and Hopwood, D. A., 1973, Genetic control of fertility in *Streptomyces coelicolor* A3(2): new kinds of donor strains, *J. Gen. Microbiol.* **76:**147–162.

74. Wellington, E. M. H, Cresswell, N., and Saunders, V. A., 1990, Growth and survival of streptomycete inoculants and extent of plasmid transfer in sterile and nonsterile soil, *Appl. Environment. Microbiol.* **56:**1413–1419.

74a. Wu, X., and Roy, K. L., 1993, Complete nucleotide sequence of a linear plasmid from *Streptomyces clavuligerus*, and characterization of its RNA transcripts, *J. Bacteriol.* **175:**37–52.

75. Yamamoto, H., Maurer, K. H., and Hutchinson, C. R., 1986, Transformation of *Streptomyces erythreus*, *J. Antibiot.* **39:**1304–1313.

76. Zaman, S., Richards, H., and Ward, J., 1992, Expression and characterization of the *korB* gene product from the *Streptomyces lividans* plasmid pIJ101 in *Escherichia coli* and determination of its binding site on the *korB* and *kilB* promoters, *Nucl. Acids Res.* **20:**3693–3700.

Chapter 12

Conjugation and Broad Host Range Plasmids in Streptococci and Staphylococci

FRANCIS L. MACRINA and GORDON L. ARCHER

1. Introduction

1.1. Overview

Interest in conjugal genetic exchange in gram-positive bacteria has centered on two major themes. First, such gene transfer systems offer comparative models for the study of the mechanisms and consequences of genetic exchange in this group of organisms. Second, the role of conjugative gene exchange in the dissemination of antimicrobial resistance has prompted investigation in order to understand such systems and, eventually, to prevent or control the spread of resistance phenotypes.

   Gene transmission systems in gram-positive bacteria may be categorized into three types. One system is represented by the pheromone-responding plasmids (20, 21) (see Chapter 14 in this volume). These plasmids are confined to the enterococci and their close relatives. Such plasmids tend to be larger than 25 kb in size and transfer with high efficiency in liquid media following the mixing of appropriate donor and recipient cells. Transfer occurs under the control of an elaborate circuitry in which recipient cells produce a small peptide sex pheromone that triggers the donor cell to produce surface components that induce cell clumping. This enhances cell–cell contact between donors and recipients, which in turn gives rise to the necessary cellular and molecular events that promote plasmid

FRANCIS L. MACRINA • Department of Microbiology and Immunology Medical College of Virginia, Virginia Commonwealth University, Richmond, Virginia 23298. GORDON L. ARCHER • Departments of Microbiology and Immunology and Medicine, Medical College of Virginia, Virginia Commonwealth University, Richmond, Virginia 23298.
Bacterial Conjugation, edited by Don B. Clewell. Plenum Press, New York, 1993.

transfer. Such pheromone-responding plasmids often carry genes encoding hemolysins, bacteriocins, or antibiotic resistance (19).

Another class of transfer systems is composed of conjugative transposons (22) (see Chapter 15 in this volume). These elements meet the criteria for procaryotic transposable elements except that, in addition to encoding the factors needed for their serial transposition, they also mediate their own conjugal transfer. Conjugative transposons usually carry determinants conferring antibiotic resistance, a feature that continues to encourage and to facilitate their study. Many of the characterized conjugative transposons are in the size range of 15 to 20 kb (22), but some are considerably larger (>50 kb) (88) and may be genetically complex in structure (6). Unlike the case for the pheromone-responding plasmids, conjugative transposon transfer generally does not occur in liquid. Donor and recipient cells must be cocultivated on a solid surface in order for transfer to proceed. Moreover, the transfer frequencies seen with conjugative transposons are quite low compared with those seen with the pheromone-responding plasmids (10^{-6} versus 10^{-2} or higher).

A third type of transfer system is composed of the so-called broad host range conjugative plasmids (21, 73). Their transfer occurs at variable frequency (generally in the 10^{-6} to 10^{-3} range), depending on plasmid type and mating-pair genotype, and requires that donor and recipient cells to be cocultivated on a solid surface. Such plasmids identified in the streptococci possess considerable latitude in the range of their transfer and inheritance. On the other hand, analogous plasmids thus far found in the staphylococci are considerably more restricted in their host range. Both groups of plasmids carry drug resistance markers and have as a lower size limit 15 to 20 kb. They do not respond to pheromone signals. Plasmids that fall into this category are the subject matter of this chapter. We will limit our discussion to plasmids of certain streptococci and staphylococci to reflect the more well-developed systems. Plasmid-mediated conjugation systems in lactic acid bacteria (e.g., *Lactococcus* and *Lactobacillus*) and in certain bacilli (e.g., *Bacillus anthracis*) have been reported but will not be covered because of their lesser state of development and apparent narrow host range (8, 33, 36, 68).

1.2. Definition of Conjugation by Broad Host Range Plasmids in the Staphylocci and Streptococci

Conjugal transfer mediated by broad host range plasmids of gram positives requires direct contact of donors and recipients, which is usually achieved by collecting cells on some type of a solid surface, for example a nitrocellulose filter. The conjugal transfer process does not involve bacteriophage or prophage and is resistant to the inhibitory effects of deoxyribonuclease. For the most part, the cellular and molecular basis of conjugative transfer mediated by the gram-positive broad host range plasmids remains unclear. Dunny (27) points out that in general bacterial conjugation can be divided into four steps: (a) initial contact of donors and recipients; (b) formation of an effective mating cell pair (or aggregate); (c) DNA transfer; and (d) genetic recombination and resolution of progeny. Although these criteria provide a useful framework for hypothesis formulation and testing, certain aspects of plasmid transfer confound the critical biochemical analysis of these phenomena. The relatively low frequency of transfer seen in these systems severely limits the analysis of mating-pair or aggregate formation as well as the processing and transfer of DNA during mating. The fact that cells must be brought into contact by artificial means raises questions about cell-cell infections that have not been addressed. How this requirement is met in nature is unclear, and this complicates the evaluation of data obtained in vitro, especially in regard to the biochemistry of cell "communication" (e.g., receptor-ligand interactions) likely to be operative at the initial stages of mating. It has been reported that for one broad host range plasmid, pAMβ1, the type and pore size of filter as well as one filter side (front or back) greatly influenced transfer frequency between *Enterococcus faecalis* and *Lactobacillus plantarum* (71). Such variations could result in 100-fold differences and, in some instances, reduced transfer frequencies to barely detectable levels (71). Passing sterile water through the membrane during the mating process could increase transfer by 10-fold (71). In another study (80), adding 10 to 40% polyethylene glycol (PEG) to liquid mating mixtures allowed mating to proceed at frequencies of up to 10^{-5}–10^{-4}. No transfer was observed at PEG concentration of less than 10%. The PEG is believed to facilitate pAMβ1 transfer by promoting aggregation of donor and recipient cells.

Although plasmid-encoded surface components on donors and recipients can be invoked as playing a role in the mating process, direct evidence to support this is missing at present. Identifiable surface structures analogous to the pili seen in gram-negative transfer systems have not been described for gram-positive broad host range plasmid-containing donor cells. Little is known about the nature of the exchange process; for example, is there unidirectional transfer of DNA through a channel or cytoplasmic bridge specifically formed in this process, or is the transfer event the result of the genetic equivalent of protoplast fusion? What is the nature of the DNA transferred in this event (e.g., is it single stranded or double stranded?), and what macromolecular processes are required for transfer? With very few exceptions, little information is known along these lines. However, as will be seen below, the stage is set to pursue studies to answer these kinds of questions with model plasmid systems of both streptococci and staphylococci.

2. Streptococcal Broad Host Range Plasmids

2.1. General Characteristics and Epidemiology

Most of the broad host range plasmids of the streptococci encode resistance to macrolides, lincosamides, and streptogramin B antibiotics (MLS[r]). Such resistance determinants encode a RNA methylase that is capable of the N^6 dimethylation of specific adenine residues in the 16s ribosomal RNA. Such an alteration prevents the binding of MLS antibiotics to the bacterial ribosome. This class of determinant is found widely in the gram-positive cocci and bacilli along with upstream regulatory transcriptional attenuation systems (58, 59). This determinant, called *erm*, has also been found in gram-negative bacteria such as the *Bacteroides* (63). Some streptococcal broad host range plasmids carry resistance determinants to antibiotics other than MLS drugs, for example chloramphenicol.

Broad host range plasmids have been demonstrated in a variety of clinically important streptococci and geographically are found worldwide (20, 21, 73). Besides sharing a common MLS[r] determinant, members of this group range approximately from 20 to 35 kb in size and often have similar restriction enzyme digest patterns. Epidemiological studies have revealed considerable DNA sequence similarities as determined by hybridization studies (37). The evolutionary relatedness of these plasmids has only begun to be systematically approached at the molecular level, however (14, 79). The MLS-type broad host range plasmid seems to be widely distributed in the streptococci, most notably Lancefield types A, B, C, and G (13, 16, 37, 42, 49, 56). Their transfer to multiple Lancefield types (16) suggests their broad natural presence within this genus. It is difficult to speculate on the origin of these plasmids, although both nucleotide and amino acid sequence data suggest that their MLS gene is ancestrally related to the one present in other gram positives, such as the staphylococci (26, 58, 89). Transfer of the streptococcal broad host range plasmids to a wide range of gram-positive organisms is illustrated in Table 12.1.

Because of the common presence of the MLS resistance gene, incompatibility classification by simple plasmid coinheritance studies has not been practical. However, as the genetic basis of replication of these plasmids has been pursued, incompatibility grouping using custom plasmid constructs with differing markers has been possible. Also contributing to such classification schema have been studies relating to the replication regions of certain of these plasmids (14, 79).

2.2. Model Broad Host Range Plasmids of Streptococcal Origin

The two most extensively studied examples of conjugative broad host range streptococcal plasmids are pIP501 and pAMβ1. pIP501, originally isolated from a strain of *Streptococcus agalactiae*, is 30.2 kb in size and confers inducible resistance to erythromycin (MLS[r]) and chloramphenicol (39). The latter resistance is conferred by a determinant (*cat*) that encodes chloramphenicol acetyltransferase. pIP501 is able to promote its self-transfer at frequencies similar to pAMβ1 (ca. 10^{-4} progeny/donor cell). pAMβ1 was initially isolated from *Enterococcus faecalis* (23), confers constitutively expressed MLS resistance, and is 26.5 kb in size. A number of similar plasmids have been identified, but pIP501 and pAMβ1 have served as the models for genetic and biochemical study in terms of their replication, resistance determinants, and transfer properties. pAMβ1 conjugative functions have been demonstrated to work in gram-positive to gram-negative transfer (85). A pBR322-pAMβ1 chimeric plasmid could be conjugally transferred from *E. faecalis* to *Escherichia coli* at frequencies of 5×10^{-9}. These data raise interesting

TABLE 12.1   Host Range of Streptococcal Plasmids

pIP501	pAMβ1
Streptococcus	*Streptococcus*
Staphylococcus	*Staphylococcus*
Clostridium	*Clostridium*
Listeria	*Lactobacillus*
Pediococcus	*Bacillus*

See references 20, 21, 27, 73.

questions (see below) about the biology of gene transfer in procaryotes and, equally important, have serious implications regarding the spread of drug resistance genes among evolutionarily diverse bacteria.

pIP501 (10, 30, 31) and pAMβ1 (47) have been characterized at the molecular level, and restriction maps have been generated. Resistance genes and origins of replication necessary for maintenance in streptococci have been located and characterized (14, 79). The copy number for pIP501 and pAMβ1 is estimated to be three to five replicons per bacterial chromosome (30). Several useful gene cloning and genetic mobilization systems have been developed using pIP501 and pAMβ1 (9, 11, 12, 51, 52, 54, 55, 69).

2.3. pIP501

Physical analysis of pIP501 (Figure 12.1) has resulted in a useful restriction enzyme site map for this plasmid (30, 31, 42). Several derivatives of this plasmid have been prepared by in vitro recombinant DNA techniques (31, 42). These and spontaneously occurring

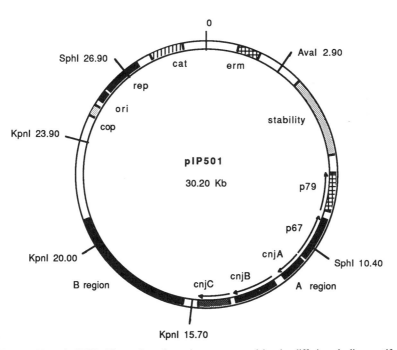

FIGURE 12.1. Map of pIP501. The various determinants are noted by the differing shading motifs. Chloramphenicol resistance, cat; erythromycin resistance, erm (30, 31); cop, region inferred to be involved in copy number (42). The origin of replication, ori; location of the replication protein gene, rep (14). Stability: the region thought to be involved in plasmid stability in *S. sanguis* (43). Two genes encoding 67 and 79 kdal proteins in *E. coli* minicells are noted as p67 and p79, with the arrows indicating the deduced direction of transcription of these genes (42). The three genes involved in transfer, identified by transposon mutagenesis, are designated cnjA, B, and C (42). The area designated as B region also appears to encode functions needed for self-transfer based on transposon mutagenesis (42).

deletion derivatives of pIP501 have helped in the localization of replication and copy control sequences (30). Partial digestion of pIP501 with HindIII, followed by ligation and transformation of *Streptococcus sanguis*, resulted in the creation of pVA798 (30). This plasmid contained the normally contiguous 6.3-kb HindIII A fragment (contains *cat*) and the 1.5-kb HindIII F fragment. pVA798 displayed an elevated copy number compared to pIP501 in *S. sanguis* (95 copies/cell versus 2 to 3 copies/cell), but it was unstable (30). This suggested an alteration in copy control and in partitioning, but the nature of the defects is unclear.

A second important construct derived from pIP501 was pVA797 (31, 54). This plasmid was constructed in vitro by substituting a streptococcal replicon, pVA380-1, for a 3-kb HpaII-AvaI fragment of pIP501. The fragment that was removed contained the MLS resistance gene. This pVA380-1::pIP501 construct replicated stably and allowed flexibility in certain genetic experiments (e.g., mutagenesis with Tn917, a transposon carrying an MLS resistance gene). Removal of the MLS^r gene and insertion of the pVA380-1 replicon did not affect conjugation frequency (30) or host range (69). The presence of the pVA380-1 replicon also provided a unique EcoRI site that enabled cloning of the entire pVA797 plasmid in *E. coli* using the positive selection vector pOP203A2$^+$ (31). This recombinant plasmid, pVA904, segregated into *E. coli* minicells, and the pVA797 sequences directed the synthesis of 13 polypeptides (as measured by the incorporation of ^{35}S-methionine into proteins detectable on polyacrylamide gels subjected to autoradiography). The pVA797 construct set the stage for a genetic analysis of pIP501, and the ability of pVA904 to synthesize proteins in *E. coli* minicells provided encouragement for identifying the function of pIP501 gene products, especially those that might be involved in conjugal transfer.

Krah and Macrina (42, 43) used the pVA797 replicon to begin an analysis of the genetic basis of conjugal transfer. As a starting point, they made a smaller replicon by removal of specific restriction fragments, and, in the process, they replaced the chloramphenicol resistance gene with a kanamycin resistance gene. The resultant plasmid, pVA1702, was 25.2 kb in size but still transferred conjugally at wild-type (pVA797/pIP501) levels. This 5-kb reduction in plasmid size compared with pVA797/pIP501 facilitated the search for genes involved in conjugative transfer. The pVA1702 plasmid was used as a target for transposon mutagenesis using both Tn917 and Tn917lac (42). The latter transposon allowed the monitoring of a promoterless β-galactosidase gene that could be activated if the transposon inserted downstream from a transcriptional unit. Transposon insertions enabled the localization of conjugal transfer determinants to two separate regions of pVA1702: Region A was 7.5 kb, and region B was 8.8 kb. Both of these regions corresponded to the lower half of the original pIP501 circular restriction map (30, 42) (see Figure 12.1). In vitro constructed deletions of the A region were evaluated for protein-encoding ability in *E. coli* minicells, and three proteins presumed to be involved in conjugal transfer were identified. These proteins were 25, 45, and 76 kDal in size. Their function has yet to be described, but this system in general should afford the necessary tools to explore the biochemical genetic basis of pIP501-conferred conjugal transfer proficiency. In a further examination of the pVA1702 plasmid, Krah and Macrina (43) discovered that this plasmid was not efficiently maintained in *S. sanguis* when transferred from *E. faecalis*. These authors concluded that a region deleted from pIP501 in constructing pVA1702 was responsible for stable maintenance and/or establishment in *S. sanguis*. These observations prompt the hypothesis that broad host range plasmids like pIP501 employ specific mechanisms enabling their maintenance in certain gram-positive bacteria. Figure 12.1, adapted from Krah and Macrina (42), illustrates our understanding of the organization of pIP501 at present.

2.4. pAMβ1

The MLS resistance gene of pAMβ1 was identified in conjugal cotransmission studies. The conjugal proficiency of pAMβ1 was discovered after it was introduced by genetic transformation into a Lancefield group F *Streptococcus* (*S. anginosus* [48]). The pAMβ1-containing transformant was shown to conjugally transfer MLS resistance, coincident with the intact plasmid, to a wide variety of streptococcal species (49). The extended host transfer range beyond streptococcal species is apparent in Table 12.1. LeBlanc and Lee (47) studied the molecular organization of pAMβ1, localizing its MLS resistance gene and a region involved in plasmid replication. These studies were facilitated by the use of spontaneously occurring deletion derivatives (53). Analysis of such deletion mutants suggested that one region of the plasmid (ca. 8 kb in size) was involved in conjugal transfer functions. Strains containing deletions could maintain the plasmid and express MLS resistance normally but were not able to act as conjugal donors (47). Using such deletion analysis results, early attempts to clone suspected conjugal transfer genes of pAMβ1 were not successful (47). Although there have been no additional reports on the analysis of conjugal transfer genes from pAMβ1, available cloning techniques, in vitro deletion methodologies, transposon mutagenesis, and nucleotide sequencing now provide the logical tools for undertaking a systematic investigation of the genetic basis of conjugal transfer conferred by this plasmid.

The MLS resistance gene carried by pAMβ1 has been analyzed by nucleotide sequencing (15, 57). This determinant, called *ermAM*, encodes an adenine methylase (see Section 2.1), and its sequence is strongly homologous to *erm* determinants found on at least one other streptococcal plasmid (38) and on the streptococcal transposon Tn*917* (74). The *ermAM* gene of pAMβ1 has never been reported to be transposable itself. Despite the strong homology of the pAMβ1 *ermAM* to other *erm* determinants, its phenotypic expression differs from many *erm* genes in that it is constitutively expressed. The sequencing of the pAMβ1 *ermAM* gene explained this phenotype by revealing that regulatory sequences immediately upstream of the gene were deleted (15, 57). Martin et al (57) have proposed that full *erm* gene expression in *E. coli* is related in part to the presence or absence of control sequences. The *ermAM* gene from pAMβ1 cloned in *E. coli* confers clear-cut resistance to erythromycin assayed in wild-type *E. coli* or in erythromycin-hypersensitive *E. coli* mutants (51, 57); this has made it very useful as a marker on *E. coli*-streptococcal shuttle plasmids (25, 51, 55).

Swinfield et al (79) have analyzed the nature of the replication region of pAMβ1. The nucleotide sequence of a 5.1-kb *Eco*RI restriction fragment carrying the replication region was determined by these authors. Seven open reading frames were identified, and two of these encoded proteins were tested in an *E. coli* in vitro transcription/translation assay. One of these proteins, the product of the orfC reading frame, contained reiterated amino acids, and the authors drew attention to the fact that such motifs are known to function as membrane attachment loci in other systems (e.g., *Staphylococcus aureus* protein A). Indeed the carboxyl terminus of the orfC protein showed strong homology to the membrane-anchoring sequences of several gram-positive proteins, including *S. aureus* protein A. Of further interest was what appeared to be a signal sequence at the amino terminus of the orfC protein, suggesting that this protein was excreted. Taken together, these data lead to the hypothesis that the pAMβ1 orfC protein is localized at the cell surface. Hence, Swinfield et al (79) suggested that a role for this protein in conjugal transfer be considered. An extension of these studies, using specific mutants, blocking antibodies, and complementation analysis should shed light on function of the orfC protein.

Swinfield et al (79) also established that a reading frame, designated orfE, was essential for pAMβ1 replication. The 57-kDal protein of the orfE gene was expressed in the *E. coli* in vitro transcription/translation assay. Plasmid constructs carrying defective orfE sequences were unable to replicate. A similar 57-kDal replicator protein encoded by pIP501 has been identified (14).

2.5. Conjugative Mobilization

Conjugative plasmids of streptococcal origin are capable of mobilizing nonconjugative plasmids. Both pAMβ1 and pIP501 can mobilize pMV158, a small streptococcal plasmid conferring tetracycline resistance. Mechanisms underlying such conjugative mobilization are unclear, although studies with pMV158 failed to reveal cointegrates between the conjugative plasmid and pMV158 (67, 87). pAMβ1 mobilization of a streptococcal-*E. coli* shuttle plasmid, pAM610 (76), also did not involve cointegrate formation. On the other hand, pVA797 was seen to mobilize the streptococcal-*E. coli* shuttle plasmid pVA838 by a mechanism that involved cointegrate formation between homologous sequences on pVA797 and pVA838 (77). Resolution of the cointegrate occurred in the progeny cell, with both plasmids being re-formed intact. The design of streptococcal broad host range transfer systems that exploit cointegrate formation has proved useful in the mobilization of genes in other systems, for example, the clostridia (64, 65).

Priebe and Lacks (67) have analyzed the mobilization of pMV158 by pIP501 and have shown that an open reading frame, encoding a putative polypeptide of 58 kDal, is essential. This determinant on pMV158 has been termed *mob* (67). The Mob protein shares homology with the amino-terminal half of a protein involved in site-specific cointegrative recombination involving certain staphylococcal plasmids (67). Priebe and Lacks (67) have also identified four palindromic sequences on pMV158. One of these sequences, called *palD*, was common to other small plasmids of gram-positive origin (e.g., pE194, pT181, and pUB110 of staphylococci). This sequence appeared to be necessary for pMV158 mobilization, and the authors have postulated that it contains a sequence that is nicked as part of the transfer process, analogous to the *oriT* sequences present in gram-negative systems (40). Direct evidence of this is lacking, but the implications of such thinking are important, as they give rise to the hypothesis that single-stranded DNA is transferred in this gram-positive system. These studies have provided important information regarding mechanisms of mobilization in the streptococcal broad host range systems. Equally important, they set the stage for similar approaches aimed at analyzing the molecular mechanisms of transfer of the conjugative plasmids themselves.

3. Staphylococcal Conjugative Plasmids

3.1. Overview

The earliest studies assessing mechanisms of gene exchange among natural populations of *Staphylococcus aureus* looked at transfer of antibiotic resistance phenotypes. Resistance to erythromycin and penicillin (62), as well as neomycin (44) and tetracycline (45), could be transferred among mixed cultures of naturally occurring *S. aureus* in broth. Transfer occurred at high frequency, was associated with the acquisition of plasmid DNA by recipients, and required bacteriophage. While some transfer appeared to be by conventional transduction, plasmid DNA could also be transferred by a mechanism that required the early steps of phage propagation but did not need fully intact phage particles (90). Differences between some of the characteristics of the transfer of antibiotic resistance phenotypes in mixed cultures of staphylococci and classic transduction prompted one investigator to name the process "phage-mediated conjugation" (46). The demonstration that the same antibiotic resistance markers that transferred in mixed cultures in vitro also transferred among *S. aureus* on mouse and human skin confirmed the relevance of phage-related gene exchange in natural settings (60, 62). True conjugation (plasmid transfer requiring cell-to-cell contact and not needing phage-encoded products in either donor or recipient) was not specifically described in any of these early reports.

The first demonstrations that true conjugative transfer of antibiotic resistance plasmids occurred between staphylococci were made at several institutions that had noticed outbreaks of gentamicin-resistant staphylococci in their hospitals (3, 32, 59). While Naidoo and Noble (60) had also noted transfer of gentamicin resistance between skin isolates of *S. aureus* in 1978 (60), recipients became lysogenized following transfer, incriminating phage in the transfer process. In contrast, conjugative transfer of gentamicin resistance had the characteristics of classic conjugation as described in gram-negative mating systems (3, 32, 59). These included (a) a requirement for cell-to-cell contact; (b) transfer between nonlysogenized donor and recipient pairs; (c) the finding of identical plasmids in all donors and gentamicin-resistant transconjugants; (d) the demonstration that there were genes on the conjugative plasmid required for plasmid self-transfer; and (e) the observation that self-transmissible plasmids could mobilize the transfer of small, nontransmissible plasmids. An even more convincing demonstration of true conjugative transfer was the observation that the same plasmids thought to be responsible for gentamicin resistance transfer between *S. aureus* mating pairs could also transfer resistance between *S. aureus* and *Staphylococcus epidermidis* by cell-to-cell contact (84). Because DNA could not be introduced into wild *S. epidermidis* isolates in vitro by using any of the techniques mediating DNA introduction into *S. aureus* (transduction, transformation, or "phage-mediated" conjugation), true conjugation appeared to be the only possible operative transfer mechanism. The demonstrations that interspecies transfer occurred on human skin (41, 82) and that conjugative plasmids with identical restriction endonuclease digestion profiles were seen in hospital isolates of *S. epidermidis* and *S. aureus* from the same patient (3) confirmed the likely epidemiological relevance of conjugative transfer.

Additional plasmids that appeared to transfer between *S. aureus* mating pairs have also been reported. One class encoded diffusible pigment and resistance to MLS antibiotics and spectinomycin; it does not mobilize nonconjugative plasmids (82). A second class has no detectable phenotype but was shown to transfer in filter matings by its ability to mobilize nonconjugative antibiotic resistance plasmids (86). Neither of these two classes of plasmids appeared to be related by DNA hybridization to the conjugative gentamicin resistance plasmids mentioned previously. However, there are, as yet, no molecular studies confirming true conjugation as the mechanism of transfer for these plasmids or showing their relationship to the pAMβ1 or pIP501 host plasmids resident in streptococci (28). Therefore, these plasmids will not be considered further in this chapter.

3.2. Epidemiology

The appearance of conjugative plasmids in staphylococci coincided with reports of the emergence of gentamicin resistance among staphylococci in U.S. hospitals in the mid-1970s. The earliest *S. aureus* isolate known to contain a conjugative plasmid was a gentamicin-resistant isolate recovered in Alaska in 1974 (72). Conjugative plasmids have also been found in gentamicin-resistant isolates from Australia (83), Great Britain, and Germany (61), but they were of minimal epidemiological significance in these countries. In Melbourne, Australia (75), and Dublin, Ireland (78), where multiresistant *S. aureus* isolates predominate in many hospitals, gentamicin resistance genes are borne largely on nonconjugative plasmids and transposable elements. In contrast, in a recent survey of staphylococci from various geographic regions of the United States, 63% of gentamicin-resistant *S. aureus* carried resistance genes on a conjugative plasmid (4). In this same study, conjugative plasmids were identified not by resistance phenotype but by hybridization of cell lysates with a DNA probe containing conjugative transfer genes. Of the 74 staphylococcal isolates containing conjugative plasmids identified by DNA hybridization, all had homologous gentamicin resistance genes encoded on the conjugative replicon. This confirms earlier observations noting the invariant link between conjugative transfer and the appearance of gentamicin resistance in transconjugants. It also supports the previous suggestion that all staphylococcal conjugative gentamicin resistance plasmids were cognate replicons, as determined by restriction endonuclease mapping and/or whole-plasmid DNA hybridization (2, 35, 83).

The original source for staphylococcal conjugative plasmids is not clear. Conjugative replicons may have existed for years in staphylococci without antibiotic resistance markers. The association of gentamicin resistance genes with the conjugative transfer apparatus and the selective pressure of increased antibiotic use in hospitals possibly facilitated the dissemination of both resistant isolates and conjugative plasmids. Alternatively, both conjugative replicons and gentamicin resistance genes may have been acquired from another bacterial species at about the same time. Analysis of staphylococcal isolates that predated the emergence of gentamicin resistance would help to clarify this issue.

Staphylococcal conjugative plasmids appear to be remarkably stable. In hospitals into which they were introduced in the early 1980s, those plasmids are still the predominant carrier of gentamicin resistance genes in *S. aureus* (4). This may be due both to the stability of the replicon itself and to its broad host range among staphylococcal strains and species. Homologous plasmid-encoded conjugative transfer genes have been found in at least seven different species of nosocomial coagulase-negative staphylococci as well as *S. aureus* (4) and can be transferred in vitro among four different staphylococci species (G. Archer, unpublished observations). The ability of these plasmids to transfer among and to be stably maintained in coagulase-negative staphylococci may provide the reservoir for their nosocomial maintenance (1).

3.3. Model Conjugative Replicons

The 52-kb plasmid pGO1 is a prototypical conjugative staphylococcal replicon (5) (Figure 12.2). This plasmid is transferred among staphylococcal species (e.g., *aureus*, *epidermidis*) at frequencies of 10^{-7} to 10^{-5} progeny/input donor cell. It determines the bifunctional 6′ AAC/2″ APH enzyme that mediates resistance to the aminoglycosides gentamicin, tobramycin, kanamycin, and amikacin (17). It also carries genes encoding resistance to quaternary ammonium compounds (*qac*) and ethidium bromide (50) as well as tandem genes encoding thymidylate synthetase and dihydrofolate reductase that mediate trimethoprim resistance (24, 70). pGO1 appears to contain an integrated copy of pUB110 that encodes 4′4″ APH, mediating kanamycin, tobramycin, neomycin, and paromomycin resistance (18). All of these genes mediate the appropriate resistance phenotype when present on conjugative plasmids, with the exception of pUB110 aminoglycoside resistance sequences. Although most conjugative plasmids examined in a recent survey contained the restriction enzyme site signature of pUB110, only 45% were resistant to neomycin (4). In addition to the resistance genes, there are eight directly repeated copies and one indirectly repeated copy of a 780- to 800-bp IS-like element, designated either IS*431* (7) or IS*257* (34). The major genes mediating conjugative transfer are found within a 14.5-kb *Bgl*II fragment between *qac* and the integrated pUB110, flanked by directly repeated copies of IS*431/257*. Transposon insertion mutagenesis showed that genes within this segment were required for transfer; cloning of the segment onto pE194 (pGO220) showed that the genes were also sufficient for transfer (81). However, subsequent studies have shown that pGO220, the "mini-*tra*" conjugative replicon, became unstable after several rounds of transfer. In some cases, gross deletions of pGO220 were found in colonies unable to transfer the plasmid, while in others there was no obvious alteration (G. Archer, unpublished data). The reason for this instability is not clear but may be related to genes in the conjugative transfer region that appear to be lethal to the cell when cloned on vectors with copy numbers higher than those of the parent plasmid (see below).

A second region of pGO1 involved in conjugative transfer that is distant from the major concentration of transfer genes has also been identified. Evans and Dyke (29) and Archer et al (unpublished data) identified similar regions of two different conjugative plasmids that, when mutagenized by transposon insertion, abolished transfer. Archer et al (unpublished observations) found that an approximately 2-kb area could be identified by transposon insertion on the opposite side of the plasmid from the other clustered transfer genes. The area was also found to be transcriptionally active by using Tn*917* lac fusions (Figure 12.2).

The role of individual transfer genes in conjugation has not yet been established. A gene or genes contained within a cloned 4.6-kb segment of the major transfer region cannot be introduced alone into *S. aureus* by either protoplast transformation or electroporation. However, if the parent replicon (pGO1) or a clone containing approximately 2 kb of DNA at one end of the transfer region is present within the cell, the 4.6-kb clone can be transferred

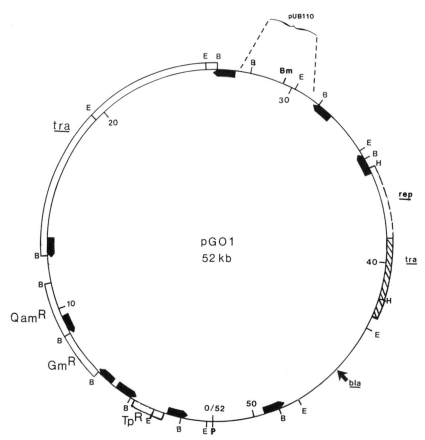

FIGURE 12.2. Map of pG01, a model staphylococcal conjugative plasmid. Designations are as follows: E = *Eco*RI; B = *Bgl*II; P = *Pst*I; H = *Hind*III; Bm = *Bam*Hi. The two *Hind*III sites indicated are not the only sites on the plasmid but are included to show possible boundaries for the secondary *tra* region. *Tra* (larger characters) is the area containing primary conjugative transfer genes; *tra* (smaller characters) is the area containing accessory transfer gene(s). Qam^R = genes encoding resistance to quaternary ammonium compounds; Gm^R = gentamicin resistance; Tp$^$ = trimethoprim resistance; *bla* = the site at which a transposon encoding β-lactamase has been shown to insert in some conjugative plasmids (but not pG01); pUB110 = the site of insertion of sequences derived from plasmid pUB110, including the gene encoding the aminoglycoside adenyl transferase ADH 4′,4″, mediating resistance to neomycin, paromycin, and tobramycin; *rep* = the deduced site of sequence(s) encoding replication functions. The numbers are kb coordinates beginning at the *Pst*I site and moving clockwise. The black arrowed boxes with the circle designate copies of the IS-like element IS431/257; the arrow shows the direction of transcription of the single internal open reading frame.

and maintained (80). The nature of the *trans*-acting regulatory region and its target are both under active investigation, since each area was identified by transposon insertional mutagenesis as being required for transfer.

The mechanism of transfer has also not been defined. However, two pieces of evidence suggest that single-stranded transfer is occurring. First, novobiocin treatment of donor, but not recipient, cells inhibits transfer at a drug concentration that does not decrease donor viability (5). This is consistent with a requirement of the enzyme for DNA unwinding after a single-stranded nick has occurred. Second, mobilization of a small, non-self-transmissible plasmid proceeds by creation of a single-stranded nick in the mobilized plasmid (66). Two genes (*mobA* and *mobB*) encoded by the mobilized plasmid and transfer genes on the conjugative plasmid are both required for mobilization. However, although these data suggest that transfer of a single DNA strand occurs, we have been unable to locate an origin of transfer by either assessing the mobilization of fragments cloned from the transfer region or observing the conversion of covalently closed circular plasmids containing the cloned fragments to open circular forms in the presence of the parent replicon (G. Archer, unpublished data).

Finally, the genetic organization of the transfer region is poorly understood. However, several observations argue against a single polycistronic message being present. First, plasmids containing transposon inserts that abolish conjugation can be complemented for transfer by cloned transfer region fragments. Second, a clone containing one end of the transfer region can be complemented for transfer by an overlapping clone containing the other half of the transfer region (81). Both clones can be found intact as individual replicons in transconjugants. Third, Northern analysis revealed the presence of multiple transcripts from various transfer region clones (G. Archer, unpublished data).

3.4. Concluding Thoughts

There are, as yet, no firm data that the F system of *E. coli* is a general paradigm for bacterial conjugation. Plasmids that transfer among gram-positive bacteria are likely to have evolved unique mechanisms for conjugative transfer that compensate for lack of filamentous surface appendages, for the presence of a thick peptidoglycan cell wall, for the frequent existence of surface exopolymers, and for the absence of an outer membrane. The careful elucidation of a pheromone mating system that is unique to *E. faecalis* is an example of a specialized transfer system that operates among gram-positive bacteria. Identification of the mechanism of plasmid DNA transfer, the presence and function of surface-localized transfer proteins, the organization and regulation of contiguous transfer genes, and the effect of host functions on transfer frequency all will be essential for understanding the mechanisms that have evolved among gram-positive bacteria for exchanging genetic information.

ACKNOWLEDGMENTS. The author's work was supported by U.S.P.H.S. grants DE04224 and DE09035 (to F.L.M.) and AI21722 (to G.L.A.).

References

1. Archer, G. L., and Armstrong, B., 1983, Alteration of staphylococcal flora in cardiac surgery patients receiving antibiotic prophylaxis, *J. Infect. Dis.* **147:**642–649.
2. Archer, G. L., Dietrick, D. R., and Johnston, J. L., 1985, Molecular epidemiology of transmissible gentamicin resistance among coagulase-negative staphylococci in a cardiac surgery unit, *J. Infect. Dis.* **151:**243–251.
3. Archer, G. L., and Johnston, J. L., 1983, Self-transmissible plasmids in staphylococci that encode resistance to aminoglycosides, *Antimicrob. Agents Chemother.* **24:**70–77.

4. Archer, G. L., and Scott, J., 1991, Conjugative transfer genes in staphylococcal isolates from the United States, *Antimicrob. Agents Chemother.* **35:**2500–2504.
5. Archer, G. L., and Thomas, W. D., 1990, Conjugative transfer of antimicrobial resistance genes between staphylococci, in: *Molecular Biology of the Staphylococci* (R. P. Novick, ed.), VCH Publishers, New York, pp. 112–122.
6. Ayoubi, P., Kilic, A. O., and Vijayakumar, M. N., 1991, Tn*5253*, the pneumococcal Ω(*cat tet*) BM6001 element, is a composite structure of two conjugative transposons, TN*5251* and Tn*5252*, *J. Bacteriol.* **173:** 1617–1622.
7. Barberis-Maino, L., Berger-Bachi, B., Weber, H., Beck, W. D., and Kayser, F. H., 1987, IS*431*, a staphylococcal insertion sequence-like element related to IS*26* from *Proteus vulgaris*, *Gene* **59:**107–113.
8. Battisti, L., Green, B. D., and Thorne, C. B., 1985, Mating system for transfer of plasmids among *Bacillus anthracis*, *Bacillus cereus*, and *Bacillus thuringiensis*, *J. Bacteriol.* **162:**543–550.
9. Behnke, D., and Ferretti, J. J., 1980, Physical mapping of plasmid pDB101: a potential vector plasmid molecular cloning in streptococci, *Plasmid* **4:**130–138.
10. Behnke, D., and Gilmore, M. S., 1981, Location of antibiotic resistance determinants, copy control, and replication functions on the double-selective streptococcal cloning vector pGB301, *Mol. Gen. Genet.* **184:** 115–120.
11. Behnke, D., Gilmore, M. S., and Ferretti, J. J., 1981, Plasmid pGB301, a new multiple resistance streptococcal cloning vehicle and its use in cloning of a gentamicin/kanamycin resistance determinant, *Mol. Gen. Genet.* **182:**414–421.
12. Behnke, D., Gilmore, M. S., and Ferretti, J. J., 1981, Plasmid pGB301, a new multiple resistance streptococcal cloning vehicle and its use in cloning of a gentamicin/kanamycin resistance determinant, *Mol. Gen. Genet.* **182:**414–421.
13. Behnke, D., Golubkov, V. J., Malke, H., Boitsov, A. S., and Totolian, A. A., 1979. Restriction endonuclease analysis of group A streptococcal plasmids determining resistance to macrolides, lincosamides, and streptogramin-B antibiotics, *FEMS Microbiol. Lett.* **6:**5–7.
14. Brantl, S., Behnke, D., and Alonso, J. C., 1990, Molecular analysis of the replication region of the conjugative *Streptococcus agalactiae* plasmid pIP501 in *Bacillus subtilis*. Comparison with plasmids pAMβ1 and pSM19035, *Nucl. Acids Res.* **18:**4783–4790.
15. Brehm, J., Salmond, G., and Minton, N., 1987, Sequence of the adenine methylase gene of *Streptococcus faecalis* plasmid pAMβ1. *Nucl. Acids Res.* **15:**3177.
16. Buu-Hoi, A., Bieth, G., and Horaud, T., 1984, Broad host range of streptococcal macrolide resistance plasmids, *Antimicrob. Agents Chemother.* **25:**289–291.
17. Byrne, M. E., Gillespie, M. T., and Skurray, R. A., 1990, Molecular analysis of a gentamicin resistance transposon-like element on plasmids isolated from North American *Staphylococcus aureus* strains, *Antimicrob. Agents Chemother.* **34:**2016–2113.
18. Byrne, M. E., Gillespie, M. T., and Skurray, R. A., 1991, 4′, 4″ adenyltransferase activity on conjugative plasmids isolated from *Staphylococcus aureus* is encoded on an integrated copy of pUB110, *Plasmid* **25:** 70–75.
19. Christie, P. J., and Dunny, G. M., 1986, Identification of regions of the *Streptococcus faecalis* plasmid pCF10 that encode antibiotic resistance and pheromone response, *Plasmid* **15:**230–241.
20. Clewell, D. B., 1981, Plasmids, drug resistance, and gene transfer in the genus *Streptococcus*, *Microbiol. Rev.* **45:**409–436.
21. Clewell, D. B., 1990, Movable genetic elements and antibiotic resistance in enterococci, *Eur. J. Clin. Microbiol. Infect. Dis.* **9:**90–102.
22. Clewell, D. B., and Gawron-Burke, C, 1986, Conjugative transposons and the dissemination of antibiotic resistance in streptococci, *Annu. Rev. Microbiol.* **40:**635–659.
23. Clewell, D. B., Yagi, Y., Dunny, G. M., and Schultz, S. K., 1974, Characterization of three plasmid deoxyribonucleic acid molecules in a strain of *Streptococcus faecalis*: identification of a plasmid determining erythromycin resistance, *J. Bacteriol.* **117:**283–289.
24. Coughter, J. P., Johnston, J. L., and Archer, G. L., 1986, Characterization of a staphylococcal trimethoprim resistance gene and its product, *Antimicrob. Agents Chemother.* **31:**1027–1032.
25. Dao, M. L., and Ferretti, J. J., 1985, *Streptococcus-Escherichia coli* shuttle vector pSA3 and its use in the cloning of streptococcal genes, *Appl. Environ. Microbiol.* **49:**115–119.

26. Dubnau, D., and Monod, M., 1986, The regulation and evolution of MLS resistance, in: *Banbury Report 24: Antibiotic Resistance Genes: Ecology Transfer, and Expression* (S. B. Levy and R. P. Novick, eds.), Cold Spring Harbor Laboratory, Cold Spring Harbor, New York, pp. 369–387.

27. Dunny, G. M., 1991, Mating interactions in gram-positive bacteria, in: *Microbial Cell-Cell Interactions* (M. Dworkin, ed.), pp. 9–33, American Society for Microbiology, Washington, DC.

28. Engel, H. W. B., Soedirman, N., Rost, J. A., van Leeuwen, W. J., and Van Embden, J. D. A., 1980, Transferability of macrolide, lincosamide and streptogramin resistance between group A, B, and D streptococci, *Streptococcus pneumoniae* and *Staphylococcus aureus*, *J. Bacteriol.* **142:**407–413.

29. Evans, J., and Dyke, K. G. H., 1988, Characterization of the conjugation system associated with the *Staphylococcus aureus* plasmid pJE1, *J. Gen. Microbiol.* **134:**1–8.

30. Evans, R. P., Jr., and Macrina, F. L., 1983, Streptococcal R Plasmid pIP501: endonuclease site map, determinant location, and construction of novel derivatives, *J. Bacteriol.* **154:**1347–1355.

31. Evans, R. P., Winter, R. B., and Macrina, F. L., 1985, Molecular cloning of a pIP501 derivative yields a model replicon for the study of streptococcal conjugation, *J. Gen. Microbiol.* **131:**145–153.

32. Forbes, B. A., and Schaberg, D. R., 1983, Transfer of resistance plasmids from *Staphylococcus epidermidis* to *Staphylococcus aureus*: evidence for conjugative exchange of resistance, *J. Bacteriol.* **153:**627–634.

33. Gasson, M. J., 1990, In vivo genetic systems in lactic acid bacteria, *FEMS Microbiol. Rev.* **87:**43–60.

34. Gillespie, M. T., Lyon, B. R., Loo, L. S. L., Mathews, P. R., Stewart, P. R., and Skurray, R. A., 1987, Homologous direct repeat sequences associates with mercury, methicillin, tetracycline and trimethoprim resistance determinants in *Staphylococcus auerus*, *FEMS Microbiol. Lett.* **43:**165–171.

35. Goering, R. V., Teeman, B. A., and Ruff, E. A., 1985, Comparative physical and genetic maps of conjugal plasmids mediating aminoglycoside resistance in *Staphylococcus aureus* in the United States, in: *The Staphylococci* (J. Jelijaszevicz, ed.), Zbl. Bakt. Suppl. 14. Gustav Fischer Verlag, New York.

36. Heemskerk, D. D., and Thorne, C. B., 1990, Genetic exchange and transposon mutagenesis in *Bacillus anthracis*, *Salisbury Med. Bull.* **68:**(Sp. Suppl.):63–67.

37. Horaud, T., Bouguenec, C. L., and Pepper, K., 1985, Molecular genetics of resistance to macrolides, lincosamides, and streptogramin B (MLS) in streptococci, *J. Antimicrob. Chemother.* **16**(Suppl. A):111–135.

38. Horinouchi, S., Byeon, W.-H., and Weisblum, B., 1983. A complex attenuator regulates inducible resistance to macrolides, lincosamides, and streptogramin B antibiotics in *Streptococcus sanguis*, *J. Bacteriol.* **154:**1252–1262.

39. Horodniceanu, T., Bouanchaud, D., Biet, G., and Chabbert, Y., 1976, R plasmids in *Streptococcus agalactiae* (group B), *Antimicrob. Agents Chemother.* **10:**795–801.

40. Ippen-Ihler, K. A., and Minkley, E. G., 1986, The conjugation system of F, the fertility factor of *Escherichia coli*, *Ann. Rev. Genet.* **20:**593–624.

41. Jaffe, H. W., Sewwney, H. M., Nathan, R. A., Weinstein, R. A., Kabins, S. A., and Cohen, S., 1980, Identity and interspecies transfer of gentamycin resistance plasmids in *Staphylococcus aureus* and *Staphylococcus epidermidis*, *J. Infect. Dis.* **141:**738–747.

42. Krah, E. R., III, and Macrina, F. L., 1989, Genetic analysis of the conjugal transfer determinants encoded by the streptococcal broad-host-range plasmid pIP501, *J. Bacteriol.* **171:**6005–6012.

43. Krah, E. R., III, and Macrina, F. L., 1991, Identification of a region that influences host range of the streptococcal conjugative plasmid pIP501, *Plasmid* **25:**64–69.

44. Lacey, R. W., 1971, High frequency transfer of neomycin resistance between naturally occurring strains of *Staphylococcus aureus*, *J. Gen. Microbiol.* **4:**73–84.

45. Lacey, R. W., 1971, Transfer of tetracycline resistance between strains of *Staphylococcus aureus* in mixed culture, *J. Gen. Microbiol.* **69:**229–237.

46. Lacey, R. W., 1980, Evidence for two mechanisms of plasmid transfer in mixed cultures of *Staphylococcus aureus*, *J. Gen. Microbiol.* **119:**423–435.

47. LeBlanc, D., and Lee, L., 1984, Physical and genetic analyses of streptococcal plasmid pAMβ1 and cloning of its replication region, *J. Bacteriol.* **157:**445–453.

48. LeBlanc, D. J., Cohen, L., and Jensen, L., 1978, Transformation of group F streptococci by plasmid DNA, *J. Gen. Microbiol.* **106:**49–54.

49. LeBlanc, D. J., Hawley, R. J., Lee, L. N., and St. Martin, E. J., 1978, "Conjugal" transfer of plasmid DNA among oral streptococci, *Proc. Natl. Acad. USA* **75:**3484–3487.

50. Lyon, B. R., and Skurray, R., 1987, Antimicrobial resistance of *Staphylococcus aureus*: genetic basis, *Microbiol. Rev.* **51:**88–134.

51. Macrina, F. L., Evans, R. P., Tobian, J. A., Hartley, D. L., and Jones, K. R., 1983, Novel shuttle plasmid vehicles for *Escherichia-Streptococcus* transgeneric cloning, *Gene* **25**:145–150.

52. Macrina, F. L., Jones, K. R., and Wood, P. H., 1980, Chimeric streptococcal plasmids and their use as molecular cloning vehicles in *Streptococcus sanguis* (Challis), *J. Bacteriol.* **143**:1425–1435.

53. Macrina, F. L., Keeler, C. L., Jr., Jones, K. R., and Wood, P. H., 1980, Molecular characterization of unique deletion mutants of the streptococcal plasmid, pAMβ1, *Plasmid* **4**:8–16.

54. Macrina, F. L., Tobian, J. A., Jones, K. R., and Evans, R. P., 1982, Molecular cloning in the Streptococci, *Basic Life. Sci.* **19**:195–210.

55. Macrina, F. L., Tobian, J. A., Jones, K. R., Evans, R. P., and Clewell, D. B., 1982, A cloning vector able to replicate in *Escherichia coli* and *Streptococcus sanguis*, *Gene* **19**:345–353.

56. Malke, H., 1979, Conjugal transfer of plasmids determining resistance to macrolides, lincosamides, and streptogramin-B type antibiotics among group A, B, D, and H streptococci, *FEMS Microbiol. Lett.* **5:** 335–338.

57. Martin, B., Alloing, G., Mejean, V., and Claverys, J.-P., 1987, Constitutive expression of erythromycin resistance mediated by the ermAM determinant of pAMβ1 results from deletion of 5′ leader peptide sequences, *Plasmid* **18**:250–253.

58. Mayford, M., and Weisblum, B., 1989, Conformational alterations in the *ermC* transcript in vivo during induction, *EMBO J.* **8**:4307–4314.

59. McDonnell, R. W., Sweeney, H. M., and Cohen, S., 1983, Conjugational transfer of gentamycin resistance plasmids intra- and interspecifically in *Staphylococcus aureus* and *Staphylococcus epidermidis*, *Antimicrob. Agents Chemother.* **23**:151–160.

60. Naidoo, J., and Noble, W. C., 1978, Transfer of gentamycin resistance between strains of *Staphylococcus aureus* on skin, *J. Gen. Microbiol.* **107**:391–393.

61. Naidoo, J., Noble, W. C., Weissman, A., and Dyke, K. G. H., 1983, Gentamicin resistant staphylococci: genetics of an outbreak in a dermatology department, *J. Hyg.* **91**:7–16.

62. Novick, R. P., and Morse, S. I., 1967, In vivo transmission of drug resistance factors between strains of *Staphylococcus aureus*, *J. Exp. Med.* **125**:45–59.

63. Odelson, D. A., Rasmussen, J. L., Smith, C. J., and Macrina, F. L., 1987, Extrachromosomal systems and gene transmission in anaerobic bacteria, *Plasmid* **17**:87–109.

64. Oultram, J. D., Davies, A., and Young, M., 1987, Conjugal transfer of a small plasmid from *Bacillus subtilis* to *Clostridium acetobutylicum* by cointegrate formation with plasmid pAMbeta1, *FEMS Microbiol. Lett.* **42**:113–119.

65. Oultram, J. D., Peck, H., Brehm, J. K., Thompson, D. E., Swinfield, T. J., and Minton, N. P., 1988, Introduction of genes for leucine biosynthesis from *Clostridium pasteurianum* into *C. acetobutylicum* by cointegrate conjugal transfer, *Mol. Gen. Genet.* **214**:177–179.

66. Projan, S. J., and Archer, G. L., 1989, Mobilization of the relaxable *Staphylococcus aurens* plasmid pC221 by the conjugative plasmid pG01 involves three pC221 loci, *J. Bacteriol.* **171**:1841–1845.

67. Priebe, S. D., and Lacks, S. A., 1989, Region of the streptococcal plasmid pMV158 required for conjugative mobilization, *J. Bacteriol.* **171**:4778–4784.

68. Reddy, A., Battisti, L., and Thorne, C. B., 1987, Identification of self-transmissible plasmids in four *Bacillus thuringiensis* subspecies, *J. Bacteriol.* **169**:5263–5270.

69. Romero, D., Slos, P., Roberts, C., Castellino, I., and Mercenier, A., 1987, Conjugative mobilization as an alternative vector delivery system for lactic streptococci, *Appl. Environ. Microbiol.* **29**:807–813.

70. Rouch, D. A., Messerotti, L. J., Loo, L. S. L., Jackson, C. A., and Skurray, R., 1989, Trimethoprim resistance transposon Tn*4003* from *Staphylococcus aureus* encodes genes for a dihydrofolate reductase and thymidylate synthetase flanked by three copies of IS*257*, *Mol. Microbiol.* **3**:161–175.

71. Sasaki, Y., Taketomo, N., and Sasaki, T., 1988, Factors affecting transfer frequency of pAMB1 from *Streptococcus faecalis* to *Lactobacillus plantarum*, *J. Bacteriol.* **170**:5939–5942.

72. Schaberg, D. R., Power, G., Betzold, J., and Forbes, B. A., 1985, Conjugative R plasmids in antimicrobial resistance of *Staphylococcus aureus* causing nosocomial infections, *J. Infect. Dis.* **152**:43–49.

73. Schaberg, D. R., and Zervos, M. J., 1986, Intergeneric and interspecies gene exchange in gram-positive cocci, *Antimicrob. Agents Chemother.* **30**:817–822.

74. Shaw, J. H., and Clewell, D. B., 1985, Complete nucleotide sequence of macrolide-lincosamide-streptogramin B-resistance transposon Tn*917* in *Streptococcus faecalis*, *J. Bacteriol.* **164**:782–796.

75. Skurray, R. A., Rouch, D. A., Lyon, B. R., Gillespie, M. J., Tennent, J. M., Byrne, M. E., Messerotti, L. J., and May, J. W., 1988, Multi-resistant *Staphylococcus aureus*: genetics and evolution of epidemic Australian strains, *J. Antimicrob. Chemother.* **21**(Suppl. C):19–38.

76. Smith, M. D., 1985, Transformation and fusion of *Streptococcus faecalis* protoplasts, *J. Bacteriol.* **162**:92–97.

77. Smith, M. D., and Clewell, D. B., 1984, Return of *Streptococcus faecalis* DNA cloned in *Escherichia coli* to its original host via transformation of *Streptococcus sanguis* followed by conjugative mobilization, *J. Bacteriol.* **160**:1109–1114.

78. Storrs, M. J., Courvalin, P., and Foster, T. J., 1988, Genetic analysis of gentamicin resistance in methicillin and gentamicin resistant strains of *Staphylococcus aureus* isolated in Dublin hospitals, *Antimicrob. Agents. Chemother.* **32**:1174–1181.

79. Swinfield, T.-J., Oultram, J. D., Thompson, D. E., Brehm, J. K., and Minton, N. P., 1990, Physical characterisation of the replication region of the *Streptococcus faecalis* plasmid pAMβ1, *Gene* **87**:79–90.

80. Taketomo, N., Sasaki, Y., and Sasaki, T., 1989, A new method for conjugal transfer of plasmid pAMB1 to *Lactobacillus plantarum* using polyethylene glycol, *Agric. Biol. Chem.* **53**:3333–3334.

81. Thomas, W. D., and Archer, G. L., 1989, Identification and cloning of the conjugative transfer region of *Staphylococcus aureus* plasmid pGO1, *J. Bacteriol.* **171**:684–691.

82. Townsend, D. E., Ashdown, N., Annear, D. I., and Grubb, W. B., 1985, A conjugative plasmid encoding production of a diffusible pigment and resistance to aminoglycosides and macrolides in *Staphylococcus aureus*, *Aust. J. Exp. Biol. Med.* **63**:573–586.

83. Townsend, D. E., Bolton, S., Ashdown, N., and Grubb, W. B., 1985, Transfer of plasmid-borne aminoglycoside resistance determinants in staphylococci, *J. Med. Microbiol.* **20**:169–185.

84. Townsend, D. E., den Hollander, L., Bolton, S., and Grubb, W. B., 1986, Clinical isolates of staphylococci conjugate on contact with dry absorbant surfaces, *Med. J. Austr.* **144**:166.

85. Trieu-Cuot, P., Carlier, C., and Courvalin, P., 1988, Conjugative plasmid transfer from *Enterococcus faecalis* to *Escherichia coli*, *J. Bacteriol.* **170**:4388–4391.

86. Udo, T., Townsend, D. E., and Grubb, W. B., 1987, A conjugative staphylococcal plasmid with no resistance phenotype, *FEMS Microbiol. Lett.* **40**:279–283.

87. Van der Lelie, D., Wösten, H. A. B., Bron, S., Oskam, L., and Venema, G., 1990, Conjugal mobilization of streptococcal plasmid pMV158 between strains of *Lactococcus lactis* subsp. *lactis*, *J. Bacteriol.* **172**:47–52.

88. Vijayakumar, M. N., Priebe, S. D., and Guild, W. R., 1986, Structure of a conjugative element in *Streptococcus pneumoniae*, *J. Bacteriol.* **166**:978–984.

89. Weisblum, B., 1985, Inducible resistance to macrolides, lincosamides and streptogramin type B antibiotics: the resistance phenotype, its biological diversity, and structural elements that regulate expression—a review, *J. Antimicrob. Chemother.* **16**:63–90.

90. Witte, W., 1981, Zur rolle des phagen bei der plasmid ubertragung in mischkulturen von *Staphylococcus aureus*, *Zantralbl. Bakteriol. Hyg. I. Abst. Orig.* **249**:195–202.

Chapter 13

Conjugal Transfer in Anaerobic Bacteria

FRANCIS L. MACRINA

1. Introduction

This chapter will review the biology and genetics of conjugative exchange in selected anaerobic bacteria. Specifically, conjugation in two genera, *Bacteroides* and *Clostridium*, will be the central theme. Conjugative systems in these two genera have been discovered and developed over the past 15 years. Although parallel systems are likely to emerge in other anaerobic bacteria, the advanced state of such systems in *Bacteroides* and *Clostridium* justifies limiting discussion in these two genera. There are multiple lines of interest driving the study of anaerobic bacteria. First, these organisms include important human and animal pathogens (11). Second, anaerobes, including *Bacteroides* and *Clostridium*, make up the bulk of the indigenous microflora of human beings and other mammals (21). This ecological position provides them with the opportunity to initiate infection at remote sites. Equally important, these organisms may contribute to pathology without leaving their normal niche. For example, both the *Bacteroides* and the clostridia have been implicated in the generation of potentially cocarcinogenic compounds in the human colon (10, 21). Colonic anaerobes also have been implicated in useful physiological process such as the production of nutrients and the degradation of complex dietary carbohydrates (42). Finally, anaerobic bacteria are of considerable interest to the biotechnologist. The metabolic diversity of the clostridia has been exploited in such processes as biodegradation, production of fuels and solvents, and biotransformations (23, 29, 70).

The means to genetically manipulate *Bacteroides* and *Clostridium* is desirable. Our ability to move genes from cell to cell can assist in the dissection of process of both medical

FRANCIS L. MACRINA • Department of Microbiology and Immunology, Medical College of Virginia, Virginia Commonwealth University, Richmond, Virginia 23298.

Bacterial Conjugation, edited by Don B. Clewell. Plenum Press, New York, 1993.

and biotechnological importance. The implications range from the facile production of novel vaccines to the construction of strains that efficiently overproduce valuable chemicals. The discovery and development of conjugal genetic exchange systems in these two genera has afforded new avenues of biological exploration in anaerobic bacteria.

2. *Bacteroides*

The *Bacteroides* are a large, heterogeneous group of non-spore-forming, nonmotile, obligately anaerobic gram-negative bacteria. Colonic species include *fragilis, uniformis, vulgatus, thetaiotaomicron, distasonis*, and *ovatus*. The *Bacteroides* species that live in the colon often represent up to one-third of the cultivable bacteria found there (30). Another group of human indigenous *Bacteroides* species reside exclusively in the oral cavity and include species such as *denticola, buccae, loescheii*, and *intermedius*. The oral species, *B. gingivalis*, believed to be important in the pathogenesis of periodontal disease, was recently renamed *Porphyromonas gingivalis*; also now included under this genus name are the former *Bacteroides* species *asaccharolyticus* and *endodontalis*. The *Bacteroides* are evolutionarily distant from the facultative gram-negative bacteria. It is believed they diverged from the eubacterial line before the appearance of gram-positive organisms (65). Furthermore, the oral and intestinal *Bacteroides* are not closely related to each other when compared by DNA hybridization. The aerotolerance and simple nutritional requirements of the colonic *Bacteroides* as compared with the oral species have facilitated the search for and development of conjugal genetic systems in the former group of organisms.

2.1. Extrachromosomal Elements and Conjugative Plasmids

Conjugal genetic exchange in *Bacteroides* initially was described in 1979 with multiple laboratories demonstrating transmissible macrolide-lincosamide-streptogramin (MLS) resistance in this genus (34, 61, 66). This phenotype was important because it conferred resistance to the clinically effective and widely used lincosamide antibiotic, clindamycin. This resistance was similar to that seen in gram-positive bacteria (e..g, streptococci, staphylococci, and bacilli), where the MLS gene (designated *erm*) encoded an RNA methylase that modified specific adenine residues in the 23S-rRNA, thus preventing MLS drugs from binding to the ribosome and inhibiting protein synthesis. One important difference is that *Bacteroides* MLS resistance described thus far has been constitutively expressed, whereas this phenotype is often inducibly expressed in gram-positive organisms.

The genetic basis of conjugal transfer was initially attributed to plasmids. However, it is now clear that nonplasmid elements that carry genes conferring conjugal transfer also are common in the *Bacteroides* (see Section 2.2.). In general, conjugal transfer in *Bacteroides* has been defined on the basis of DNase resistance and the necessity of donor and recipient cells to be in close contact. Transfer requires cell-to-cell contact on a nitrocellulose or polycarbonate filter or on an agar matrix. Transfer frequencies are low (ca. 10^{-4} to 10^{-6} transconjugants/donor cell), but they can be elevated one to two orders of magnitude by using isogenic mating pairs (66). The transfer process appears to reach maximal levels after 2 to 3 hours of cocultivation of cells on filter disks (66).

The cellular or molecular events of conjugal transfer in *Bacteroides* remain unclear. Pili have not been associated with the conjugative process. The number of genes needed for conjugal transfer are not known for any of the well-characterized systems. The existence of specific surface receptors on either donor or recipient cells has not been described. Information on whether surface exclusion is operative in matings is unavailable. There have been no reports of either donor- or recipient-specific bacteriophage. At the cellular level, important unanswered questions about *Bacteroides* conjugal transfer include whether an *oriT*-like sequence is needed for transfer or mobilization, whether single- or double-stranded DNA is transferred, and what macromolecular processes are required for transfer. Evidence supporting the genetic exchange between strains of *Bacteroides* in vivo has been reported (5), and it is likely that the conjugal transfer of drug resistance plasmids occurs in the colon, resulting in the spread of antibiotic resistance genes. Transfer of drug resistance plasmids and nonplasmid elements at the interspecies level among the colonic *Bacteroides* is well documented (31, 42). However, only one transfer system (nonplasmid) has been described for the oral *Bacteroides* (12), and a full appreciation of its host range is not yet available. Interestingly, transfer of the element from oral species to the colonic species *fragilis* was observed (see Section 2.2.).

Three different resistance plasmids of the colonic *Bacteroides* have been intensely studied: pBF4, a 41-kb plasmid from *B. fragilis*; pBFTM10, a 15-kb plasmid from *B. fragilis* (and its indistinguishable counterpart pCP1 from *Bacteroides thetaiotaomicron*); and pBI136, an 80-kb plasmid from *B. ovatus*. All three plasmids encode high-level MLS resistance conferred by a conserved *erm* gene. The location of the *erm* gene on each of these plasmids initially was deduced by the analysis of spontaneously occurring plasmid deletion mutants (27, 53, 62, 66). The nucleotide sequence of two of the *erm* genes (*ermF* from pBF4 and *ermFS* from pBI136) has been determined, and the genes have been found to differ at only one nucleotide position. Based on restriction maps and the DNA hybridization under high stringency, it is likely that the *erm* gene of pBFTM10 is strongly homologous to *ermF* and *ermFS*. It is believed that all three of these genes shared a common ancestor and that they have not diverged significantly from one another since entering *Bacteroides* (31). Physical analyses of both wild-type and mutant plasmids revealed that the *erm* gene of each plasmid was bordered by a similar, if not identical, insertion sequence (IS) element that flanked the gene in a directly repeated fashion. Other than their *erm* genes and the associated insertion sequence elements, pBF4, pFBTM10, and pBI136 shared no detectable homology. The flanking IS elements (IS*4400* on pBFTM10, IS*4351* on pBF4, and iso-IS*4351* on pBI136) form compound transposons with their associated *erm* genes. Genetic evidence for their transposition in both *B. fragilis* and *Escherichia coli* has been obtained (39, 46, 48, 56). In all three of these transposons, the *erm* genes were juxtaposed to one copy of the IS element, suggesting some interaction between these two sequences. Indeed, the analysis of nucleotide sequence data, S1 nuclease transcription mapping studies, and the construction of gene fusions to evaluate promoter activity (36, 37, 54) indicated that the *erm* genes are under the transcriptional control of the promoters contained in the IS element. Outward-firing promoters exist at both ends of this class of element, so it was able to drive *erm* transcription regardless of its orientation with respect to the drug resistance gene. Examples of the IS element being in opposite orientations with respect to the *erm* gene were found in the comparative examination of pBF4 and pBFTM10 with pBI136. Odelson et al (31) have proposed a model to explain the formation of theses transposons. It predicts that all three transposons evolved independent of one another. Following the entry of a similar, if not identical, *erm* gene into *Bacteroides*, transposition of the IS element next to the newly acquired gene resulted in the transcriptional control of *erm* by the IS element. This may have been required in the face of selective pressure because the invading *erm* gene could not be expressed in *Bacteroides* for any number of reasons (e.g., unrecognizable promoter, nonfunctional upstream regulatory sequences on the *erm* mRNA [31]). A second proposed transposition event then resulted in the *erm* gene being flanked by two copies of the IS element, generating a compound transposon. The different orientation of the IS element with respect to the *erm* gene on pBF4/pBFTM10 versus pBI136, along with the qualitatively and quantitatively different intervening DNA sequences in the three transposons, argues that these elements were formed as the result of three separate events. Figure 13.1 illustrates the genetic organization of these three *Bacteroides* transposons.

The control of drug resistance determinants by insertion elements in *Bacteroides* has been reported to occur with genes other than *erm*. Rasmussen and Kovacs (35) have cloned a *B. fragilis* chromosomal metallo-β-lactamase gene, designated *ccrA*. They observed that sequences upstream of the *ccrA* gene appeared to be present in multiple copies on the chromosome of the host strain from which this determinant was cloned. Nucleotide sequence analysis revealed this upstream region to be a 1598-base-pair (bp) element bearing all of the molecular characteristics of an insertion sequence; serial transposition of this sequence was not reported by these authors (35). One terminus of the element was 19 bp from the start codon of the *ccrA* gene, leading these authors to conclude that the transcriptional control signals for the resistance determinant were contained within this insertion-sequence-like element. The insertion-sequence-like element found upstream of the *ccrA* gene did appear to be novel and did not display similarity with other insertion sequence elements, including the *Bacteroides* element IS4351 (37). The findings of Rasmussen and Kovacs (35) suggest that drug resistance gene activation by insertion elements may be a common theme in *Bacteroides*.

The existence of drug resistance transposons on three unrelated conjugative plasmids in *Bacteroides* suggests that such plasmids play a role in the dissemination of antibiotic resistance in this genus. Moreover, these studies indicate the presence of multiple, different conjugative plasmids. The genetic fine structure of their conjugative apparatus must be further probed at both the molecular and cellular levels before definitive statements can be made about the relatedness of their transfer genes. Interestingly, a plasmid (pBI106) sharing strong similarity to pBF4, but without a copy of Tn4351, has been isolated from *B. ovatus* (55). This plasmid might be a progenitor of the pBF4 class of plasmid (i.e., a conjugative plasmid minus any detectable drug resistance genes). Examination of the conjugal transfer proficiency of pBI106 was not examined, but its relatedness to pBF4 suggests the existence of cryptic conjugative plasmids in *Bacteroides* that are analogous to the F plasmid of *E. coli*.

The ongoing characterization of *Bacteroides* conjugative R-plasmids like pBF4 sets the stage for a systematic analysis of the genetic basis of conjugal transfer in this genus. The use of *E. coli-Bacteroides* conjugative shuttle systems (see Section 2.4.) in concert with deletion and transposon mutagenesis provides a plausible way to at least identify the regions encoding essential genes for transfer. This information then could be logically used to begin

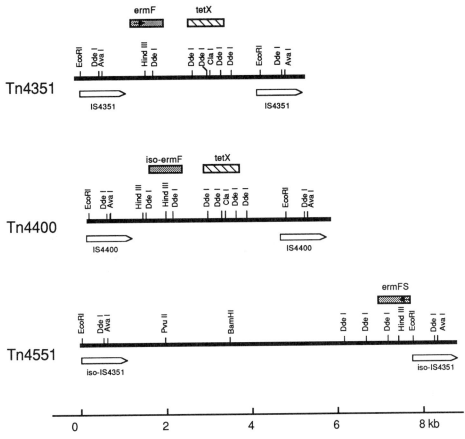

FIGURE 13.1. *Bacteroides* transposons. Restriction maps of the three transposons that are discussed in the text are illustrated here. A size scale in kilobase (kb) pairs is shown at the bottom of the figure. The shaded rectangles indicate the locations of their *erm* genes (confer clindamycin resistance), and the arrows within the rectangles indicate the direction of transcription of the gene (not definitively established for iso-*ermF* in Tn*4400*). The open arrow bars correspond to the related, flanking insertion sequences that border each of these transposons in a directly repeated fashion. The *tetX* locus (cross-hatched rectangle) is located adjacent to the *erm* genes seen on Tn*4400* and Tn*4351*. This determinant confers Tc^r in aerobically grown *E. coli*; it does not confer Tc^r in *Bacteroides* (13).

an identification of proteins needed for conjugal transfer. These and similar studies are urgently needed to expand our understanding of plasmid-specified conjugative transfer in *Bacteroides*.

2.2. Conjugative Nonplasmid Elements

Certain donor strains of *B. fragilis* are capable of transferring their antibiotic resistance without the involvement of detectable plasmid DNA (25, 28, 60). Such transfer elements have fallen into two categories with respect to antibiotic resistance phenotypes:

those that transfer tetracycline resistance alone and those that transfer linked tetracycline
and clindamycin resistances. All of these systems have met the operational definition of
conjugal transfer (see Section 2.1.). Mating occurred only on a solid matrix (filter or agar
surface) similar to the conditions seen for plasmid transfer. Donor strains carrying many of
the Tc^r elements require that cells first be grown in low but inhibitory levels of tetracycline
in order to display maximal conjugal transfer efficiency (50, 57). Exceptions to this do
exist, and one well-characterized donor has been shown to transfer Tc^r at comparable
frequencies regardless of exposure to tetracycline (25). The mechanism of tetracycline
enhancement of conjugal transfer proficiency remains unexplained. Usually, the Tc^r
determinant itself is inducibly expressed. These elements are capable of mobilizing
nonconjugative plasmids, presumably operating *in trans* (see Section 2.3.). Those elements
examined shared no cross-hybridizing sequences with conjugative plasmids, although
many of the MLS resistance conferring elements carry a gene that is related to the
Bacteroides erm gene family (e.g., *ermF, ermFS* [25, 28]). The presence of insertion
sequences has not been associated with these elements based on either hybridization with
known IS sequences or genetic structure analysis (i.e., detectable repeated sequences
associated with the elements) (18). Evidence suggests that nonplasmid conjugal elements
exist that do not carry antibiotic resistance markers (51, 56). Moreover, the presence of such
elements has been demonstrated in clinical isolates of *Bacteroides* already shown to be
carrying a conjugative resistance plasmid (56).

Shoemaker and Salyers (50) reported an interesting observation associated with the
transfer of several of these types of elements, all of which conferred Tc^r and displayed
greatly increased transfer efficiency following exposure to tetracycline. Specifically, when
these elements were transferred to *Bacteroides uniforms* and the resulting strain was
exposed to inhibitory tetracycline, two circular DNA forms (ca. 10 and 11 kb) could be
detected as extrachromosomal elements. Southern analyses indicated that these circular
molecules were normally integrated into the host cell chromosome. Certain of the Tc^r
nonplasmid elements examined by these authors did not mediate the excision of the circular
DNA forms. Stevens et al (59) have recently presented evidence suggesting that the control
of production of these circular DNA forms resides in a region contained within the Tc^r
transfer element called ERL (confers Tc^r and MLS resistance). They were able to clone
regions of a similar element (termed DOT) and create insertion mutations in the subcloned
fragments. When these insertion mutations were introduced back into the ERL element and
their transfer was evaluated, these authors found reduced conjugal transfer efficiencies. Two
such mutants did not display the formation of the circular DNA molecules (termed
plasmidlike forms [Plf]). One of these mutants also showed reduced efficiency of conjugative mobilization of coresident plasmids. These authors also identified an additional mutant,
prepared by insertion mutagenesis, which displayed the Plf phenotype in a constitutive
fashion. Although these studies clearly established the need for this Tc^r element to produce
the Plf phenotype, the significance of this phenomenon remains unexplained.

Indirect evidence first suggested that nonplasmid transfer elements were large (ranging
from 40 to 50 kb). Shoemaker et al (45) used a novel shuttle to clone segments of the DOT
element (Tc^r, MLS^r). They obtained sequences that contained one or both drug resistance
markers. Some of the clones that carried only Tc^r were capable of conjugal transfer in
intergeneric *Bacteroides* matings. They used cloned overlapping sequences to construct a
linear restriction endonuclease map. This map indicated the size of this element to be at

least 50 kb. These authors could not determine whether they had cloned a chromosome-DOT element junctions sequence that would enable them to ascribe a chromosomal location to this element. Nonetheless, their work was the first to report on the physical isolation of functional segments of one of these prototype elements. Recently, Bedzyk et al (2) have employed transverse alternating field electrophoresis to approximate the size of certain of the nonplasmid elements that they have characterized, including the DOT element. These studies revealed the DOT element (as well as other phenotypically similar elements) to range from 70 to 80 kb in size. The electrophoretic analyses of Bedzyk et al (2) also suggested that elements such as DOT had a limited number of insertion sites (three to eight) on the *Bacteroides* genome. Halula and Macrina (18) have also reported the partial cloning of a Tc^r, MLS^r element, called Tn*5030*, from *B. fragilis*. They estimated the size of this element to be in the 40- to 50-kb range based on the analysis of presumed overlapping cloned fragments. A considerably larger estimate (150 to 200 kb) for this element has been obtained from DNA fragment sizes observed on transverse alternating field electrophoresis gels (2). This difference in the size estimates of this particular element remains unreconciled at present. Interpretation of the results of Halula and Macrina (18) was complicated by the unexpected finding that sequences related to Tn*5030* were present in the recipient strain from which their genomic libraries were prepared. Further studies are needed to fully define the molecular structure of Tn*5030*.

Shoemaker and Salyers (51) have isolated a 65-kb conjugal transfer element devoid of drug resistance determinants. This element was believed to have transposed to a *Bacteroides-Escherichia* shuttle plasmid. This element was designated XBU4422 and was shown by DNA hybridization experiments to be integrated into the chromosome of the host *B. uniforms* strain from which it was initially isolated. The availability of an apparent complete element of this type on a plasmid has opened avenues for extending studies using both physical and genetic means. In this vein, Bedzyk et al (2) have performed nucleotide sequence analysis on the junction regions of the XBU4422 element and its site of insertion. Sequencing of several insertions did not indicate any type of target site duplication as is commonly observed with other types of mobile genetic elements in procaryotes. However, they did observe that XBU4422 did carry 4 to 5 bp of adjacent chromosomal DNA when it transposed from the host chromosome to a plasmid. Excision of the XBU4422 element from a plasmid replicon usually resulted in a single base (an A) being left behind at the target site; however, precise excision was also observed in one case. These results set the stage for more detailed studies on the molecular mechanisms of transposition and conjugation employed by this class of elements.

Malamy and coworkers (19, 20) have described a 9.6-kb element designated Tn*4399* that is capable of conjugally mobilizing plasmids when it is present in *cis*. Tn*4399* was discovered in experiments where a Tc^r element was being studied in terms of its ability to restore conjugal proficiency to a transfer-defective plasmid. Transposition of Tn*4399* to the transfer-defective plasmid as well as to one other plasmid in the cell occurred at several sites. One copy of this element was found on the chromosome of the donor strain used in their studies; however, a transconjugant strain isolated by Hecht et al contained three copies of Tn*4399* at novel chromosomal sites. These workers have further characterized the molecular events associated with the insertion of Tn*4399* into a target site. Tn*4399* contains 13-bp inverted repeat sequences at its termini, with an extra base present within the left inverted repeat sequence. Their results suggested some specificity with a majority of independent insertions occurring in one of two locations in a single A +s T rich 14-bp target site. In contrast to other conjugative transposons (7, 8), Tn*4399* generated a novel 5-bp sequence at its right terminus target junction in addition to creating a 3-bp repeated sequence of the target site. The basis of mobilization conferred by Tn*4399* presently is under study by this group. This element could be carrying an *oriT*-like sequence that is working in concert with yet additional conjugal elements within the donor cell. Tn*4399* may also encode all or part of the conjugal transfer apparatus, as is the case with conjugative transposons like Tn*916* (7, 8). Tn*4399* was capable of mobilizing plasmids in intergeneric *Bacteroides* matings as well as in *Bacteroides-E. coli* matings.

Recently, nonplasmid-associated conjugative transfer of linked tetracycline and penicillin resistance has been demonstrated in species of oral *Bacteroides* by Guiney and Bouic (12). The transfer process appeared similar in all respects to that seen for nonplasmid elements in the colonic *Bacteroides*. The host range of these elements included three oral *Bacteroides* species (*buccae*, *denticola*, and *intermedius*). Of equal importance was the transfer of this class of element to the colonic species *B. fragilis*. Moreover, a cloned Tc[r] gene from *B. fragilis* was found to hybridize to the transferred tetracycline resistance locus from the oral *Bacteroides*. This report was the first to establish the presence of conjugal elements in the oral *Bacteroides*. Furthermore, it provided compelling evidence that conjugal genetic exchange occurs between oral and colonic *Bacteroides*. Intergroup conjugative transfer of drug resistance elements has also been observed between anaerobes of the bovine rumen and the human colon (52). A naturally occurring, conjugative plasmid carrying a tetracycline resistance gene conserved in the colonic *Bacteroides* (*tetQ*) has been discovered in *Prevotella* (formerly *Bacteroides*) *ruminicola*. This plasmid could be conjugally transferred from *P. ruminicola* to *B. fragilis*. Moreover, conjugative chromosomal elements from *B. fragilis* could be transferred to *P. ruminicola*. *Bacteroides* shuttle plasmids (see below) for doing recombinant DNA work have also been successfully mobilized into *Prevotella ruminicola*, as recently reported by Shoemaker et al (52).

Table 13.1 lists some of the conjugative plasmids or systems under study in the *Bacteroides*.

2.3. Conjugative Mobilization

Available evidence supports the notion that presumed nonconjugative *Bacteroides* plasmids are capable of being mobilized by conjugative elements. Mays et al (28) first reported that the phenotypically cryptic 5.7-kb plasmid of *B. fragilis* V503 frequently was cotransferred to recipients along with a nonplasmid conjugative element conferred MLS and tetracycline resistance (about 80% of the drug resistant transconjugants received the 5.7-kb plasmid). The lack of a selectable marker on this plasmid prevented the unequivocal establishment of its own self-transmissibility, but its small size argued against it being conjugative. Valentine and coworkers (63) confirmed and extended these observations using a phenotypically cryptic plasmid as well as the *Bacteroides* resistance plasmids pBFTM10 and pCP1. Using an intergeneric conjugal transfer system, these workers demonstrated that an IncP conjugative plasmid could mobilize these elements in both inter- (*Escherichia-Bacteroides*) and intra- (*Escherichia-Escherichia*) generic matings. These results led the authors to suggest that these *Bacteroides* plasmids contained a site that operationally

TABLE 13.1   Some Conjugative Elements or Systems of *Bacteroides*

Name	Size (kb)	Host of Origin	Genotype or Phenotype[a]	Comments	Reference
pBF4	41	*B. fragilis*	*ermF*, *tetX*, Tra+	Carries Tn*4351*[b]	(66, 67)
pBI136	82	*B. ovatus*	*ermFS*, Tra+	Carries Tn*4551*[b]	(53, 55)
pBFTM10	14.6	*B. fragilis*	*erm*, *tetX*, Tra+	Carries Tn*4400*[b]	(39, 62)
pBI106	46	*B. ovatus*	Tra+[c]	Appears similar to pBF4 without Tn*4351* sequences	(55)
DOT	50[d]	*B. thetaiotaomicron*	*erm*, *tet*, Tra+	Chromosomal element[e]	(45)
XBU4422	65	*B. uniformis*	Tra+	Conjugative, transposonlike; originating from host chromosome	(51)
Tn*4399*	9.6	*B. fragilis*	Tra+	Conjugative transposon; mobilizes plasmids *in cis*	(19, 20)
pDP1	19	*Bacteroides-E. coli* shuttle plasmid	Ap^r, Tc^r (*E. coli*) *erm* (*Bacteroides*)	Can be mobilized from *E. coli* to *Bacteroides* by incP plasmid	(13)
pE5-2	17	*Bacteroides-E. coli* shuttle plasmid	Su^r, Tc^r (*E. coli*) *erm* (*Bacteroides*)	Can be mobilized from *E. coli* to *Bacteroides* by incP plasmid	(47, 48)

[a]Abbreviations: *ermF*, *ermFS*, *erm*, resistance to macrolide (erythromycin), lincosamide (clindamycin), and streptogramin antibiotics; Tc^r, tetracycline; *tetX*, tetracycline resistance in aerobically grown *E. coli* (see 13); *tet*, tetracyline resistance in *Bacteroides*; Ap^r, ampicillin resistance; Su^r, sulfonamide resistance; Tra+, conjugal transfer proficiency.
[b]See Figure 13.1 for transposon maps.
[c]Transfer proficiency assumed based on physical similarity to pBF4.
[d]The size 50 kb represents a lower limit based on the characterization of overlapping cosmid clones of this element (45). Revised estimates using transverse alternating field electrophoresis indicate this element's size to be in the 70- to 80-kb range (2).
[e]Chromosomal location proposed based on inability to detect plasmid DNA involvement in the transfer process.

functioned as an *oriT* in DNA transfer. Physical evidence supporting the existence of such a site is presently unavailable. Interestingly, shuttle plasmids made from either pB8-51 or pBFTM10 could be mobilized from *Bacteroides* donors containing a nonplasmid Tc^r element designated ERL. Whether the same region was being recognized by both the IncP plasmid and the ERL elements is still open to question. Shoemaker and Salyers (49) also have provided evidence that the ERL element is capable of mobilizing nonreplicating, cytoplasmic DNA molecules from *Bacteroides* to *E. coli*. An integrated copy of the IncB plasmid R*751* can excise molecules from the *Bacteroides* chromosome, presumably because of the recombination between Tn*4351* and IS*4351* sequences. Although this molecule is unable to replicate in *Bacteroides*, the evidence suggests that it is mobilized by the ERL element into *E. coli*, where it is stably maintained as a plasmid replicon.

As mentioned previously, the presence of nonplasmid (i.e., chromosomal) conjugative elements in the *Bacteroides* resembling conjugative transposons appears to be a rather common occurrence. Such elements potentially complicate studies aimed at defining conjugal transfer involving plasmids, since integrated elements could mobilize *in trans*. In sum, the mechanism of mobilization of nonconjugative elements remains ill-defined, although available evidence suggests that the process involves mechanisms similar to those operative in other bacterial systems. The existence of a *cis*-acting region appears necessary for mobilization to occur.

2.4. Conjugation as a Genetic Tool

The discovery of conjugative antibiotic resistance in *Bacteroides* has led to a rapid development of genetic systems suitable for manipulation of these organisms. Early work in this area was hampered by the fact that indigenous *Bacteroides* plasmids were neither maintained nor expressed in *E. coli*. Likewise, no *E. coli* plasmid or antibiotic resistance gene has been found to function in the *Bacteroides*. However, using indigenous R-plasmids (e..g, pBF4), it has been possible to construct the plasmid vectors required for routine genetic analysis.

Conjugation and plasmid mobilization have been useful as methods for the introduction of DNA into the *Bacteroides*. The essential elements required for transfer between *E. coli* and *Bacteroides* were established by Guiney et al (13) in the construction of pDP1. This hybrid plasmid contains pBR322 sequences for replication and selection in *E. coli*, and it contains the transfer origin (*oriT*) from RK2 for high-frequency mobilization (14). This was fused to the 15-kb MLS^r plasmid, pCP1 (pBFTM10), which contained a *Bacteroides* replication region and a selective marker. Mobilization of pDP1 from *E. coli* to *Bacteroides* required the transfer functions of a helper plasmid such as RK2 or RK231, and these can be provided *in trans* in the pDP1 host cell or by a second helper strain in triparental matings. No transfer was observed in the absence of an appropriate helper plasmid. Genetic and physical analyses of the *Bacteroides* transconjugants revealed that pDP1 was transferred. However, the helper plasmid was not found in transconjugants, presumably because of its inability to replicate (46).

Two improvements in the construction and use of mobilization vectors have occurred since the initial description of pDP1. The first of these was an improvement in the transfer frequencies of *E. coli* to *Bacteroides* matings. Shoemaker and coworkers (46) found that incubation of mating mixtures under aerobic conditions resulted in a 20- to 100-fold increase in the number of transconjugants as compared with identical matings performed under an anaerobic atmosphere. Under aerobic conditions, transfer frequencies with typical shuttle vectors reached 1.1×10^{-3}. The aerotolerant *Bacteroides* recipients did not suffer a loss in viability. This strategy enabled the authors to detect the transposition of Tn*4351* in *Bacteroides* following its conjugal entry from an *E. coli* donor.

The second improvement resulted from an observation that many indigenous *Bacteroides* plasmids contained regions that promoted their mobilization by *E. coli* conjugative plasmids (47). Shuttle vectors containing *Bacteroides* replicons such as pBFGM10 (pCP1) and the cryptic plasmid pB8-51 could be mobilized to *B. uniformis* recipient by IncP (e.g., RK231, R751) plasmids at frequencies of 10^{-8} to 10^{-5}. The IncIa plasmid R64*drd*-11 also mobilized (albeit inefficiently) the shuttle vectors, but no mobilization was observed with the IncN, W, or F1 plasmids. Transfer also occurred between *E. coli* strains but was 100 times greater than seen with *Bacteroides* recipients. By using *E. coli* to *E. coli* matings with various derivatives of the shuttle vectors, it was possible to localize the mobilization (*mob*) regions on both pBFTM10 and pB8-51 (47).

The indigenous cryptic plasmids of *Bacteroides* were readily mobilized by the conjugative antibiotic resistance transposons such as Tn*5030* (28). Based on studies with derivatives of the pBFTM10 and pB8-51 shuttle plasmids, it appears that the mobilization regions identified above in *E. coli* are in fact the same as those recognized by the *Bacteroides* conjugative transposons (47). Transfer between *Bacteroides* strains occurred at relatively high frequencies for plasmids containing intact mobilization regions but at greatly reduced or undetectable frequencies when the region was disrupted. Similar results were seen in experiments using *E. coli* recipients. No transfer was observed from a *Bacteroides* host strain unless the shuttle plasmids were coresident with a conjugative transposon. Further, plasmids containing a functional *E. coli oriT* but a disrupted *Bacteroides* mobilization region were not mobilized by the conjugative transposons, suggesting that *oriT* was not recognized by this system. Although these mobilization regions are not as efficient for transfer as the *E. coli* transfer origins, they can be used to achieve genetic transfer.

The work of Salyers and coworkers serves as an example of the application of conjugative gene transfer systems to the study of *Bacteroides*. These authors have cleverly used an intergeneric conjugative shuttle system to construct specific *Bacteroides* mutants. *B. thetaiotaomicron* contains two enzymes that degrade chondroitin sulfate, a mucopolysaccharide present in the colon. The gene encoding chondroitin lyase II (CSase II) has been cloned and characterized (17). The CSase II gene was found to be adjacent to the gene encoding chondro-4-sulfatase, the next enzyme in the degradation pathway (16). A DNA fragment corresponding to an internal (incomplete) portion of the CSase II gene was cloned into an *E. coli-Bacteroides* vector pE3-1 and mobilized into *Bacteroides* with IncP plasmid R751. Selection for the MLS[r] carried on the vector resulted in a recombination event that interrupted chromosomal copy of the CSase II gene (16). The contribution of the CSase II to total chondroitin sulfate use then was investigated in vitro (15) and in vivo in germ-free mice (41). The evaluation of specific, characterized, insertion mutations that were identical to the parent strain except for the single alteration demonstrated that CSaseII was not required for chondroitin sulfate use. The CSaseII mutant also displayed no selective disadvantage in the intestinal tract of the germ-free mouse (41). These as well as other studies illustrate the use of conjugative mobilization systems to introduce cloned sequences back into their original host cells in order to construct specific mutants by insertion mutagenesis or allelic exchange (58).

3. *Clostridium*

The genus *Clostridium* consists of obligately anaerobic, spore-forming, gram-positive rods. Species of this genus are of considerable economic importance from two standpoints: infectious diseases (11) and commercial biotechnology 23, 29, 70). Significant pathogens that belong to this genus include *C. perfringens* (gas gangrene), *C. tetani* (tetanus), and *C. botulinum* (botulism). The clostridia produce some of the most potent exotoxins known (e.g., the neurotoxins produced by *C. botulinum*). On the other hand, many species of this genus provide considerable benefit to humankind. The clostridia possess a wide array of metabolic capabilities. Some species are of commercial importance in the production of organic solvents (29). The clostridia are composed of many species that are free-living in nature. However, several species normally occupy the intestinal tract of mammals and birds (21, 31). Like the *Bacteroides*, the clostridia are a genetically heterogeneous group that appears to have ancient procaryotic origins. Many species are nutritionally versatile and aerotolerant, making them amenable to genetic studies. Their ability to sporulate provides a model developmental system in this unique procaryote that can be studied comparatively by genetic and biochemical approaches. Much of the current work aimed at clostridial species should provide the means to rationally use classic and recombinant genetics to manipulate metabolic pathways in order to improve existing commercial applications or to create novel ones. Conjugal genetic exchange systems in the clostridia will make significant contributions toward achieving this goal and will be the prime focus of this discussion. Those wishing a general treatment of genetic systems in the clostridia are urged to consult any one of several recent reviews on the subject (24, 29, 40, 70).

3.1. Clostridial Plasmids

The study of conjugal genetic exchange in the clostridia had it beginnings in the seminal work by Sebald and coworkers (4, 43, 44). Specifically, this group reported the existence of large conjugative plasmids conferring resistance to tetracycline and chloramphenicol. These studies established the existence of plasmid-mediated conjugal transfer in the clostridia and, furthermore, piqued interest in the study of extrachromosomal elements in this group of anaerobes. Since that time, there have been numerous reports of the existence of plasmids in clostridia of both medical and biotechnological significance. Included in this list are species such as *butyricum, paraputrificum, botulinum, perfringens, difficile, novyi, tetani, acetobutylicum,* and *cochlearum*. Although many of the reported plasmids remain phenotypically cryptic, some carry observable genetic markers, including conjugal transfer proficiency. The reviews by Odelson et al (31) and Young et al (70) provide tabular catalogues of plasmids that have been identified in this genus. A number of cryptic plasmids indigenous to *C. acetobutylicum* or *C. perfringens* have been used to construct shuttle plasmids that replicate in clostridial species as well as in other hosts (e..g, *E. coli, B. subtilis*). For the most part, such vectors have been introduced back into the clostridial host by some form of genetic transformation (e.g., protoplast transformation, electrotransformation) rather than by conjugation (29, 70). However, one family of shuttle plasmids has recently been described that uses the RK2 *oriT*-based conjugative mobilization systems to transfer the shuttle vector (68).

3.2. Conjugative Plasmids

The conjugative antibiotic resistance plasmids of the clostridia were the first genetic elements to be described in this genus (4, 44). The 54-kb plasmid pIP401 carried resistance determinants for both chloramphenicol (Cm^r) and tetracycline (Tc^r). Abraham and Rood (1) have since shown that the Cm^r gene is carried by a 6-kb transposon (Tn*4451*). Reports of R-plasmid transfer occurring presumably by conjugation in vivo has been made (4). Interspecies conjugative R-plasmid transfer has been demonstrated (e.g., *C. perfringens* to *C. difficile*), but systematic studies to define host range of such plasmids have not been reported. The operational definition of conjugal transfer has revolved around the standard parameters examined for the *Bacteroides* systems (see Section 2.1.). Generally, conjugal transfer is inefficient in liquid culture, and most mating procedures are performed using cells that have been collected on a solid filter or agar matrix. Similarly, there is no available information as to the cellular or molecular basis of conjugal DNA transfer in this genus. However, available genetic tools in this genus should facilitate progress in this area in the near future.

Conjugative plasmids appear able to mobilize nonconjugative, coresident plasmids, but the molecular basis of these observations remains unclear (4). Such mobilization could be the result of interaction of a *trans*-acting factor with an *oriT*-like site or, alternatively, it might be the result of cointegrate formation. In this vein, plasmid mobilization via cointegrate formation has been demonstrated at both the intrageneric (33) and intergeneric (32) levels. These systems basically involved the construction of shuttle-type plasmids that carry a copy of the *erm* gene present on the broad host range streptococcal plasmid pAMβ1 (6). Mobilizable plasmids in these systems contain appropriate cloning sites; accordingly, sequences of interest can be cloned in a host like *E. coli*. The resultant plasmid can then be transformed into an intermediate host, *Bacillus subtilis*, where the plasmid is recombinationally rescued by a resident pAMβ1 plasmid. This event creates cointegrate where the cloned sequences are flanked by a copy of the duplicated *erm* gene. This cointegrate can then be conjugationally transferred to *Clostridium*. In this fashion, Oultram et al (33) were able to identify cloned clostridial sequences that complemented a leucine deficiency in an auxotrophic strain of *C. acetobutylicum*.

Recently, Williams et al (68) have developed a system that uses IncP plasmids to mobilize DNA molecules in the clostridia in a manner similar to that used in the *Bacteroides* (see Section 2.4.). A shuttle plasmid family carrying the RK2 origin of transfer (*oriT*) has been constructed. The plasmids have differing origins of replication on them for maintenance in a variety of hosts (e.g., *C. acetobutylicum*, *C. butyricum*, *Enterococcus faecalis*). The system permits the direct RK2 mobilization of cloned sequences contained on the *oriT* shuttle plasmids from *E. coli* to gram-positive recipients. This system establishes a useful alternative to cointegrate-based transfer using an intermediate *Bacillus* host (see above) and to electroporation of shuttle plasmids. It also underscores the range of operation of the transfer apparatus encoded by the broad host range plasmid RK2. Indeed, the system developed by Williams et al (68) may open avenues to the genetic manipulation of heretofore genetically recalcitrant anaerobic organisms such as the methanogens.

3.3. Broad Host Range Gram-Positive Plasmids

As mentioned previously, several broad host range plasmids, primarily of streptococcal origin, have been shown to be conjugally transferred to and maintained in certain clostridial species. Some of the more widely used replicons are listed in Table 13.2. As pointed out previously, these plasmids show promise in being able to mobilize other extrachromosomal elements into the clostridia using gram-positive donors. This can be accomplished by using constructs that result in deliberate cointegrate formation. Alternatively, direct mobilization of plasmids (presumably without cointegrate formation) may also be possible (71), but more work is needed to elucidate mechanisms operative in such systems.

3.4. Nonplasmid-Mediated Conjugal Genetic Exchange

The conjugative transfer of drug resistance in the absence of plasmid involvement have been made. Initial reports involved *C. difficile* donors that could transfer Tc[r] by a mechanism that operationally resembled conjugation (22). Progeny strains from such matings in turn became conjugative donors of Tc[r]. Similar transfer systems involving MLS[r]

TABLE 13.2   Streptococcal Conjugative Elements that Function in
the Clostridia[a]

Name	Size (kb)	Host of origin	Genotype or Phenotype[b]	Reference
pAMβ1	26.5	*S. faecalis*	*erm*, Tra$^+$	(9, 32, 71)
pIP501	30.2	*S. agalactiae*	*erm*, Cm^{r}Tra$^+$	(38)
Tn916	16	*S. faecalis*	Tc^r, Tra$^+$	(64)
Tn1545	25	*S. pneumoniae*	*erm*, Tc^r, Km^r, Tra$^+$	(3)

[a]Experiment done with *C. tetani, C. acetylbutylicum*.
[b]Abbreviations: *erm*, resistance to macrolide (erythromycin), lincosamide (clindamycin), and streptogramin antibiotics; *tetX*, Tc^r, tetracycline resistance; Km^r, kanamycin resistance; Cm^r, chloramphenicol resistance; Tra$^+$, conjugal transfer proficiency.

have been reported in the clostridia as well (26, 69). Very little is known about such transfer elements at present. Although it is likely that they represent a class of naturally occurring conjugative transposonlike elements, much more work is needed to elucidate their nature. The relationship of their conjugative apparatus to that present on conjugative plasmids remains unclear.

Streptococcal conjugative transposons can be conjugally transferred from gram-positive hosts to the clostridia. These include Tn916 (16 kb; confers Tc^r), Tn925 (similar to Tn916), and Tn1545 (25 kb; confers Tc^r, MLS^r, Km^r [kanamycin resistance]) (3, 9, 64, 70). For example, Tn916 can be conjugally transferred from *Enterococcus faecalis* to both *C. tetani* (64) and *C. acetobutylicum* (3). Such systems seem to assure the means to directly perform transposon mutagenesis in the clostridia, without the need for developing suitable plasmid delivery vectors. Because these transposons reside on the chromosome in the *Enterococcus* donor, they transpose and transfer in the absence of plasmid DNA. Thus, when they enter the clostridial recipient host, they must integrate into the chromosome in order to be stably maintained. Because this integration is likely to be random, one should be able to collect desired mutants owing to transposon insertions at various sites on the clostridial chromosome.

In sum, the precise cellular and molecular mechanisms associated with conjugal transfer in the clostridia remain unclear. This genus contains its own conjugative factors, including plasmids and nonplasmid transfer elements. Considerable work is needed to gain a comparative understanding of the genetics and biochemistry of conjugal transfer in these systems. Broad host range plasmids and conjugative transposons of streptococcal origin can be transferred to the clostridia, where they are maintained and expressed. Progress is being made to exploit such systems in order to be able to genetically manipulate the clostridia. Such work has significant implications for the improvement of strains for commercial purposes.

4. Concluding Thoughts

The development of conjugative genetic exchange systems in the *Bacteroides* and the clostridia has played a major role in advancing the molecular genetic study of these anaerobes. The study of conjugative systems has largely been focused on R-plasmid systems, and the clinical significance of such elements is of great importance. Moreover, such systems have provided the means to develop cloning vectors and to use conjugal mobilization to ask molecular genetic questions in these genera. Such systems allow the performance of experiments in homologous host backgrounds, thus obviating problems with gene expression in hosts like *E. coli*. Equally important, conjugative genetic exchange will provide the means to modify and/or construct strains for novel biotechnological applications. Finally, the study of the biology and genetics of conjugation in these genera offers a comparative opportunity for studying mechanisms of gene exchange in phylogenetically unique procaryotes.

ACKNOWLEDGMENTS. The contributions of C. Jeff Smith and Madelon C. Halula to the preparation of the manuscript are gratefully acknowledged. Work from the author's lab was supported by U.S. Public Health Service grants nos. DE09035 and DE04224.

References

1. Abraham, L.J., and Rood, J.I., 9187, Identification of Tn*4451* and Tn*4452*, chloramphenicol resistance transposons from *Clostridium perfringens*, *J. Bacteriol*. 169:1579–1584.
2. Bedzyk, L.A., Shoemaker, N.B., Young, K.E., and Salyers, A.A., 1992, Insertion and excision of *Bacteroides* conjugative chromosomal elements, *J. Bacteriol*. 174:166–172.
3. Bertram, J., and Dürre, P., 1989, Conjugal transfer and expression of streptococcal transposons in *Clostridium acetobutylicum*, *Arch. Microbiol*. 151:551–557.
4. Brefort, G., Magot, M., Ionesco, H., and Sebald, M., 1977, Characterization and transferability of *Clostridium perfringens* plasmids, *Plasmid* 1:52–66.
5. Butler, E., Joiner, K.A., Malamy, M., Bartlett, J.C., and Tally, F.P., 1984, Transfer of tetracycline or clindamycin resistance among strains *Bacteroides fragilis* in experimental abscesses, *J. Infect. Dis*. 150: 20–24.
6. Clewell, D.B., 1981, Plasmids, drug resistance, and gene transfer in the genus *Streptococcus*, *Microbiol. Rev*. 45:409–436.
7. Clewell, D.B., 1990, Movable genetic elements and antibiotic resistance in enterococci, *Eur. J. Clin. Microbiol. Infect. Dis*. 9:90–102.
8. Clewell, D.B., and Gawron-Burke, C., 1986, Conjugative transposons and the dissemination of antibiotic resistance in streptococci, *Annu. Rev. Microbiol*. 40:635–659.
9. Davies, A., Oultram, J., Pennock, A., Williams, D., Richards, D., Minton, N., and Young, M., 1988, Conjugal gene transfer in *Clostridium acetobutylicum*, in: *Genetics and Biotechnology of Bacilli*, (A.T. Ganesan and J.A. Hoch, eds.), Academic Press, London, pp. 391–395.
10. Fears, R., and Sabine, J.R., 1986, Cholesterol 7-alpha-hydroxylase 7-alpha-monooxygenase), CRC Press, Boca Raton, Florida.
11. Finegold, S.M., and George, W.L., 1989, *Anaerobic Infections in Humans*, Academic Press, San Diego.
12. Guiney, D.G., and Bouic, K., 1990, Detection of conjugal transfer systems in oral, black-pigmented *Bacteroides* spp., *J. Bacteriol*. 172:495–497.
13. Guiney, D.G., Hasegawa, P., and Davis, C.E., 1984, Plasmid transfer from *Escherichia coli* to *Bacteroides fragilis*: differential expression of antibiotic resistance phenotypes, *Proc. Natl. Acad. Sci. USA* 81:7203–7206.
14. Guiney, D.G., and Yakobson, E., 1983, Location and nucleotide sequence of the transfer origin of the broad host range plasmid RK2, *Proc. Natl. Acad. Sci. USA* 80:3595–3598.
15. Guthrie, E.P., and Salyers, A.A., 1986, Use of targeted insertional mutagenesis to determine whether chondroitin lyase II is essential for condroitin sulfate utilization by *Bacteroides thetaiotaomicron*, *J. Bacteriol*. 166:966–971.

16. Guthrie, E.P., and Salyers, A.A., 1987, Evidence that the *Bacteroides thetaiotaomicron* chondroitin lyase gene is adjacent to the chondro-4-sulfatase gene and may be part of the same operon, *J. Bacteriol.* 169:1192–1199.

17. Guthrie, E.P., Shoemaker, N.B., and Salyers, A.A., 1985, Cloning and expression in *Escherichia coli* of a gene coding for chondroitin lyase from *Bacteroides thetaiotaomicron*, *J. Bacteriol.* 164:510–515.

18. Halula, M., and Macrina, F.L., 1990, Tn*5030*: a conjugative transposon conferring clindamycin resistance in *Bacteroides* species, *Rev. Infect. Dis.* 12(Suppl. 2):S235–S242.

19. Hecht, D.W., and Malamy, M.H., 1989, Tn*4399* a conjugal mobilizing transposon of *Bacteroides fragilis*, *J. Bacteriol.* 171:3603–3608.

20. Hecht, D.W., Thompson, J.S., and Malamy, M.H., 1989, Characterization of the termini and transposition products of Tn*4399*, a conjugal mobilizing transposon of *Bacteroides fragilis*, *Proc. Natl. Acad. Sci. USA* 86:5340–5344.

21. Hentges, D., 1983, *Human Intestinal Microflora in Health and Disease*, Academic Press, New York.

22. Ionesco, H., 1980, Transfert de la resistance a la tetracycline chez *Clostridium difficile*, *Ann. Microbiol. Inst. Pasteur* 131A:171–179.

23. Jones, D.T., and Woods, D.R., 1986, Acetone-butanol fermentation revisited, *Microbiol. Rev.* 50:484–524.

24. Jones, D.T., and Woods, D.R., 1986, Gene transfer, recombination and gene cloning in *Clostridium acetobutylicum*, *Microbiol. Sci.* 3:19–22.

25. Macrina, F.L., Mays, T.K., Smith, C.J., and Welch, R.A., 1981, Non-plasmid associated transfer of antibiotic resistance in *Bacteroides*, *J. Antimicrob. Chemother.* 8:77–86.

26. Magot, M., 1983, Transfer of antibiotic resistances from *Clostridium innocuum* to *Clostridium perfringens* in the absence of detectable plasmid DNA, *FEMS Microbiol. Lett.* 18:149–151.

27. Magot, M., Fayolle, F., Privitera, G., and Sebald, M., 1981, Transposon-like structures in the *B. fragilis* MLS plasmid pIP410, *MGG* 181:559–561.

28. Mays, T.D., Smith, C.J., Welch, R.A., Delfini, C., and Macrina, F.L., 1982, Novel antibiotic resistance transfer in *Bacteroides*, *Antimicrob. Agents Chemother.* 21:110–118.

29. Minton, N.J., and Oultram, J.D., 1988, Host:vector systems for gene cloning in *Clostridium*, *Microbiol. Sci.* 5:310–315.

30. Moore, W.E.C., and Holdeman, L., 1974, Human fecal flora: the normal flora of 20 Japanese-Hawaiians, *Appl. Microbiol.* 27:961–979.

31. Odelson, D.A., Rasmussen, J.L., Smith, C.J., and Macrina, F.L., 1987, Extrachromosomal systems and gene transmission in anaerobic bacteria, *Plasmid* 17:87–109.

32. Oultram, J.D., Davies, A., and Young, M., 1987, Conjugal transfer of a small plasmid from *Bacillus subtilis* to *Clostridium acetobutylicum* by cointegrate formation with plasmid pAMβ1, *FEMS Microbiol. Lett.* 42:113–119.

33. Oultram, J.D., Peck, H., Brehm, J.K., Thompson, D.E., Swinfield, T.J., and Minton, N.P., 1988, Introduction of genes for leucine biosynthesis from *Clostridium pasteurianum* into *C. acetobutylicum* by cointegrate conjugal transfer, *Mol. Gen. Genet.* 214:177–179.

34. Privitera, G., Dublanchet, A. and Sebald, M., 1979, Transfer of multiple antibiotic resistance between subspecies of *Bacteroides fragilis*, *J. Infect. Dis.* 139:97–101.

35. Rasmussen, B.A., and Kovacs, E., 1991, Identification and DNA sequence of a new *Bacteroides fragilis* insertion sequence-like element, *Plasmid* 25:141–144.

36. Rasmussen, J.L., Odelson, D.A., and Macrina, F.L., 1986, Complete nucleotide sequence and transcription of *ermF*, a macrolide-lincosamide-streptogramin B resistance determinant *Bacteroides fragilis*, *J. Bacteriol.* 168:523–533.

37. Rasmussen, J.L., Odelson, D.A., and Macrina, F.L., 1987, Complete nucleotide sequence of insertion element IS*4351* from *Bacteroides fragilis*, *J. Bacteriol.* 169:3573–3580.

38. Reysset, G., and Sebald, M., 1985, Conjugal transfer of plasmid mediated antibiotic resistance from streptococci to *Clostridium acetylbutylicum*, *Ann. Microbiol. Inst. Pasteur* 136:275–282.

39. Robillard, N.J., Tally, F.P., and Malamy, M.H., 1985, Tn*4400*, a compound transposon isolated from *Bacteroides* functions in *Escherichia coli*, *J. Bacteriol.* 164:1248–1255.

40. Rogers, P., 1986, Genetics and biochemistry of *Clostridium* revelant to development of fermentation processes, *Adv. Appl. Microbiol.* 31:1–60.

41. Salyers, A.A., and Guthrie, E.P., 1988, A deletion in the chromosome of *Bacteroides thetaiotaomicron*

abolishes production of chondroitinase II does not affect of the organisms in gastrointestinal tracts of germfree mice, *Appl. Environ. Microbiol.* 54:1964–1969.

42. Salyers, A.A., Shoemaker, N.B., and Guthrie, E.P., 1987, Recent advances in *Bacteroides* genetics, *CRC Crit. Rev. Microbiol.* 14:49–71.

43. Sebald, M., Bouanchaud, D., Bieth, G., and Prevot, A.R., 1975, Plasmids controlling the resistance to several antibiotics in *C. perfringens* type A, strain 659, *C. R. Acad. Sci.* [D] (Paris) 280:2401–2404.

44. Sebald, M., and Brefort, M.G., 1975, Transfer of the tetracycline-chloramphenicol plasmid in *Clostridium perfringens*, *C. R. Acad. Sci.* [D] (Paris) 281:317–319.

45. Shoemaker, N.B., Barber, R.D., and Salyers, A.A., 1989, Cloning and characterization of a *Bacteroides* conjugal tetracycline-erythromycin resistance element by using a shuttle cosmid vector, *J. Bacteriol.* 171:1294–1302.

46. Shoemaker, N.B., Getty, C., Gardner, J.F., and Salyers, A.A., 1986, Tn*4351* transposes in *Bacteroides* spp. and mediates the integration of plasmid R751 into the *Bacteroides* chromosome, *J. Bacteriol.* 165:929–936.

47. Shoemaker, N.B., Getty, G., Guthrie, E.P., and Salyers, A.A., 1986, Regions in *Bacteroides* plasmids pBFTM10 and pB8-51 that allow *Escherichia coli-Bacteroides* shuttle vectors to be mobilized by plasmids and by a conjugative *Bacteroides* tetracycline element, *J. Bacteriol.* 166:959–965.

48. Shoemaker, N.B., Guthrie, E.P., Salyers, A.A., and Gardner, J.F., 1985, Evidence that the clindamycin-erythromycin resistance gene of *Bacteroides* plasmid pBF4 is on a transposable element, *J. Bacteriol.* 162:626–632.

49. Shoemaker, N.B., and Salyers, A.A., 1987, Facilitated transfer of IncP beta R751 derivatives from the chromosome of *Bacteroides uniformis* to *Escherichia coli* by a conjugative *Bacteroides* tetracycline resistance element, *J. Bacteriol.* 169:3160–3167.

50. Shoemaker, N.B., and Salyers, A.A., 1988, Tetracycline-dependent appearance of plasmidlike forms in *Bacteroides uniformis* 0061 mediated by conjugal *Bacteroides* tetracycline resistance elements, *J. Bacteriol.* 170:1651–1657.

51. Shoemaker, N.B., and Salyers, A.A., 1990, A cryptic 65-kilobase-pair transposonlike element isolated from *Bacteroides uniformis* has homology with *Bacteroides* conjugal tetracycline resistance elements, *J. Bacteriol.* 172:1694–1702.

52. Shoemaker, N.B., Wang, G., and Salyers, A.A., 1991, Evidence for natural transfer of a tetracycline resistance gene between the bacteria from the human colon and bacteria from the bovine rumen, *Appl. Environ. Microbiol.* 58:1313–1320.

53. Smith, C.J., 1985, Characterization of *Bacteroides ovatus* plasmid pBI136 and of its clindamycin resistance region, *J. Bacteriol.* 161:1069–1073.

54. Smith, C.J., 1987, Nucleotide sequence analysis of Tn*4551*: use of *ermFS* operon fusions to detect promoter activity in *Bacteroides fragilis*, *J. Bacteriol.* 169:4589–4596.

55. Smith, C.J., and Macrina, F.L., 1984, Large transmissible clindamycin resistance plasmid in *Bacteroides ovatus*, *J. Bacteriol.* 158:739–741.

56. Smith, C.J., and Spiegel, H., 1987, Transposition of Tn*4551* in *Bacteroides fragilis*: identification properties of a new transposon from *Bacteroides* spp., *J. Bacteriol.* 169:3450–3457.

57. Smith, C.J., Welch, R.A., and Macrina, F.L., 1982, Two independent conjugal transfer systems operating in *Bacteroides fragilis* V479-1, *J. Bacteriol.* 151:281–287.

58. Smith, K.A., and Salyers, A.A., 1989, Cell-associated pullulanase from *Bacteroides thetaiotaomicron*: cloning, characterization, and insertional mutagenesis to determine role in pullulan utilization, *J. Bacteriol.* 171:2116–2123.

59. Stevens, A.M., Shoemaker, N.B., and Salyers, A.A., 1990, The region of a *Bacteroides* conjugal chromosomal tetracycline resistance element which is responsible for production of plasmidlike forms from unlinked chromosomal DNA might also be involved in transfer of the element, *J. Bacteriol.* 172:4271–4279.

60. Tally, F., Shimell, M., Carson, G., and Malamy, M., 1981, Chromosomal and plasmid-mediated transfer of clindamycin resistance in *Bacteroides fragilis*, in: *Molecular Biology, Pathogenicity and Ecology of Bacterial Plasmids* (S.B. Levy and R.C. Clowes, eds.), Plenum, New York, p. 51.

61. Tally, F., Snydman, D., Gorbach, S., and Malamy, M., 1979, Plasmid mediated, transferable resistance to clindamycin and erythromycin in *B. fragilis*, *J. Infect. Dis.* 139:83–88.

62. Tally, F., Snydman, D., Shimell, M., and Malamy, M., 1982, Characterization of pBFTM10, a clindamycin-erythromycin resistance transfer factor from *Bacteroides fragilis*, *J. Bacteriol.* 151:686–689.

63. Valentine, P.J., Shoemaker, N.B., and Salyers, A.A., 1988, Mobilization of *Bacteroides* plasmids by *Bacteroides* conjugal elements, *J. Bacteriol.* 170:1319–1324.
64. Volk, W.A., Bizzini, B., Jones, K.R., and Macrina, F.L., 1988, Inter- and intrageneric transfer of Tn*916* between *Streptococcus faecalis* and *Clostridium tetani*, *Plasmid* 19:255–259.
65. Weisburg, W.G., Oyaizu, Y., Oyaizu, H., and Woese, C.R., 1985, Natural relationship between *Bacteroides* and flavobacteria, *J. Bacteriol.* 164:230–236.
66. Welch, R.A., Jones, K.R., and Macrina, F.L., 1979, Transferable lincosamide-macrolide resistance in *Bacteroides*, *Plasmid* 2:261–268.
67. Welch, R.A., and Macrina, F.L., 1981, Physical characterization of *Bacteroides fragilis* R plasmid pBF4, *J. Bacteriol.* 145:867–872.
68. Williams, D.R., Young, D.I., and Young, M., 1990, Conjugative plasmid transfer from *Escherichia coli* to *Clostridium acetobutylicum*, *J. Gen. Microbiol.* 136:819–826.
69. Wust, J., and Hardegger, U., 1983, Transferable resistance to clindamycin, erythromycin and tetracycline in *Clostridium difficile*, *Antimicrob. Agents Chemother.* 23:784–786.
70. Young, M., Minton, N.P., and Staudenbauer, W.L., 1989, Recent advances in the genetics of clostridia, *FEMS Microbiol. Rev.* 63:301–326.
71. Yu, P., and Pearse, L., 1986, Conjugal transfer of streptococcal antibiotic resistance plasmids into *Clostridium acetobutylicum*, *Biotech. Lett.* 8:469–474.

Chapter 14

Sex Pheromones and the Plasmid-Encoded Mating Response in *Enterococcus faecalis*

DON B. CLEWELL

1. Introduction

Plasmid-related conjugation in gram-positive, nonstreptomycete bacteria was first reported in 1974 by Jacob and Hobbs (38). They identified an R-plasmid (pJH1) and a hemolysin/bacteriocin plasmid (pJH2) in a clinical isolate of *Enterococcus* (formerly *Streptococcus*) *faecalis* and found both capable of transfer in broth matings (38, 39). Independently, and at about the same time, Tomura et al (67), in Japan, reported on the transferability of a hemolysin/bacteriocin determinant in *E. faecalis*. Although physical evidence for plasmid involvement was not presented, it is highly likely that this was indeed the case, as the frequency of transfer was quite high (10^{-2} per donor in broth matings); and subsequent studies showed that the hemolysin/bacteriocin trait exhibited by many *E. faecalis* strains usually involved a conjugative plasmid (4, 6).

As more conjugative plasmids were identified in *E. faecalis*, it became evident that some would transfer efficiently in broth (e.g., 10^{-2} per donor), whereas others appeared to transfer poorly, if at all, but could transfer reasonably well on solid surfaces such as filter membranes. It was eventually found that those plasmids that transferred in broth encoded a response to an extracellular peptide that induced a series of biosynthetic events resulting in conjugation. Although "aggregates" were frequently observed visually during matings, the first indication that a pheromone was involved came when donor cells were found to self-

DON B. CLEWELL • Department of Biologic and Materials Sciences, School of Dentistry and Department of Microbiology and Immunology, School of Medicine, The University of Michigan, Ann Arbor, Michigan 48109-0402.
Bacterial Conjugation, edited by Don B. Clewell. Plenum Press, New York, 1993.

aggregate, or clump, after exposure to a culture filtrate of recipient cells (18, 19). Induced donors transferred plasmid DNA to recipients in very short (10-minute) matings at frequencies up to 10^5 times that of controls not preexposed to pheromone. In a mating mixture, without preexposure to pheromone, it generally takes about 30 minutes before significant transfer becomes detectable (32). Figure 14.1 illustrates the kinetics of the induction process for the hemolysin/bacteriocin plasmid pAD1 and illustrates how donor potential reaches a maximum after exposure for 60 to 90 minutes.

Transfer actually occurs between donors if they are induced to clump by exposure to a recipient culture filtrate containing pheromone (7). In an experiment where two homologous donor cultures bearing distinguishable derivatives of pAD1 (i.e., marked with the tetracycline resistance transposon Tn916 or the erythromycin resistance transposon Tn917) were both induced and then mated briefly (20 minutes) with each other, plasmid transfer occurred in both directions (7). If only one of the donors was induced, significant transfer occurred only from the induced to the uninduced cells. This result supported the view that pheromone induced functions necessary for DNA transfer as well as aggregation. Were this not the case, one would have expected transfer in both directions between the two donors, regardless of which had been induced prior to the mating. There was also evidence that entry (surface) exclusion functions may be inducible, since transfer was significantly reduced if both donors were induced, compared with the case where only one was induced.

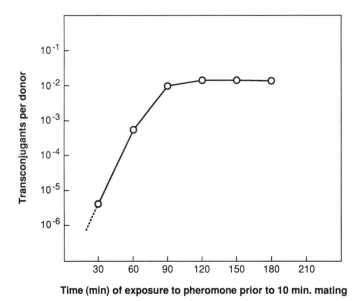

Time (min) of exposure to pheromone prior to 10 min. mating

FIGURE 14.1. Transfer of pAM714 (pAD1::Tn917) during brief matings, as a function of length of prior exposure to pheromone. (Redrawn from reference 30.) An overnight culture of *E. faecalis* FA2-2(Pam714) was diluted 1:10 into a 1:1 mixture of cAD1 (culture filtrate of strain JH2-2) and fresh broth. At the indicated times, a sample of donors was mixed with recipients (strain JH2SS; an overnight culture similarly diluted) at a ratio of 1 donor per 10 recipients. After 10 minutes the mixture was plated on medium selective for recipients that had acquired the Tn917-associated erythromycin resistance.

2. Plasmid Specificity of Sex Pheromones

Plasmid-free recipients secrete multiple sex pheromones (10, 19, 52), each one exhibiting specificity for a given plasmid type. When a plasmid is acquired, secretion of the related pheromone is prevented; however, unrelated pheromones continue to be produced. As many as five different sex pheromones are known to be produced by a single recipient cell, and it is likely that this number is actually much larger.

Enterococcal plasmids reported to exhibit a pheromone response are listed in Table 14.1 along with their corresponding pheromones. Determinants carried by these plasmids include those encoding hemolysin, bacteriocins, and resistance to antibiotics and ultraviolet light. In other cases no associated phenotype, other than the ability to respond to a pheromone, has been observed. A given donor may carry more than one pheromone-responding plasmid. As noted in Table 14.1, two clinical isolates (*E. faecalis* strains DS5 and HH22) were found to contain three such plasmids, with each plasmid conferring a response to a different pheromone. When present together, the specific induction of one pheromone-responding plasmid does not necessarily result in mobilization (*trans-activation*) of another. In the case of cells constructed to carry both pAD1 and pPD1, exposure to just one or the other of the corresponding pheromones (i.e., cAD1 or cPD1, respectively) induced transfer of only the related plasmid (23).

Some plasmids determine the production of peptides that act as competitive inhibitors of the corresponding pheromone (13, 34). For example, cells harboring pAD1 secrete a substance designated iAD1, which specifically inhibits the ability to respond to the pheromone cAD1. The purpose of such compounds may be to prevent induction by pheromone levels too low to result in the generation of mating aggregates (i.e., when recipients are too far away to encounter by random collision). They might also prevent self-

TABLE 14.1 Plasmids Known to Encode a Pheromone Response

Plasmid	Size (kb)	Original Host[a]	Phenotype Encoded[b]	Related Pheromone	Related References
pAD1	60	DS16	Hly/Bac, UV^r	cAD1	(9, 12, 19, 65)
pPD1	56	39-5	Bac	cPD1	(18, 77)
pAM373	36	RC73	?	cAM373	(11)
PCF10	54	SF-7	Tc^r(Tn925)	cCF10	(20)
pAMγ1	60	DS5	Hly/Bac, UV^r	cAD1	(10)
pAMγ2	~60	DS5	Bac	cAMγ2	(10)
pAMγ3	~60	DS5	?	cAMγ3	(10)
pOB1	71	5952	Hly/Bac	COB1	(4, 56, unpublished data)
pJH2	59	JH1	Hly/Bac	cAD1	(4, 39, unpublished data)
pBEM10	70	HH22	Pen^r, Gm^r, Km^r, Tm^r	cAD1	(52)
pAM323	66	HH22	Em^r	cAM323	(52)
pAM324	53	HH22	?	cAM324	(52)
pHKK100	55	228[a]	Hly, Vm^r	cHKK100	(29, unpublished data)

[a]All strains are *E. faecalis* except 228, which is *E. faecium*.
[b]Hly, hemolysin; Bac, bacteriocin; UV^r, ultraviolet light resistance; Tc^r, tetracycline resistance; Pen^r, penicillin resistance; Gm^r, gentamicin resistance; Km^r, kanamycin resistance; Tm^r, tobramycin resistance; Em^r, erythromycin resistance; Vm^r, vancomycin resistance.

induction by small amounts of endogenous pheromone that escape the shutdown process or by other peptides that might cross-react with the related receptor. In a situation where there are equal numbers of donors and recipients, the pheromone produced by the recipients will be in excess of the donor's inhibitor and will successfully induce a mating response.

3. Pheromone and Inhibitor Structure

Several pheromones and related inhibitors have been isolated and characterized by Suzuki and coworkers (47–51, 63). Their structures (and related references) are shown in Table 14.2. All are hydrophobic octa- or heptapeptides. Interestingly, each pheromone contains one hydroxyamino acid residue. Conceivably, these could be sites for modification (inactivation) as part of the plasmid-determined shutdown process, although there is as yet no evidence for this. cAD1 and iAD1 (both octapeptides) exhibit identity at four of the eight positions, whereas cPD1 and the corresponding inhibitor iPD1 exhibit identity at only two positions. The two inhibitors iAD1 and iPD1 both contain the sequence Thr-Leu-Val, as does the pheromone cCF10. A greater similarity is seen between cCF10 and iAD1, both of which contain Val-Thr-Leu-Val. cCF10 exhibited no detectable inhibitor activity in the appropriate assays, however, iAD1 did exhibit a very weak pheromone activity when exposed to cells harboring pCF10 (51).

Synthetic pheromone and inhibitor peptides exhibit full activity and specificity for the related plasmid system. The pheromones are active at concentrations as low as 5×10^{-11} M, and a report dealing with cCF10 showed that a donor cell harboring pCF10 could detect as few as one or two molecules (51). The latter study was based on measurements of plasmid transfer frequencies in 15-minute matings after 1-hour exposures to dilutions of synthetic cCF10.

In the case of cPD1, removal of one amino acid residue from the amino terminus reduced activity to 0.5%, and further eliminations from this end resulted in no detectable activity (43). Removal of the carboxy-terminal residue eliminated activity completely. Omission of the amino-terminal residue from cAD1 resulted in no detectable activity, whereas deletion of the carboxy-terminal residue reduced activity to 0.02% (43). The full lengths of cPD1 and cAD1, therefore, are important to pheromone activity.

A synthetic hybrid peptide consisting of cPD1 on the amino-terminal half and cAD1 on

TABLE 14.2 Structures of Sex Pheromones
and Some Related Inhibitors

Pheromone or Inhibitor (M)	Peptide Structure	Reference
cPD1 (912)	H-Phe-Leu-Val-Met-Phe-Leu-Ser-Gly-OH	(63)
cAD1 (818)	H-Leu-Phe-Ser-Leu-Val-Leu-Ala-Gly-OH	(47)
cAM373 (733)	H-Ala-Ile-Phe-Ile-Leu-Ala-Ser-OH	(49)
cCF10 (789)	H-Leu-Val-Thr-Leu-Val-Phe-Val-OH	(51)
iPD1 (828)	H-Ala-Leu-Ile-Leu-Thr-Leu-Val-Ser-OH	(50)
iAD1 (846)	H-Leu-Phe-Val-Val-Thr-Leu-Val-Gly-OH	(48)

the carboxy-terminal half exhibited 10% cPD1 activity but no cAD1 activity (43). The converse structure had the opposite result, with 1% cAD1 activity and no cPD1 activity. The specificity of the peptide is therefore determined by the amino-terminal sequences.

It has been possible to generate mutants in pheromone production by chemical mutagenesis or by using the conjugative transposon Tn916 (5, 34). Screening for such variants was possible by overlaying colonies with a "lawn" of donor cells harboring pPD1 and looking for colonies that failed to give rise to a halolike appearance that represented donors responding (aggregating) to the diffusing pheromone. Two such variants were reduced to about 2 to 3% of wild-type production of both cAD1 and cPD1, but both secreted unaltered levels of cAM373. These strains were relatively poor recipients in broth, but when plasmid DNA was harbored, they behaved normally as donors. The nature of these mutants is not known, but the fact that two pheromones are affected suggests some form of linkage or a defect in transport or processing.

4. Aeration and Pheromone Production

Production of cAD1 is significantly reduced as a result of aeration during cell growth (15, 75). Cultures grown with vigorous shaking secrete only 5 to 7% as much cAD1 as cultures grown without shaking. Shaking cultures that were inside an anaerobic jar did not give rise to the lower pheromone titer. Interestingly, the activities of the pheromones cPD1 and cAM373 in the same culture filtrates were not affected by aeration; thus, it would appear that oxygen has a specific regulatory effect on the production of cAD1. The level of iAD1 produced by cells harboring pAD1 does not appear to be affected by oxygen.

cAD1 is also expressed at different levels by different bacterial strains. In the case of the *E. faecalis* strain OG1X, culture filtrates have a cAD1 activity that is about eightfold higher than that from the nonisogenic *E. faecalis* strain FA2-2 grown under the same conditions. In the same culture filtrates, however, cPD1 and cAM373 activities are similar for both hosts.

5. The Aggregation Event

In the case of the plasmids pAD1 and pPD1, the pheromone-induced mating response results in synthesis of a proteinaceous microfibrillar substance on the cell surface that can be resolved by immunoelectron microscopy (27, 70, 77). This "fuzzy" material has been referred to as "aggregation substance" (AS); it binds to a substance referred to as "binding substance" (BS) on the recipient surface. There is evidence that BS may in part by made up of lipoteichoic acid (LTA), since very low concentrations of purified LTA (e.g., 0.1 to 1.0 μg/mL) were able to inhibit aggregation (23). This is also consistent with the fact that induced donor cells clump in the absence of recipients; LTA corresponds to the Lancefield group D antigen on the *E. faecalis* surface (76) and is therefore likely to be on the surface of donors as well as recipients. Trotter and Dunny (69) recently reported the generation of plasmid-free, bacteriophage-resistant mutants that were defective in binding to induced donors and were also poor recipients in broth matings. These variants were altered in LTA fatty acid composition.

Early studies showed that aggregates could be dissociated by a chelating agent such as EDTA (18), and it was subsequently found that phosphate and divalent cations were necessary for induced cells to aggregate (77). When induced cells were exposed to the nonphysiological pH of 2.5, which would be expected to remove the negative charge on LTA, aggregation occurred in the absence of phosphate and Mg^{++}. The role of phosphate and Mg^{++} at physiological pHs, may therefore be to help stabilize the interaction of AS and the negatively charged BS (LTA).

Western blot analyses using antiserum raised against induced pAD1-containing donors fixed with gluteraldehyde have revealed a number of surface antigens that appear only after induction. The size of these proteins range from about 52,000 to 190,000 Da (23, 74). In the case of pAD1, these proteins have been referred to as AD52, AD74, AD130, AD153, and AD157. The designations were originally intended to reflect the relative sizes (in kilodaltons). However, more recent data have shown that AD153 and AD157 are larger than previously thought and appear closer to 170,000 and 190,000 Da, respectively (74). Induced pPD1-containing cells exhibited a similar pattern, giving rise to surface antigens designated PD78, PD130, PD153, and PD157 (23, 42). The latter three proteins migrated at rates essentially identical to the corresponding pAD1 proteins and strongly cross-reacted immunologically. A barely detectable band migrating at M_r about 140,000 (AD140) was also seen in preparations from induced donors and appeared elevated in certain clumping-defective mutants (22). In the case of pAD1, the dominant protein was AD74, whereas in the case of pPD1, the dominant protein was PD78. Several proteins were also identified in the pCF10 system, with a dominant protein being designated SA73 (68). It is noteworthy that different banding patterns have appeared in Western blots where different extraction procedures (e.g., using lysozyme and/or heat) have been used (28, 55, K. Tanimoto and D. Clewell, unpublished), and some caution should therefore be exercised when comparing data generated in laboratories using different procedures.

Wirth and colleagues isolated AD74 and raised polyclonal antibodies that enabled detection of surface "hairs" (the microfibrillar substance noted above) by immunogold labeling and electron microscopy (27, 70). Interestingly, the hairs were not evenly distributed on the donor surface during relatively short induction times (<40 minutes) and seemed to appear on "old" regions of the cell envelope (i.e., regions not close to where septum formation was occurring and cell wall was actively being synthesized). The purpose of this type of distribution on the surface is not clear, but it may be useful in allowing dissociation of donors and recipients after mating. That is, the absence of AS on newly synthesized wall would mean that cell division within a mating aggregate could lead to cells that could more easily dissociate after plasmid transfer.

The pAD1 structural gene for AS has been identified and sequenced (28). The determinant, designated *asa1* (stands for *a*ggregation *s*ubstance of p*AD1*), was found to encode a protein of 1296 amino acids, with a signal peptide consisting of 43 amino acid residues. The mature peptide would represent a protein with a molecular weight of 137,429. In its carboxyl-terminal region is a segment corresponding to an "anchor," as defined by similar regions found in other surface proteins of gram-positive bacteria (e.g., M protein of *Streptococcus pyogenes* [31]). A proline-rich segment, probably associated with cell wall (28), is close by. The protein also exhibited the amino acid motifs Arg-Gly-Asp-Ser and Arg-Gly-Asp-Val (one of each), which in other bacterial systems are believed to play a role in adherence to the integrin family of proteins on the surface of eukaryotic cells. Because the protein used to raise the antiserum against the *asa1* product was the lower molecular weight AD74, the latter was presumed to be a degradation product. Indeed, it represented the amino-terminal half of the protein. Under different extraction conditions, a larger protein (137 kDa) could be resolved in a Western blot analysis (28, 55). The carboxyl terminus of AD74 was recently shown to be a lysine at position 510 (54).

The AS determined by pCF10 appears to be very similar to that of pAD1, based on sequence comparisons (40, 45), and hybridization studies using a segment of pAD1 DNA containing part of the *asa1* sequence showed that a number of other pheromone plasmids tested, with the exception of pAM373, exhibited significant homology (26). A recent report by Nakayama et al (53) described a sequence on pPD1 relating to the PD78 surface protein (M_r of 78,000); however, a smaller protein of about 54 kDa, and exhibiting some homology with the TraD protein (thought to be involved in DNA transport) of the gram-negative plasmid R100, was deduced from the sequence. The sequence also revealed X-X-Pro repeated 15 times, similar to the sequences contained in the carboxyl terminus of TraD. In this case, though, the X-X-Pro sequences were located in the middle of the deduced polypeptide. It was suggested that the slower migration in SDS polyacrylamide gel (i.e., as 78 kDa rather than 54 kDa) could be due to unusually high levels of glutamic acid (13.8%) and lysine (11.5%). The fact that this pPD1 determinant does not exhibit homology with *asa1* of pAD1, while another region of pPD1 apparently does exhibit homology (26), implies that Asa1 and PD78 have different functions.

6. Entry Exclusion

Uninduced donor cells act as good recipients. Evidence for a pheromone-inducible entry exclusion function in the case of pAD1 was noted on the basis of a reduced transfer frequency between donor cells induced to clump by a culture-filtrate of recipients (7). Studies with pCF10 also showed that induced donors were 10- to 100-fold reduced in "recipient potential" when mated with homologous donor cells (21). The reduction was plasmid specific, as induction did not affect the ability to take up a nonhomologous pAD1 derivative. An inducible surface protein identifiable as a 130-kDa protein by Western blot analyses was reported to be associated with surface exclusion. A monoclonal antibody preparation specific for this protein reduced exclusion in mating mixtures and allowed for a 4- to 10-fold enhancement of transfer frequencies (21). A pheromone-induced surface exclusion would seem important to minimize exchange of plasmid DNA between homologous donors that are in close proximity to one another during a mating response to a plasmid-free (pheromone-producing) recipient.

7. Genetic Structure of pAD1

7.1. General

Genetic analyses of the pheromone response have focused primarily on the plasmids pAD1 and pCF10. The transposon Tn917 (encodes erythromycin resistance; a member of the Tn3 family) has been extensively employed in these endeavors. Tn917 was originally identified on a nonconjugative plasmid (pAD2) in *E. faecalis* DS16—the same clinical isolate in which pAD1 was first revealed (66). Interestingly, it exhibits a 20-fold preference for pAD1 over the bacterial chromosome as a transposition target (71), making it relatively easy to generate pAD1::Tn*917* insertions. Youngman (78) and colleagues have constructed transposon derivatives including constructs having a promoterless *lacZ* gene (Tn*917lac*) able to report transcriptional read-in from outside the transposon and carried by temperature-sensitive plasmid delivery vectors. These have been extremely useful in analyses of the pheromone-responding conjugation systems.

In the case of pAD1, transposon insertions over a segment representing more than half of the 60-kb plasmid have been found to affect the mating response (15, 22). Various functions have been identified as depicted in the map shown in Figure 14.2. Insertions in the regions designated E, F, G, and H greatly reduced or eliminated the ability of pAD1 to transfer in broth. Insertions in region H did not prevent inducible aggregation, but transfer was not detectable in broth and was less than 0.7% of the wild-type in filter matings. It is likely that the H region contains genes specifically devoted to DNA transfer. Aggregation was also not affected by insertions in the adjacent G region; however, plasmid transfer occurred at about 5% of wild-type levels in broth. Conceivably, this region is involved in the stabilization of aggregates and/or the formation of the conjugation bridge. The four inducible surface proteins AD74, AD130, AD153, and AD157 (see above) were all induced normally in the case of mutations in the G and H region. Region F insertions did not undergo induced aggregation, and/or were altered in the appearance of one or more of the above inducible surface proteins. The AD74 protein was not detectable in most cases. The AS determinant *asa1* (see above) is included in this region. F region mutants exhibited a reduced transfer (about 4% of wild-type) in broth, but wild-type transfer on filter membranes, indicating that the ability to transfer DNA was not affected. It will be interesting to correlate the protein patterns of the specific insertion mutations with their precise locations on the map in the context of the accumulating nucleotide sequence information in this region.

Tn*917lac* insertions in regions F, G, and H have revealed pheromone-inducible LacZ

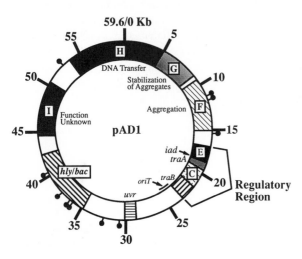

FIGURE 14.2. Map of pAD1. Regions associated with various functions are indicated as shaded areas. The hemolysin/bacteriocin determinant is indicated as *hly/bac*, and *uvr* indicates a region that confers increased resistance to UV. *traA*, *traB*, and regions C and E are related to regulation of the mating process. This region also contains the determinant *iad* for the pheromone inhibitor peptide iAD1. Regions F, G, H, and I include various structural genes that are induced as a result of exposure to pheromone. (The determinants and their functions are discussed in the text.) The markers on the outer circle indicate *Eco*RI restriction sites.

expression in the case of all inserts oriented where *lacZ* could be transcribed in a counterclockwise direction (15, 58) (Figure 14.2). Although it is conceivable that the genes in all three regions are on a single large operon, secondary promoters would still be likely; insertional mutations are not necessarily polar. (For example, insertion in the F region blocked aggregation but allowed plasmid DNA transfer if matings were performed on filters). Interestingly, Tn*917lac* insertions counterclockwise to the H region and defining a region designated I (Figure 14.2) exhibit induction of LacZ (again counterclockwise) on exposure to pheromone (58). A phenotype has not been ascribed to this region yet, as aggregation and transfer in most cases appear reasonably normal.

7.2. The Regulatory Region

The remaining regions involved in plasmid transfer are located within a 7-kb segment (from about 17 to 24 kb on the map) and are involved in regulation of the pheromone response (22, 32, 72) (Figures 14.2 and 14.3). Region E is believed to include one or more positive regulator genes, since some insertions here knock out the ability to aggregate, produce surface proteins, and transfer plasmid DNA. As noted in Figure 14.3 an open reading frame, designated *traE1*, has been recently identified in this region (60). Tn*917lac* inserts here exhibit a pheromone-inducible expression of LacZ (counterclockwise). This induction was also observed using miniplasmids of E region pAD1::Tn*917lac* mutants (73, 74) consisting only of the *Eco*RI fragment B (about 15.5 to 31.0 kb on the pAD1 map; see Figure 14.2) containing the insert. Thus, all the information necessary for the detection and regulation of the pheromone response must be in this segment, presumably in the above-

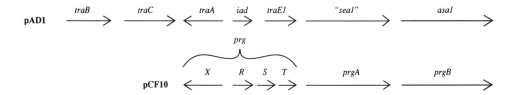

	Determinant	Protein M.W. (10^3)	Function
pAD1	*traA*	38	• neg. reg., sensing
	traB	44	• cAD1 shutdown
	traC	61	• sensing of cAD1
	traE1	14	• pos. reg.
	iad		• iAD1 precursor
	sea1	100	• surface exclusion?
	asa1	137	• aggr. substance
pCF10	*prgR, S, T, X*		• pos. reg. of *prgB*
	prgA (sec10)	98	• surface exclusion
	prgB (asc10)	137	• aggr. substance

FIGURE 14.3. Key reading frames/determinants relating to the pheromone responses of pAD1 and pCF10. The location, orientation, and relative size of the determinants are indicated by the arrows. Related information is also tabulated. Portions of the sequence information reflected here are based on work presently being prepared for publication, or recently submitted, by the laboratories of R. Wirth, G. Dunny, and D. Clewell.

noted 7-kb region. At least one product from the E region, presumably TraE1, is necessary for expression of regions F through I.

7.2.1. *traA*

The E region (*traE1*), in turn, is under negative control of the product of *traA*. Insertions in *traA* result in constitutive clumping and expression of surface proteins, and plasmid transfer occurs at high frequency in short (10-minute) matings (32). Mutants defective in TraA production have a characteristic colony morphology on solid media. These colonies, which are easily identifiable, were found to exhibit a "dry" appearance resembling that of wild-type cells plated on media containing pheromone. When a *traA* mutation was combined (on the same plasmid) with a *lac* fusion in the E region (NR5), LacZ was expressed constitutively (74), as would be expected if the E region were negatively controlled by TraA.

There is also evidence that TraA is involved in transduction of the pheromone signal. Western blot analyses revealed that the surface proteins produced by *traA* mutants are at levels only 10 to 50% (depending on the bacterial host) of that observed for induced wild-type cells (74). These levels were not increased by exposure to cAD1, indicating that their expression was insensitive to pheromone. A few insertions in *traA* resulted in much lower expression of surface proteins and even transferred at a frequency a few orders of magnitude lower than the other *traA* derivatives; however, they remained insensitive to cAD1. It is possible, therefore, that TraA interacts directly with, or even corresponds to, the pheromone receptor. The *traA* determinant has recently been sequenced (59) and found to encode a protein with an inferred molecular weight of 37,856 (Figure 14.3). It has limited homology with several DNA binding proteins.

7.2.2. *traB*

The determinant referred to as *traB* is involved in the shutdown of endogenous pheromone (74). Insertions resulted in a constitutive aggregation believed to be caused by a self-induction by endogenous cAD1; however, because pheromone was barely (if at all) detectable in filtrates of these mutants, one or more other factors must help the *traB* product in shutting down cAD1. Insertions in *traB* gave rise to a "ringed" colony morphology with a "dry" center but became uniformly dry when synthetic pheromone was added to the medium or when the plates were grown anaerobically (74, 75). As noted earlier, plasmid-free cells secreted significantly more cAD1 if grown anaerobically. It was deduced that the ringed appearance of the *traB* colonies was due to differences in the local environment within the colony—with the central part being more anaerobic and therefore producing enough cAD1 to result in self-induction. In addition, short (10-minute) matings using donors with inserts in *traB* gave rise to relatively high levels of pAD1 transfer but at a lower frequency than the case for *traA* mutants. When the cells (*traB* mutants) were grown anaerobically, transfer occurred at a higher frequency. Experiments in which a *traB* mutation was combined with a *lacZ* fusion in the E region gave results that were consistent with these observations (74).

Additional support for a *traB* role in cAD1 shutdown came from the comparison of mutants in two different (nonisogenic) bacterial hosts (74). In contrast to the case for *E*.

faecalis OG1X carrying *traB* mutants, colonies looked normal (i.e, not ringed) when the corresponding pAD1::Tn*917* derivative was harbored by an *E. faecalis* FA2-2 host; colonies were dry, however, if cAD1 was provided in the medium. A reasonable explanation for the difference stems from the fact that OG1X secretes about eightfold more cAD1 than FA2-2 and would therefore be more likely to cause a self-induction when harboring a pAD1 *traB* mutation. Unlike the case for *traB*, *traA* mutants appeared dry in both hosts. Sequence analysis has recently found *traB* to encode a 44-kDa protein (F. An and D. Clewell, manuscript in preparation) (Figure 14.3).

7.2.3. *traC*

The C region is believed to contribute to pheromone sensing, as insertions in this region resulted in a decreased sensitivity to cAD1 (22, 74). The mutants were characterized by a four- to eightfold increase in iAD1 detectable in culture filtrates. The increase in iAD1 appears to be related to a decrease in binding affinity, which affected the ability of both cAD1 and iAD1 to bind. The C-region-related function exhibited some host specificity, as the differences in iAD1 levels could be easily observed in OG1X but not in FA2-2. Interestingly, FA2-2 carrying wild-type pAD1 exhibited levels of iAD1 that were four- to eightfold higher than those for OG1X(pAD1); these levels, however, were similar to those found in the case of C-region mutants in OG1X. When a C-region mutation was combined on the same plasmid with a *lacZ* fusion in the E region, the production of LacZ was relatively insensitive to cAD1 in the OG1X host but remained sensitive in FA2-2 (74). Thus, the C-region product (TraC, see below) may not play as significant a role in the FA2-2 host as it apparently does in OG1X. This raises the interesting possibility that FA2-2 may have a chromosome-borne receptor determinant, while OG1X may not. Thus the plasmid-encoded receptor (C-region product?) would be important for a cAD1 response in OG1X but might not be necessary in FA2-2. That is, the TraA protein (see above) might interact with either receptor. The possibility that a receptor could be determined by the bacterial chromosome has been considered previously (74); it could serve to regulate the level of extracellular pheromone via a feedback mechanism. The absence of a chromosome-determined receptor in OG1X might also explain the eightfold elevated levels of cAD1 produced by this strain (i.e., perhaps the peptide cannot feedback-repress its synthesis). The C region has recently been sequenced (K. Tanimoto, personal communication) and has been found to encode a 61-kDa product containing a typical signal sequence at its amino terminus. The protein exhibits strong homology with known oligopeptide binding proteins in *E. coli* and *S. typhimurium* (30, 41). The determinant has been designated *traC* (Figure 14.3).

7.3. *oriT*

The precise location of the transfer origin (*oriT*) has not yet been determined; however, recent experiments (F. An and D. Clewell, unpublished) support the view that *oriT* is located close to *traB*, as indicated on Figure 14.2. A chimera consisting of the shuttle vector pAM401 carrying a 3.1-kb *Hin*dIII fragment beginning about a third of the way into the 5′ end of *traB* and extending back (clockwise; Figure 14.2) was efficiently mobilized by pAM7607 (a constitutively conjugative derivative of pAD1) from the Rec⁻ *E. faecalis* strain UV202. pAM401 clones containing other segments of pAD1 failed to be mobilized.

8. The iAD1 Determinant

Nucleotide sequence analyses (14) have revealed the determinant for iAD1 (designated *iad*) between the E region and *traA*. The determinant, depicted in Figure 14.4, shows a short open reading frame corresponding to what appears to encode a 22 amino acid residue precursor with the last eight residues representing iAD1. The first of two methionines was chosen as the likely translational initiation site because of the optimally located Shine-Dalgarno (SD) site. Two potential promoters (overlapping) were also observed with spacings between their −10 and −35 hexamers of 18 and 20 nucleotides. The promoter region appears functional in an *E. coli* host, as expression of iAD1 is detectable in culture filtrates of appropriate clones in that host but not in clones where the promoter was removed (14). Three sets of inverted repeats were observed downstream of *iad* and have been designated A, B, and C (Figure 14.4). The farthest (over 300 nucleotides away) downstream (C) exhibits structural properties of a potential transcription terminator in that it corresponds to 17-bp inverted repeats ($\Delta G = -34.4$ kcal) followed by 5 T's. Possible roles for the other repeats remain unknown.

There is a possible SD sequence preceding a short open reading frame (15 amino acid residues) within the putative transcription terminator. If this is indeed real, its level of translation might have an influence on transcription termination and affect readthrough into *traE1*. It is noteworthy that iAD1 activity in culture filtrates is similar, although not identical, for uninduced wild-type and *traA* mutants; thus iAD1 expression appears to be constitutive. The behavior of the Tn*917lac* fusion indicated as AA12 (Figure 14.4) is consistent with this notion, as LacZ expression is high in the absence of pheromone (15). In contrast, the fusion designated AA43 immediately downstream of the transcription terminator exhibits expression (high level) only in the presence of cAD1, as do fusions farther downstream in *traE1* (60). Because these are negatively regulated by TraA, expression of TraE1 depends on readthrough of the transcription terminator of *iad*. Thus, TraA may exercise its negative control by binding to DNA at or near the transcription terminator and enhancing termination.

It is interesting that iAD1 appears to be part of its own signal sequence (14). Four of the first eight residues in the amino terminus are positively charged, while the remaining residues, including the iAD1 moiety, are hydrophobic. An α-helical wheel projection (not shown) has indicated hydrophilic and hydrophobic sides. Although N-terminal amphipathic signal peptides are known to occur in prokaryotes, their significance is not yet understood; in eukaryotes, such structures appear to be important for targeting to mitochondria (61). The processing that results in the mature iAD1 may involve a somewhat "general" signal peptidase, since this event seems to occur normally in an *E. coli* host (14). Analysis of the iAD1 activity in filtrates of *E. coli* harboring an *iad*-containing plasmid chimera showed that the retention time on reverse phase HPLC chromatography was essentially identical to that of authentic iAD1.

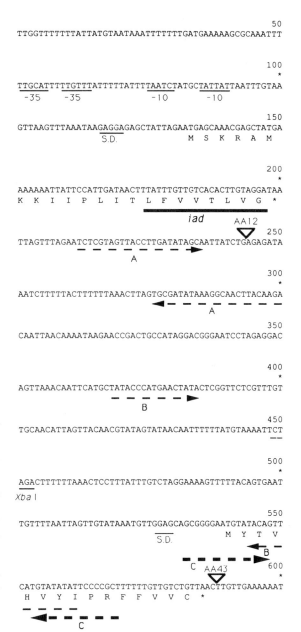

                                                                                50
                                    TTGGTTTTTTTATTATGTAATAAATTTTTTTGATGAAAAAGCGCAAATTT

                                                                                100
                                                                                 *
                                    TTGCATTTTTGTTTATTTTTATTTTAATCTATGCTATTATTAATTTGTAA
                                      -35   -35   -10   -10

                                                                                150
                                    GTTAAGTTTAAATAAGAGGAGAGCTATTAGAATGAGCAAACGAGCTATGA
                                                   S.D.   M S K R A M

                                                                                200
                                                                                 *
                                    AAAAAATTATTCCATTGATAACTTTATTTGTTGTCACACTTGTAGGATAA
                                    K K I I P L I T L F V V T L V G *
                                                                    _____
                                                                    iad   AA12
                                                                            ▽   250
                                    TTAGTTTAGAATCTCGTAGTTACCTTGATATAGCAATTATCTGAGAGATA
                                    - - - - - - - - - - - - - - - - ►
                                                       A

                                                                                300
                                                                                 *
                                    AATCTTTTTACTTTTTTAAACTTAGTGCGATATAAAGGCAACTTACAAGA
                                                          ◄ - - - - - - - -
                                                                    A

                                                                                350
                                    CAATTAACAAAATAAGAACCGACTGCCATAGGACGGGAATCCTAGAGGAC

                                                                                400
                                                                                 *
                                    AGTTAAACAATTCATGCTATACCCATGAACTATACTCGGTTCTCGTTTGT
                                                      - - - - - ►
                                                           B
                                                                                450
                                    TGCAACATTAGTTACAACGTATAGTATAACAATTTTTTATGTAAAATTCT
                                                                                ─ ─

                                                                                500
                                                                                 *
                                    AGACTTTTTTAAACTCCTTTATTTGTCTAGGAAAAGTTTTTACAGTGAAT
                                    Xba I

                                                                                550
                                    TGTTTTAATTAGTTGTATAAATGTTGGAGCAGCGGGGAATGTATACAGTT
                                                             S.D.   M Y T V
                                                                                 ◄ -
                                                                                 B
                                                                        - - - - - ►
                                                                    C AA43   600
                                                                           ▽   *
                                    CATGTATATATTCCCCGCTTTTTTGTTGTCTGTTAACTTGTTGAAAAAAT
                                    H V Y I P R F F V V C *
                                    ◄ - - - - -
                                                      C

FIGURE 14.4. Nucleotide sequence of region including the *iad* determinant. Two open reading frames based on the position of Shine-Dalgarno (S.D.) sequences are indicated with the inferred amino acid residues. The first corresponds to *iad* and is preceded by two possible promoters (−10 and −35 hexamers), whereas the second reflects a short reading frame beginning within a putative transcription termination site (indicated by the dashed arrows designated C) downstream from *iad*. Other arrows, designated A and B, show inverted repeats of unknown function. Two Tn917lac insertions are designated as AA12 and AA43. The AA12 fusion expressed β-galactosidase constitutively, whereas AA43 expressed it only in the presence of cAD1. (From reference 14.)

9. Phase Variation

As previously noted, colonies of *E. faecalis* harboring pAD1 have a characteristic dry colony morphology if cAD1 is provided in the medium. If pheromone is not present, the colonies are smooth; however, dry colonies appear spontaneously at a frequency of 10^{-4} to 10^{-2} (15, 58). Subcultures of these variants continue to give rise to dry colonies, but reversion to normal occurs—again at a frequency of 10^{-4} to 10^{-2}. Cells from the dry colonies exhibit clumping in liquid cultures and donate plasmid at a high frequency in short (10-minute) broth matings. They behave somewhat like *traA* mutants.

The phase variation affects expression in the regions E through I, as colonies representing corresponding transcriptional fusions oriented counterclockwise in these regions appear blue at similar frequencies on X-gal plates devoid of cAD1 (58). The phenomenon was genetically linked to the plasmid, and a pAD1 derivative with a transcriptional fusion in the E region exhibited the variation even when all but the *Eco*RI B fragment (containing the Tn917lac insertion) was missing. The specific nature of the reversible change is not yet known, but it probably involves expression of *traA* or its ability to regulate *traE1*; however, sequence analysis of phase variants has not detected structural differences in *traA* (L. Pontius and D. Heath, personal communication). The ability to undergo such a phase variation should allow for transfer of plasmid DNA to bacterial strains/species that do not secrete cAD1.

10. Genetic Analyses of pCF10

pCF10 has been analyzed making use of Tn917 as well as by cloning fragments on an *E. coli-E. faecalis* shuttle plasmid and introducing Tn5 insertions while in an *E. coli* background (2, 3, 17, 40, 57). Conjugation-related determinants appear confined to a 25-kb contiguous segment. Some emphasis has been placed on a 7-kb segment that incudes all or part of two surface protein determinants. The proteins are designated Sec10 (Tra130) and Asc10 (Tra150) and are believed to represent surface exclusion protein and aggregation substance, respectively. The corresponding determinants are respectively designated *prgA* (*sec10*) and *prgB* (*asc10*) (*prg* stands for *pheromone responsive gene*), which map close together (Figure 14.3). Significant homologies with equivalent determinants on pAD1 have recently been identified (40, 57, G. Dunny and R. Wirth, personal communication). On the *prgA* proximal side of these determinants are four genes designated *prgX*, *prgR*, *prgS*, and *prgT*. There is evidence that all four are involved in positive regulation of *prgB*. As seen in Figure 14.3, it would appear that *prgX-prgR-prgS-prgA-prgB* may correspond roughly to the segment on pAD1 bounded by the E region and *asa1*. It will be interesting to compare the genetic organization of the regulatory regions of pAD1 and pCF10 as work progresses.

11. Concluding Remarks

A given plasmid-free strain of *E. faecalis* secretes at least five, and probably many more, different pheromones, and the corresponding plasmids are generally in different incompatibility groups. Those plasmids that respond to cAD1 exhibit extensive homology and appear closely related (8, 33, 36, 46). Most of those identified are hemolysin/bacteriocin plasmids similar to pAD1 and represent incompatibility group IncHly (16). An exception is pBEM10 (52), which does not encode hemolysin/bacteriocin (it encodes a β-lactamase and resistance to gentamicin) but is IncHly (F. An and D. Clewell, unpublished). Hemolysin/bacteriocin is expressed by a significant percentage of *E. faecalis* clinical isolates and is particularly common among isolates associated with human parenteral infections. One study found that over 50% of hemolytic strains carried transferable hemolysin determinants (36). Considering the reported evidence that the hemolysin/bacteriocin determinant of pAD1 reduced the LD50 in a mouse model (35), there would appear to be a role in virulence. The fact that pAD1 aggregation substance (Asa1) contains motifs that are known in other systems to be involved in binding to receptors (integrins) on eukaryotic cells (28) brings to mind the possibility that this protein could play a role in pathogenesis. Recent observations that cells expressing Asa1 exhibit adherence to cultured renal tubular cells (44) makes this a reasonable consideration. It also raises the question of whether the observed phase variation could be involved (in the absence of pheromone) in a manner similar to that for type I pili, which contribute to the pathogenesis of certain *E. coli* strains (24).

The highly evolved transfer system encoded by pAD1 has enabled its wide dissemination. The interesting connection between the associated hemolytic and bacteriolytic activities, which are exhibited by the same protein product, is somewhat unique in its own right (1, 37). The bacteriocin activity has a broad host range and is self-selective in relation to conjugal transfer. That is, potential recipients are in danger of being killed by nearby donors if they do not take up the plasmid that can protect them with its related immunity function.

The production of multiple pheromones by plasmid-free strains of *E. faecalis* raises the interesting question as to how these peptides evolved as key participants in mating phenomena. It has been speculated (4, 5) that these compounds might have functions independent of conjugation and that certain plasmids evolved to take advantage of their potential use as intercellular signals. There have been reports recently that certain pheromones exhibit potent neutrophil chemotaxis activity (25, 62), thus raising the possibility that some of these substances can influence the course of a related bacterial infection by enhancing or modifying the host response. Conceivably, some pheromone/inhibitor peptides could competitively inhibit the activity of other (unrelated) chemotactic factors at neutrophil receptor sites and might thereby contribute to virulence. (In the case of the pAD1-related peptides, it was found that iAD1 exhibited chemotactic activity with human neutrophils, whereas cAD1 did not.) It is interesting that essentially all strains of *Staphylococcus aureus*, an important human pathogen, produce an activity similar to the *E. faecalis* pheromone cAM373, whereas, strains of coagulase negative staphylococci (less pathogenic) did not (11). Although the enterococcal peptide exhibited some neutrophil chemotaxis, the related staphylococcal activity, which has recently been found to differ in structure only at the C-terminal residue (A. Suzuki, personal communication), has not yet been tested. There is no evidence that the *S. aureus* activity is a pheromone in that species. In fact the enterococcal plasmid pAM373 (which responds to cAM373) does not appear able to establish in *S. aureus* (11).

It is clear more work is necessary to gain an understanding of the mechanism and regulation of pheromone-induced plasmid transfer in *E. faecalis*. It is surprising that in the bacterial world the involvement of pheromones in the interbacterial transfer of plasmid

DNA is something still known only in enterococci—even though well over a decade has passed since discovery of the phenomenon. The closest thing to sex pheromones are probably the competence factors associated with transformation phenomena in species such as *Streptococcus pneumoniae* and *Streptococcus sanguis* (45, 64). There are examples, however, of certain tumor-inducing (Ti) plasmids in *Agrobacterium* that transfer T-DNA to plants in response to specific plant-produced phenolic compounds (see Chapter 9). Tumors resulting from the acquired T-DNA then secrete opines able to induce the transfer of the conjugative Ti plasmid from donor to recipient (plasmid-free) bacteria (see Chapter 10). It will be interesting to compare the various inducible systems as more is learned about their regulation.

ACKNOWLEDGMENTS. Work conducted in my laboratory relating to material discussed in this review was supported by Public Health Service Grant GM33956. I thank R. Wirth, G. Dunny, A. Suzuki, D. Galli, and J. Nakayama for providing unpublished information helping to make this review as up to date as possible.

References

1. Brock, T.D., and Davie, J.M., 1963, Probable identity of group D hemolysin with bacteriocin, *J. Bacteriol.* 86:708–712.
2. Christie, P.J., and Dunny, G.M., 1986, Identification of regions of the *Streptococcus faecalis* plasmid pCF-10 that encode antibiotic resistance and pheromone response functions, *Plasmid* 15:230–241.
3. Christie, P.J., Kao, M., Adsit, J.C., and Dunny, G.M., 1988, Cloning and expression of genes encoding pheromone-inducible antigens of *Enterococcus (Streptococcus) faecalis*, *J. Bacteriol.* 170:5156–5168.
4. Clewell, D.B., 1981, Plasmids, drug resistance, and gene transfer in the genus *Streptococcus*, *Microbiol. Rev.* 45:409–436.
5. Clewell, D.B., 1985, Sex pheromones, plasmids, and conjugation in *Streptococcus faecalis*, in: *The Origin and Evolution of Sex* (H.O. Halvorson and A. Monroy, eds.), MBL lecture Series in Biology, Vol. 7, Alan R. Liss, New York, pp. 13–28.
6. Clewell, D.B., 1990, Movable genetic elements and antibiotic resistance in enterococci, *Europ. J. Clinic. Microbiol. Infect. Dis.* 9:90–102.
7. Clewell, D.B., and Brown, B.L., 1980, Sex pheromone cAD1 in *Streptococcus faecalis*: induction of a function related to plasmid transfer, *J. Bacteriol.* 143:1063–1065.
8. Clewell, D.B., and Weaver, K.E., 1989, Sex pheromones and plasmid transfer in *Enterococcus faecalis* (a review), *Plasmid* 21:175–184.
9. Clewell, D.B., Tomich, P.K., Gawron-Burke, M.C., Franke, A.E., Yagi, Y., and An, F.Y., 1982, Mapping of *Streptococcus faecalis* plasmids pAD1 and pAD2 and studies relating to transposition of Tn*917*, *J. Bacteriol.* 152:1220–1230.
10. Clewell, D.B., Yagi, Y., Ike, Y., Craig, R.A., Brown, B.L., and An, F., 1982, Sex pheromones in *Streptococcus faecalis*: multiple pheromone systems in strain DS5, similarities of pAD1 and pAMγ1, and mutants of pAD1 altered in conjugative properties, in: *Microbiology–1982* (D. Schlessinger, ed.), American Society for Microbiology, Washington, DC, pp. 97–100.
11. Clewell, D.B., An, F.Y., White, B.A., and Gawron-Burke, C., 1985, *Streptococcus faecalis* sex pheromone (cAM373) also produced by *Staphylococcus aureus* and identification of a conjugative transposon (Tn*918*), *J. Bacteriol.* 162:1212–1220.
12. Clewell, D.B., Ehrenfeld, E.E., Kessler, R.E., Ike, Y., Franke, A.E., Madion, M., Shaw, J.H., Wirth, R., An, F., Mori, M., Kitada, C., Fujino, M., and Suzuki, A., 1986, Sex-pheromone systems in *Streptococcus faecalis*, in: *Banbury Report 24: Antibiotic Resistance Genes: Ecology, Transfer and Expression*, Cold Spring Harbor, Cold Spring Harbor, New York, pp. 131–142.
13. Clewell, D.B., An, F.Y., Mori, M., Ike, Y., and Suzuki, A., 1987, *Streptococcus faecalis* sex pheromone

(cAD1) response: evidence that the peptide inhibitor excreted by pAD1-containing cells may be plasmid determined, *Plasmid* 17:65–68.

14. Clewell, D.B., Pontius, L.T., An, F.Y., Ike, Y., Suzuki, A., and Nakayama, J., 1990, Nucleotide sequence of the sex pheromone inhibitor (iAD1) determinant of *Enterococcus faecalis* conjugative plasmid pAD1, *Plasmid* 24:156–161.

15. Clewell, D.B., Pontius, L.T., Weaver, K.E., An, F.Y., Ike, Y., Suzuki, A., and Nakayama, J., 1991, *Enterococcus faecalis* hemolysin/bacteriocin plasmid pAD1: regulation of the pheromone response, in: *Genetics and Molecular Biology of Streptococci, Lactococci, and Enterococci* (G. Dunny, P. Cleary, and L. McKay, eds.), American Society for Microbiology, Washington, DC, pp. 3–8.

16. Colmar, I., and Horaud, T., 1987, *Enterococcus faecalis* hemolysin-bacteriocin plasmids belong to the same incompatibility group, *Appl. Environ. Microbiol.* 53:567–570.

17. Dunny, G.M., 1990, Genetic functions and cell-cell interactions in the pheromone-inducible plasmid transfer system of *Enterococcus faecalis*, *Molec. Microbiol.* 4:689–696.

18. Dunny, G.M., Brown, B.L., and Clewell, D.B., 1978, Induced cell aggregation and mating in *Streptococcus faecalis*: evidence for a bacterial sex pheromone, *Proc. Natl. Acad. Sci. USA* 75:3479–3483.

19. Dunny, G.M., Craig, R.A., Carron, R.L., and Clewell, D.B., 1979, Plasmid transfer in *Streptococcus faecalis*: production of multiple sex pheromones by recipients, *Plasmid* 2:454–465.

20. Dunny, G.M., Funk, C., and Adsit, J., 1981, Direct stimulation of the transfer of antibiotic resistance by sex pheromones in *Streptococcus faecalis*, *Plasmid* 6:270–278.

21. Dunny, G.M., Zimmerman, D.L., and Tortorello, M.L., 1985, Induction of surface exclusion (entry exclusion) by *Streptococcus faecalis* sex pheromones: use of monoclonal antibodies to identify an inducible surface antigen involved in the exclusion process, *Proc. Natl. Acad. Sci. USA* 82:8582–8586.

22. Ehrenfeld, E.E., and Clewell, D.B., 1987, Transfer functions of the *Streptococcus faecalis* plasmid pAD1: organization of plasmid DNA encoding response to sex pheromone, *J. Bacteriol.* 169:3473–3481.

23. Ehrenfeld, E.E., Kessler, R.E., and Clewell, D.B., 1986, Identification of pheromone-induced surface proteins in *Streptococcus faecalis* and evidence for a role for lipoteichoic acid in formation of mating aggregates, *J. Bacteriol.* 168:6–12.

24. Eisenstein, B.I., 1981, Phase variation of type I fimbriae in *Escherichia coli* is under transcriptional control, *Science* 214:337–339.

25. Ember, J.A., and Hugli, T.E., 1989, Characterization of the human neutrophil response to sex pheromones from *Streptococcus faecalis*, *Am. J. Pathol.* 134:797–805.

26. Galli, D., and Wirth, R., 1991, Comparative analysis of *Enterococcus faecalis* sex pheromone plasmids identifies a single homologous DNA region which codes for aggregation substance, *J. Bacteriol.* 173:3029–3033.

27. Galli, D., Wirth, R., and Wanner, G., 1989, Identification of aggregation substances of *Enterococcus faecalis* cells after induction by sex pheromones. An immunological and ultrastructural investigation, *Arch. Microbiol.* 151:486–490.

28. Galli, D., Lottspeich, F., and Wirth, R., 1990, Sequence analysis of *Enterococcus faecalis* aggregation substance encoded by the sex pheromone plasmid pAD1, *Molec. Microbiol.* 4:895–904.

29. Handwerger, S., Pucci, M.J., and Kolokathis, A., 1990, Vancomycin resistance is encoded on a pheromone response plasmid in *Enterococcus faecium* 228, *Antimicrob. Ag. Chemother.* 34:358–360.

30. Hiles, I.D., Higgins, C.F., 1986, Peptide uptake by *Salmonella typhimurium*. The periplasmic oligopeptide-binding protein, *Europ. J. Biochem.* 158:561–567.

31. Hollingshead, S.K., Fischetti, V.A., and Scott, J.R., 1986, Complete nucleotide sequence of type 6M protein of the group A *Streptococcus*—repetitive structure and membrane anchor, *J. Biol. Chem.* 261:1677–1686.

32. Ike, Y., and Clewell, D.B., 1984, Genetic analysis of the pAD1 pheromone response in *Streptococcus faecalis*, using transposon Tn917 as an insertional mutagen. *J. Bacteriol.* 158:777–783.

33. Ike, Y., and Clewell, D.B., 1987, High incidence of hemolysin production by *Streptococcus faecalis* strains associated with human parenteral infections: structure of hemolysin plasmids, in: *Streptococcal Genetics* (J. Ferretti and R. Curtiss III, eds.), American Society for Microbiology, Washington, DC, pp. 159–164.

34. Ike, Y., Craig, R.C., White, B.A., Yagi, Y., and Clewell, D.B., 1983, Modification of *Streptococcus faecalis* sex pheromones after acquisition of plasmid DNA, *Proc. Natl. Acad. Sci. USA* 80:5369–5373.

35. Ike, Y., Hashimoto, H., and Clewell, D.B., 1984, Hemolysin of *Streptococcus faecalis* subsp. *zymogenes* contributes to virulence in mice, *Infect. Immun.* 45:528–530.

36. Ike, Y., Hashimoto, H., and Clewell, D.B., 1987, High incidence of hemolysin production by *Enterococcus (Streptococcus) faecalis* strains associated with human parenteral infections, *J. Clin. Microbiol.* 25:1524–1528.

37. Ike, Y., Clewell, D.B., Segarra, R.A., and Gilmore, M.S., 1990, Genetic analysis of the pAD1 hemolysin/bacteriocin determinant in *Enterococcus faecalis*: Tn*917* insertional mutagenesis and cloning, *J. Bacteriol.* 172:155–163.

38. Jacob, A., and Hobbs, S.J., 1974, Conjugal transfer of plasmid-borne multiple antibiotic resistance in *Streptococcus faecalis* var. *zymogenes*, *J. Bacteriol.* 117:360–372.

39. Jacob, A., Douglas, G.I., and Hobbs, S.J., 1975, Self-transferable plasmids determining the hemolysin and bacteriocin of *Streptococcus faecalis* var. *zymogenes*, *J. Bacteriol.* 121:863–872.

40. Kao, S., Olmsted, S.B., Viksnins, A.S., Gallo, J.C., and Dunny, G.M., 1991, Molecular and genetic analysis of a region of the plasmid pCF10 containing positive control genes and structural genes encoding surface proteins involved in pheromone-inducible conjugation in *Enterococcus faecalis*, *J. Bacteriol.* 173:7650–7664.

41. Kashiwagi, K., Yamaguchi, Y., Sakai, Y., Kobayashi, H., and Igarashi, K., 1990, Identification of the polyamine-induced protein as a periplasmic oligopeptide binding protein, *J. Biol. Chem.* 265:8387–8391.

42. Kessler, R., and Yagi, Y., 1983, Identification and partial characterization of a pheromone-induced adhesive surface antigen of *Streptococcus faecalis*, *J. Bacteriol.* 155:714–721.

43. Kitada, C., Fujino, M., Mori, M., Sakagami, Y., Isogai, A., Suzuki, A., Clewell, D., and Craig, R., 1985, Synthesis and structure-activity relationships of *Streptococcus faecalis* sex pheromones, cPD1 and cAD1, in: *Peptide Chemistry 1984* (N. Izumiya, ed.), Protein Research Foundation, Osaka, pp. 43–48.

44. Kreft, B., Marre, R., Schramm, U., and Wirth, R., 1992, Aggregation substance of *Enterococcus faecalis* mediates adhesion to cultured renal tubular cells, *Infect. Immun.* 60:25–30.

45. Lacks, S., 1977, Binding and entry of DNA in bacterial transformation, in: *Microbiol Interactions Series B. Receptors and Recognition*, Vol. 3 (J.L. Reissig, ed.), Chapman and Hall, London, pp. 177–232.

46. LeBlanc, D.J., Lee, L.N., Clewell, D.B., and Behnke, D., 1983, Broad geographical distribution of a cytotoxin gene mediating β-hemolysis and bacteriocin activity among *Streptococcus faecalis* strains, *Infect. Immun.* 40:1015–1022.

47. Mori, M., Sakagami, Y., Narita, M., Isogai, A., Fujino, M., Kitada, C., Craig, R., Clewell, D., and Suzuki, A., 1984, Isolation and structure of the bacterial sex pheromone, cAD1, that induces plasmid transfer in *Streptococcus faecalis*, *FEBS Lett.* 178:97–100.

48. Mori, M., Isogai, A., Sakagami, Y., Fujino, M., Kitada, C., Clewell, D.B., and Suzuki, A., 1986, Isolation and structure of *Streptococcus faecalis* sex pheromone inhibitor, iAD1, that is excreted by donor strains harboring plasmid pAD1, *Agric. Biol. Chem.* 50:539–541.

49. Mori, M., Tanaka, H., Sakagami, Y., Isogai, A., Fujino, M., Kitada, C., White, B.A., An, F.Y., Clewell, D.B., and Suzuki, A., 1986, Isolation and structure of the *Streptococcus faecalis* sex pheromone, cAM373, *FEBS Lett.* 206:69–72.

50. Mori, M., Tanaka, H., Sakagami, Y., Isogai, A., Fujino, M., Kitada, C., Clewell, D.B., and Suzuki, A., 1987, Isolation and structure of the sex pheromone inhibitor, iPD1, excreted by *Streptococcus faecalis* donor strains harboring plasmid pPD1, *J. Bacteriol.* 169:1747–1749.

51. Mori, M., Sakagami, Y., Ishii, Y., Isogai, A., Kitada, C., Fujino, M., Kitada, C., Adsit, J.C., Dunny, G.M., and Suzuki, A., 1988, Structure of cCF10, a peptide sex pheromone which induces conjugative transfer of the *Streptococcus faecalis* tetracycline-resistance plasmid, pCF10, *J. Biol. Chem.* 263:14,574–14,578.

52. Murray, B.E., An, F., and Clewell, D.B., 1988, Plasmids and pheromone response of the β-lactamase producer *Streptococcus (Enterococcus) faecalis* HH22, *Antimicrob. Agents Chemother.* 32:547–551.

53. Nakayama, J., Nagasawa, H., Isogai, A., Clewell, D.B., and Suzuki, A., 1990, Amino acid sequence of pheromone-inducible surface protein in *Enterococcus faecalis* that is encoded on the conjugative plasmid pPD1, *FEBS Lett.* 268:245–248.

54. Nakayama, J., Watarai, H., Isogai, A., Clewell, D.B., and Suzuki, A., 1992, C-terminal identification of AD74, a proteolytic product of *Enterococcus faecalis* aggregation substance: application of liquid chromatography/mass spectrometry, *Biosci. Biotech. Biochem.* 56:127–131.

55. Nakayama, J,. Watarai, H., Nagasawa, H., Isogai, A., Clewell, D.B., and Suzuki, A., 1992, Immunological characterization of pheromone-induced proteins associated with sexual aggregation in *Enterococcus faecalis*, *Biosci. Biotech. Biochem.* 56:264–269.

56. Oliver, D.R., Brown, B.L., and Clewell, D.B., 1977, Characterization of plasmids determining hemolysin and bacteriocin production in *Streptococcus faecalis* 5952, *J. Bacteriol.* 130:948–950.

57. Olmsted, S.B., Kao, S., van Putte, L.J., Gallo, J.C., Dunny, G.M., 1991, Role of pheromone-inducible surface protein Ascl0 in mating aggregate formation and conjugal transfer of the *Enterococcus faecalis* plasmid pCF10, *J. Bacteriol.* 173:7665–7672.

58. Pontius, L.T., and Clewell, D.B., 1991, A phase variation event that activates conjugation functions encoded by the *Enterococcus faecalis* plasmid pAD1, *Plasmid* 26:172–185.

59. Pontius, L.T., and Clewell, D.B., 1992, Regulation of the pAD1-encoded sex pheromone response in *Enterococcus faecalis*: nucleotide sequence analysis of *traA*, *J. Bacteriol.* 174:1821–1827.

60. Pontius, L.T., and Clewell, D.B., 1992, Conjugative transfer of *Enterococcus faecalis* plasmid pAD1: nucleotide sequence and transcriptional fusion analysis of a region involved in positive regulation, *J. Bacteriol.* 174:3152–3160.

61. Saier, M.H., Jr., Werner, P.K., and Muller, M., 1989, Insertion of proteins into bacterial membranes: mechanism, characteristics, and comparisons with the eucaryotic process, *Microbiol. Rev.* 53:333–366.

62. Sannomiya, P., Craig, R.A., Clewell, D.B., Suzuki, A., Fujino, M., Till, G.O., and Marasco, W.A., 1990, Characterization of a new class of *Enterococcus faecalis* derived neutrophil chemotactic peptides: the sex pheromones, *Proc. Natl. Acad. Sci. USA* 87:66–70.

63. Suzuki, A., Mori, M., Sakagami, Y., Isogai, A., Fujino, M., Kitada, C., Craig, R.A., and Clewell, D.B., 1984, Isolation and structure of bacterial sex pheromone cPD1, *Science* 226;849–850.

64. Tomasz, A., 1969, Some aspects of the competent state in genetic transformation, *Annu. Rev. Genet.* 3:217–232.

65. Tomich, P.K., An, F.Y., Damle, S.P., and Clewell, D.B., 1979, Plasmid related transmissibility and multiple drug resistance in *Streptococcus faecalis* subs. *zymogenes* strain DS16, *Antimicrob. Agents Chemother.* 15: 828–830.

66. Tomich, P.K., An, F.Y., and ClewelL, D.B., 1980, Properties of erythromycin-inducible transposon Tn*917* in *Streptococcus faecalis*, *J. Bacteriol.* 141:1366–1374.

67. Tomura, T., Hirano, T., Ito, T., and Yoshioka, M., 1973, Transmission of bacteriocinogenicity by conjugation in group D streptococci, *Japan J. Microbiol.* 17:445–452.

68. Tortorello, M.L., and Dunny, G.M., 1985, Identification of multiple surface antigens associated with the sex pheromone response of *Streptococcus faecalis*, *J. Bacteriol.* 162:131–137.

69. Trotter, K.M., and Dunny, G.M., 1990, Mutants of *Enterococcus faecalis* deficient as recipients in mating with donors carrying pheromone-inducible plasmids, *Plasmid* 24:57–67.

70. Wanner, G., Formanek, H., Galli, D., and Wirth, R., 1989, Localization of aggregation substances of *Enterococcus faecalis* after induction by sex pheromones. An ultrastructural comparison using immunolabeling, transmission and high resolution scanning electron microscopic techniques, *Arch. Microbiol.* 151:491–497.

71. Weaver, K.E., and Clewell, D.B., 1987, Transposon Tn*917* delivery vectors for mutagenesis in *Streptococcus faecalis*, in: *Streptococcal Genetics* (J. Ferretti and R. Curtiss III, eds.), American Society for Microbiology, Washington, DC, pp. 17–21.

72. Weaver, K.E., and Clewell, D.B., 1988, Regulation of the pAD1 sex pheromone response in *Enterococcus faecalis*: construction and characterization of *lacZ* transcriptional fusions in a key control region of the plasmid, *J. Bacteriol.* 170:4343–4352.

73. Weaver, K.E., and Clewell, D.B., 1989, Construction of *Enterococcus faecalis* pAD1 miniplasmids: identification of a minimal pheromone response regulatory region and evaluation of a novel pheromone-dependent growth inhibition, *Plasmid* 22:106–119.

74. Weaver, K.E., and Clewell, D.B., 1990, Regulation of the pAD1 sex pheromone response in *Enterococcus faecalis*: effects of host strain and *traA*, *traB*, and C region mutants on expression of an E region pheromone-inducible lacZ fusion, *J. Bacteriol.* 172:2633–2641.

75. Weaver, K.E., and Clewell, D.B., 1991, Control of *Enterococcus faecalis* sex pheromone cAD1 elaboration: effect of culture aeration and pAD1 plasmid-encoded determinants, *Plasmid* 25:177–189.

76. Wicken, A.J., Elliott, S.D., and Baddiley, J., 1963, The identity of streptococcal group D antigen with teichoic acid, *J. Gen. Microbiol.* 31:231–239.

77. Yagi, Y., Kessler, R.E., Shaw, J.H., Lopatin, D.E., An, F.Y., and Clewell, D.B., 1983, Plasmid content of *Streptococcus faecalis* strain 39-5 and identification of a pheromone (cPD1)-induced surface antigen, *J. Gen. Microbiol.* 129:1207–1215.

78. Youngman, P.J., 1987, Plasmid vectors for recovering and exploiting Tn*917* transposition in *Bacillus* and other gram-positives, in: *Plasmids: A Practical Approach* (K. Hardy, ed.), IRL Press, Oxford, pp. 79–103.

Chapter 15

The Conjugative Transposons of Gram-Positive Bacteria

DON B. CLEWELL and SUSAN E. FLANNAGAN

1. Introduction

Conjugative transposons are characterized by their ability to move from one bacterial cell to another by a process requiring cell-to-cell contact. Evidence for such elements became apparent about 13 years ago from studies of *Enterococcus* (formerly *Streptococcus*) *faecalis* strain DS16, a clinical isolate obtained in 1975 from St. Joseph's Mercy Hospital in Ann Arbor, Michigan (38, 39). DS16 was of interest at that time because of its multiple antibiotic resistance and hemolytic properties. It was found to harbor a hemolysin/bacteriocin plasmid, pAD1, and a resistance plasmid, pAD2 (26, 112). pAD1 was conjugative and conferred a mating response to a peptide sex pheromone secreted by potential recipient (pAD1-free) cells, whereas pAD2 conferred resistance to erythromycin, streptomycin, and kanamycin and was nonconjugative. The erythromycin resistance determinant (*erm*) of pAD2 was associated with a transposon designated Tn917 (111). Derivatives of DS16 cured of both pAD1 and pAD2 maintained a resistance to tetracycline (Tc), indicating that a Tc-resistance determinant (*tet*) was located on the bacterial chromosome. A series of filter membrane mating experiments designed to examine transfer of the various resistance determinants of DS16 showed that *tet* could be mobilized at frequencies of 10^{-5} per donor (38). The majority of transconjugants (about 90%) harbored pAD1 and had *tet* on the chromosome. Among most of the remaining transconjugants, *tet* was linked with pAD1, and this correlated with insertion of a 16-kb segment of DNA. A surprising result arose

DON B. CLEWELL • Department of Biologic and Materials Sciences, School of Dentistry and Department of Microbiology and Immunology, School of Medicine, The University of Michigan, Ann Arbor, Michigan 48109-0402. SUSAN E. FLANNAGAN • Department of Biologic and Materials Sciences, School of Dentistry, The University of Michigan, Ann Arbor, Michigan 48109-0402.
Bacterial Conjugation, edited by Don B. Clewell. Plenum Press, New York, 1993.

when certain "control" experiments were performed using a plasmid-free (cured) derivative of DS16 as a donor. As reported by Franke and Clewell (38), *tet* was able to transfer from such strains at a frequency of about 10^{-8} per donor, and transconjugants could pass on the trait in subsequent matings. Intercellular transfer was DNase resistant and was not affected by the presence of a Rec$^-$ allele in either the donor or the recipient. In addition, donor filtrates did not transfer *tet* to recipients, nor did donor cells exposed to chloroform prior to mating. Because cell contact appeared necessary, the term "conjugative transposon," was adopted, and the element was designated Tn916. Additional studies showing interspecies transfer added strong support that the transposon encoded its own fertility functions.

During the latter part of the 1970s, when the emergence of multiple-drug resistant strains of *Streptococcus pneumoniae* was becoming a clinical concern, investigations in several laboratories were unable to identify related plasmids (10, 31, 97, 103, 105, 130). Shoemaker et al (104) and Buu-Hoi and Horodniceanu (10), however, reported that resistance determinants could transfer to recipient strains on membrane filters by a DNase-resistant process. Horodniceanu et al (56) reported similar observations in *Streptococcus pyogenes*, *Streptococcus agalactiae*, and streptococcal Lancefield groups F and G. Details of these early observations have been previously reviewed (25). Some of these systems were found to involve large segments (more than 60 kb) of chromosomal DNA containing two or more resistance determinants and eventually proved to be relatively complex conjugative transposons. However, one pneumococcal element, designated Tn1545, was found by Courvalin and associates (11, 27, 28) to be only about 25 kb; it carried *tet*, *erm*, and a determinant for resistance to kanamycin (*aphA-3*). Both Tn1545 and Tn916 have been studied in some detail and are considered to be members of a closely related family of transposons that includes Tn918 (22), Tn920 (79), Tn925 (20), and Tn3702 (53) from *E. faecalis*; Tn5031, Tn5032, and Tn5033 from *Enterococcus faecium* (37); Tn919 (35) from *Streptococcus sanguis*; and Tn5251 (2) from *Streptococcus pneumoniae*.

Table 15.1 lists conjugative transposons that have been identified to date. In most cases the elements include a *tet* determinant, which is generally of the *tet*(M) variety (9, 75, 110). In the discussion below, some emphasis will be given to investigations of Tn916 and Tn1545, as their relatively small sizes have facilitated their characterization. Thus far, the transposition properties of these two systems seem virtually identical.

Certain nonplasmid elements bearing both conjugative and transpositional properties have been identified in species of *Bacteroides*. These appear somewhat different from the type of transposons addressed below and are discussed in Chapter 13.

2. Properties of Tn916

2.1. Frequency of Transposition and Generation of Multiple Inserts

Under conditions where transfer of chromosomal mutational markers could not be detected, a plasmid-free derivative of *E. faecalis* DS16 was able to donate Tn916 at a low frequency (10^{-8} per donor). However, when Tn916-containing transconjugants were subsequently tested for donor potential, a wide range of values was observed (10^{-9} to 10^{-5} per donor) (41). While donor potential differed from strain to strain, each individual strain exhibited a characteristic value.

Southern blot hybridization studies showed that Tn916 was able to insert at many

TABLE 15.1 Conjugative Transposons of Gram-Positive Bacteria

Element	Source	Size	Determinant[a]	Reference
Tn916	*Enterococcus faecalis* DS16	16.4 kb	*tet*(M)	38, 41
Tn918	*Enterococcus faecalis* RC73	16 kb	*tet*(M)	22
Tn919	*Streptococcus sanguis* FC1	16 kb	*tet*(M)	35, 50
Tn920	*Enterococcus faecalis* HH22	23 kb	*tet*(M)	79
Tn925	*Enterococcus faecalis* SF-7 (plasmid pCF10)[b]	16 kb	*tet*(M)	19, 20
Tn1545	*Streptococcus pneumoniae* BM4200	25.3 kb	*tet*(M) *aphA-3* *ermAM*	11, 27
Tn3701 (Tn3703)[c]	*Streptococcus pyogenes* A454	>50 kb	*erm* *tet*(M)	69, 71
Tn3702	*Enterococcus faecalis* D434	18.5 kb	*tet*(M)	53
Tn3951	*Streptococcus agalactiae* B109	67 kb	*cat* *tet*(M) *erm*	58, 106
Tn5031 Tn5032 Tn5033	*Enterococcus faecium* CH1, CH3, and SF8	16.5 kb	*tet*(M)	37
Tn5253 (Tn5252)[c] (Tn5251)[c]	*Streptococcus pneumoniae* BM6001	65.5 kb	*cat* *tet*(M)	2, 117, 118
Tn5276	*Lactococcus lactis* NIZO R5	70 kb	*nisA* *sacA*	92
Tn5301	*Lactococcus lactis* NCFB894	70 kb	*nis* *sac*	55

[a]*aph*, kanamycin resistance; *cat*, chloramphenicol resistance; *erm*, erythromycin resistance; *nis*, nisin biosynthesis; *sac*, sucrose fermentation; *tet*, tetracyline resistance.
[b]Tn925 was originally detected on plasmid pCF10; all others were originally found in chromosomal locations.
[c]Contained within another transposon.

different sites in the recipient chromosome and often appeared in several sites simultaneously (41). Usually greater than 50% of the transconjugants contained more than one insert, and there could be as many as five or six (24, 41). There was no correlation between the number of Tn916 copies present and the transfer potential. A positive correlation was observed, however, between the frequency of conjugative transposition and transposition to a resident plasmid (subsequently introduced). For example, two strains whose conjugative transfer potential differed by a factor of 100 also exhibited a 100-fold difference in the ability to generate insertions in pAD1 (41). Excision of the element, most probably giving rise to a nonreplicative circular molecule, was suggested as the common basis for these two events (41; also see below). In the case of conjugative transposition, the excised DNA could be viewed as transferring to the recipient cell by a "plasmidlike" process followed by insertion into the target chromosome.

2.2. Zygotic Induction

When a conjugative plasmid carrying Tn916 (e.g., pAM81 or pAD1) transferred to recipients, a "zygotic induction" was observed in which the transposon excised from the plasmid and either inserted into the recipient chromosome or was segregated (41). Restriction enzyme analyses indicated that the excision regenerated a plasmid fragment indistinguishable from that present prior to the initial insertion. It was this observation that provided early support for the concept of transposition being an excision/insertion process involving a circular intermediate incapable of autonomous replication (see below). It is noted that the uptake of an intermediate of Tn916 from the donor is significantly different from the uptake of a conjugative replicon carrying the transposon. In the former case, the system is already activated and probably committed to events necessary to complete transposition, whereas in the latter the plasmid simply "carries" the transposon into the recipient.

2.3. Absence of Immunity or Entry Exclusion Functions

The presence of Tn916 in a recipient cell does not reduce the uptake of additional homologous elements (24, 86). When cells harboring a transposon derivative called Tn916ΔE (contains an *erm* determinant substituted for *tet* [98]) were mated with a strain harboring Tn916, transfer occurred in both directions and at frequencies similar to those using control recipients having no transposon (24). These observations implied the absence of immunity or entry exclusion functions. Southern blot analyses showed that in the majority of cases transconjugants had Tn916 and Tn916ΔE inserted at different locations.

A resident element in recipient cells also failed to reduce "zygotic induction" of a transposon being carried in on a conjugative plasmid (24). This implied the absence of a strong *trans*-acting inhibitor (repressor) being provided by the resident element in the recipient. It is interesting that when the same Tn916-carrying plasmid (pAM81::Tn916) that gave rise to zygotic induction after conjugal plasmid transfer was introduced into *E. faecalis* cells by protoplast transformation, zygotic induction was not observed. Thus, the mechanism by which the DNA was introduced into the cell appeared to influence the occurrence of this phenomenon. It was speculated that this difference might be related to DNA entering the cell as a single strand in the case of conjugation, whereas protoplast transformation involved uptake of duplex DNA. (In the few cases where it has been examined, plasmid DNA [gram-negative systems] entered recipients as a single strand during conjugation [125; see also Chapter 5 in this volume]). In the case of single-stranded entry, perhaps a transient state (e.g., hemimethylated site[s]) occurring after the complementary strand is replicated is an important activation factor. In this regard it is noted that the *dam* methylation system is known to play a significant role in the regulation of transposable elements such as IS10 (93) and IS50 (128).

2.4. Mobilization Properties

As already noted when present in plasmid-free cells, conjugative transfer of Tn916 occurred without detectable cotransfer of chromosomal mutational markers (38). In addition, mobilization of the nonconjugative plasmid pAD2 by chromosome-borne Tn916 was not detected (25). More recently it was shown (36) that if the nonconjugative shuttle vector pAM401 or the nonconjugative pVA749 was present in *E. faecalis* donor cells harboring Tn916 in the chromosome, about 0.6% of *E. faecalis* transconjugants acquiring the transposon also received plasmid DNA. In addition, Naglich and Andrews (81) reported that Tn916 was able to mobilize the nonconjugative plasmids pC194 and pUB110 from *Bacillus subtilis* to *Bacillus thuringiensis*. The transfer of plasmid DNA occurred at least two orders of magnitude less frequently than that of Tn916. Whereas both plasmids could be mobilized, coestablishment with Tn916 was detected only in the case of pC194.

When Tn916 was located on a nonconjugative plasmid, matings resulted in transfer of the transposon without detection of transfer of an outside marker on the plasmid (24, 36). This is consistent with the view that excision is a key step and that nontransposon DNA located *in cis* is rarely, if ever, mobilized. In contrast, when two conjugative transposons were present on the bacterial chromosome at different locations, the transfer of one element was accompanied by transfer of the other at a relatively high frequency. For example, when Tn916 and Tn916ΔE were located at different sites on the chromosome, selection for transfer of either element (i.e., with Tc or Em) resulted in significant cotransfer (e.g., 50%) of the other (24). A similar result was observed (36) when the two transposons were in the Rec⁻ host UV202 and on different replicons (i.e., a nonconjugative plasmid and the chromosome). It was proposed that movement of one element results in a *trans*-activation of other homologous elements in the same cell.

In experiments using Tn925, Torres et al (113) reported that mobilization of chromosomal mutational markers occurred at frequencies similar to the transfer frequency of the transposon, and Guffanti et al (43) reported a relatively high frequency of plasmid mobilization by the transposon. It was suggested (113) that Tn925 transfer involves a cell-fusion event where genomic DNA from both donor and recipient are pooled. The data differed significantly from the observations of others using Tn916, where mobilization of nonconjugative plasmids generally occurs at two or more orders of magnitude lower than that of the transposon (24, 36, 81). It should be kept in mind, however, that nothing is known about the conjugation bridge generated by these elements. It is possible that once a conjugative "pore" is established between the donor and recipient cell, the size of that pore might vary—with one extreme corresponding to a total fusion of the two cells. Conceivably, the level of expression of a key protein that could also be influenced by different conditions (such as media) might affect the size of the pore and, thus, affect the apparent frequency of mobilization.

2.5. On the Occurrence of Multiple Inserts

The common presence of multiple inserts among transconjugants raises the question of how more than one element is acquired from a donor with only one copy of the element. One possibility is that one circular intermediate structure repeatedly transfers a copy of itself simply by repeated donation events that may even resemble multiple rounds of a rolling-circle process. As the copies enter, they target different sites. Alternatively, the *trans*-activation phenomenon might activate a similarly located sibling on a sister (perhaps partially replicated) chromosome with passage of the latter into the same recipient cell that received the initially excised transposon.

Another possibility is that insertions derive from intracellular transposition (excision/insertion) of the original insert in the recipient after passage of the chromosomal replication fork through the element. Movement of one of the daughter transposons to a new site in the sister chromosome or an as-yet unreplicated portion of the same chromosome would ultimately result in copies at two sites. It is speculated that the stability of a given insert may be inversely related to the "attractiveness" of the target site. Thus, sites that are hot targets could result in the least stable insertions, and insertions here are more likely to give rise to subsequent intracellular transpositions.

The fact that insertions of Tn916 at different sites in the bacterial chromosome exhibit donor potentials ranging over more than 1000-fold implies that a cell with multiple inserts probably has certain transposons that would, if alone, move at frequencies very different than others in the same cell. The insertion with the greatest potential would be the most likely to excise first, and *trans*-activation would stimulate movement of the other elements. If this is indeed the case, a donor with more than one transposon might give rise to transconjugants more likely to have multiple inserts than the case where the donor has a single insertion. There is currently no evidence, however, that addresses this notion.

2.6. Host Range

Tn916 and closely related transposons have an extremely broad host range and have been introduced into numerous species by conjugation or transformation. A list of bacteria in which these elements have been established is shown in Table 15.2. Movement of Tn916 appears to occur in most species tested, although conjugative transposition has been observed primarily among the gram positives. There have been recent reports (8, 83) that Tn916 could transfer between gram-positive and gram-negative bacteria. In addition, a Tn916-like element recently identified in a strain of *Fusobacterium nucleatum* (a gram-negative anaerobe) (94) was able to conjugatively transfer to other strains of the same species as well as to the gram-positive organisms *E. faecalis* and *Peptostreptococcus anaerobius* (96). The broad host range implies that the functions related to transposition and conjugation that are determined by the element are quite autonomous and that any host involvement is likely to reflect components that are highly conserved in the bacterial world. It is also consistent with the absence of many restriction endonuclease recognition sequences that might otherwise be expected within a DNA segment this size (see legend to Figure 15.1).

3. Structure and Genetics of Tn916/Tn1545

3.1. Behavior in *E. coli*

Tn916 and Tn1545 have each been cloned on plasmid vectors in *E. coli* (11, 25, 28, 42). This has facilitated structural analyses of the elements that are shown in Figure 15.1. Clones carrying the transposon on a high-copy-number plasmid replicon are generally unstable, and under nonselective conditions, the chimera is readily eliminated (24) or the transposon excises and is lost. Excision is RecA independent, and the DNA that formerly flanked the transposon becomes spliced together (28, 42). Both Tn1545 and Tn916 are fully capable of inserting into the *E. coli* chromosome, where they are relatively stable, and subsequent

TABLE 15.2 Host Range of Tn916/Tn1545 Family[a]

Strain	Reference
Acetobacterium woodii	108
Acholeplasma laidlawii	34
Alcaligenes eutrophus	8
Bacillus anthracis	59
Bacillus firmus	43
Bacillus pumilus	45
Bacillus stearothermophilus	83
Bacillus subtilis	8, 20, 43, 59, 78, 81, 100, 101
Bacillus thuringiensis	80, 81
Butyrivibrio fibrisolvens	46
Citrobacter freundii	8
Clostridium acetobutylicum	6, 7, 8, 126
Clostridium botulinum	74
Clostridium difficile	44, 78
Clostridium tetani	119
Enterococcus durans	3
Enterococcus faecalis	20, 22, 38, 53, 79
Enterococcus faecium	4, 37
Enterococcus hirae	3, 22
Escherichia coli	8, 24, 28, 42, 51
Eubacterium limosum	5
Fusobacterium nucleatum	94, 96
Haemophilus influenzae	51, 64
Haemophilus parainfluenzae	64
Lactobacillus curvatus	65
Lactobacillus plantarum	48
Lactococcus lactis	6, 8, 17, 27, 48, 49
Leuconostoc cremoris	48
Listeria monocytogenes	14, 27, 32, 40, 62
Mycoplasma hominis	95
Mycoplasma hyorhinis	33
Mycoplasma mycoides	124
Mycoplasma pulmonis	33, 34
Neisseria gonorrhoeae	82
Neisseria meningitidis	63, 82
Peptostreptococcus anaerobius	96
Staphylococcus aureus	22, 27, 61, 66, 129
Streptococcus agalactiae	99, 120, 123
Streptococcus anginosus	21
Streptococcus bovis	47
Streptococcus gordonii	22, 27, 42
Streptococcus mutans	18, 91
Streptococcus pneumoniae	27, 121
Streptococcus pyogenes	15, 84, 100
Streptococcus sanguis	35
Thermus aquaticus	101
Veillonella parvula	94, 96

[a]Bacterial species were considered to be hosts if they supported establishment of a fully characterized conjugative transposon, or if they harbored a naturally occurring element that showed significant homology to a previously characterized conjugative transposon and that also exhibited evidence of transposition/transfer abilities.

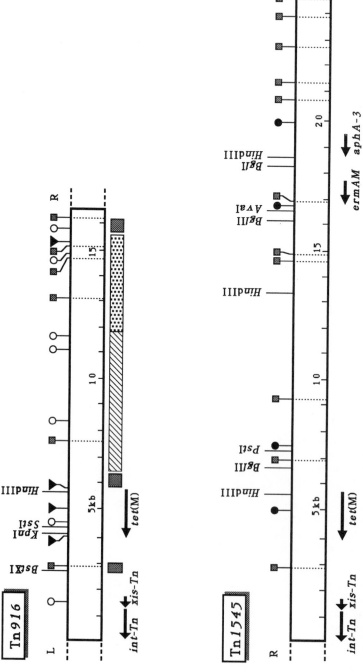

FIGURE 15.1. Linear restriction and functional maps of Tn916 and Tn1545 (based on references 9, 11, 13, 24, 25, 75, 85, 90, 102, 110, 115). Transposon DNA is bounded by heavy solid lines. Bordering DNA is indicated by dashed lines. The ends of each transposon designated "left" and "right" are indicated by L and R, respectively (these are opposite for the two transposons). Known open reading frames and direction of transcription are shown by arrows. (For diagrammatic clarity, *xis-Tn* is shown slightly larger than its actual size of 0.2 kb. A Tn5 insert between *int-Tn* and *xis-Tn* did not impair excision, transposition, or transfer functions.) Functional areas of Tn916 defined by insertional mutagenesis are indicated as follows (areas shown are defined by at least two insertion points): ▨, necessary for transfer; ⬚, necessary for transfer and transposition; ■, nonessential for transfer or transposition. Restriction endonuclease recognition sites are shown as follows: ●, *Cla*I; ■, *Hinc*II; ▼, *Hpa*II; ○, *Sau*3AI. Enzymes that cleave at three or fewer sites within the element are illustrated without symbols. *Hinc*II fragments are separated by dotted lines. Enzymes shown for each element are limited to those for which data are available throughout the entire length of the element. There are no *Ava*I, *Bam*HI, *Bgl*II, *Cla*I, *Eco*RI, *Nco*I, *Nru*I, *Pst*I, *Pvu*I, *Sal*I, *Sma*I, *Sph*I, *Xba*I, or *Xho*I sites in Tn916. There are no *Bam*HI, *Eco*RI, *Sal*I, or *Sma*I sites in Tn1545.

transposition has been observed via movement to a conjugative plasmid (24, 28). Transposition from an enterococcal plasmid (acting as a suicide vector) to the *E. coli* chromosome has also been noted upon introduction by transformation (24). Attempts by some to observe conjugative transfer between *E. coli* strains or from *E. coli* to gram-positive recipients have been unsuccessful (28, J. Jones, personal communication, Flannagan, unpublished), although more recent accounts (8, 51, 83) report success in such efforts.

3.2. Nucleotide Sequence of End-Regions

The ends of Tn*916* and Tn*1545* have been sequenced (12, 23) and are essentially identical over the lengths compared. As seen in Figure 15.2, they are very AT-rich and contain imperfect inverted repeats referred to as IRL and IRR. The inverted repeat sequences could have identity at 19 of 25 nucleotides or 20 of 26 nucleotides, depending on variability in the number of T residues (4 or 5) on the right terminus. (Alternatively, if it is found that the junction regions extend for 7 bases instead of 6 [see Figure 15.3], the inverted repeats actually contained in the transposon DNA would correspond to 18 of 24, and 19 of 25). The right end also contains two sets of short direct repeats, designated DR-1 and DR-2. The two DR-1 sequences are 9 bp long and separated by 11 bp, while the DR-2 sequences are 11 nucleotides long and are contiguous. Another set of contiguous DR-2 repeats appears near the left end of the transposon; however, one segment (indicated as "DR-2") differs by 2 bp. Overlapping the DR-2 repeats at both the right and left ends are longer (27 bp) directly repeated sequences designated DR-3 with differences at only three positions. The left DR-3 sequence is 136 nucleotides from the terminus; the right DR-3 is about 72 nucleotides from the terminus. The significance of these repeats remains unknown.

Another point of interest is the presence of potential outward-reading promoter sites (-10 and -35 boxes) in the right end of the transposon. It is not yet clear whether these are functional promoters capable of influencing the expression of outside genes. (Note: The outward-reading open reading frames [*int-Tn* and *xis-Tn*] discussed below for Tn*1545* and Tn*916* are located at the opposite end of the transposon and do not appear to have transcription termination sites; thus it is conceivable that expression of these regions also could exhibit some influence beyond the transposon terminus.)

Potential integration host factor (IHF) binding sites were found scattered throughout each end and on both DNA strands. When the regions near the inverted repeats on the extreme ends of the element were analyzed, six potential binding sites were found clustered within the first 34 bases on the left end. Significantly, none were found in the corresponding region on the right end, despite the inverted repeat symmetry. This leads to speculation that specific interaction of IHF with the left end may be involved in the transposition process.

3.3. Junctions

Insertion of Tn*916* did not give rise to a duplication of the target sequence (23), in contrast to the characteristic behavior of a number of "nonconjugative" transposons. Surprisingly, upon excision of the transposon in *E. coli*, the regenerated "target" sequence could differ from the original target sequence by a few base pairs. Of significance was that

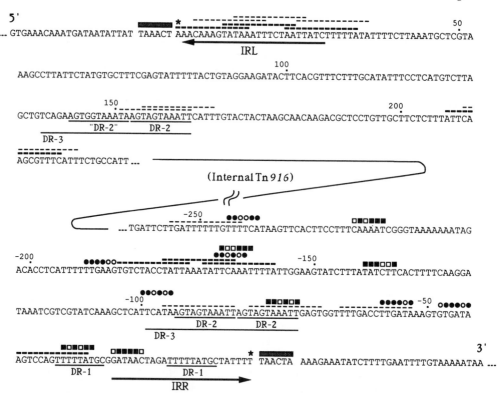

FIGURE 15.2. Nucleotide sequence of the ends of Tn916. The sequences shown are of the insert in pAM160 (23). Numbering proceeds forward from the first nucleotide of Tn916 on the left and backward from the last nucleotide of Tn916 on the right. Each end of the transposon is designated by an asterisk. Junction sequences (shown as the minimum length of 6 nucleotides) are highlighted by a shaded box. Transposon, junction, and bordering DNA are separated by spaces. Inverted repeat sequences on the left and right ends are designated IRL and IRR, respectively. Direct repeat sequences DR-1, DR-2 (and "DR-2"), and DR-3 are underlined. Potential outward-reading −10 and −35 promoter sequences on the right end are indicated by squares and circles, respectively. Filled symbols represent agreement with the *E. coli* consensus −10 (TATAAT) and −35 (TTGACA) sequences. (Because of the A + T richness of this region, many potential promoter sequences can be found. Those shown here were selected for optimum spacing [16 to 20 bp] between each −10 and −35 sequence.) Potential IHF binding sites were identified by comparison to the consensus A/TATCAANNNNTTA/G (77), allowing for two mismatches, and are shown by thick dashed lines (on the DNA strand illustrated) and thin dashed lines (for the opposite strand).

the "altered" sequence was identical to one of the junctions that bordered the inserted element, while the other junction corresponded to the original target. This suggested that excision involved a recombinational event between heterologous junctions that could lead to alternative products. It also suggested that the excised intermediate form of the transposon (see 3.4) contained one or the other of these sequences at the region joining the two ends of the circularized element and that this sequence should appear as one of the transposon/target junctions after a subsequent insertion. As discussed below (see Section 4.), however, both junction sequences from a given insert probably end up in the intermediate as a heteroduplex.

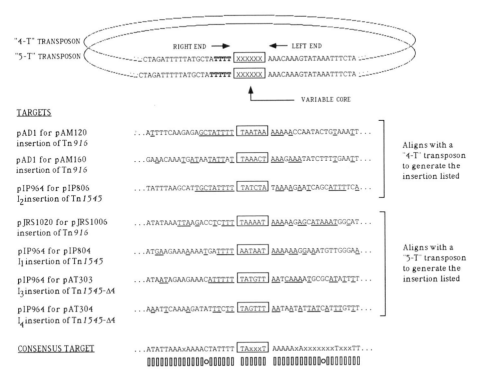

FIGURE 15.3. Targets for insertion of the Tn*916*/Tn*1545* family of conjugative transposons. Data for the Tn*916* sequence and the pAM120 and pAM160 insertion targets are from Clewell et al (23) and Flannagan (unpublished data). The pJRS1006 insertion target is from Caparon and Scott (16) and Perez-Casal et al (88). The Tn*1545* sequence and I₁ and I₂ insertion targets were compiled from Caillaud and Courvalin (12) and Poyart-Salmeron et al (89, 90). I₃ and I₄ insertion targets are from Poyart-Salmeron et al (90). Analysis of data for the inserted transposon and targets both before insertion and after excision allowed determination of the sequences in the targets (boxed) that participate in insertional recombination with the variable core of the transposon (shown boxed; each variable base of the core sequence is represented by a capital X). The target and core regions are illustrated as minimal sequences of 6 nucleotides. Because a T residue can be found on the left of each core sequence and each target sequence, this position could also be involved in recombination and not be detected. Current data allow a size estimate of 6 or 7 nucleotides for the recombination regions (assuming a standard number is involved in each case). The illustrated regions on the ends of the transposons are identical for Tn*916* and Tn*1545*. The two naturally occurring variants are shown with either 4 or 5 T residues (in bold) in the intermediate on the right end. The target consensus was derived from comparison among the seven examples presented. A letter designation of A, C, G, or T was assigned to each position when it occurred in the majority of cases (≥4/7); a lowercase x indicates a position for which a consensus could not be determined. The high A + T content of these regions is indicated below the consensus (□, A or T in the majority of cases; ○, C or G in the majority of cases). The consensus presented here differs slightly from that derived similarly by Poyart-Salmeron et al (90) and will no doubt be modified as more data become available. A salient feature of the consensus is a clustering of A residues just to the right side of the recombination region (similar to the left end of the transposon) and T residues on the left (similar to the right end of the transposon). The targets and transposons are illustrated in the way they are believed to align prior to insertional recombination. Underlined bases in each target represent homologies with the relevant transposon.

Changes in the target sequence after Tn916 excision were also observed by Caparon and Scott (16), and similar occurrences were reported by Poyart-Salmeron et al (90) in the case of Tn1545. Figure 15.3 compiles the profiles of all reported target sequences, along with a view of the transposon shown as an intermediate with a variable core sequence joining the right and left ends.

Variability of nucleotide sequence occurring in the junction regions can, in some cases, have a significant phenotypic effect. For instance, insertions of conjugative transposons into enterococcal hemolysin-encoding plasmids such as pAD1 or pIP964 have been found to produce altered hemolytic phenotypes (25, 27, 38, 72). Analysis of such mutants by Ike et al (57) has shown that isolates that have Tn916 inserted at precisely or nearly the same position can differ widely in phenotypic profile (i.e., nonhemolytic versus hyperhemolytic). It is believed that the variation in hemolysin expression is brought about by translational changes that reflect, simply, the variation of nucleotides in the junction region that joins the transposon to the interrupted hemolysin gene.

3.4. The Intermediate

Investigation of a plasmid (cosmid) chimera in *E. coli* containing a fragment of *S. pyogenes* chromosomal DNA bearing Tn916 led Scott et al (100) to identify the presence of small amounts of a covalently closed monomeric circular molecule with a size (16.4 kb) corresponding to the transposon. Purified preparations of this molecule, which represented about 1 to 5% (J. Scott, personal communication) of the plasmid preparations, were able to transform protoplasts of *Bacillus subtilis* (selecting for Tc resistance), giving rise to insertions of Tn916 into different chromosomal sites. The transformants were able to conjugatively transfer Tn916 to *S. pyogenes* recipients, indicating the element was fully functional. The Tn916-sized circular molecules established as inserts in the *B. subtilis* chromosome by protoplast transformation at frequencies much higher than those observed when transforming with plasmid chimeras containing the transposon. The former, therefore, exhibited properties one would predict for an intermediate in the transposition process, and the data supported the earlier suggestion (41, 42) that excision was the rate-limiting step. Senghas et al (102) also provided genetic evidence for this view in that mutants defective in the ability to excise when on a plasmid chimera in *E. coli* were unable to transform *E. faecalis* protoplasts (see Section 3.5.).

The presence of "Tn916 circles" has been confirmed in studies of *E. coli* plasmid chimeras pAM120, pAM160, and pAM170 (S.E. Flannagan and D.B. Clewell, unpublished). The latter represent clones using the vector pGL101 ([68]; a derivative of pBR322) ligated with different enterococcal DNA fragments containing the transposon. The general presence of these molecules in plasmid preparations from *E. coli* strongly suggests their involvement as the primary transforming activity when used to introduce Tn916 into other bacteria by transformation mechanisms that allow uptake of circular duplex molecules. (This is presumed to be the case during protoplast transformation, in contrast to natural processes where only a single strand enters the cell.)

Sequence analyses of the "joint" region connecting the two transposon ends in intermediate structures were conducted by Caparon and Scott (16) after first cloning this region in a pUC18 vector. Unexpectedly, they found that over a 5-base-pair span, some of the clones had two different nucleotides at equivalent positions for several base pairs in the joint. This proved to represent mixtures of two chimeras that could be separated by subsequent transformation. The two sequences corresponded to those found at the two transposon junctions of the parent insertion. That is, one joint sequence corresponded to the junction at one end of the previous insertion, whereas the other resembled that found at the opposite end. The data suggested that the joint region between the ends was actually a heteroduplex that gave rise to a mixture of two homoduplexes after cloning. Support for this view came from studies in which the sensitivity of the joint region of the intermediate to *Dra*I was analyzed. It was reasoned that a heteroduplex structure would destroy the integrity of a *Dra*I site (TTTAAA), and this was indeed found to be the case.

3.5. Genetic Analyses

A genetic analysis of Tn*916* (60, 102) made use of the plasmid vector pVA891, which is unable to replicate in *E. faecalis* but has an *erm* determinant able to express in that host. A chimeric derivative (pAM620) carrying Tn*916* served as a useful vehicle for the introduction of the transposon into *E. faecalis* by protoplast transformation (102, 127). Selected Tc-resistant transformants were devoid of *erm* and represented Tn*916* insertions in the bacterial chromosome. This provided a means for genetic analyses whereby mutations in the transposon could be generated in *E. coli* and the effects could be examined in *E. faecalis*.

Tn*5* insertions were generated over the length of Tn*916*, and most derivatives could be introduced into the *E. faecalis* chromosome. However, Tn*5* insertions near the left end of Tn*916* (Figure 15.1) eliminated the ability to establish in *E. faecalis*. These mutants also did not undergo the Tn*916* excision event characteristic of plasmid clones in *E. coli*; however, excision occurred readily when the wild-type region was provided *in trans* on a different plasmid. The inactivation of an "excisase" was therefore suggested, and a role for such an activity in transposition was implicated (see below). The majority of Tn*916*::Tn*5* insertions that could be established in *E. faecalis* were unable to undergo subsequent conjugative transposition, and for some of these, intracellular movement (to pAD1) was also not detectable. Insertions close to *tet*, and two inserts near the right end, did not appear to affect the transposon. The map in Figure 15.1 shows the location of the related regions. That insertions within a portion of the right half of Tn*916* resulted in a failure to detect intracellular transposition, even though transposition was necessary to establish these derivatives in *E. faecalis*, suggested that Tn*5*-affected steps were bypassed during, or prior to, introduction by transformation. For example, the plasmid preparation from *E. coli* probably already contained excised intermediate structures stemming from the instability of location on a high-copy-number plasmid (see Section 3.1). When located on the chromosome, initiation of the excision event occurs at a much lower frequency and may depend on additional factors.

In the case of Tn*1545*, Poyart-Salmeron et al (89) constructed a series of deletion derivatives devoid of large segments of the internal portion of the transposon but maintaining the determinant *aphA-3* (kanamycin resistance) and at least 185 bp at each end. It was only when a region close to the right end of the transposon (equivalent to the left end of Tn*916*) was deleted that the element no longer exhibited a high tendency to excise from the plasmid. Excision of this derivative, designated Tn*1545*-Δ*4*, could be complemented *in* *trans* by coresident plasmid chimeras containing the appropriate portion of the transposon. Sequence analyses led to the identification of two open reading frames corresponding to proteins designated ORF1 and ORF2 (subsequently called Xis-Tn and Int-Tn, respectively; see Figure 15.1), which played a role in excision. ORF2 appeared essential for excision, but its activity was greatly stimulated by the additional presence of ORF1. ORF1 and ORF2 had deduced masses of 8,100 and 46,925 daltons, respectively, and bands approximating these sizes could be detected in in vitro protein synthesis assays (89).

A plasmid derivative containing Tn*1545*-Δ*4* was used to construct a replicative transposon intermediate using the following approach (90). A compatible replicon, pACYC184, was first inserted into the transposon, after which excision from the original replicon was accomplished by providing ORF1 and ORF2 *in trans*. Subsequent transformation and appropriate selection yielded stable "intermediate" structures. These "replicative intermediates" were then used to determine the requirement for insertion (integration) by providing a second plasmid containing a target fragment (segment of a gram-positive plasmid) and a third plasmid containing transposon DNA to be tested for *trans*-acting function. Insertion was determined by screening for cointegration of the replicative intermediate and the plasmid carrying the target. Using this approach, ORF2 was found to be sufficient for insertion.

The generation of stable "intermediates" also allowed for examination of joint sequences as well as the sequences regenerated in the plasmid from which excision occurred. These sequences were found to vary in a manner that was completely consistent with the data described above for Tn*916* (see Figure 15.3).

Determinants of ORF1 and ORF2 have also been sequenced for Tn*916* (24, 109) and are essentially identical to those of Tn*1545*; Su and Clewell (109) showed that both ORF1 and ORF2 were essential for excision in *E. coli*. Using an allelic replacement approach, Storrs et al (107) constructed a mutant of Tn*916*, designated Tn*916-int1*, which was defective in both excision and integration. They reported that conjugative transfer of this derivative from the *B. subtilis* chromosome to *E. faecalis* was detected only when the transposon-encoded integrase (ORF2) was provided *in trans* in both donor and recipient. Bringel et al (8a), however, recently reported that for transfer to occur it was only necessary to provide ORF2 in the donor.

3.6. Similarities with λ *xis* and *int*

Poyart-Salmeron et al (89) made the important observation that the Tn*1545* ORF1 and ORF2 noted above exhibit homology with the excisionases (Xis) and integrases (Int), respectively, of lambdoid bacteriophages. Considering also the similar functions played by these products, it seems clear that the corresponding determinants evolved from common ancestors. The transposon determinants have therefore been designated *xis-Tn* and *int-Tn* (90).

As pointed out by Trieu-Cuot et al (116), Xis-Tn is a slightly basic protein (net charge of +3) with an amino-terminal half similar to Xis of bacteriophage P22 and to a lesser extent λ and φ80. Int-Tn is highly basic (net charge is +17) and exhibits homology with the Int-related family of site-specific recombinases (29, 89). These include phage-encoded integrases (as noted above), transposon-encoded transposases and resolvases, and invertases. Members of this group have two domains in their carboxy-terminal halves that are highly conserved and thought to be part of their active sites (1, 73); Int-Tn contains these domains. An invariant tyrosine residue present in one of these domains is probably linked to the DNA during recombination (67, 87).

4. Model for Transposition

Gawron-Burke and Clewell (41) hypothesized that an excision event giving rise to a nonreplicating circular intermediate is the first step in Tn916 transposition. The later demonstration that the junctions at opposite ends of the element could differ and that either could be maintained in the DNA from which Tn916 excised (23) implicated an excision process involving recombination between the two dissimilar sequences. A circular intermediate containing one or the other "junction," designated now as "variable core" sequence, was hypothesized, and the core (joint) sequence was predicted to appear at one of the new junctions of a subsequent insertion. The junction sequence data of Caparon and Scott (16) and Poyart-Salmeron et al (for Tn1545 [90]) extended the originally suggested minimal size of the variable core (23) from 4 to 6 (or possibly 7) base pairs.

The evidence reported by Caparon and Scott (16) for a heteroduplex within circular intermediates strongly supports a recombination mechanism proposed both by them and by Poyart-Salmeron et al (89). This is illustrated in Figure 15.4 and involves the generation of staggered double-stranded cuts across each of the two junction sequences to generate single-stranded overhangs that are then paired and ligated to form a covalently closed circular molecule. It is not known at this time whether the staggered cleavage generates $5'$ or $3'$ overhangs. Because the two junctions are not identical, a heteroduplex is formed. The site from which the transposon excises also results in a heteroduplex, but subsequent replication or mismatch repair processes would generate two daughter molecules having one or the other of the junction sequences. In the intermediate the heteroduplex is maintained (see Section 3.4.)—a view consistent with the notion that it does not replicate—and the apparent absence of mismatch repair suggests that the region may be protected by interaction with one or more proteins. The heteroduplex in the intermediate is believed to be maintained as long as the circular element remains in the original host cell, and therefore would be present to interact with a new target in the case of intracellular transposition. Concerning conjugative transposition (transfer to a new cell) however, events are not well understood. If the element transfers via a single-stranded mechanism, the intermediate would presumably be regenerated in the new host as a homoduplex.

Insertion of the intermediate at a new site can be visualized as the reverse of the excision mechanism, after the region containing the joint pairs up with a suitable target sequence. Because the target site need not be homologous with the joint sequence, heteroduplexes could be generated at the new junctions, and replication or mismatch repair would then give rise to homoduplexes. That a former junction can be carried by the transposon to a new site was shown by Caparon and Scott (16), who reported that the joint sequence of an intermediate generated in E. coli could be identified as a junction in a subsequent insertion in the B. subtilis chromosome.

It should be pointed out that excision frequently occurs in such a way that the number of thymyl residues representing the right end of Tn916 (corresponding to adenyl residues in the left end of Tn1545)in the intermediate may vary by $+/-$ one (16, 90). This difference is reflected in the DNA from which excision occurred as well as in the subsequent insertion. A

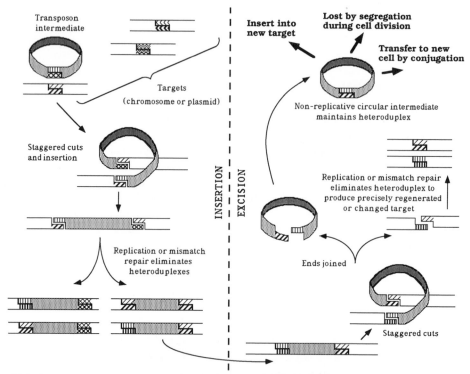

FIGURE 15.4. Model for insertion and excision of the Tn*916*/Tn*1545* family of conjugative transposons. Transposon DNA is shaded; bordering sequences are clear. Junction DNA is patterned, with homologous pairing in the junction regions illustrated by thick and thin representations of similar pattern. The illustration shows the transposon intermediate before insertion as a heteroduplex; however it is not known at this time whether the intermediate maintains a heteroduplex during transfer to a new host. It is conceivable (perhaps probable) that the conjugation process involves transfer of a single strand, with subsequent regeneration of the double-stranded (homoduplex) molecule in the new host. See text for details.

likely explanation for this is that the right end sometimes undergoes a staggered cut using a different sequence (offset by 1 bp) than was used on insertion, while the left end is cleaved normally. Slippage of factors that bind to the run of T's at this end might facilitate this event. Thus, an insertion/excision cycle can generate not only base substitutions but also frameshift mutations.

Sequence analyses of junction regions for both Tn*916* and Tn*1545* revealed that the target on either side of the region that recombines with the variable core of the intermediate usually has significant homology with the ends of the transposon (23). This is illustrated in Figure 15.3. By comparing several targets, a possible consensus target site (Figure 15.3) has been deduced (90). It was speculated (23) that the widely differing donor (excision?) potentials of strains harboring Tn*916* insertions in different sites might relate to different degrees of homology between each end of the transposon and the region just outside the opposite end (which would be expected to align prior to excision—see Figure 15.4). However, recent experiments (D. Jaworski, personal communication) have indicated that independently obtained Tn*916* insertions in the same site but with different junctions can exhibit widely different transposition frequencies. This implies that junction sequences per se are probably recognized with different efficiencies by the excisase/integrase system.

The strong similarities of the above model with the lysogenic aspects of lambdoid bacteriophages have been recognized and discussed in some detail by Poyart-Salmeron et al (89, 90). Of particular relevance is the fact that integration of λ into the bacterial chromosome involves a reciprocal recombination between 15 bp homologous core regions on the phage DNA (*attP*) and bacterial DNA (*attB*). Int-λ binds to the *attB* and *attP* sites and introduces 7-bp staggered nicks within the core sequence (30, 67, 76). The resulting overlapping structures interact, giving rise to recombinants. In contrast to Tn*916*/Tn*1545*, the λ system uses homologous sites; however, it is important to note that λ is also capable of integrating at nonhomologous sites (i.e., in absence of *attB*) with much lower efficiencies (122). It appears likely that the transposition mechanism of Tn*916*/Tn*1545* resembles the process by which lambdoid bacteriophages integrate into and excise from the bacterial chromosome. The regions flanking the "variable core" region of the transposon in the intermediate are probably intimately involved, much the same way equivalent regions relating to *attP* participate in the recombination process (67).

5. Complex Conjugative Transposons

Horaud (formerly Horodniceanu) and colleagues (10, 52, 54, 56, 70) have accumulated information on genetic aspects of antibiotic resistance with respect to a large number of clinical isolates of pathogenic streptococci. Examination of 152 antibiotic resistant strains, belonging to Lancefield groups A, B, C, and G, as well as *S. pneumoniae*, *S. bovis*, and *S. anginosus* (*milleri*), found that most resistance determinants were chromosome-borne (52). Indeed, only 7 of 71 (10%) β-hemolytic streptococci of groups A, B, C, and G harbored resistance plasmids; these usually encoded resistance to erythromycin and in some cases also chloramphenicol. (All 10% also carried a nonconjugative *tet* determinant on the chromosome.) Among the larger group of 152 strains, 25% were able to transfer multiple chromosomal resistance determinants en bloc at frequencies of 10^{-9} to 10^{-6} per donor in filter matings, and transferability usually extended over a broad host range. Although elements resembling Tn*916*/Tn*1545* appeared to be present in a number of these strains, many seemed to represent larger and more complex elements. Three of these systems were investigated in some detail in the laboratories of Guild, Burdett, Horaud, and Vijayakumar, and these are now identified as Tn*5253* (previously referred to as Ω *cat tet*) (2, 10, 103, 117, 118); Tn*3951* (56, 58, 106); and Tn*3701* (69, 70, 71). (See Table 15.1). A comparison of these elements is shown in Figure 15.5.

All three transposons are large (50 kb or more), and Le Bouguenec et al (70) found them to have significant homology. Interestingly, each contained a region very similar to Tn*916*. In the case of Tn*3701*, the Tn*916*-like region carried *erm*, as well as *tet*, and exhibited independent transposition. It was therefore designated Tn*3703*; however, unlike Tn*916*, it did not appear to have conjugative properties.

In the case of Tn*5253* (65.5 kb), Ayoubi et al (2) recently found that the internal Tn*916*-like region is capable of independent conjugative transposition, and this 18-kb internal element bearing *tet* has been designated Tn*5251*. It was possible to delete Tn*5251*

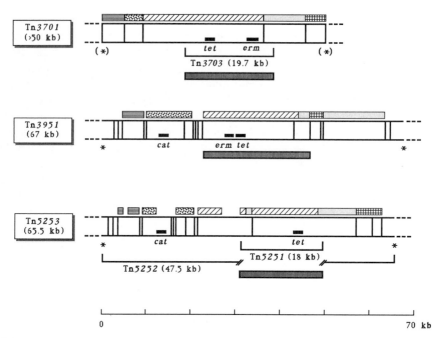

FIGURE 15.5. Comparison of complex conjugative transposons Tn*3701* of *S. pyogenes* A454, Tn*3951* of *S. agalactiae* B109, and Tn*5253* of *S. pneumoniae* BM6001. The ends of each transposon are indicated by asterisks. (The exact positions of the ends of Tn*3701* are not known.) Bold vertical lines within each element represent *Eco*RI cleavage sites. Transposon DNA (solid lines) and bordering sequences (dashed lines) are represented. Transposons that are contained within larger elements are indicated (⌐⌐). The approximate positions of antibiotic resistance determinants are illustrated (▮); *tet*, tetracycline resistance; *erm*, erythromycin resistance; *cat*, chloramphenicol resistance. Homology among the three elements has been determined by Le Bouguenec et al (70). Patterned segments above each element correspond to homologies with the *Eco*RI fragments of Tn*3701*. Sequences that bear much similarity to Tn*916* are indicated (▦) below each element. Transposon maps were compiled based on the following references: Tn*3701* (69, 70), Tn*3951* (58), and Tn*5253* (2, 117).

from the larger element in vivo, leaving an element (bearing *cat*) still capable of conjugative transposition. Designated Tn5252 and having a size of 47.5 kb, it was able to insert at multiple sites in the chromosome of *E. faecalis* recipients, and establishment appeared independent of host homologous recombination functions. However in pneumococcus, Tn*5252*, unlike Tn*5251*, preferred to insert at a unique target site, which is similar to the behavior of the parental element Tn*5253*. Tn*5251*, therefore, did not play a mechanistic role in the conjugation of the larger transposon.

It is evident that Tn*5253* is a composite structure consisting of one conjugative transposon (Tn*5251*) inserted within another (Tn*5252*). The two elements do not exhibit significant homology and therefore correspond to distinct classes of conjugative transposons. Although both Tn*5251* and Tn*5252* are capable of independent transfer, the separation of these elements was not observed when the two were associated as Tn*5253*. Conceivably, sequences outside of Tn*5251* influence the ability to excise independently.

Insofar as Le Bouguenec et al (70) showed that significant homology exists among Tn*3701*, Tn*5253*, and Tn*3951*, with respect to regions outside of the internal Tn*916*-like segments, it would appear that all three contain Tn*5252*-like sequences. Because similar homologies exist on the chromosome of many antibiotic-resistant clinical isolates of various streptococcal species (70), it is evident that Tn*5252*-like elements (many containing Tn*916*-like sequences within them) are widespread.

6. Concluding Remarks

Conjugative transposons appear to be widespread in the bacterial world. While the majority of the elements have been identified in species of streptococci and enterococci, they clearly are not limited to these groups. Indeed their very broad host range, which includes the gram negatives, suggests that they are highly ubiquitous. (They even occur naturally in gram-negative bacteria [94, 96]). The degree to which they have been found in *Streptococcus* probably relates to the fact that plasmids, which in many other groups of bacteria commonly bear resistance determinants, are not particularly common in this genus. Indeed, in streptococci and perhaps some other groups, conjugative transposons are probably more responsible for the dissemination of antibiotic resistance than are plasmids.

It is now evident that there are at least two basic classes of conjugative transposons in gram-positive bacteria: those resembling Tn*916*/Tn*1545* and those exhibiting homology with Tn*5252*. Perhaps the recently described (55, 92) lactococcal elements encoding nisin biosynthesis and sucrose fermentation represent a third class. It is likely that other groups will soon be identified; many will no doubt be composites consisting of more than one element, bearing similarity, in this respect, to Tn*5253* (and probably Tn*3701* and Tn*3951*). It will be interesting to compare the various types with respect to their structures and mechanisms of movement.

The Tn*916*-like elements appear to be related to the lambdoid phages in that they excise and integrate via a reciprocal recombination (Campbell-like) mechanism and use proteins related to Int and Xis. An important difference is that the transposon target sites are not highly specific, and recombination occurs readily between heterologous sequences. Indeed, a short sequence brought in by the transposon may even remain behind after subsequent excision; thus, a gene from which the transposon excises can remain mutated. Conceivably the heterologous junctions could act to stabilize the transposon. That is, insertions into sites resulting in significant homology around the junctions could be relatively unstable, leading to loss of the element.

Little is known of the actual process of cell-to-cell transfer of conjugative transposons and their establishment in the recipient. The failure of a resident transposon to prevent uptake of a second element again distinguishes these systems from the lambdoid phages. The frequent appearance of multiple copies (inserts) in transconjugants (at least in the case of the Tn*916*-like elements) may represent a means by which ultimate survival of the element is ensured.

The very broad host range of conjugative transposons has made elements like Tn*916* attractive "tools" for doing genetic analyses on bacterial species in which genetic studies have been severely limited. For example, one simply can select for transconjugants (or transformants) that are drug resistant and then screen for the desired phenotypic change. DNA fragments from appropriate derivatives can then be cloned in *E. coli* (or any desired host), again selecting for the resistance trait (42). If the vector is a multicopy plasmid, the resulting instability of the transposon allows excision of the transposon and splicing together of the flanking DNA. The latter may represent an intact (or possibly slightly altered) form of the original gene of interest. Also, the development of systems for introducing specific genes into the transposon for subsequent insertion into the bacterial chromosome (83, 85, 114) should prove to be productive.

ACKNOWLEDGMENTS. Work conducted in the authors' laboratory relating to material discussed in this review was supported by Public Health Service Grant AI10318. We wish to thank Y. Su, D. Jaworski, L. Zitzow, J. Scott and G. Dunny for helpful discussions and/or for providing unpublished information.

References

1. Argos, P., Landy, A., Abremski, K., Egan, J.B., Haggard-Ljungquist, E., Hoess, R.H., Kahn, M.L., Kalionis, B., Narayana, S.V.L., Pierson, L.S., III, Sternberg, N., and Leong, J.M., 1986, The integrase family of site-specific recombinases: regional similarities and global diversity, *EMBO J.* 5:433–440.
2. Ayoubi, P., Kilic, A.O., and Vijayakumar, M.N., 1991, Tn*5253*, the pneumococcal Ω(*cat tet*) BM6001 element, is a composite structure of two conjugative transposons Tn*5251* and Tn*5252*, *J. Bacteriol.* 173:1617–1622.
3. Bentorcha, F., Clermont, D., de Cespedes, G., and Horaud, T., 1992, Natural occurrence of structures in oral streptococci and enterococci with DNA homology to Tn*916*, *Antimicrob. Agents Chemother.* 36:59–63.
4. Bentorcha, F., de Cespedes, G., and Horaud, T., 1991, Tetracycline resistance heterogeneity in *Enterococcus faecium*, *Antimicrob. Agents Chemother.* 35:808–812.
5. Berman, M.H., and Frazer, A.C., 1991, Program Abstr. 91st General Meeting, American Society for Microbiology, Abstr. Q-21.
6. Bertram, J., and Durre, P., 1989, Conjugal transfer and expression of streptococcal transposons in *Clostridium acetobutylicum*, *Arch. Microbiol.* 151:551–557.
7. Bertram, J., Kuhn, A., and Durre, P., 1990, Tn*916*-induced mutants of *Clostridium acetobutylicum* defective in regulation of solvent formation, *Arch. Microbiol.* 153:373–377.
8. Bertram, J,. Stratz, M., and Durre, P., 1991, Natural transfer of conjugative transposon Tn*916* between gram-positive and gram negative bacteria, *J. Bacteriol.* 173:443–448.
8a. Bringel, F., von Alstine, G. L., and Scott, J. R., 1992, Conjugative transposition of Tn*916*: the transposon *int* gene is required only in the donor. *J. Bacteriol.* 174:4036–4041.
9. Burdett, V., 1990, Nucleotide sequence of the *tet*(M) gene of Tn*916*, *Nucl. Acids Res.* 18:6137.
10. Buu-Hoi, A., and Horodniceanu, T., 1980, Conjugative transfer of multiple antibiotic resistance markers in *Streptococcus pneumoniae*, *J. Bacteriol.* 143:313–320.
11. Caillaud, F., Carlier, C., and Courvalin, P., 1987, Physical analysis of the conjugative shuttle transposon Tn*1545*, *Plasmid* 17:58–60.
12. Caillaud, F., and Courvalin, P., 1987, Nucleotide sequence of the ends of the conjugative shuttle transposon Tn*1545*, *Mol. Gen. Genet.* 209:110–115.
13. Caillaud, F., Trieu-Cuot, P., Carlier, C., and Courvalin, P., 1987, Nucleotide sequence of the kanamycin resistance determinant of the pneumococcal transposon Tn*1545*: evolutionary relationships and transcriptional analysis of *aphA-3* genes, *Mol. Gen. Genet.* 207:509–513.
14. Camilli, A., Paynton, C.R., and Portnoy, D.A., 1989, Intracellular methicillin selection of *Listeria monocytogenes* mutants unable to replicate in a macrophage cell line, *Proc. Natl. Acad. Sci. USA* 86:5522–5526.
15. Caparon, M.G., and Scott, J.R., 1987, Identification of a gene that regulates expression of M protein, the major virulence determinant of group A streptococci, *Proc. Natl. Acad. Sci. USA* 84:8677–8681.
16. Caparon, M.G., and Scott, J.R., 1989, Excision and insertion of the conjugative transposon Tn*916* involves a novel recombination mechanism, *Cell* 59:1027–1034.
17. Casey, J., Daly, C., and Fitzgerald, G.F., 1991, Chromosomal integration of plasmid DNA by homologous

recombination in *Enterococcus faecalis* and *Lactococcus lactis* subsp. *lactis* hosts harboring Tn919, *Appl. Environ. Microbiol.* 57:2677–2682.

18. Caufield, P.W., Shah, G.R., and Hollingshead, S.K., 1990, Use of transposon Tn916 to inactivate and isolate a mutacin-associated gene from *Streptococcus mutans, Infec. Immun.* 58:4126–4135.

19. Christie, P.J., and Dunny, G.M., 1986, Identification of regions of the *Streptococcus faecalis* plasmid pCF-10 that encode antibiotic resistance and pheromone response functions, *Plasmid* 15:230–241.

20. Christie, P.J., Korman, R.Z., Zahler, S.A., Adsit, J.C., and Dunny, G.M., 1987, Two conjugation systems associated with *Streptococcus faecalis* plasmid pCF10: identification of a conjugative transposon that transfers between *S. faecalis* and *Bacillus subtilis, J. Bacteriol.* 169:2529–2536.

21. Clermont, D., and Horaud, T., 1990, Identification of chromosomal antibiotic resistance genes in *Streptococcus anginosus* ("*S. milleri*"), *Antimicrob. Agents Chemother.* 34:1685–1690.

22. Clewell, D.B., An, F.Y., White, B.A., and Gawron-Burke, C., 1985, *Streptococcus faecalis* sex pheromone (cAM373) also produced by *Staphylococcus aureus* and identification of a conjugative transposon (Tn918), *J. Bacteriol.* 162:1212–1220.

23. Clewell, D.B., Flannagan, S.E., Ike, Y., Jones, J.M., and Gawron-Burke, C., 1988, Sequence analysis of termini of conjugative transposon Tn916, *J. Bacteriol.* 170:3046–3052.

24. Clewell, D.B., Flannagan, S.E., Zitzow, L.A., Su, Y.A., He, P., Senghas, E., and Weaver, K.E., 1991, Properties of conjugative transposon Tn916, in: *Genetics and Molecular Biology of Streptococci, Lactococci, and Enterococci* (G.M. Dunny, P.P. Cleary, and L.L. McKay, eds.), American Society for Microbiology, Washington, DC, pp. 39–44.

25. Clewell, D.B., and Gawron-Burke, C., 1986, Conjugative transposons and the dissemination of antibiotic resistance in streptococci, *Ann. Rev. Microbiol.* 40:635–659.

26. Clewell, D.B., Tomich, P.K., Gawron-Burke, M.C., Franke, A.E., Yagi, Y., and An, F.Y., 1982, Mapping of *Streptococcus faecalis* plasmids pAD1 and pAD2 and studies relating to transposition of Tn917, *J. Bacteriol.* 152:1220–1230.

27. Courvalin, P., and Carlier, C., 1986, Transposable multiple antibiotic resistance in *Streptococcus pneumoniae, Mol. Gen. Genet.* 205:291–297.

28. Courvalin, P., and Carlier, C., 1987, Tn1545: a conjugative shuttle transposon, *Mol. Gen. Genet.* 206:259–264.

29. Craig, N.L., and Kleckner, N., 1987, Transposition and site-specific recombination, in: Escherichia coli *and* Salmonella typhimurium, (F.C. Neidhardt, J.L. Ingraham, K.B. Low, B. Magasanik, M. Schaechter, and H.E. Umbarger, eds.), American Society for Microbiology, Washington, DC, pp. 1054–1070.

30. Craig, N.L., and Nash, H.A., 1983, The mechanism of phage λ site-specific recombination: site-specific breakage of DNA by Int topoisomerase, *Cell* 35:795–803.

31. Dang-Van, A., Tiraby, G., Acar, J.F., Shaw, W.V., and Bouanchaud, D.H., 1978, Chloramphenicol resistance in *Streptococcus pneumoniae*: enzymatic acetylation and possible plasmid linkage, *Antimicrob. Agents Chemother.* 13:577–583.

32. Doucet-Populaire, F., Trieu-Cuot, P., Dosbaa, I., Andremont, A., and Courvalin, P., 1991, Inducible transfer of conjugative transposon Tn1545 from *Enterococcus faecalis* to *Listeria monocytogenes* in the digestive tracts of gnotobiotic mice, *Antimicrob. Agents Chemother.* 35:185–187.

33. Dybvig, K., and Alderete, J., 1988, Transformation of *Mycoplasma pulmonis* and *Mycoplasma hyorhinis*: transposition of Tn916 and formation of cointegrate structures, *Plasmid* 20:33–41.

34. Dybvig, K., and Cassell, G.H., 1987, Transposition of gram-positive transposon Tn916 in *Acholeplasma laidlawii* and *Mycoplasma pulmonis, Science* 235:1392–1394.

35. Fitzgerald, G.F., and Clewell, D.B., 1985, A conjugative transposon (Tn919) in *Streptococcus sanguis, Infect. Immun.* 47:415–420.

36. Flannagan, S.E., and Clewell, D.B., 1991, Conjugative transfer of Tn916 in *Enterococcus faecalis: trans* activation of homologous transposons, *J. Bacteriol.* 173:7136–7141.

37. Fletcher, H.M., Marri, L., and Daneo-Moore, L., 1989, Transposon-916-like elements in clinical isolates of *Enterococcus faecium, J. Gen. Microbiol.* 135:3067–3077.

38. Franke, A.E., and Clewell, D.B., 1981, Evidence for a chromosome-borne resistance transposon (Tn916) in *Streptococcus faecalis* that is capable of "conjugal" transfer in the absence of a conjugative plasmid, *J. Bacteriol.* 145:494–502.

39. Franke, A.E., and Clewell, D.B., 1981, Evidence for conjugal transfer of a *Streptococcus faecalis* transposon (Tn916) from a chromosomal site in the absence of plasmid DNA, *Cold Spring Harbor Symp. Quant. Biol.* 45:77–80.

40. Gaillard, J.L., Berche, P., and Sansonetti, P., 1986, Transposon mutagenesis as a tool to study the role of hemolysin in the virulence of *Listeria monocytogenes*, *Infec. Immun.* 52:50–55.

41. Gawron-Burke, C., and Clewell, D.B., 1982, A transposon in *Streptococcus faecalis* with fertility properties, *Nature* (London) 300:281–284.

42. Gawron-Burke, C., and Clewell, D.B., 1984, Regeneration of insertionally inactivated streptococcal DNA fragments after excision of transposon Tn*916* in *Escherichia coli*: strategy for targeting and cloning of genes from gram-positive bacteria, *J. Bacteriol.* 159:214–221.

43. Guffanti, A.A., Quirk, P.G., and Krulwich, T.A., 1991, Transfer of Tn*925* and plasmids between *Bacillus subtilis* and alkaliphilic *Bacillus firmus* OF4 during Tn*925*-mediated conjugation, *J. Bacteriol.* 173:1686–1689.

44. Hachler, H., Kayser, F.H., and Berger-Bachi, B., 1987, Homology of a transferable tetracycline resistance determinant of *Clostridium difficile* with *Streptococcus* (*Enterococcus*) *faecalis* transposon Tn*916*, *Antimicrob. Agents Chemother.* 31:1033–1038.

45. Hendrick, C.A., Johnson, L.K., Tomes, N.J., Smiley, B.K., and Price, J.P., 1991, Insertion of Tn*916* into *Bacillus pumilus* plasmid pMGD302 and evidence for plasmid transfer by conjugation, *Plasmid* 26:1–9.

46. Hespell, R.B., and Whitehead, T.R., 1991, Conjugal transfer of Tn*916*, Tn*916ΔE*, and pAMβ1 from *Enterococcus faecalis* to *Butyrivibrio fibrisolvens* strains, *Appl. Environ. Microbiol.* 57:2703–2709.

47. Hespell, R.B., and Whitehead, T.R., 1991, Introduction of Tn*916* and pAMβ1 into *Streptococcus bovis* JB1 by conjugation, *Appl. Environ. Microbiol.* 57:2710–2713.

48. Hill, C., Daly, C., and Fitzgerald, G.F., 1985, Conjugative transfer of the transposon Tn*919* to lactic acid bacteria, *FEMS Microbiol. Lett.* 30;115–119.

49. Hill, C., Daly, C., and Fitzgerald, G.F., 1987, Development of high-frequency delivery system for transposon Tn*919* in lactic streptococci: random insertion in *Streptococcus lactis* subsp. *diacetylactis* 18–16, *Appl. Environ. Microbiol.* 53:74–78.

50. Hill, C., Venema, G., Daly, C., and Fitzgerald, G.F., 1988, Cloning and characterization of the tetracycline resistance determinant of and several promoters from within the conjugative transposon Tn*919*, *Appl. Environ. Microbiol.* 54:1230–1236.

51. Holland, J., Towner, K.J., and Williams, P., 1992, Tn*916* insertion mutagenesis in *Escherichia coli* and *Haemophilus influenzae* type b following conjugative transfer, *J. Gen. Microbiol.* 138:509–515.

52. Horaud, T., de Cespedes, G., Clermont, D., David, F., and Delbos, F., 1991, Variability of chromosomal genetic elements in streptococci, in: *Genetics and Molecular Biology of Streptococci, Lactococci, and Enterococci* (G.M. Dunny, P.P. Cleary, and L.L. McKay, eds.), American Society for Microbiology, Washington, DC, pp. 16–20.

53. Horaud, T., Delbos, F., and de Cespedes, G., 1990, Tn*3702*, a conjugative transposon in *Enterococcus faecalis*, *FEMS Microbiol. Lett.* 72:189–194.

54. Horaud, T., Le Bouguenec, C., and Pepper, K., 1985, Molecular genetics of resistance to macrolides, lincosamides and streptogramin B (MLS) in streptococci, *J. Antimicrob. Chemother.* 16A (Suppl.):111–135.

55. Horn, N., Swindell, S., Dodd, H., and Gasson, M., 1991, Nisin biosynthesis genes are encoded by a novel conjugative transposon, *Mol. Gen. Genet.* 228:129–135.

56. Horodniceanu, T., Bougueleret, L., and Bieth, G., 1981, Conjugative transfer of multiple-antibiotic resistance markers in beta-hemolytic group A, B, F, and G streptococci in the absence of extrachromosomal deoxyribonucleic acid, *Plasmid* 5:127–137.

57. Ike, Y., Flannagan, S.E., and Clewell, D.B., 1992, Hyperhemolytic phenomena associated with insertions of Tn*916* into the hemolysin determinant of *Enterococcus faecalis* plasmid pAD1, *J. Bacteriol.* 174:1801–1809.

58. Inamine, J.M., and Burdett, V., 1985, Structural organization of a 67-kilobase streptococcal conjugative element mediating multiple antibiotic resistance, *J. Bacteriol.* 161:620–626.

59. Ivins, B.E., Welkos, S.L., Knudson, G.B., and Leblanc, D.J., 1988, Transposon Tn*916* mutagenesis in *Bacillus anthracis*, *Infec. Immun.* 56:176–181.

60. Jones, J.M., Gawron-Burke, C., Flannagan, S.E., Yamamoto, M., Senghas, E., and Clewell, D.B., 1987, Structural and genetic studies of the conjugative transposon Tn*916*, in: *Streptococcal Genetics* (J.J. Ferretti, and R. Curtiss III, eds.), American Society for Microbiology, Washington, DC, pp. 54– 60.

61. Jones, J.M., Yost, S.C., and Pattee, P.A., 1987, Transfer of the conjugal tetracycline resistance transposon Tn*916* from *Streptococcus faecalis* to *Staphylococcus aureus* and identification of some insertion sites in the staphylococcal chromosome, *J. Bacteriol.* 169:2121–2131.

62. Kathariou, S., Metz, P., Hof, H., and Goebel, W., 1987, Tn916-induced mutations in the hemolysin determinant affecting virulence of *Listeria monocytogenes*, *J. Bacteriol.* 169:1291–1297.

63. Kathariou, S., Stephens, D.S., Spellman, P., and Morse, S.A., 1990, Transposition of Tn916 to different sites in the chromosome of *Neisseria meningitidis*: a genetic tool for meningococcal mutagenesis, *Molec. Microbiol.* 4:729–735.

64. Kauc, L., and Goodgal, S.H., 1989, Introduction of transposon Tn916 into *Haemophilus influenzae* and *Haemophilus parainfluenzae*, *J. Bacteriol.* 171:6625–6628.

65. Knauf, H.J., Vogel, R.F., and Hammes, W.P., 1989, Introduction of the transposon Tn919 into *Lactobacillus curvatus* Lc2-c, *FEMS Microbiol. Lett.* 65:101–104.

66. Kuypers, J.M., and Proctor, R.A., 1989, Reduced adherence to traumatized rat heart valves by a low-fibronectin-binding mutant of *Staphylococcus aureus*, *Infec. Immun.* 57:2306–2312.

67. Landy, A., 1989, Dynamic, structural, and regulatory aspects of λ site-specific recombination, *Annu. Rev. Biochem.* 58:913–949.

68. Lauer, G., Pastrana, R., Sherley, J., and Ptashne, M., 1981, Construction of overproducers of the bacteriophage 434 repressor and cro proteins, *J. Molec. Appl. Genet.* 1:139–147.

69. Le Bouguenec, C., de Cespedes, G., and Horaud, T., 1988, Molecular analysis of a composite chromosomal conjugative element (Tn3701) of *Streptococcus pyogenes*, *J. Bacteriol.* 170:3930–3936.

70. Le Bouguenec, C., de Cespedes, G., and Horaud, T., 1990, Presence of chromosomal elements resembling the composite structure Tn3701 in streptococci, *J. Bacteriol.* 172:727–734.

71. Le Bouguenec, C., Horaud, T., Bieth, G., Colimon, R., and Dauguet, C., 1984, Translocation of antibiotic resistance markers of a plasmid-free *Streptococcus pyogenes* (group A) strain into different streptococcal hemolysin plasmids, *Mol. Gen. Genet.* 194:377–387.

72. Le Bouguenec, C., Horaud, T., Geoffroy, C., and Alouf, J.E., 1988, Insertional inactivation by Tn3701 of pIP964 hemolysin expression in *Enterococcus faecalis*, *FEMS Microbiol. Lett.* 49:455–458.

73. Leong, J.M., Nunes-Duby, S., Oser, A.B., Lesser, C.F., Youderian, P., Susskind, M.M., and Landy, A., 1986, Structural and regulatory divergence among site-specific recombination genes of lambdoid phage, *J. Mol. Biol.* 189:603–616.

74. Lin, W., and Johnson, E.A., 1991, Transposon Tn916 mutagenesis in *Clostridium botulinum*, *Appl. Environ. Microbiol.* 57:2946–2950.

75. Martin, P., Trieu-Cuot, P., and Courvalin, P., 1986, Nucleotide sequence of the *tetM* tetracycline resistance determinant of the streptococcal conjugative shuttle transposon Tn1545, *Nucl. Acids Res.* 14:7047–7058.

76. Mizuuchi, K., Weisberg, R., Enquist, L., Mizuuchi, M., Buraczynska, M., Foeller, C., Hsu, P.L., Ross, W., and Landy, A., 1981, Structure and function of the phage λ *att* site: size, Int-binding sites, and location of the crossover point, *Cold Spring Harbor Symp. Quant. Biol.* 45:429–437.

77. Morisato, D., and Kleckner, N., 1987, Tn10 transposition and circle formation in vitro, *Cell* 51:101–111.

78. Mullany, P., Wilks, M., and Tabaqchali, S., 1991, Transfer of Tn916 and Tn916ΔE into *Clostridium difficile*: demonstration of a hot-spot for these elements in the *C. difficile* genome, *FEMS Microbiol. Lett.* 79: 191–194.

79. Murray, B.E., An, F.Y., and Clewell, D.B., 1988, Plasmids and pheromone response of the β-lactamase producer *Streptococcus (Enterococcus) faecalis* HH22, *Antimicrob. Agents Chemother.* 32:547–551.

80. Naglich, J.G., and Andrews, R.E., Jr., 1988, Introduction of the *Streptococcus faecalis* transposon Tn916 into *Bacillus thuringiensis* subsp. *israelensis*, *Plasmid* 19:84–93.

81. Naglich, J.G., and Andrews, R.E., Jr., 1988, Tn916-dependent conjugal transfer of pC194 and pUB110 from *Bacillus subtilis* into *Bacillus thuringiensis* subsp. *israelensis*, *Plasmid* 20:113–126.

82. Nassif, X., Puaoi, D., and So, M., 1991, Transposition of Tn1545-Δ3 in the pathogenic neisseriae: a genetic tool for mutagenesis, *J. Bacteriol.* 173:2147–2154.

83. Natarajan, M.R., and Oriel, P., 1991, Conjugal transfer of recombinant transposon Tn916 from *Escherichia coli* to *Bacillus stearothermophilus*, *Plasmid* 26:67–73.

84. Nida, K., and Cleary, P.P., 1983, Insertional inactivation of streptolysin S expression in *Streptococcus pyogenes*, *J. Bacteriol.* 155:1156–1161.

85. Norgren, M., Caparon, M.G., and Scott, J.R., 1989, A method for allelic replacement that uses the conjugative transposon Tn916: deletion of the *emm*6.1 allele in *Streptococcus pyogenes* JRS4, *Infec. Immun.* 57:3846–3850.

86. Norgren, M., and Scott, J.R., 1991, The presence of conjugative transposon Tn916 in the recipient strain does not impede transfer of a second copy of the element, *J. Bacteriol.* 173:319–324.

87. Pargellis, C.A., Nunes-Duby, S.E., Moitoso de Vargas, L., and Landy, A., 1988, Suicide recombination substrates yield covalent λ integrase-DNA complexes and lead to identification of the active site tyrosine, *J. Biol. Chem.* 263:7678–7685.

88. Perez-Casal, J., Caparon, M.G., and Scott, J.R., 1991, Mry, a *trans*-acting positive regulator of the M protein gene of *Streptococcus pyogenes* with similarity to the receptor proteins of two-component regulatory systems, *J. Bacteriol.* 173:2617–2624.

89. Poyart-Salmeron, C., Trieu-Cuot, P., Carlier, C., and Courvalin, P., 1989, Molecular characterization of two proteins involved in the excision of the conjugative transposon Tn*1545*: homologies with other site-specific recombinases, *EMBO J.* 8:2425–2433.

90. Poyart-Salmeron, C., Trieu-Cuot, P., Carlier, C., and Courvalin, P., 1990, The integration-excision system of the conjugative transposon Tn*1545* is structurally and functionally related to those of lambdoid phages, *Mol. Microbiol.* 4:1513–1521.

91. Procino, J.K., Marri, L., Shockman, G.D., and Daneo-Moore, L., 1988, Tn*916* insertional inactivation of multiple genes on the chromosome of *Streptococcus mutans* GS-5, *Infec. Immun.* 56:2866–2870.

92. Rauch, P.J.G., and de Vos, W.M., 1992, Characterization of the novel nisin-sucrose conjugative transposon Tn*5276* and its insertion in *Lactococcus lactis*, *J. Bacteriol.* 174:1280–1287.

93. Roberts, D., Hoopes, B.C., McClure, W.R., and Kleckner, N., 1985, IS*10* transposition is regulated by DNA adenine methylation, *Cell*, 43:117–130.

94. Roberts, M.C., 1990, Characterization of the Tet M determinants in urogenital and respiratory bacteria, *Antimicrob. Agents Chemother.* 34:476–478.

95. Roberts, M.C., and Kenny, G.E., 1987, Conjugal transfer of transposon Tn*916* from *Streptococcus faecalis* to *Mycoplasma hominis*, *J. Bacteriol.* 169:3836–3839.

96. Roberts, M.C., and Lansciardi, J., 1990, Transferable Tet M in *Fusobacterium nucleatum*, *Antimicrob. Agents Chemother.* 34:1836–1838.

97. Robins-Brown, R.M., Gaspar, M.N., Ward, J.I., Wachsmuth, I.K., Koornhof, H.J., Jacobs, M.R., and Thornsberry, C., 1979, Resistance mechanisms of multiply resistant pneumococci: antibiotic degradation studies, *Antimicrob. Agents Chemother.* 15:470–474.

98. Rubens, C.E., and Heggen, L.M., 1988, Tn*916*ΔE: A Tn*916* transposon derivative expressing erythromycin resistance, *Plasmid* 20:137–142.

99. Rubens, C.E., Wessels, M.R., Heggen, L.M., and Kasper, D.L., 1987, Transposon mutagenesis of type III group B *Streptococcus*: correlation of capsule expression with virulence, *Proc. Natl. Acad. Sci. USA* 84: 7208–7212.

100. Scott, J.R., Kirchman, P.A., and Caparon, M.G., 1988, An intermediate in the transposition of the conjugative transposon Tn*916*, *Proc. Natl. Acad. Sci. USA* 85:4809–4813.

101. Sen, S., and Oriel, P., 1990, Transfer of transposon Tn*916* from *Bacillus subtilis* to *Thermus aquaticus*, *FEMS Microbiol. Lett.* 67:131–134.

102. Senghas E., Jones, J.M., Yamamoto, M., Gawron-Burke, C., and Clewell, D.B., 1988, Genetic organization of the bacterial conjugative transposon Tn*916*, *J. Bacteriol.* 170:245–249.

103. Shoemaker, N.B., Smith, M.D., and Guild, W.R., 1979, Organization and transfer of heterologous chloramphenicol and tetracycline resistance genes in pneumococcus, *J. Bacteriol.* 139:432–441.

104. Shoemaker, N.B., Smith, M.D., and Guild, W.R., 1980, DNase-resistant transfer of chromosomal *cat* and *tet* insertions by filter mating in pneumococcus, *Plasmid* 3:80–87.

105. Smith, M.D., and Guild, W.R., 1979, A plasmid in *Streptococcus pneumoniae*, *J. Bacteriol.* 137:735–739.

106. Smith, M.D., and Guild, W.R., 1982, Evidence for transposition of the conjugative R determinants of *Streptococcus agalactiae* B109, in: *Microbiology–1982* (D. Schlessinger, ed.), American Society for Microbiology, Washington, DC, pp. 109–111.

107. Storrs, M.J., Poyart-Salmeron, C., Trieu-Cuot, P., and Courvalin, P., 1991, Conjugative transposition of Tn*916* requires the excisive and integrative activities of the transposon-encoded integrase, *J. Bacteriol.* 173:4347–4352.

108. Stratz, M., Gottschalk, G., and Durre, P., 1990, Transfer and expression of the tetracycline resistance transposon Tn*925* in *Acetobacterium woodii*, *FEMS Microbiol. Lett.* 68:171–176.

109. Su, Y.A., and Clewell, D.B., Characterization of the left four kilobases of conjugative transposon Tn*916*. Determinants involved in excision, submitted for publication.

110. Su, Y.A., He, P., and Clewell, D.B., 1992, Characterization of the *tet*(M) determinant of Tn*916*: evidence for regulation by transcription attenuation, *Antimicrob. Agents Chemother.* 36:769–778.

111. Tomich, P.K., An, F.Y., and Clewell, D.B., 1980, Properties of erythromycin-inducible transposon Tn917 in *Streptococcus faecalis*, *J. Bacteriol.* 141:1366–1374.

112. Tomich, P.K., An, F.Y., Damle, S.P., and Clewell, D.B., 1979, Plasmid-related transmissibility and multiple drug resistance in *Streptococcus faecalis* subsp. *zymogenes* strain DS16, *Antimicrob. Agents Chemother.* 15:828–830.

113. Torres, O.R., Korman, R.Z., Zahler, S.A., and Dunny, G.M., 1991, The conjugative transposon Tn925: enhancement of conjugal transfer by tetracycline in *Enterococcus faecalis* and mobilization of chromosomal genes in *Bacillus subtilis* and *E. faecalis*, *Mol. Gen. Genet.* 225:395–400.

114. Trieu-Cuot, P., Carlier, C., Poyart-Salmeron, C., and Courvalin, P., 1991, An integrative vector exploiting the transposition properties of Tn1545 for insertional mutagenesis and cloning of genes from gram-positive bacteria, *Gene* 106:21–27.

115. Trieu-Cuot, P., Poyart-Salmeron, C., Carlier, C., and Courvalin, P., 1990, Nucleotide sequence of the erythromycin resistance gene of the conjugative transposon Tn1545, *Nucl. Acids Res.* 18:3660.

116. Trieu-Cuot, P., Poyart-Salmeron, C., Carlier, C., and Courvalin, P., 1991, Molecular dissection of the transposition mechanism of conjugative transposons from gram-positive cocci, in: *Genetics and Molecular Biology of Streptococci, Lactococci, and Enterococci* (G.M. Dunny, P.P. Cleary, and L.L. McKay, eds.), American Society for Microbiology, Washington, DC, pp. 21–27.

117. Vijayakumar, M.N., Priebe, S.D., and Guild, W.R., 1986, Structure of a conjugative element in *Streptococcus pneumoniae*, *J. Bacteriol.* 166:978–984.

118. Vijayakumar, M.N., Priebe, S.D., Pozzi, G., Hageman, J.M., and Guild, W.R., 1986, Cloning and physical characterization of chromosomal conjugative elements in streptococci, *J. Bacteriol.* 166:972–977.

119. Volk, W.A., Bizzini, B., Jones, K.R., and Macrina, F.L., 1988, Inter- and intrageneric transfer of Tn916 between *Streptococcus faecalis* and *Clostridium tetani*, *Plasmid* 19:255–259.

120. Wanger, A.R., and Dunny, G.M., 1985, Development of a system for genetic and molecular analysis of *Streptococcus agalactiae*, *Res. Vet. Sci.* 38:202–208.

121. Watson, D.A., and Musher, D.M., 1990, Interruption of capsule production in *Streptococcus pneumoniae* serotype 3 by insertion of transposon Tn916, *Infec. Immun.* 58:3135–3138.

122. Weisberg, R.A., and Landy, A., 1983, Site-specific recombination in phage lambda, in: *Lambda II* (R.W. Hendrix, J.W. Roberts, F.W. Stahl, and R.A. Weisberg, eds.), Cold Spring Harbor Laboratory, Cold Spring Harbor, New York, pp. 211–250.

123. Weiser, J.N., and Rubens, C.E., 1987, Transposon mutagenesis of group B streptococcus beta-hemolysin biosynthesis, *Infec. Immun.* 55:2314–2316.

124. Whitley, J.C., and Finch, L.R., 1989, Location of sites of transposon Tn916 in the *Mycoplasma mycoides* genome, *J. Bacteriol.* 171:6870–6872.

125. Willetts, N., and Wilkins, B., 1984, Processing of plasmid DNA during bacterial conjugation, *Microbiol. Rev.* 48:24–41.

126. Woolley, R.C., Pennock, A., Ashton, R.J., Davies, A., and Young, M., 1989, Transfer of Tn1545 and Tn916 to *Clostridium acetobutylicum*, *Plasmid* 22:169–174.

127. Yamamoto, M., Jones, J.M., Senghas, E., Gawron-Burke, C., and Clewell, D.B., 1987, Generation of Tn5 insertions in streptococcal conjugative transposon Tn916, *Appl. Environ. Microbiol.* 53:1069–1072.

128. Yin, J.C.P., Krebs, M.P., and Reznikoff, W.S., 1988, Effect of *dam* methylation on Tn5 transposition, *J. Mol. Biol.* 199:35–45.

129. Yost, S.C., Jones, J.M., and Pattee, P.A., 1988, Sequential transposition of Tn916 among *Staphylococcus aureus* protoplasts, *Plasmid* 19:13–20.

130. Young, F.E., and Mayer, L., 1979, Genetic determinants of microbial resistance to antibiotics, *Rev. Infect. Dis.* 1:55–63.

Index